# Agricultural Crop Production

# Agricultural Crop Production

Edited by **Alabaster Jenkins**

R CALLISTO REFERENCE

New York

Published by Callisto Reference,
106 Park Avenue, Suite 200,
New York, NY 10016, USA
www.callistoreference.com

**Agricultural Crop Production**
Edited by Alabaster Jenkins

International Standard Book Number: 978-1-63239-661-7 (Hardback)

Printed in the United States of America.

# Contents

# Preface

Agriculture is the pillar of human existence as it provides us with food, timber, fibre, etc. Crop production is the branch of agriculture which specifically deals with growing plants to convert them into usable items such as food, clothing, medicine, etc. This book traces the progress of this field and highlights some of its key concepts and applications. The various studies that are constantly contributing towards advancing technologies and evolution of this field are examined in detail. The topics covered in this book offer the readers new insights in the area of agricultural crop production. It will prove to be a valuable source of reference for agriculturists, environmentalist, researchers and students alike.

The researches compiled throughout the book are authentic and of high quality, combining several disciplines and from very diverse regions from around the world. Drawing on the contributions of many researchers from diverse countries, the book's objective is to provide the readers with the latest achievements in the area of research. This book will surely be a source of knowledge to all interested and researching the field.

In the end, I would like to express my deep sense of gratitude to all the authors for meeting the set deadlines in completing and submitting their research chapters. I would also like to thank the publisher for the support offered to us throughout the course of the book. Finally, I extend my sincere thanks to my family for being a constant source of inspiration and encouragement.

**Editor**

# Foliar Sprays of Citric Acid and Malic Acid Modify Growth, Flowering, and Root to Shoot Ratio of *Gazania* (*Gazania rigens* L.): A Comparative Analysis by ANOVA and Structural Equations Modeling

**Majid Talebi,[1] Ebrahim Hadavi,[2] and Nima Jaafari[1]**

[1] *Graduate School of Horticulture, Islamic Azad University, Karaj Branch, Karaj 31876 44511, Iran*
[2] *Department of Horticulture, Islamic Azad University, Karaj Branch, Karaj 31876 44511, Iran*

Correspondence should be addressed to Ebrahim Hadavi; hadavi@kiau.ac.ir

Academic Editor: Pia Parolin

Foliar application of two levels of citric acid and malic acid (100 or 300 mg L$^{-1}$) was investigated on flower stem height, plant height, flower performance and yield indices (fresh yield, dry yield and root to shoot ratio) of *Gazania*. Distilled water was applied as control treatment. Multivariate analysis revealed that while the experimental treatments had no significant effect on fresh weight and the flower count, the plant dry weight was significantly increased by 300 mg L$^{-1}$ malic acid. Citric acid at 100 and 300 mg L$^{-1}$ and 300 mg L$^{-1}$ malic acid increased the root fresh weight significantly. Both the plant height and peduncle length were significantly increased in all applied levels of citric acid and malic acid. The display time of flowers on the plant increased in all treatments compared to control treatment. The root to shoot ratio was increased significantly in 300 mg L$^{-1}$ citric acid compared to all other treatments. These findings confirm earlier reports that citric acid and malic acid as environmentally sound chemicals are effective on various aspects of growth and development of crops. Structural equations modeling is used in parallel to ANOVA to conclude the factor effects and the possible path of effects.

## 1. Introduction

*Gazania* is a small genus of 16 species, all endemic to southern Africa belonging to the Asteraceae (Compositae) family. It is a member of the tribe *Arctoteae*, subtribe *Gorteriinae*, which includes seven other genera: *Berkheya, Gorteria, Cuspidia, Didelta, Heterorhachis, Cullumia,* and *Hirpicium* [1]. Besides other flower species, the *Gazania rigens* L., a member of annual flowers, represents an important decorative material for public greens, gardens, flower stands, and so forth [2]. It is propagated easily and has attractive flowers with a long flowering period, from middle spring to autumn; *Gazania* is used extensively in urban decorations nowadays. However, *Gazania* is considered among those landscape groundcovers that are susceptible to water stress [3]. Increasing of ornamental performance is a major concern, especially in harsh urban condition, which demands an increased tolerance to a variety of environmental stresses. Therefore, easy-to-apply practices that can improve the plant performance would be of interest.

Endogenous organic acids are the source of both carbon skeleton and energy for cells and are used in the respiratory cycle and other biochemical pathways. Therefore, they can influence the cut flower's vase life [4]. Malic acid is metabolized in plant mitochondria by reaction of malic enzyme [5]. Malate is a common reserve anion playing a role in the plant vacuole as counter ion for K and Ca [6], especially in nitrate-dependent plants [7]. Roots of calcicole plants (plants growing in alkaline soils) exudate citrate and malate, which enables them to extract P and Fe from such soils [8]. Eidyan et al. reported that citric acid spray (0.1%, w/v) increased the vase life of cut tuberose plants and increased the size of bulblets in a synergism with foliar Fe [9]. Preharvest spray of citric acid (0.15%, w/v) increased the mean vase life of cut lilium flowers from 11.8 in control treatment to 14 days

[10]. Basil plants (*Ocimum basilicum* L.) treated with citric acid (0.1%, w/v) produced higher biomass and essential oil yield [11]. In another study on dill, it was revealed that plant vegetative parameters as well as tolerance to powdery mildew were improved by the combination of 0.3% citric acid and 0.1% malic acid [12]. They suggested elevated exudation of organic acids by roots in response to foliar organic acids, which was confirmed recently by An et al. [13]. El-Tohamy et al. reported that citric acid spray induced drought tolerance in bean [14]. Increased yield caused by positive interactions between citric acid spray and seed inoculation with mycorrhiza, *Azotobacter*, and *Azospirillum* is reported [15].

The present study aimed to evaluate the use of citric acid and malic acid as biostimulants for increasing the ornamental performance of *Gazania*. In this study, we aimed to test whether applying foliar sprays of citric acid and malic acid could increase quality and performance of an ornamental landscape decorative plant. We investigated the effect of these organic acids on both canopy and root system to check the possible change in allocation pattern between shoot and root.

## 2. Materials and Methods

The experiment was conducted in the experimental greenhouse of Qom Municipal (Qom, Iran). Citric acid and malic acid (Sigma-Aldrich, 99% Pure) were applied as foliar sprays each at two concentrations (100 or 300 mg L$^{-1}$). The control treatment consisted of foliar sprays of distilled water. The experiment was conducted in a completely randomized design. Sample size was 18 pots per treatment, except for the control treatment where 9 pots were used.

*Gazania* seeds were planted in plugs in Nov. and then after 30 days were transplanted to plastic pots containing a (1:1:1 by volume) mixture of garden soil, leaf compost, and peat moss. The pH was determined (pH ≈ 7) on a 1:1 medium/water mixture with a Jenway pH meter model 3030 (Felsted, UK). Plants were placed in a greenhouse at a mean day/night temperature of 26/18°C and a soil moisture of 60–70%. A total of six sprays were applied starting from 45 days after emergence of seeds when the leaves were about seven cm long. The sprays were repeated biweekly. A hand-held sprayer was used and each spraying was continued until the liquid runs off freely from the foliage. A commercial adjuvant was added to all sprays (Citogate). After the onset of flowering, the flower display time was recorded in all treatments. Other traits were recorded 150 days after emergence including the flower peduncle length, plant height, flower performance, and yield indices, including fresh yield, dry yield, and R:S (root to shoot ratio).

*2.1. Data Analysis.* Analysis of variance was performed by SPSS software (ver. 16.0, SPSS Inc., Chicago, USA). Preliminary analyses (examination of normal probability plots and scatterplots) were performed to insure no violation of assumptions of normality and homoscedasticity. Means were separated by Duncan's test when the $F$ test proved significant at $\alpha < 0.05$. Structural equations modeling (SEM) analysis was carried out using AMOS extension of SPSS.

## 3. Results

*Shoot Parameters.* Plants treated with 300 mg L$^{-1}$ malic acid produced the highest level of shoot dry weight per pot, in comparison with control (Table 1). The same pattern was noticeable in total dry biomass production by plants. However, the effect on the shoot fresh weight was less prominent, and no significant difference was noticed between treatments. Water content of *Gazania* plants was slightly decreased by malic acid when compared to citric acid containing treatments; however, none was considered significantly different compared to control treatment (Table 1).

*Root Parameters.* Both citric acid treatments and 300 mg L$^{-1}$ malic acid increased the root fresh weight significantly (Table 1). By an increase in concentration of both applied organic acids a larger root dry weight resulted (Table 1). However, there was no difference between 100 and 300 mg L$^{-1}$ malic acid concentrations, while in citric acid treatment, the root fresh weight was significantly larger when the concentration increased from 100 to 300 mg L$^{-1}$. Malic acid in 300 mg L$^{-1}$ level and both citric acid levels increased the R:S ratio compared to control.

*Flowering Parameters.* All concentration of organic acids significantly lengthened the flower display time. Among treatments, only 100 mg L$^{-1}$ of citric acid increased it significantly (Table 1). All applied levels of organic acids created significantly taller plants (Table 1). In all treatments of organic acids, *Gazania* plants had significantly longer peduncle than untreated control. Any significant difference among the applied concentrations of organic acids was missing (Table 1).

As it could be seen in the developed path diagram from experimental data, citric acid is nominated as the only responsible factor for increased flower display time; however, a small portion of variance is explained (7%), and the related mechanism remains unclear (Figure 1). SEM suggests that citric acid primarily increased the plant height and, consequently, the peduncle length. In the case of malic acid, the effect on peduncle length was considered direct, while the effect on plant height is suggested to be mediated indirectly. The positive effect of malic acid on peduncle length is concurrent with a negative effect on the flower count. The observed effect on root allocation is mediated via the increase in shoot dry weight induced by both citric acid and malic acid (52% of variance explained).

## 4. Discussion

Our results reveal a positive effect by applied organic acids on root parameters, plant height, and flowering parameters. This indicates that these organic acids could improve the ornamental value of *Gazania* as an outdoor decorative plant. A shift in allocation pattern of assimilates was noted as suggested before. The increase in flower display time in our experiment could be considered relatively similar with the previous findings of increased vase life by citric acid sprays reported on cut tuberose [9], lilium [10], and increased shelf

TABLE 1: Mean comparison table.

| Treatment | Shoot fresh weight (g) | Shoot dry weight (g) | Root fresh weight (g) | Root dry weight (g) | Root to shoot ratio (DW) | Dry biomass (g) | Water content (%) | Peduncle length (cm) | Plant height (cm) | Flower count | Flower display time (day) |
|---|---|---|---|---|---|---|---|---|---|---|---|
| Control | 223 | 30 | 2.1[b§] | 0.96[c] | 0.035 | 31.4 | 85.9[ab] | 16.3[b] | 19.1[c] | 16.3 | 7.4[c] |
| MA 100 mg L$^{-1}$ | 231 | 34.5 | 2.8[ab] | 1.2[bc] | 0.034 | 35.6 | 84.9[b] | 21.2[a] | 22.5[b] | 16.2 | 8.6[b] |
| MA 300 mg L$^{-1}$ | 279 | 43.8 | 3.6[a] | 1.6[ab] | 0.042 | 45.5 | 84.4[b] | 22.4[a] | 23.4[ab] | 13.5 | 8.4[b] |
| CA 100 mg L$^{-1}$ | 254 | 30.4 | 3.7[a] | 1.1[bc] | 0.039 | 31.1 | 87.6[a] | 20.7[a] | 24[ab] | 17.7 | 9.4[a] |
| CA 300 mg L$^{-1}$ | 266 | 38.6 | 3.7[a] | 1.8[a] | 0.049 | 40.4 | 85.5[ab] | 22[a] | 25[a] | 16.3 | 8.7[b] |
| *F* test | | | | | | | | | | | |
| Treatment effect | ns[†] | ns | ** | ** | ns | ns | * | ** | ** | ns | ** |
| CV (%, mean) | 38.9 | 45.7 | 39.3 | 48.8 | 41.5 | 42.9 | 3.4 | 15.8 | 12.8 | 44.7 | 6.3 |

§Values in the same column that are followed by the same letter do not differ significantly according to Douncan's multiple range test ($P < 0.05$). Letters were not shown for those considered non-significant by *F* test.

† ns: non significant, * significant ($\alpha < 0.05$), and ** significant ($\alpha < 0.01$).

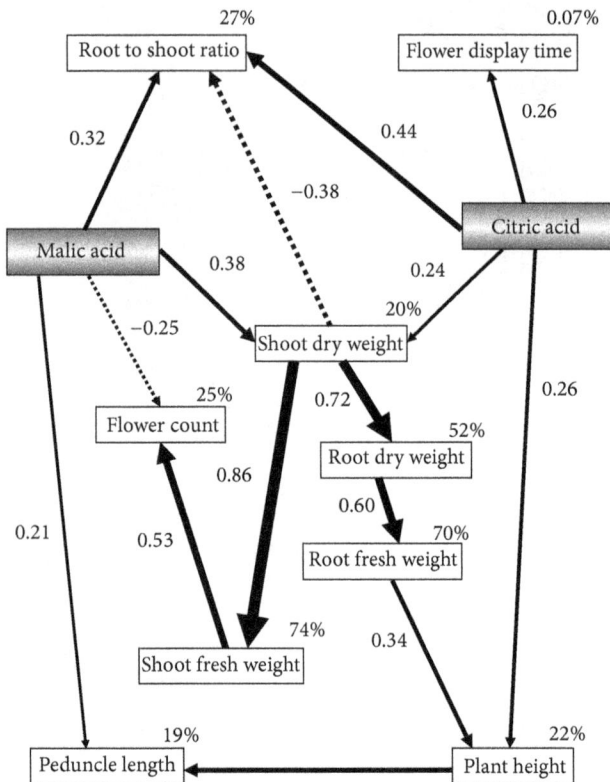

FIGURE 1: The path analysis of experimental data. Dashed lines indicate negative coefficients, and the width of the arrow indicates the strength of the effect. The lines represent significant effect and the numbers on arrows are standardized regression coefficients (when factor goes up or down by 1 standard deviation, the variable goes up or down by the multiplying the mentioned number in its standard deviation). The numbers on the upper corner of each variable are determination coefficient ($R^2$) (e.g., it is estimated that the predictors of the root dry weight explain 52 percent of its variance). Variation due to error is not included for simplicity.

life of fresh cut dill [12]. However, here we are reporting for the first time that both organic acids increased the lifespan of attached flowers to the plant.

The taller plants in response to spray of both organic acids are consistent with an earlier report on dill [12], where it is observed that both organic acids and their combination created taller dill plants. SEM analysis suggests that increased peduncle length is at least partially regulated by the same factors that cause increased plant height.

Eidyan et al. [9] noted a synergism between foliar application of Fe and citric acid in increasing of bulblet size in tuberose. A reduction in aerial bulbil weight by citric acid and some combinations of citric acid and malic acid sprays in lilium was reported. They suggested that this could be due to a carbohydrate partitioning effect by these organic acids [10]. Our observation of increased carbohydrate allocation to roots is in consistency with these reports. We can postulate that in those experiments a similar pattern of allocation to root system, like what we noticed here, existed that was responsible for larger underground bulblets in tuberose and reduced aerial bulbil weights in lilium.

Andrews considered Leaf N level [16] and leaf protein concentration [17] as factors that explain the major proportion of the variation in S : R (shoot to root) ratio. Later, he concluded that all environmental influences on S : R are often related mechanistically to their effects on leaf protein concentration [18]. However, while these conclusions may be applied to well fertilized plants, there are other insights to R : S ratio from the stress management viewpoint. The absolute amounts of below- and above-ground environmental stress, by the below/above-ground environmental stress ratio, and the growth potential of the species itself are suggested to determine R : S ratio [19]. In another study seedling survival of the different species was positively correlated with root allocation [20]. In our experiment, we may consider increasing R : S ratio by applied organic acid as plant adaptation to the potential environmental stresses. Vegetative indexes like plant height and peduncle length and in some cases the dry biomass accumulation are positively affected along with an increased R : S ratio, so we can conclude that these treatments helped the plant to adapt to and perform better in response to environmental situation by a more effective root system which has supported a better shoot growth.

We observed a range of R : S ratio (3.5% to 4.9% DW), which was lower than what was reported earlier by Niu and Rodriguez [21]. They reported R : S ratios between 7 to 8% by planting the plants in a lighter medium (Canadian sphagnum peat moss, perlite + wetting agent) which could induce more root proliferation than what we used (mixture of garden soil, leaf compost, and peat moss). Using a lighter medium can increase the recovery rate of fine roots, which hold a large share in *Gazania*. In our experiment, possibly the heavier applied medium caused losing part of extra fine roots and root hairs during the root isolation process for weighting. However, as this potential loss has affected all of treatments in a same manner, it cannot be considered as a source of variation in our study. Our longer experiment period (150 versus 90 days) could be another point explaining this difference. Decrease in R : S ratio by age/size in herbaceous plants is well documented [22]. Nevertheless, more data on R : S ratio of *Gazania* plant in different culture conditions may be needed to reach a final conclusion.

In the root, dry weight, and flower display time, we observed significant differences in response to applied citric acid concentrations. In malic acid, while there were no significant differences between applied concentrations, but in the root, fresh weight, and root dry weight, we see that only the higher concentration is considered significantly different from the control. When comparing with earlier reports, here we observe more similarity between responses of selected traits to organic acids. In some traits, the lower concentration gave good results that give us the idea of possible distinct patterns of response to concentration of applied organic acids. Therefore, we suggest testing both lower and higher concentrations of these organic acids to reach a better understanding in this regard.

SEM analysis proposes a controlling effect by shoot dry weight on R : S ratio. On the other side, it suggests that root dry and fresh weights are positively affected by shoot biomass. This may seem contrasting and could be a matter of question. We can conclude that the observed controlling effect of shoot dry weight on R : S ratio could result from decrease in R : S ratio by increasing of plant size/age. This would be in accordance with a previous conclusion in a review of literature by Wilson regarding decrease in R : S ratio of herbaceous plants with age/size, with the exception of being in root crops [22]. Observation of the dependence of the root dry and fresh weight on the shoot dry weight could be a representative of the dependence of root growth on allocated carbohydrates by shoot. Therefore, it would be natural to expect a larger root system supported by a larger shoot system.

## 5. Conclusion

Citric acid and malic acid possibly affected carbohydrate partitioning toward root allocation in our study. The stronger root system supported better shoot system in response. Considering the accumulated data regarding effects of citric acid and malic acid and the way they affect plants in relatively low concentrations and the consistency among plants from different families, a distinct pattern of regulative effects could be noted. The fact that they affect plant in low concentration along with inhibitory effect by higher concentrations of these organic acids, which is reported earlier, enables us to conclude that their carbohydrate content may not play an important role in creating the observed responses. As these are both environment-friendly and low cost agents, it seems promising to continue work to summarize the distinct effects of these organic acids on plants and to understand the mechanism by which they manifest their effects on plants.

In addition, we may draw attention to the use of structural equations modeling (SEM) in this kind of studies; as we can see here, it can help to uncover and visualize the key important points of interest out of the results in a concise manner. However, there are some differences visible between outputs of ANOVA and SEM that are due to the difference in algorithm of calculation. SEM analysis gives more weight to the correlation between variables than ANOVA. By putting both analyses together, we can obtain a more comprehensive output. To confirm the suggested path of effects and proposed model, further work in different conditions on the same plant would be of interest. Mitchell [23] recommended use of SEM as a powerful tool for applying path analysis to observational data sets in ecology and evolution, and here we can see a positive perspective for application of this modeling tool in plant studies, as well.

## Conflict of Interests

The authors declare that there is no conflict of interests regarding the publication of this paper.

## References

[1] S. Howis, N. P. Barker, and L. Mucina, "Globally grown, but poorly known: Species limits and biogeography of Gazania Gaertn. (Asteraceae) inferred from chloroplast and nuclear DNA sequence data," *Taxon*, vol. 58, no. 3, pp. 871–882, 2009.

[2] A. Vujošević, N. Lakić, D. Beatović, and S. Jelačić, "Influence of slow disintegrating fertilizer rates on quality of gazania (*Gazania rigens* L.) seedlings," *Journal of Agricultural Sciences*, vol. 52, pp. 121–130, 2007.

[3] D. R. Pittenger, D. A. Shaw, D. R. Hodel, and D. B. Holt, "Responses of landscape groundcovers to minimum irrigation," *Journal of Environmental Horticulture*, vol. 19, pp. 78–84, 2001.

[4] J. A. T. da Silva, "The cut flower: postharvest considerations," *Journal of Biological Sciences*, vol. 3, no. 4, pp. 406–442, 2003.

[5] D. A. Day and J. B. Hanson, "Pyruvate and malate transport and oxidation in corn mitochondria," *Plant Physiology*, vol. 59, no. 4, pp. 630–635, 1977.

[6] I. P. Ting, "Towards a model for malate accumulation in plant tissues," *Plant Science Letters*, vol. 21, no. 3, pp. 215–221, 1981.

[7] C. Osmond and G. G. Laties, "Compartmentation of malate in relation to ion absorption in beet," *Plant Physiology*, vol. 44, no. 7, 1969.

[8] J. Lopez-Bucio, M. F. Nieto-Jacobo, V. Ramírez-Rodríguez, and L. Herrera-Estrella, "Organic acid metabolism in plants: from adaptive physiology to transgenic varieties for cultivation in extreme soils," *Plant Science*, vol. 160, no. 1, pp. 1–13, 2000.

[9] B. Eidyan, E. Hadavi, and N. Moalemi, "Pre-harvest foliar application of iron sulfate and citric acid combined with urea fertigation affects growth and vase life of tuberose (*Polianthes tuberosa* L.) 'Por-Par'," *Horticulture Environment and Biotechnology*, vol. 55, no. 1, pp. 9–13, 2014.

[10] N. Darandeh and E. Hadavi, "Effect of pre-harvest foliar application of citric acid and malic acid on chlorophyll content and post-harvest vase life of Lilium cv. Brunello," *Frontiers in Plant Science: Crop Science and Horticulture*, vol. 2, article 106, 2012.

[11] N. Jaafari and E. Hadavi, "Growth and essential oil yield of Basil (ocimum basilicum L.) as affected by foliar spray of citric acid and salicylic acid," *Zeitschrift fur Arznei- und Gewurzpflanzen*, vol. 17, no. 2, pp. 80–83, 2012.

[12] N. Jafari and E. Hadavi, "Growth and essential oil yield of dill (Anethum graveolens) as affected by foliar sprays of citric acid and malic acid," *Acta Horticulturae*, vol. 955, pp. 287–290, 2012.

[13] Y. An, P. Zhou, Q. Xiao, and D. Shi, "Effects of foliar application of organic acids on alleviation of aluminum toxicity in alfalfa," *Journal of Plant Nutrition and Soil Science*, vol. 177, no. 3, pp. 421–430, 2014.

[14] W. El-Tohamy, H. El-Abagy, M. Badr, and N. Gruda, "Drought tolerance and water status of bean plants (*phaseolus vulgaris* l.) as affected by citric acid application," *Journal of Applied Botany and Food Quality*, vol. 86, pp. 212–216, 2013.

[15] V. Maleki, M. R. Ardakani, F. Rejali, and A. A. Taherpour, "Physiological responses of sweet basil (*Ocimum basilicum* L.) to triple inoculation with azotobacter, azospirillum, glomus intraradices and foliar application of citric acid," *Annals of Biological Research*, vol. 4, pp. 62–71, 2013.

[16] M. Andrews, J. I. Sprent, J. A. Raven, and P. E. Eady, "Relationships between shoot to root ratio, growth and leaf soluble protein concentration of Pisum sativum, *Phaseolus vulgaris* and *Triticum aestivum* under different nutrient deficiencies," *Plant, Cell and Environment*, vol. 22, no. 8, pp. 949–958, 1999.

[17] M. Andrews, J. A. Raven, and J. I. Sprent, "Environmental effects on dry matter partitioning between shoot and root of crop plants: relations with growth and shoot protein concentration," *Annals of Applied Biology*, vol. 138, no. 1, pp. 57–68, 2001.

[18] M. Andrews, J. A. Raven, P. J. Lea, and J. I. Sprent, "A role for shoot protein in shoot-root dry matter allocation in higher plants," *Annals of Botany*, vol. 97, no. 1, pp. 3–10, 2006.

[19] R. Hunt and A. O. Nicholls, "Stress and the coarse control of growth and root-shoot partitioning in herbaceous plants.," *Oikos*, vol. 47, no. 2, pp. 149–158, 1986.

[20] F. Lloret, C. Casanovas, and J. Peñuelas, "Seedling survival of Mediterranean shrubland species in relation to root:shoot ratio, seed size and water and nitrogen use," *Functional Ecology*, vol. 13, no. 2, pp. 210–216, 1999.

[21] G. Niu and D. S. Rodriguez, "Relative salt tolerance of selected herbaceous perennials and groundcovers," *Scientia Horticulturae*, vol. 110, no. 4, pp. 352–358, 2006.

[22] J. B. Wilson, "A review of evidence on the control of shoot: root ratio, in relation to models," *Annals of Botany*, vol. 61, no. 4, pp. 433–449, 1988.

[23] R. J. Mitchell, "Testing evolutionary and ecological hypotheses using path analysis and structural equation modelling," *Functional Ecology*, vol. 6, no. 2, pp. 123–129, 1992.

# 2

# Maize Residue as a Viable Substrate for Farm Scale Cultivation of Oyster Mushroom (*Pleurotus ostreatus*)

**Abena O. Adjapong, Kwame D. Ansah, Faustina Angfaarabung, and Henry O. Sintim**

*Department of General Agriculture, School of Applied Science & Technology, Sunyani Polytechnic, Sunyani, Ghana*

Correspondence should be addressed to Henry O. Sintim; hosintim@gmail.com

Academic Editor: Albino Maggio

In the search for alternatives to sawdust as growing media in commercial mushroom cultivation, three organic substrates obtainable as crop residue, maize husk, maize cob, and maize stalk, with each being supplemented with rice bran, were evaluated as growth media for the oyster mushroom, *Pleurotus ostreatus* (Kummer). For the tested alternatives to sawdust, the harvested weight of fruiting bodies that sprouted on a kilogram maize husk media per crop (32.99 g) was the highest. Sawdust media supported significantly ($P < 0.001$) heavier fruiting bodies (42.18) than the maize residues. The peak mushroom harvests for the various substrates were obtained between the first and seventh fruiting body flushes. The biological efficiency of the substrates, which measured usable nutrients indicated that maize stalk supplemented with rice bran, was 39% compared to that of the sawdust media (60%). The maize husk media and the maize cob media had biological efficiencies of 32% and 9.5%, respectively. These results indicate that two of the tested growing media (maize stalk or husk) produced mushrooms with yield characteristics that were comparable to the well-used sawdust in the cultivation of oyster mushrooms. The environmental and economic parameters involved in the use and carting of sawdust make these on-farm crop residues a viable alternative for mushroom cultivation in especially nonforest zones of Ghana.

## 1. Introduction

Edible mushrooms include fungi that thrive on damp decaying organic matter alone or in combination with soil [1] as it depends on nutrients obtained from dead and decaying materials. Mushrooms have been used for food and medicine and have often been considered a luxurious food reserved for the elite [2]. Global mushroom production has increased tremendously, from about 0.3 million tons in 1961 to about 3.41 million tons in 2010 [3, 4]. China is the leading producer of mushrooms worldwide, producing about 65% of global mushrooms and 85% of oyster mushroom worldwide [3]. Africa produces only 1% of the total world output of oyster mushroom [5].

Oyster mushroom (*Pleurotus ostreatus*, Kummer) is the second largest commercially produced and important edible mushroom in the world market [6] after *Agaricus* mushrooms. It is the most popular mushroom cultivated in Ghana and it also does well in other tropical and subtropical regions [7]. *P. ostreatus* has a unique adaptation to a wide range of lignocellulosic substrates [2, 8]; hence, it is not uncommon to find wild oyster mushrooms sprouting naturally in clusters on dead trees. It is also an easy and a cost effective mushroom to grow [8]. Oyster mushroom is consumed for its taste and medicinal and nutritional properties [8]. It contains proteins, vitamins, and crude fibre and has been recommended for its lovastatin properties [6] to patients with cholesterol related ailments [4, 7, 9]. It is believed that oyster mushroom can prevent high blood pressure, constipation, and hangovers and can assist in the recovery from fatigue and again it is used as a soil conditioner in agriculture [2].

Oyster mushrooms have been cultivated using tree logs or containers such as shelves, boxes, bags, and bottles containing growing media [2]. Various crop residues have been used in producing oyster mushrooms either as main substrates or in combination with supplements [10]. Oyster mushroom can be grown on various substrates including paddy straw, maize stalks/cobs, vegetable plant residues, bagasse [11, 12], sawdust, wheat straw, cotton waste, waste paper, and cotton stalks [13, 14]. The preferred method of cultivation is dependent on the mushroom variety, market demand, farmer's preferences, and availability of growing media [15].

In Ghana, oyster mushroom is cultivated using plastic bags filled with decomposed sawdust [16]. However, as wood (timber) becomes scarce, the unavailability of appropriate sawdust substrate becomes a limiting factor for mushroom cultivation in Ghana. In addition, the timber industry which provides the sawdust as a by-product from the timber mills is restricted to the forest zones. Mushroom growers outside the forest zones are therefore challenged with the carting cost of the sawdust which increases production cost. The development of cultivation techniques that utilizes other sustainable sources of substrates for oyster mushroom cultivation can fill the sawdust void and is likely to be adopted by mushroom growers.

The utilization of maize residues as substrate for oyster mushroom under controlled conditions has been reported by Atikpo et al. [15], Obodai et al. [16], and Onyango et al. [5]. Maize is one of the major crops grown in Ghana and its residues (husks, cobs, and stalks) are abundant and available during the year and could be exploited as a sustainable substrate for mushroom growing. The extrapolation of laboratory growing successes [5, 15, 16] of mushroom on maize residues need to be tested under field conditions using home-grown crop varieties before it can be recommended to local farmers. This study investigated the relative performance of indigenous maize residues (cobs, husks, and stalks) in oyster mushroom cultivation under farmer field conditions. The potential of these substrates and with other supplements to support oyster mushroom cultivation was also determined.

## 2. Materials and Methods

*2.1. Source of Substrates and Spawn.* The study was conducted at BenCom farm at Techiman in the Brong Ahafo Region of Ghana, between March and July 2011. BenCom farm located at 7.5772°N, 1.9292°W is the largest mushroom farm in Ghana that produces about 2500 tons of mushrooms annually. *Pleurotus ostreatus* spawn was obtained from BenCom farms. Sawdust of unknown timber species was collected from a local sawmill at the Techiman timber market. The maize residues (husk, cob, and stalk) were abundantly available from local farmers.

*2.2. Preparation of Growing Substrates.* Ten (10) kg of sawdust was mixed with 0.5% lime to maintain pH. The mixture was moistened to attain a moisture content of 70%. The mixture was then heaped into a pyramidal shape and remixed at 4-day intervals to ensure good aeration. The mixture was adequately fermented and ready for bagging after 28 days. The maize husks and stalks were cut into approximately 4 cm lengths and the cobs were crushed to approximately 0.5–4 cm. Each of the residues was then soaked in excess water for 12 hours. The excess water was drained out and the wet residue was sun-dried for one hour to obtain the maize residue substrate. In treatments with supplements, 840 g rice bran was admixed. Each substrate was bagged into one-kilogram lots using 33 × 18 cm heat resistant propylene bags. Polyvinyl chloride (PVC)

pipes were used to stiffen the spouts of the bags and these were sterilized at 100°C for 4 hours in a metal drum over naked flame and were allowed to cool before use.

*2.3. Spawning, Incubation, and Fruiting.* The bagged substrates were placed in a sterilized room and 50 g of mushroom spawn was inoculated into each 1 kg bag making a 5% spawn. The thorough spawning technique was used to enhance mycelia growth where the spawn was completely mixed with the substrate. The bags were covered with a sterilized cotton wool and then paper, which was fastened in place with a rubber band. This was incubated at 26–28°C for about 20–34 days in a well-ventilated room. Each treatment was replicated five times. The bags were moved to the growing house for fruiting when mycelia had fully grown. The PVC rings, cotton plugs, and the papers were removed to enhance aeration and stimulate pinning when the walls of the bags had been fully colonized. Holes were also created on the bags for emergence of fruiting bodies. Temperature and humidity were regulated by wetting the concrete floor and spawning bags.

*2.4. Data Collection and Analyses.* The fruiting bodies were pulled out and the number of mushroom and its fresh weight were recorded every six days for 6 consecutive weeks. The minimum period (6.33 days) between flushes reported by Bhatti et al. [17] was adopted. The total number of fruiting bodies, total number of flushes, and yield of each bag and total yield (g) of the mushrooms were calculated after the cropping period. After the last harvest the dry weight of the spent substrate was determined by exposing it to direct sunlight until there was no further change in weight for 3 consecutive days. The biological efficiency (BE) of each treatment was calculated using the formula:

$$BE = \frac{\text{Total fresh weight of mushroom (g) per substrate} \times 100}{\text{Dry weight of spent substrate (g)}}. \quad (1)$$

The mushroom weight and numbers were square-root-transformed before statistical analyses. The data was subjected to analysis of variance (ANOVA) using GenStat 9th edition. Mean separation was done using Least Significant Difference and effects were declared significant at 5% level.

## 3. Results

The number of fruiting bodies of oyster mushroom was significantly influenced by the different substrates. The four substrates sawdust, maize husk, maize stalk, and maize cobs significantly influenced the number of fruiting bodies of mushroom (Figure 1). Maize stalk substrate recorded the highest mean number of fruiting bodies per crop (5.17) and maize husks had the lowest 3.38 which was significantly different from the other substrates (Figure 3). Each of the substrates during the 7-week harvesting period had different periods for peak mushroom fruiting body numbers. Sawdust, the traditional substrate for mushroom production in the locality, and maize husk produced the highest fruiting body flushes at the first flush harvesting week (Figure 1), while maize cobs and maize stalk had their highest peaks during the

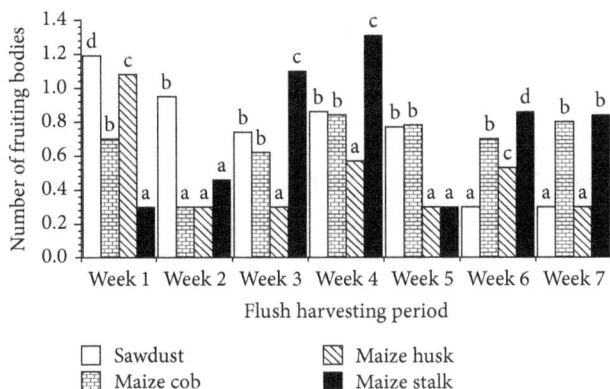

FIGURE 1: The effect of substrate material on fruiting body sprouts (flushes) over time. (1) Bars with the same Arabic numerals during a harvesting period are not significantly different $P < 0.001$. (2) 50 g of mushroom spawn was mixed with 1 kg substrate and incubated at 26–28°C for about 20–34 days. (3) The mushroom flushes were recorded over six weeks.

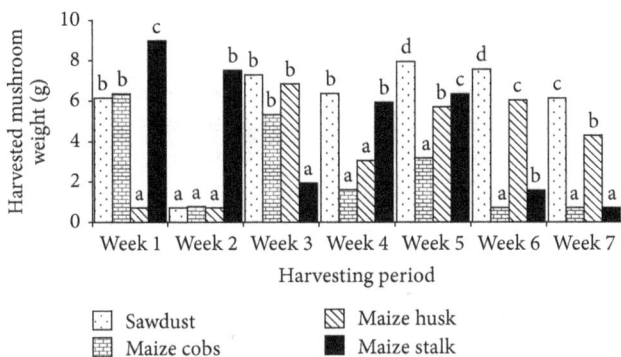

FIGURE 2: The effect of substrate material on mushroom yield over time. (1) Bars with the same Arabic numerals during a harvesting period are not significantly different $P < 0.001$. (2) 50 g of mushroom spawn was mixed with 1 kg substrate and incubated at 26–28°C for about 20–34 days. (3) The mushroom yield was recorded over six weeks.

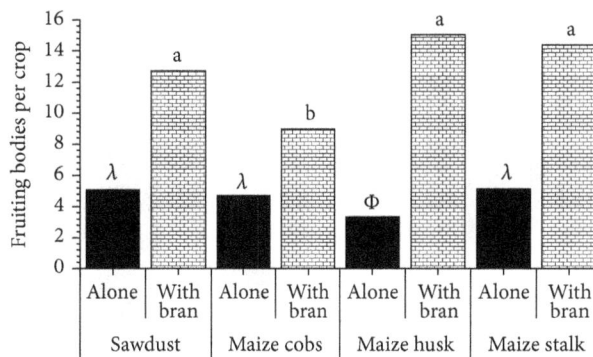

FIGURE 3: The effect of substrates on the total number of sprouted fruiting bodies in a crop. (1) Bars with the same Arabic letter for substrate material alone or same Greek letter for bran fortified substrates are not significantly different $P < 0.001$. (2) Substrates with bran contain 8.4% rice bran.

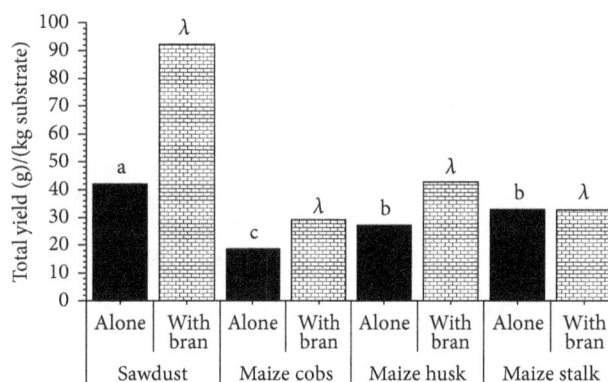

FIGURE 4: The effect of substrates on the yield of oyster mushrooms in a crop. (1) Bars with the same Arabic letter for substrate material alone or same Greek letter for bran fortified substrates are not significantly different $P < 0.001$. (2) Substrates with bran contain 8.4% rice bran.

fourth week. Supplemented substrates significantly affected the number of fruiting bodies that sprouted per crop. The average number of fruiting bodies obtained in each substrate alone was lower than when it was supplemented with rice bran. Among the substrates which were supplemented with rice bran, maize husks produced the highest mean number of fruiting bodies (2.15) per harvesting period followed by maize stalk (2.06) and the least was maize cobs which produced 1.28 fruiting bodies.

The substrates significantly influenced the fresh weight of mushroom produced (Figure 4). Sawdust recorded a significantly ($P < 0.001$) higher (42.18 g) fresh mushrooms per kilogram substrate in a crop and was followed by maize stalk (32.99 g), maize husk (27.55 g), and a significantly low mushroom weight from maize cob substrate (18.67 g). The fifth harvesting period gave the highest fresh mushroom weight and the least was during the second harvesting period (Figure 2). The highest harvested fresh mushroom weight

from a substrate was recorded on maize stalk during the first harvesting week (Figure 2). In the rice bran supplemented substrates, sawdust media resulted in the highest (92.2 g) fresh mushroom weight per kilogram substrate (Figure 4) but were not significantly different from either the maize cob (29.2 g), maize husk (42.8), or maize stalk (32.7 g).

The biological efficiency (BE) which is a function of the extent of substrate nutrient utilization was significantly different among the substrates. Sawdust had the highest biological efficiency of 60.1% and maize cobs had the lowest value of 14%. Of the supplemented substrates, maize stalk gave the highest biological efficiency of 39.2% followed by maize husk. Maize cobs recorded the lowest biological efficiency of 9.5% (Figure 5).

## 4. Discussion

Several edible mushrooms have been successfully cultivated at commercial level worldwide using lignocellulose wastes as substrates [18]. An ideal substrate should contain nitrogen

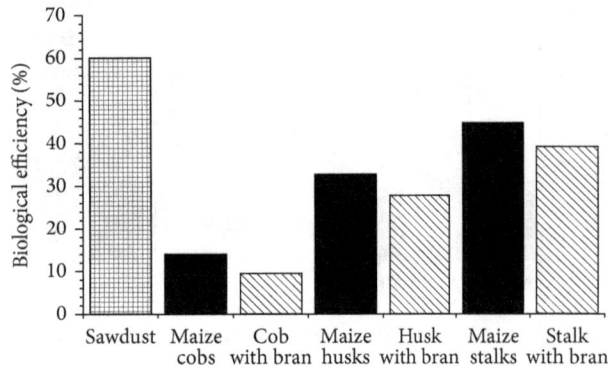

FIGURE 5: Biological efficiency (BE) of sawdust and maize residues. (1) Biological efficiency (BE) was a function of substrate usage to amount converted into harvested mushroom. (2) Substrates with bran contain 8.4% rice bran.

(supplement) and carbohydrates for rapid mushroom growth [19]. Various crop residues have been reported to have been used in producing oyster mushrooms either as main substrates or in combinations with supplements [10]. Oyster mushroom can be grown on various substrates including paddy straw, maize stalks/cobs, vegetable plant residues, bagasse [12], wheat straw [13], cotton waste, waste paper, and cotton stalks that are all suitable for high production capacity for oyster mushroom [14].

The response of the different substrates used shows differences in respect of time taken for formation of pinheads, maturation of fruiting bodies, period between flushes, number of flushes, and yield. Lozano [20] reported that seven harvestings were carried during 60 days, whereas Jiskani [21] reported 7.5 days between flushes and Bughio [22] recorded 9–15 days between flushes. There were 7 flushes within 42 days for all the substrates tested in this experiment based on our selected 5% spawning rate. The number and frequency of flushes depend on spawning rates. A minimum period of 6.33 days between flushes was taken when spawning rate was 2% and the period was 16.72 days at 10% spawning rate [17]. However, only one flush was harvested when a spawn rate of 1% on dry weight basis was used [20].

It is the general experience that the second flush will be the largest. It is not uncommon for the first flush to be the largest and with a few substrates the third flush may occasionally be the largest [23]. The flush number with the highest yield in this experiment was not definite as each substrate had its peak at different periods. On the average, the fifth harvesting period gave the highest fresh mushroom weight and the least was during the second harvesting period which flouts the general experience reported in [23]. The results showed that the tested maize by-products for the cultivation of oyster mushroom influenced the weight of mushroom produced as reported by Mendez et al. [11]. When oyster mushrooms are grown in the field, harvested fruiting bodies yield only 9 g/kg substrate [13] compared to the ~42 g/kg recorded for sawdust in this experiment. Sawdust has been reported as the best substrate for mycelia growth and fructification [24]. The high harvested mushroom

weight from the sawdust treatment could be due to good aeration as has been reported by Chang and Miles [25] that mushroom production was influenced by the composition of the substrate and the degree of aeration. The suitability of the substrates was also confirmed by their comparable biological efficiencies to the traditional sawdust. The nutritional components of mushroom substrates are known to influence mycelia growth [26]. Oyster mushrooms have the ability to utilize biomolecules for their growth [27]. A highly nutritive substrate also improves the sustenance of mycelia vegetative growth which leads to vigorous growth and late pinning [28].

The lipid contents of these residues were not tested in the current experiment; however, according to Naraian et al. [29] maize stalks have higher lipid components than cobs and husks. The lipids present in the maize stalks might have contributed to the high biological efficiency 44.4% (60.1% for sawdust) which represents the usable and available nutrients in a growing media. However, in other reports, the biological efficiency reached about 90–97% after 50–60 days with a spawn rate of 10% [30] as compared to the 5% used here. Although 25% spawn rate appeared superior, Fan et al. [30] recommended 10% spawn rate in view of the process economics while Bhatti et al. [17] also recommended 7% spawning rate on dry weight basis as this resulted in early and high yielding crop of oyster mushroom, with minimum period for maturation of fruiting bodies, maximum number of flushes and fruiting bodies. Supplementation of substrates has become one of the major techniques in mushroom cultivation. An ideal substrate should contain nitrogen (supplement) and carbohydrates for rapid mushroom growth [19]. Rice bran proved to be a suitable material for supplementing the maize residues as substrates. It increased the biological efficiency of all the substrates except for maize cobs. Ayodele and Akpaja [31] reported that supplementation of sawdust with oil palm fibres can enhance mycelia growth and sporophore yield of wood ear mushroom.

## 5. Conclusion

The study demonstrated that maize stalk is potentially suitable for use in commercial oyster mushroom production. Many locally available organic substrates have high potential for utilization as substrates and/or supplement for oyster mushroom production. In this study, even though sawdust is known to be the most suitable substrate for mushroom production, maize residues (with or without rice bran supplement) have also proven to be suitable substrates for oyster mushroom cultivation at the farm level. Maize husk and stalk were the most suitable in terms of the number of fruiting bodies and fresh weight of mushroom and are therefore recommended as potential substrate for cultivation of oyster mushroom. The differences in the results reported by other research workers should be expected since it may be due to the variation in growing requirements for cultivation of oyster mushroom such as temperature, humidity, and light regimens.

## Conflict of Interests

The authors declare that there is no conflict of interests regarding the publication of this paper.

## Acknowledgments

The authors are grateful to the management and staff of BenCom Youth Association/Enterprise in the Techiman Municipality of the Brong Ahafo Region of Ghana, where this study was conducted.

## References

[1] M. A. Belewu and K. Y. Belewu, "Cultivation of mushroom (*Volvariella volvacea*) on banana leaves," *African Journal of Biotechnology*, vol. 4, no. 12, pp. 1401–1403, 2005.

[2] T. H. Quimio, "Why grow mushroom," in *Mushroom Growers' Handbook*, pp. 1–12, Mushroom World, 1st edition, 2004.

[3] D. J. Royse, *Cultivation of Oyster Mushrooms*, College of Agricultural Sciences, Pennsylvania State University, State College, Pa, USA, 2003.

[4] G. C. Wakchaure, "Production and marketing of mushrooms: global and national scenario," in *Mushrooms-Cultivation, Marketing and Consumption*, M. Singh, B. Vijay, S. Kamal, and C. G. Wakchaure, Eds., Directorate of Mushroom Research, pp. 15–22, Solan, India, 2011.

[5] B. O. Onyango, V. A. Palapal, P. F. Arama, S. O. Wagai, and B. M. Gichimu, "Morphological characterization of Kenyan native wood ear mushroom (*Auricularia auricula* (L. ex Hook.) Underw and the effect of supplemented millet and sorghum grains in spawn production," *Agriculture and Biology Journal of North America*, vol. 3, pp. 2151–2157, 2010.

[6] S. W. Kang, "What is oyster mushroom," in *Mushroom Growers' Handbook*, pp. 48–51, Mushroom World, 1st edition, 2004.

[7] N. V. Chiejina and J. O. Olufokunbi, "Effects of different substrates on the yield and protein content of *Pleurotus tuberregium*," *African Journal of Biotechnology*, vol. 9, no. 11, pp. 1573–1577, 2010.

[8] J. C. Rajapakse, P. Rubasingha, and N. N. Dissanayake, "The potential of using cost-effective compost mixtures for oyster mushroom (*Pleurotus* spp.) cultivation in Sri Lanka," *Tropical Agriculture Research Extension*, vol. 10, pp. 29–32, 2007.

[9] G. Jonathan, A. Ajayi, I. Oku, and D. Wankasi, "Nutritive value of common wild edible mushrooms from Southern Nigeria," *Global Journal of Biotechnology & Biochemistry*, vol. 1, no. 1, pp. 16–21, 2006.

[10] J. Ashraf, M. A. Ali, W. Ahmad, C. M. Ayyub, and J. Shafi, "Effect of different substrate supplements on oyster mushroom (*Pleurotus* spp.)," *Production Food Science and Technology*, vol. 1, no. 3, pp. 44–51, 2013.

[11] L. A. Mendez, C. A. S. Castro, R. B. Casso, and C. M. C. Leal, "Effect of substrate and harvest on the amino acid profile of Oyster mushroom (*Pleurotus ostreatus*)," *Journal of Food Composition and Analysis*, vol. 18, no. 5, pp. 447–450, 2005.

[12] S. Hassan, A. Y. Mohammad, and K. Kiramat, "Cultivation of the oyster mushroom (*Pleurotus ostreatus* (Jacq.) P. Kumm) in two different agroecological zones of Pakistan," *African Journal of Biotechnology*, vol. 10, pp. 183–188, 2011.

[13] N. Badshah, N. Ur-Rehman, and M. Wahid, "Yield and quality of mushrooms grown on different substrates," *Sarhad Journal of Agricultural*, vol. 8, no. 6, pp. 631–635, 1992.

[14] T. Marimuthu, "Prospects of oyster mushroom cultivation in Tamil Nadu," *Journal of Ecobiology*, vol. 7, no. 1, pp. 27–34, 1995.

[15] M. Atikpo, O. Onokpise, M. Abazinge et al., "Sustainable mushroom production in Africa: a case study in Ghana," *African Journal of Biotechnology*, vol. 7, no. 3, pp. 249–253, 2008.

[16] M. Obodai, J. Cleland-Okine, and K. A. Vowotor, "Comparative study on the growth and yield of *Pleurotus ostreatus* mushroom on different lignocellulosic by-products," *Journal of Industrial Microbiology and Biotechnology*, vol. 30, no. 3, pp. 146–149, 2003.

[17] M. I. Bhatti, M. M. Jiskani, K. H. Wagan, M. A. Pathan, and M. R. Magsi, "Growth, development and yield of oyster mushroom, *Pleurotus ostreatus* (Jacq. Ex. Fr.) Kummer as affected by different spawn rates," *Pakistan Journal of Botany*, vol. 39, no. 7, pp. 2685–2692, 2007.

[18] A. Ingale and A. Ramteke, "Studies on cultivation and biological efficiencies of mushrooms grown on different agro residues," *Innovative Romanian Food Biotechnology*, vol. 6, pp. 25–28, 2010.

[19] K. B. Khare, J. M. Mutuku, O. S. Achwania, and D. O. Otaye, "Production of two oyster mushrooms, *Pleurotus sajorcaju* and *P. florida* on supplemented and un-supplemented substrates," *International Journal of Agriculture and Applied Sciences*, vol. 6, pp. 4–11, 2010.

[20] J. C. Lozano, "Commercial production of oyster mushroom (*Pleurotus ostreatus*) in coffee pulp," *Fitopatologia Colombiana*, vol. 14, no. 2, pp. 42–47, 1990.

[21] M. M. Jiskani, *A Brief Outline "The Fungi" Cultivation of Mushrooms*, Izhar Publisher, Tando Jam, Pakistan, 1999.

[22] I. Bughio, *Yield performance of oyster mushroom, Pleurotus ostreatus (Jacq. ex. Fr.) Kummer on combination of different straws [M.S. thesis]*, Department of Plant Pathology Sindh Agriculture University, Tando Jam, Pakistan, 2001.

[23] R. H. Kurtzman Jr., "A review Mushrooms: sources for modern western medicine," *Micologia Aplicada International*, vol. 17, no. 2, pp. 21–33, 2005.

[24] M. Kadiri and I. O. Fasidi, "Variations in chemical composition of *Chlorophyllum molybditis* (Mayerex. Fr.) Massee and Pleurotus tuber-regium (Fries) during fruit body development," *Nigerian Journal of Science*, vol. 24, pp. 86–89, 1990.

[25] S. T. Chang and P. G. Miles, *Edible Mushrooms and Their Cultivation*, CRC Press, Boca Raton, Fla, USA, 1989.

[26] B. O. Onyango, V. A. Palapala, P. F. Arama, S. O. Wagai, and B. M. Gichimu, "Suitability of selected supplemented substrates for cultivation of Kenyan native wood ear mushrooms (*Auricularia auricula*)," *American Journal of Food Technology*, vol. 6, no. 5, pp. 395–403, 2011.

[27] J. Poppe, "Use of agricultural waste materials in the cultivation of mushrooms," *Mushroom Science*, vol. 15, pp. 3–23, 2000.

[28] J. W. Kimenju, G. O. M. Odero, E. W. Mutitu, P. M. Wachira, R. D. Narla, and W. M. Muiru, "Suitability of locally available substrates for oyster mushroom (*Pleurotus ostreatus*) cultivation in Kenya," *Asian Journal of Plant Sciences*, vol. 8, no. 7, pp. 510–514, 2009.

[29] R. Naraian, R. K. Sahu, S. Kumar, S. K. Garg, C. S. Singh, and R. S. Kanaujia, "Influence of different nitrogen rich supplements during cultivation of *Pleurotus florida* on corn cob substrate," *Environmentalist*, vol. 29, no. 1, pp. 1–7, 2009.

[30] L. Fan, A. Pandey, R. Mohan, and C. R. Soccol, "Use of various coffee industry residues for the cultivation of *Pleurotus ostreatus* in solid state fermentation," *Acta Biotechnologica*, vol. 20, no. 1, pp. 41–52, 2000.

[31] S. M. Ayodele and E. O. Akpaja, "Yield evaluation of *Lentinus squarosulus* (Mont) Sing on selected sawdust of economic tree species supplemented with 20% oil palm fruit fibers," *Asian Journal of Plant Sciences*, vol. 6, pp. 1098–1102, 2007.

# 3

# Effects of Different Treatments on Seed Germination Improvement of *Calotropis persica*

**Asghar Farajollahi,[1] Bahram Gholinejad,[2] and Hamed Jonaidi Jafari[2]**

[1] *Combating Desertification, Gorgan University of Agricultural Science and Natural Resources, Gorgan 49138 15739, Iran*
[2] *Faculty of Natural Resources, University of Kurdistan, Sanandaj 31585 4314, Iran*

Correspondence should be addressed to Bahram Gholinejad; bahramgholinejad@yahoo.com

Academic Editor: Tibor Janda

The purpose of this study was to investigate the effects of different treatments on seed germination in the desert plant species *Calotropis persica* (Gand.). This species is known to have long time for seed germination considering arid region condition and short time of access moist. An experiment was performed with 13 treatments and 4 replications in a completely randomized design. Treatments included $KNO_3$ with concentrations of 0.1, 0.2, and 0.3 percent, immersion in hot water for five min, acetylsalicylic acid 100, 200, and 300 mg $L^{-1}$, ethereal sulfuric acid (60%) for 5 and 10 min, thiourea with concentrations of 0.1% and 0.3%, and prechilling for 10 days. Tap water was used as the control. Our findings indicate that $KNO_3$ 0.1% and 100 mg $L^{-1}$ acetylsalicylic acid were the most effective treatments for improvement of seed germination properties in this species. In a comparison of the two mentioned treatment, $KNO_3$ 0.1% treatments is the best.

## 1. Introduction

Germination is a critical stage in the life cycle of weeds and crop plants and often controls population dynamics, with major practical implications. Seed germination is the critical stage for species survival [1, 2]. In recent 20 years, desertification has been recognized as a major environmental problem and is a major focus of United Nations Environment Programme [3]. Vegetation is a protector of the soil against water and wind erosion as well as a casualty of soil erosion [4, 5]. Each desert-inhabiting plant has its own complex of strategies that enables it to persist in desert habitats [6]. Strategies for improving the growth and development of arid region plant species have been investigated for many years. Treated seeds with chemical compound usually would exhibit rapid germination when absorbing water under field conditions [7].

*Calotropis* is a genus of flowering plants in the dogbane family, Apocynaceae. They are commonly known as milkweeds because of the latex they produce. *Calotropis* species are considered common weeds in some parts of the world. The flowers are fragrant and are often used in making floral tassels in some mainland Southeast Asian cultures. *Calotropis persica* is growing in tropical region only. Iran is a country in the mid-latitude belt of arid and semiarid regions of the Earth. Approximately 60% of Iran is classified as arid and semiarid [8]. Based on results, the seed of full ripening fruits with scarification had the highest germination percent. [9] investigated the effects of salt stress and prime on germination improvement and seedling growth of *Calotropis procera* L. seeds and the results showed that priming improved the seedling characteristics in all samples, especially in −0.05 MPa, but a decrease with decrease in osmotic potential. The work in [10] studied the effect of temperature, light, pretreatment, and storage on seed germination of *Rhodomyrtus tomentosa* and their result showed that light significantly improved germination of fresh seeds but storage decreased the light-sensitivity of germination. Soaking for 24 hours in 250–600 mg $L^{-1}$ gibberellic acid, 5–20% potassium nitrate, or 10% hydrogen peroxide solution increased seed germination. *Calotropis* sp. is an important economic plant used for drug and other purposes. The purpose of this study was to develop methods to increase germination percentage, shorten germination time, provide

TABLE 1: Analysis of variance for treatments on germination properties of C. persica.

| Germination properties | Source of variation | Degree of freedom | Sum of squares | Mean of squares | $F$ (Fisher test) |
|---|---|---|---|---|---|
| Germination percentage | Between groups | 12 | 26678.793 | 2223.233 | 35.758** |
|  | Within groups | 39 | 2424.805 | 62.174 |  |
| Mean germination time | Between groups | 12 | 97.666 | 8.139 | 4.194** |
|  | Within groups | 39 | 75.675 | 1.940 |  |
| Germination rate | Between groups | 12 | 0.051 | 0.004 | 3.550** |
|  | Within groups | 39 | 0.047 | 0.001 |  |

**Significant difference at 1%.

more rate germination, and result in more efficient seed propagation techniques for C. persica seeds.

## 2. Material and Methods

Seeds of C. persica were collected from Jiroft arid regions in southern Iran in 2013. A preliminary germination test was performed and low germination percentage was obtained. To solve this problem, we implemented an experiment with a randomized complete design. Before the start of experiment, seeds were surface sterilized in 1% sodium hypochlorite solution for 5 min, then rinsed with sterilized water, and air-dried for 28 h before putting in petri dishes. Treatments included pretreatment with $KNO_3$ (0.1 and 0.3 percent) for 48 hours, acetylsalicylic acid to the moisture in the petri dish (100, 200 and 100 mg $L^{-1}$), prechilling (4 degrees centigrade for 10 days), hot water (70°C) for 5 min, ethereal sulfuric acid (60%) for 5 and 10 min, thiourea with concentrations of 0.1% and 0.3%, and control treatment (irrigation with distilled water). The seeds were placed on top of Whatman paper number 1 within 10 cm petri dishes containing 10 mL distilled water. Counting number of germinating seeds began from the first day and was done till the end of the experiment (19 days). Germination percentage was recorded daily during the study period. Rate of germination was estimated using modified Timpson's index of germination velocity [11]. Mean germination time (MGT) was calculated to assess the rate of germination [12]:

$$MGT = \frac{\sum D \cdot N}{n}, \qquad (1)$$

where $N$ is the number of seeds which in $D$ day grow, $n$ the total number of seeds grown, and $D$ the number of days from the date of germination and the germination rate index was obtained by reversing MGT at the end of this period; final germination percentage was recorded. There are no outliers; normality of data was checked and nonnormal data transformed by arc sin to verification of this hypothesis arc sin transformation was used for germination percentage before analysis [13]. Experimental data was analyzed by SPSS 17.0 to analyze the data and Duncan's test at 5% level was used to compare the means.

## 3. Results

The results of ANOVA (Table 1) showed that there are significant differences (at 1% level) between effective treatments on germination characteristics and the different treatments resulted in significant differences among germination properties (Table 1).

The results of this research showed that germination percentage of C. persica increases due to application of $KNO_3$ in different concentrations and acetylsalicylic acid 100 and 200 mg $L^{-1}$ and decreased germination percentage due to application of hot water for 5 min, prechilling for 10 days, sulfuric acid 5 and 10 min, and thiourea 0.3%. Acetylsalicylic acid 300 mg $L^{-1}$ and thiourea 0.1% have the same effect on germination percentage in comparison to control treatment. The increased germination percentage by $KNO_3$ 0.1, 0.2, and 0.3% and acetylsalicylic acid 100 mg $L^{-1}$ was significant (Figure 1).

The seed germination rates of C. persica increased significantly when $KNO_3$ 0.1% was used. Acetylsalicylic acid 200 and 300 mg $L^{-1}$ and thiourea 0.3% increased seed germination rate, but this increase was not significant. However, the germination rate was decreased when hot water for 5 min, prechilling for 10 days, sulfuric acid for 5 and 10 min, $KNO_3$ 0.2 and 0.3%, acetylsalicylic acid 100 mg $L^{-1}$, and thiourea 0.3% were used (Figure 2).

Mean germination time of C. persica decreased by using $KNO_3$ 0.1% but this difference was not significant. In seeds of C. persica, all treatments, except for $KNO_3$ 0.1%, caused increase in mean germination time (Figure 3).

## 4. Discussion and Conclusion

According to the obtained results, $KNO_3$ 0.1% and acetylsalicylic acid 100 mg $L^{-1}$ were the most effective treatments for improvement of seed germination properties in C. persica plant species. In a comparison of the two mentioned treatments, $KNO_3$ 0.1% treatment is the best. This technique has become a common seed treatment that can increase rate, percentage, and uniformity of germination or seedling emergence, mainly under unfavorable environmental conditions. Rapid seed germination and stand establishment are critical factors for crop production under stress conditions. Hot water for 5 min and prechilling for 10 days did not show positive effect on germination improvement. The study result of scarification of seeds of Acacia angustissima showed that seeds soaking in hot water cause seed germination induction but increasing duration of seed contact with hot water leads to decline of seed germination percentage [14]. In a research it is shown that prechilling for 10 days had

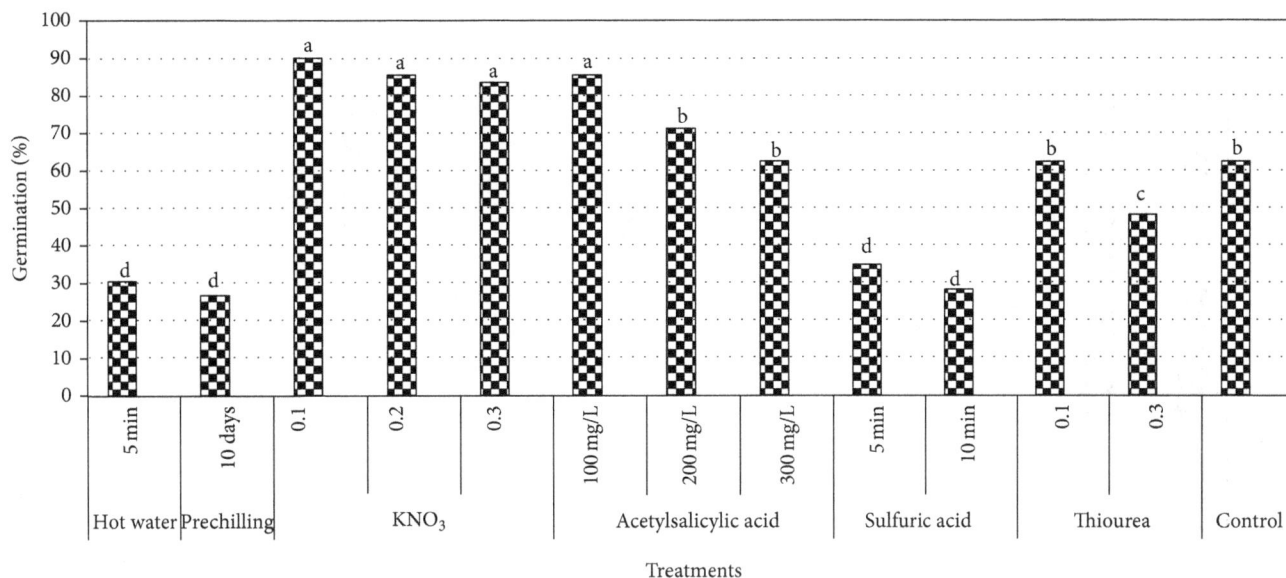

FIGURE 1: Comparison effects of various treatments on germination percentage of *C. persica*.

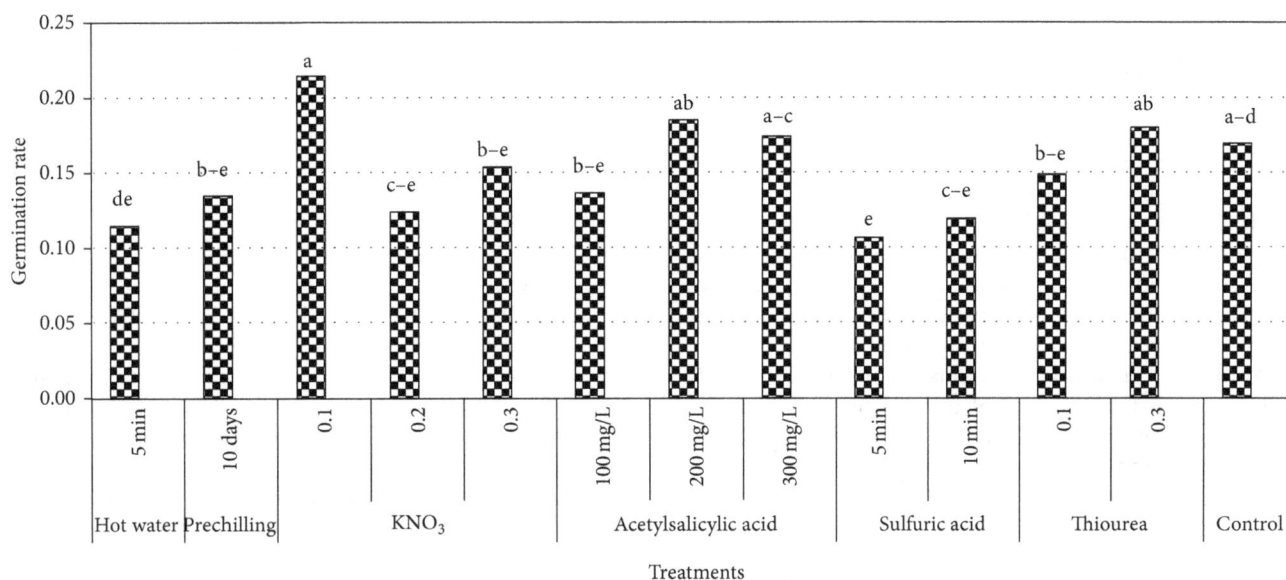

FIGURE 2: Comparison effects of various treatments on germination rate of *C. persica*.

a positive effect on germination rate and mean germination time of both medicinal species of *Foeniculum vulgare* and *Cuscuta epithymum* but germination percentage decreased due to application of prechilling [15]. Sulfuric acid for 5 and 10 min did not have positive effect on seed germination of *C. persica*; as a result, seed treatment with sulfuric acid cannot improve seed germination. This result demonstrated that above mentioned treatment had the destructive effect on embryo. It is notified that increasing in doses of sulfuric acid caused germination improvement and suggested chemical scarification in concentrated $H_2SO_4$ for 2 hours [16]. In this research thiourea did not have a positive effect on germination improvement of *C. persica*. Acetylsalicylic acid

$100\,mg\,L^{-1}$ improved mean germination time and germination percentage in comparison with amount of 200 and $300\,mg\,L^{-1}$.

In this research, $KNO_3$ 0.1% is recognized as the best treatment for improvement seed germination properties of *C. persica*. Similar results were reported in previous studies for the species of *Citrullus colocynthis* [17], *Foeniculum vulgare* and *Cuscuta epithymum* [15], *Hypericum aviculariifolium* [18], and *Avena fatua* [19]. According these results, $KNO_3$ 0.1% treatment is suggested for improvement of *C. persica* germination and this treatment is proper for propagation of studied species. Positive effect of $KNO_3$ could be due to its role in balancing hormonal portion within seed which

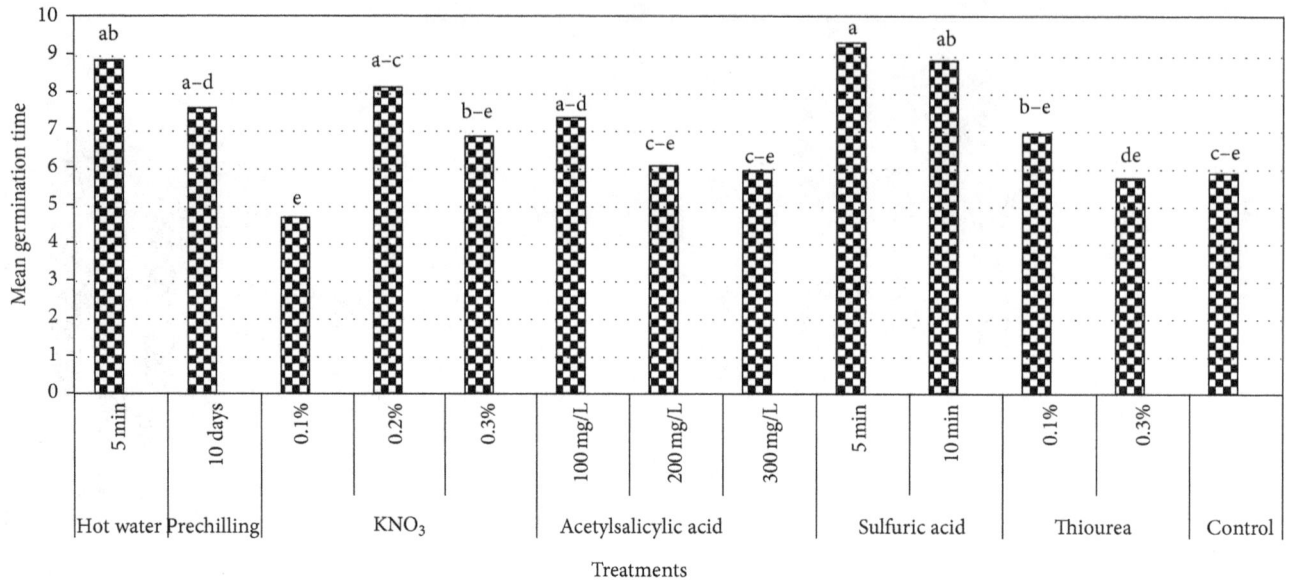

FIGURE 3: Comparison effects of various treatments on mean germination time of *C. persica*.

in turn results in germination inhibitors ratio like ABA. (abscisic acid). Virtually all of the cellular and metabolic events that are known to occur before the completion of germination of nondormant seeds also occur in imbibed dormant seeds; indeed, the metabolic activities of the latter are frequently only subtly different from those of the former [20]. The seeds of most Mediterranean and desert species have dormancy characteristics or structural properties that prevent immediate germination of at least a proportion of the seeds [21–24]. The results obtained will be useful in carrying out tree improvement and plantings of *C. persica* trees for fuel wood, local medicine, and industrial production. Rapid seedling growth is also essential for reclamation of desert. This information could ultimately help in the sustainable development of the arid zones.

## Conflict of Interests

The authors declare that there is no conflict of interests regarding the publication of this paper.

## Acknowledgments

This research was supported by Iranian Revolutionary Guards Navy. The authors would like to thank Hamid Reza Ahmadinia Ph.D. student of Fishery in Gorgan University of Agriculture and Natural Resources for his help with the creation of this work.

## References

[1] Z. Huang, X. Zhang, G. Zheng, and Y. Gutterman, "Influence of light, temperature, salinity and storage on seed germination of *Haloxylon ammodendron*," *Journal of Arid Environments*, vol. 55, no. 3, pp. 453–464, 2003.

[2] Q. H. Yang, X. Wei, X. L. Zeng et al., "Seed biology and germination ecophysiology of *Camellia nitidissima*," *Forest Ecology and Management*, vol. 255, no. 1, pp. 113–118, 2008.

[3] S. Zare, A. Tavili, and M. J. Darini, "Effects of different treatments on seed germination and breaking seed dormancy of *Prosopis koelziana* and *Prosopis Juliflora*," *Journal of Forestry Research*, vol. 22, no. 1, pp. 35–38, 2011.

[4] C. J. Yates, D. A. Norton, and R. J. Hobbs, "Grazing effects on plant cover, soil and microclimate in fragmented woodlands in south-western Australia: implications for restoration," *Austral Ecology*, vol. 25, no. 1, pp. 36–47, 2000.

[5] M. G. Manzano and J. Návar, "Processes of desertification by goats overgrazing in the *Tamaulipan thornscrub* (matorral) in North-eastern Mexico," *Journal of Arid Environments*, vol. 44, no. 1, pp. 1–17, 2000.

[6] Y. Gutterman, "Strategies of seed dispersal and germination in plants inhabiting deserts," *Botanical Review*, vol. 60, no. 4, pp. 373–425, 1994.

[7] M. Ashraf and M. R. Foolad, " Pre-sowing seed treatment-a shotgun approach to improve germination growth and crop yield under saline and none-saline conditions," *Advances in Agronomy*, vol. 88, pp. 223–271, 2005.

[8] S. J. Milton, "Spatial and temporal patterns in the emergence and survival of seedlings in arid Karoo shrubland," *Journal of Applied Ecology*, vol. 32, no. 1, pp. 145–156, 1995.

[9] M. Taghvaei, N. Khaef, and H. Sadeghi, "The effects of salt stress and prime on germination improvement and seedling growth of *Calotropis procera* L. seeds," *Journal of Ecology and Field Biology*, vol. 35, no. 2, pp. 73–78, 2012.

[10] H. L. Liang, H. N. Liu, Q. H. Yang et al., "Seed germination of *Rhodomyrtus tomentosa*," *Seed Science and Technology*, vol. 41, no. 2, pp. 188–189, 2013.

[11] M. A. Khan and I. A. Ungar, "The effect of salinity and temperature on germination of polymorphic seeds and growth of *Atriplex triangularis* wild," *The American Journal of Botany*, vol. 71, pp. 481–489, 1984.

[12] R. A. Ellis and E. H. Roberts, "The quantification of ageing and survival in orthodox seeds," *Seed Science and Technology*, vol. 9, pp. 373–409, 1981.

[13] S. M. Khan, J. Nazir, H. K. Zahoor, and M. K. Sultan, "Yield performance of oyster mushroom," *Pakistan Journal of Phytopathology*, vol. 18, pp. 89–93, 2006.

[14] R. Rincón-Rosales, N. R. Culebro-Espinosa, F. A. Gutierrez-Miceli, and L. Dendooven, "Scarification of seeds of *Acacia angustissima* (Mill.) Kuntze and its effect on germination," *Seed Science and Technology*, vol. 31, no. 2, pp. 301–307, 2003.

[15] A. Tavili, A. Farajollahi, H. Pouzesh, and E. Bandak, "Treatment induced germination improvement in medicinal species of foeniculum vulgare miller and Cuscuta epithymum(L.) L," *Journal of Modern Applied Science*, vol. 4, no. 7, pp. 163–169, 2010.

[16] T. Merou, I. Takos, E. Konstantinidou, S. Galatsidas, and G. Varsamis, "Effect of different pretreatment methods on germination of *Albizia julibris* sin seeds," *Seed Science and Technology*, vol. 39, no. 1, pp. 248–252, 2011.

[17] M. Saberi, A. Shahriari, F. Tarnian, and S. Noori, "Comparison the effect of different treatments for breaking seed dormancy of *Citrullus colocynthis*," *Journal of Agricultural Science*, vol. 3, no. 4, pp. 62–67, 2011.

[18] C. Çirak, K. Kevseroğlu, and A. K. Ayan, "Breaking of seed dormancy in a Turkish endemic Hypericum species: *Hypericum aviculariifolium* subsp. *depilatum* var. *depilatum* by light and some pre-soaking treatments," *Journal of Arid Environments*, vol. 68, no. 1, pp. 159–164, 2007.

[19] J. R. Hilton, "The influence of light and potassium nitrate on the dormancy and germination of *Avena fatua* L. (wild oat) seed and its ecological significance," *New Phytologist*, vol. 96, no. 1, pp. 31–34, 1984.

[20] J. D. Bewley, "Seed germination and dormancy," *The Plant Cell*, vol. 9, no. 7, pp. 1055–1066, 1997.

[21] C. A. Thanos, K. Georghiou, and F. Skarou, "*Glaucium flavum* seed germination—an ecophysiological approach," *Annals of Botany*, vol. 63, no. 1, pp. 121–130, 1989.

[22] E. Jurado and M. Westoby, "Germination biology of selected central Australian plants," *Australian Journal of Ecology*, vol. 17, no. 3, pp. 341–348, 1992.

[23] Y. Gutterman, *Seed Germination in Desert Plants*, Springer, Berlin, Germany, 1993.

[24] D. T. Bell, D. P. Rokich, C. J. McChesney, and J. A. Plummer, "Effects of temperature, light and gibberellic acid on the germination of seeds of 43 species native to Western Australia," *Journal of Vegetation Science*, vol. 6, no. 6, pp. 797–806, 1995.

# 4

# Impacts of Improved Switchgrass and Big Bluestem Selections on Yield, Morphological Characteristics, and Biomass Quality

Erik Delaquis,[1] Roger Samson,[2] Philippe Seguin,[1] Arif Mustafa,[3] and Huguette Martel[4]

[1] Department of Plant Science, McGill University, Macdonald Campus, 21111 Lakeshore Road, Sainte-Anne-de-Bellevue, QC, Canada H9X 3V9

[2] REAP Canada, 21111 Lakeshore Road, Sainte-Anne-de-Bellevue, QC, Canada H9X 3V9

[3] Department of Animal Science, McGill University, Macdonald Campus, 21111 Lakeshore Road, Sainte-Anne-de-Bellevue, QC, Canada H9X 3V9

[4] MAPAQ, Direction Régionale de l'Estrie, Sherbrooke, QC, Canada J1N 2A5

Correspondence should be addressed to Philippe Seguin; philippe.seguin@mcgill.ca

Academic Editor: Qiquan Wang

Switchgrass (*Panicum virgatum* L.) and big bluestem (*Andropogon gerardii* V.) are promising warm-season grasses for biomass production. Understanding the morphological and quality-related traits of these grasses can guide breeders in developing strategies to improve yield and quality for bioindustrial applications. Elite selections were made in Southern Quebec from four promising varieties of switchgrass and one of big bluestem. Biomass yield, morphological characteristics, and selected quality traits were evaluated at two sites in 2011 and 2012. Significant variation was detected for all measured characteristics, with differences varying by site and year. In some cases the selection process modified characteristics including increasing height and reducing tiller mortality. Switchgrasses reached a similar tiller equilibrium density in both years of $690\,\mathrm{m}^{-2}$ and $379\,\mathrm{m}^{-2}$ at a productive and marginal site, respectively. Differences in yield were pronounced at the marginal site, with some advanced selections having a higher yield than their parent varieties. Switchgrass yields were generally greater than those of big bluestem. A delayed spring harvest date greatly reduced yield but reduced moisture content and slightly increased cellulose concentration. Big bluestem had a higher cellulose content than switchgrass, likely due to greater stem content.

## 1. Introduction

Switchgrass (*Panicum virgatum* L.) is a perennial $C_4$ grass native to North America which has historically been used as a forage crop and for conservation plantings [1–3]. Recently interest has shifted towards the use of the crop in biomass energy and biofibre applications [4], with research indicating significant production potential in Eastern Canada [5]. Growing interest in this multipurpose crop has led to an increasing number of breeding and selection programmes and the release of new commercial switchgrass selections in the 1980s ("Trailblazer," "Forestburg," "Dacotah," and "KY1625"), 1990s ("Sunburst" and "Shawnee"), and 2000s ("High Tide," "Carthage," "BoMaster," "Performer," "Cimarron," and "Colony") [6–8]. Despite an increasing acreage dedicated to the production of this crop in Eastern Canada,

these selections all represent the efforts of plant material improvement programs located in the Atlantic Coastal, Midwest, and Northern Great Plains regions of the United States. There has been little selection effort *in situ* in Eastern Canada or in the Northeastern United States.

Big bluestem (*Andropogon gerardii* Vitman) is the most dominant grass of the tallgrass prairie ecosystem and has been demonstrated to be another high yielding warm-season grass in the Eastern North American context [9, 10]. Fully established stands of big bluestem can approximate the yields of switchgrass stands of a similar maturity [10, 11]. Big bluestem is generally taller than upland switchgrass and native big bluestem ecovars have been found to contain 62% of the biomass in the stem component [12], compared with native switchgrass with 47% [13]. Big bluestem appears to have some biomass quality advantages over switchgrass for certain fibre applications as it generally has improved fibre strength properties [11] and higher cellulose contents due to its higher stem-to-leaf ratio [14]. In an assessment of 5 switchgrasses and 4 big bluestems at heading, it was found that there were no differences in cellulose contents between big bluestem and switchgrass stems and leaves; however, stems of both species had on average 11% higher cellulose contents than leaves [14]. Early-harvested big bluestem has also been identified to have improved digestibility for livestock due to its leafy composition prior to heading [15]. Big bluestem has been used for forming polycultures with switchgrass and other native grasses and forbs in mixed prairie systems. Mixtures have many demonstrated benefits including increased indigenous avian and arthropod diversity and species richness [16, 17] and once fully established can produce similar yields to switchgrass monocultures [18, 19].

Switchgrass production in Quebec currently covers about 1500 ha, with the acreage increasing up to 30% per year (Huguette Martel, unpublished data). The vast majority of this acreage is the variety "Cave-in-Rock," a collection made from a field in Cave-in-Rock, Illinois, in 1958 [20]. Previous experiences indicate that appropriate regional selections can significantly increase yield potential in a given environment [21], but despite this there are currently no commercial varieties available that have been developed specifically for agronomic applications for Eastern Canada or the Northeastern United States. There has also been no plant breeding effort conducted in these regions for big bluestem.

Previous studies with switchgrass and tall fescue confirmed the effectiveness of seedling-stage selection for developing selections with modified tiller number [22–24] and increased yield per tiller [22, 25]. It has also been demonstrated that selection and breeding programmes have the capacity to alter fiber composition in switchgrass [26], an important characteristic for energy and fibre crops. Fiber content, including lignin and cellulose concentrations, has been demonstrated to be important parameters for multiple conversion technologies and thus important breeding targets for biomass crops [15].

The development of a spring harvest system for switchgrass is another potential avenue for improving biomass quality for energy and fibre applications. In biomass crops, the leaching of minerals from plant tissues throughout the winter has been shown to result in lower mineral concentrations [27, 28], while the loss of leaves may decrease the concentration of major ash components such as silicates and increase cellulose content [29]. Elevated levels of silica are generally considered undesirable for commercial combustion processes [30]. Spring harvesting also carries the added advantages of reducing labor demands during peak harvest time for commercial cash crops and drastically reducing moisture content at harvest which reduces or eliminates the need for drying [31] and is thus beneficial for baling and storage [27].

The goal of the present study was to compare the agronomic characteristics of several commercial varieties of upland switchgrass and big bluestem along with advanced selections made from these varieties to determine the effects on their performance and morphological characteristics in Southern Quebec. Several of these selections were also subjected to both spring and fall harvest to determine the effects of harvest date on yield, fiber, mineral, and energy concentrations of the harvested biomass.

## 2. Material and Methods

*2.1. Plant Materials.* Selection lines and parent material were supplied by REAP-Canada, based in Southern Quebec (Sainte-Anne-de-Bellevue, QC, Canada). The biomass breeding program is utilizing short breeding cycles incorporating selection at the seedling stage followed by recurrent restricted phenotypic selection (RRPS) in mature spaced-plant nurseries [32]. A total of 11 upland switchgrass selections and three big bluestem selections were evaluated. These included three switchgrass parent varieties and seven selections derived from these (Table 1). "Sandlover," a selection made by REAP-Canada from "NU-94-2" [15], was also evaluated bringing the total switchgrass selections evaluated to eleven. In addition, two selections of big bluestem were also evaluated alongside their parent material "Prairie View Indiana germplasm".

The "Blue Jacket" and "Blue Jacket II" switchgrass selections were made from "Sunburst" switchgrass using RRPS [33]. The population of "Blue Jacket II" was selected for reduced tillering at the seedling stage, with 200 superior plants selected from a population of 1000 plants [22] followed by a RRPS selection made on mature plants. The "Blue Jacket Early" selection was derived from 50 plants which exhibited early spring growth in the year after establishment in a one ha seed field of "Blue Jacket" switchgrass in Valleyfield, Quebec. The plants were dug up in the spring of 2008 and replanted and seed was collected from the population in the fall of 2009.

"Cave-in-Rock II" was derived from a seed collection in the fall of 2006 from 30 superior plants in an eight-year-old switchgrass field in Valleyfield, Quebec. The population was subsequently selected for reduced tillering at the seedling stage were transplanted in the field [22]. At the mature plant stage, the population was selected using RRPS and seed derived from superior plants from a 200-spaced-plant nursery was collected in fall 2009. The "Cave-in-Rock Early" population was derived from seed collected in fall of 2009 from 30 early maturing switchgrass plants in a 10-year-old field of "Cave-in-Rock" switchgrass in Valleyfield, Quebec.

The "Tecumseh" and "Tecumseh II" selections were made from "Summer" switchgrass using RRPS [33]. A greenhouse population of 1000 plants of "Tecumseh II" was selected for reduced tillering at the seedling stage [22]. At the mature plant stage, the population was selected using RRPS and seed derived from superior plants from the 200-spaced-plant nursery was collected in fall of 2009.

"Sandlover" was derived from 200 plants selected from a 1,000-spaced-plant nursery in Valleyfield, Quebec, using RRPS [33]. The original source population was Northern Upland "NU-942" obtained from Charles Taliaferro at the University of Oklahoma in 2006.

Finally, the selections "Prairie View Early" and "Prairie View II" were selected from an 800-spaced-plant nursery of Prairie View Indiana germplasm big bluestem originally transplanted in 2008 in Valleyfield, Quebec.

TABLE 1: Origins of the commercial switchgrass and big bluestem varieties evaluated in the present study and the selections derived from them through a local selection programme in Southern Quebec.

| Variety/selection | Year of release | Site of origin | Latitude of origin (°N) | References |
|---|---|---|---|---|
| Switchgrass | | | | |
| Cave-in-Rock | 1958 | Cave-in-Rock, Illinois | 37 | Jefferson and McCaughey, 2012 [6] |
| Cave-in-Rock II | N/A | Valleyfield, Quebec | | Present study |
| Cave-in-Rock Early | N/A | Valleyfield, Quebec | | Present study |
| Summer | 1953 | Nebraska City, Nebraska | 41 | Jefferson and McCaughey, 2012 [6] |
| Tecumseh | N/A | Valleyfield, Quebec | | Present study |
| Tecumseh II | N/A | Valleyfield, Quebec | | Present study |
| Sunburst | 1998 | Union County, South Dakota | 43 | Jefferson and McCaughey, 2012 [6] |
| Blue Jacket | N/A | Valleyfield, Quebec | | Present study |
| Blue Jacket II | N/A | Valleyfield, Quebec | | Present study |
| Blue Jacket Early | N/A | Valleyfield, Quebec | | Present study |
| Sandlover | 2009 | Northern Nebraska | 36–40 | Vermerris, 2008 [15] |
| Big bluestem | | | | |
| Prairie view | 1994 | 20 sites across Indiana | 38–41 | Samson et al., 2014 [32] |
| Prairie view II | N/A | Valleyfield, Quebec | | Present study |
| Prairie view Early | N/A | Valleyfield, Quebec | | Present study |

*2.2. Field Management and Data Collection.* Plots were arranged in a randomized complete block design with four replications at two sites in Southern Quebec, Sainte-Anne-de-Bellevue (45°25′32.45″N, 73°55′48.63″W, 37 m elevation) and Cookshire-Eaton (45°20′57.49″N, 71°47′08.40″W, 243 m elevation). Plot size was 4 × 5 m in Sainte-Anne-de-Bellevue and 2.4 × 5 m in Cookshire-Eaton. The soil at the Sainte-Anne-de-Bellevue site was a free-draining St. Bernard sandy clay loam, while the Cookshire-Eaton site was an imperfectly drained Magog stony loam. All selections of both species were seeded in spring 2010 using a Fabro precision planter (Fabro Enterprises, Swift Current, SK, Canada) with an 18 cm row spacing. The seeding rate used for all selections was 300 pure live seeds (PLS) m$^{-2}$. Nitrogen was applied at both sites as urea at a rate of 50 kg N ha$^{-1}$ in May of 2011 and 2012; no other fertilization was done.

Plant height was determined in each plot biweekly from May to September in 2011 and 2012. The height of 10 plants was measured in each of three subplots selected at random within each plot, with the height recorded as the tallest overall part of the plant. Tiller numbers were also determined biweekly in 2011 and 2012 in three 50 × 50 cm quadrats placed randomly in each plot. Tiller counts were continued until tiller dies down at the beginning of the transition to dormancy at the end of each growing season. Phenology stages were determined based on sampling 10 plants at random from each plot and was scored based on the mean stage count (MSC) as described by Moore et al. [34]. Big bluestem selections were not included in tiller counts and were measured for height once at the end of the growing season.

All plots were harvested in the fall after a first frost using a flail-type forage harvester with a cutting surface of 0.6 × 5 m and a cutting height of 7 cm. The harvest dates for 2011 and 2012 were November 1 and October 23 in Sainte-Anne-de-Bellevue and October 24 and October 17 in Cookshire-Eaton,

respectively. In addition, the selections from "Cave-in-Rock" switchgrass and "Prairie View Indiana" big bluestem were also harvested in the spring to determine yield differences related to harvest time. The 2012 spring harvest dates were April 19 and April 30 in Sainte-Anne-de-Bellevue and Cookshire-Eaton, respectively. All harvested biomass was weighed and subsamples of approximately 500 g were collected from each plot to determine biomass yield on a dry matter basis. These subsamples were also used to determine the moisture, ash, fiber, and energy concentration of the harvested biomass for selections that were harvested both in the fall and in the spring.

*2.3. Laboratory Analyses.* Subsamples taken from each plot were dried for 48 hours at 60°C and then reweighed to determine moisture concentration [35] (method number 934.01). Dry samples were then ground with a Model 4 Thomas-Wiley Laboratory Mill forage grinder (Thomas Scientific, Swedesboro, NJ, USA) to pass through a 1 mm screen. The ash fraction was determined using a Thermolyne (Dubuque, IA, USA) muffle furnace [35] (method number 924.05). Ground samples were analyzed for neutral (NDF) and acid (ADF) detergent fiber using an Ankom Fiber Analyzer (Ankom Technology Corporation, Macedon, NY, USA) by incubating the samples in neutral [35] and acid detergent solutions [35] (method number 9738.18), respectively. Acid digestible lignin (ADL) was determined by washing ADF residues with 20 N $H_2SO_4$ [35] (method number 9738.18). From the results of these analyses cellulose, hemicellulose, and lignin concentrations were calculated (hemicellulose = NDF − ADF, cellulose = ADF − ADL, and lignin = ADL − mineral ash) [36]. Energy concentration was determined using a Parr adiabatic bomb calorimeter (Parr Oxygen Bomb Model 1341EB, Calorimeter Thermometer Model 6772, Parr Instrument Company, Moline, IL, USA) to calculate higher heating value (HHV).

TABLE 2: Yield on dry matter basis of 11 selections of switchgrass and 3 selections of big bluestem seeded in 2010 and harvested in fall of 2011 and 2012 at two sites in Southern Quebec (in $Mg\,ha^{-1}$).

| Entry | Sainte-Anne-de-Bellevue | | Cookshire-Eaton | |
| --- | --- | --- | --- | --- |
| | 2011 | 2012 | 2011 | 2012 |
| Switchgrass | | | | |
| Sunburst | 7.7 | 9.1 | 3.6 | 5.6 |
| Blue Jacket | 9.0 | 9.1 | 4.2 | 4.8 |
| Blue Jacket II | 9.7 | 9.1 | 5.0 | 5.3 |
| Blue Jacket Early | 8.1 | 9.1 | 3.6 | 4.7 |
| Summer | 9.2 | 11.2 | 4.9 | 6.8 |
| Tecumseh | 8.2 | 10.4 | 2.1 | 4.2 |
| Tecumseh II | 9.3 | 10.3 | 4.6 | 6.7 |
| Cave-in-Rock | 10.0 | 10.3 | 3.5 | 7.3 |
| Cave-in-Rock II | 10.2 | 11.1 | 5.9 | 9.0 |
| Cave-in-Rock Early | 8.8 | 9.9 | 2.5 | 6.4 |
| Sandlover | 9.3 | 9.7 | 1.6 | 5.3 |
| Switchgrass mean | 9.0 | 9.9 | 3.8 | 6.0 |
| Big Bluestem | | | | |
| Prairie view | 6.0 | 7.5 | 2.1 | 5.1 |
| Prairie view II | 5.6 | 8.8 | 1.9 | 5.6 |
| Prairie view Early | 6.6 | 8.4 | 1.6 | 6.1 |
| Big bluestem mean | 6.1 | 8.2 | 1.9 | 5.6 |
| Overall mean | 8.4 | 9.6 | 3.4 | 5.9 |
| P value | <0.0001 | 0.0046 | <0.0001 | 0.0004 |
| LSD (5%) | 1.1 | 1.7 | 1.4 | 1.6 |

*2.4. Statistical Analyses.* The experiment design used at both sites was a randomized complete block design with four replications. All data were analyzed using the GLM procedure of the SAS statistical software program [37]. Data were analyzed separately for each environment due to differences in sampling time and sampling numbers. Replicates were considered random effects and treatments (i.e., selections or harvest time (fall versus spring)) were considered fixed. Unless otherwise noted, tests were considered significant at the 5% probability level. Comparisons between means were made using LSD when ANOVAs indicated significant model and treatment effects. Only significant results are discussed in the text unless otherwise noted.

## 3. Results and Discussion

*3.1. Biomass Yield.* Switchgrass biomass yield across selections, years, and sites averaged $7.2\,Mg\,ha^{-1}$ with individual site-year averages ranging between 3.8 and $9.9\,Mg\,ha^{-1}$ (Table 2). Large differences were observed between sites, with yields averaging 4.9 and $9.5\,Mg\,ha^{-1}$ across years in Cookshire-Eaton and Sainte-Anne-de-Bellevue, respectively. Yields of all selections were greater in Sainte-Anne-de-Bellevue than in Cookshire-Eaton, although differences in the ranking of selections varied at the two sites. Overall much lower yields were observed in Cookshire-Eaton which may have been due to several environmental factors. The Cookshire-Eaton site is less productive as it is at a higher elevation

and has a heavier, imperfectly drained soil. It also suffered more intense perennial weed pressure as compared to the Sainte-Anne-de-Bellevue site. The switchgrass and especially the big bluestem appeared to take longer to become established at the Cookshire-Eaton site. The mean switchgrass yield at Sainte-Anne-de-Bellevue increased by 10% in year 2 while the mean switchgrass yield at Cookshire Eaton increased by 58% in year 2. Mean big bluestem yields increased by 34% and 294% in year 2 at the Sainte-Anne-de-Belle-vue and Cookshire-Eaton sites, respectively. Overall it appeared that yields are not optimized for switchgrass until the second production year with most selections on productive sites. In the case of switchgrass and especially big bluestem on more marginal sites it may take until year three for the stands to become fully productive as very large increases were experienced in year 2. Amongst the switchgrass accessions tested "Blue Jacket II" appeared to establish very well and had similar yields in both the first and the second year at both sites. This may have been a result of Blue Jacket II's larger seed size [32, 38] and relatively early maturity. These characteristics may allow it to more fully complete its growth cycle in the establishment year. Switchgrass usually reaches 33–66% of its yield potential in the first production year, while the second production year is generally indicative of long-term biomass yield potential [21]. Overall it appeared that, in the case of a productive site in southern Quebec, a large seeded upland cultivar selected for improved establishment could achieve near full productivity

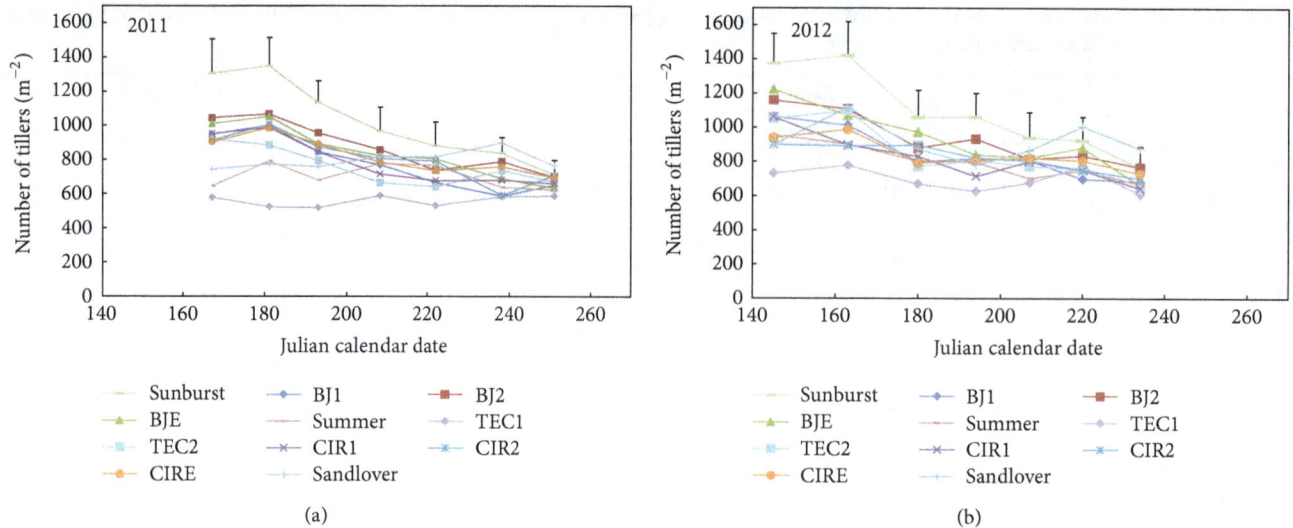

FIGURE 1: Number of tillers m$^{-2}$ of 11 selections of switchgrass at Sainte-Anne-de-Bellevue, Quebec, in 2011 and 2012. Vertical bars represent the least significant difference (LSD, $P = 0.05$) at each sampling point throughout the season. BJ1, Blue Jacket; BJ2, Blue Jacket II; BJE, Blue Jacket Early; TEC1, Tecumseh; TEC2, Tecumseh II; CIR, Cave-in-Rock; CIR2, Cave-in-Rock II; CIRE, Cave-in-Rock Early.

in the first production year. In the case of big bluestem on a marginal site, as little as 1/4 of the productive potential may be achieved in the first production year.

Variation in switchgrass biomass yield across sites, varieties, and fertilizer regimes has previously been reported [39–42], with a recent meta-analysis reporting a mean yield of $8.7 \pm 4.2$ Mg ha$^{-1}$ over 39 trials conducted across the United States [43]. Jannasch et al. [44] reported an average plot yield of 9.2 Mg ha$^{-1}$ for switchgrass in Southern Quebec, while average yields of 7.0 and 9.0 Mg ha$^{-1}$ were reported in Minnesota [45] and Iowa [39], respectively.

Big bluestem biomass yields averaged 5.5 Mg ha$^{-1}$ across all selections, sites, and years and were more variable than switchgrass, with values ranging between 1.9 and 8.2 Mg ha$^{-1}$ in individual site-years. Previous experiences have suggested that big bluestem may not reach its maximum yield potential until 3-4 years after seeding [15], a result supported by the large increases in yield and plant height observed between years in the present study.

Differences in switchgrass biomass yield were also observed between selections. Overall yields followed similar patterns as previous studies with early maturing cultivars generally being lower yielding selections than later maturing selections [46] (Table 2). However, higher yields for some of the most recent selections were observed when compared to the original parent materials. For example "Cave-in-Rock II" outyielded Cave-in-Rock in both years at Cookshire-Eaton, while "Blue Jacket II" outyielded "Sunburst" at both sites in 2011. In the case of Tecumseh II there were no significant differences in yield with "Summer" switchgrass observed. "Summer" is known to be a relatively high producing cultivar for its maturity class and breeding progress may be more difficult to achieve. The "Cave-in-Rock II" selection overall appeared to be amongst the most productive selections across sites and years. "Cave-in-Rock" selections generally performed at least

as well as or better than the other selections tested, confirming the results of other studies which have found "Cave-in-Rock" to be well adapted and high yielding in Southern Quebec [47].

Some selections from the "Sunburst" variety performed poorly in Cookshire-Eaton and in both 2011 and 2012 significantly higher stem rust (*Puccinia* spp.) incidence was noted on those selections relative to all other selections in the study (data not shown). Stem rust infections are the most common disease reported in switchgrass and may cause significant damage including stunting, early flowering, and reduced yield [39]. The prevalence of rust infections on all "Cave-in-Rock" selections was extremely low, as seen in previous research [48]. Gustafson et al. [49] evaluated the genetic variation in resistance to rust disease in switchgrass populations including "Summer," "Sunburst," and "Cave-in-Rock." "Sunburst" had the highest susceptibility to rust, while "Cave-in-Rock," having evolved under heavy rust pressure in humid Southern Illinois, was the least susceptible.

*3.2. Tiller Density.* Significant differences in tiller number were observed among selections at all measurement points in both years at Sainte-Anne-de-Bellevue (Figure 1). At the first measurement period (early June in 2011 and late May in 2012), the mean tiller density per square meter across all entries was 906 in 2011 and increased to 1033 in 2012, with "Sunburst" producing the most tillers m$^{-2}$ in both years (1307 in 2011 and 1375 in 2012). Low tiller counts were observed on "Tecumseh" and "Sandlover" switchgrass in 2011 as the plots had relatively thin plant stands. Due to the large variability between plots, the significant differences observed were often between selection lineages and not between selections within a lineage. By the end of August, tiller density in Sainte-Anne-de-Bellevue had decreased to 674 and 706 tillers per square meter in 2011 and 2012, respectively, representing average decreases of 22% and 30%, respectively. The magnitude of the decrease

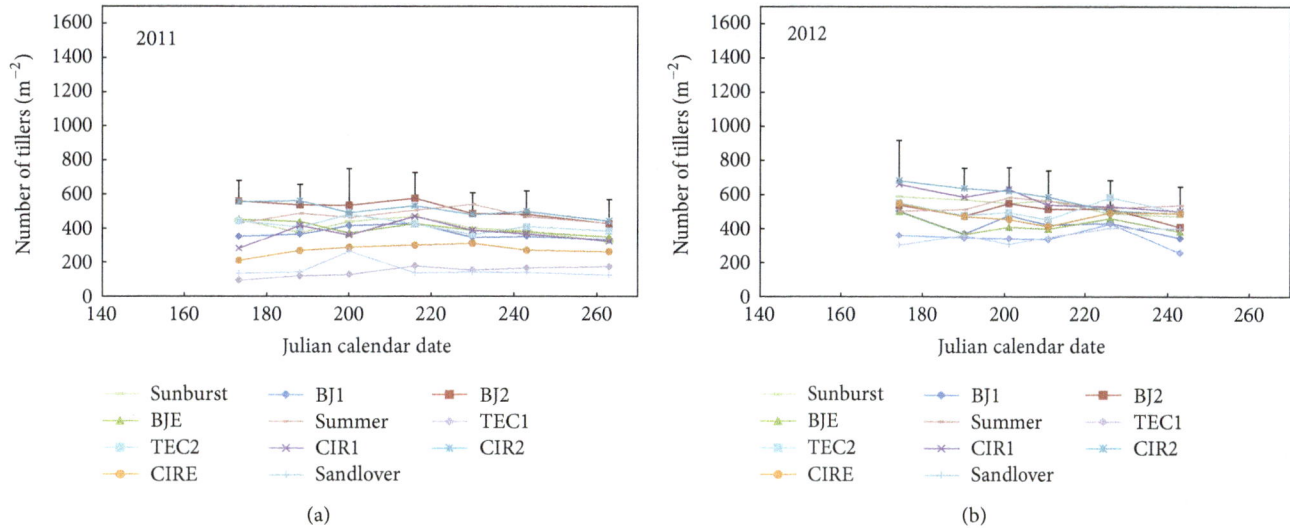

FIGURE 2: Number of tillers m$^{-2}$ of 11 selections of switchgrass at Cookshire-Eaton, Quebec, in 2011 and 2012. Vertical bars represent the least significant difference (LSD, $P = 0.05$) at each sampling point throughout the season. BJ1, Blue Jacket; BJ2, Blue Jacket II; BJE, Blue Jacket Early; TEC1, Tecumseh; TEC2, Tecumseh II, CIR; Cave-in-Rock; CIR2, Cave-in-Rock II; CIRE, Cave-in-Rock Early.

varied considerably depending on the selection lineage. The average decrease in tiller density throughout the season across both years was 38% for "Sunburst" selections, 26% for "Cave-in-Rock" selections, and 19% for "Summer" selections, while no decrease was observed for "Sandlover" which had a thin stand. All selections from the "Sunburst" lineage displayed reduced tillering early in the season. The variable degree of tiller loss across selections ultimately had an equalizing effect on tiller density between all selections.

In Cookshire-Eaton stands were considerably thinner and more variable, but significant differences were still observed between selections at all measurement periods (Figure 2). At both sites differences between selections were much larger at the beginning of the season, with only stands with the highest initial density self-thinning throughout the season. However, due to the thinner stands at Cookshire-Eaton tiller mortality during the season was greatly reduced. In 2011, the low initial tiller density in Cookshire-Eaton resulted in no significant decrease over the course of the season with mean tiller number across all selections decreasing from 360 to 324 tillers m$^{-2}$. In 2012 all stands had higher initial tiller density and a correspondingly greater decrease from 522 to 434 tillers m$^{-2}$, an average decrease of 14% over the season. Once stands had more fully established at the Cookshire-Eaton site in 2012, trends in average tiller decrease also began to align more closely with those seen at Sainte-Anne-de-Bellevue with an average decrease of 25% for "Sunburst" selections, 21% for "Cave-in-Rock" selections, and 11% for "Summer" selections, while "Sandlover" did not decrease at all due to a thin stand.

High tiller mortality as the growing season progresses is common in grasses, with crops such as ryegrass losing 60% of their tillers [50]. The large majority of these are small vegetative tillers which senesce in the sward after being shaded out by the larger reproductive tillers. In perennial grasses such as switchgrass high vegetative tiller production occurs in

the spring with subsequent mortality throughout the season [51].

The results of the present study support previous findings that indicate that there is likely an optimal tiller density for efficient capture of solar radiation which swards target through increased vegetative tiller production or self-thinning. This optimum is known as tiller equilibrium density [23]. Madakadze et al. [52] found that populations of switchgrass had differing leaf area indices associated with various heights of their canopy. The authors also found that leaf angle and leaf area index varied by variety, indicating that different populations have different canopy structures. The variability in tiller mortality among selections in the present study may indicate that tiller plasticity also plays a large role in this variability. Boe and Casler [13] reported a large degree of plasticity in tiller morphology within populations. In the present study this plasticity seemed to allow tiller density across many selections to converge towards equilibrium density. "Sunburst" had the highest initial density and displayed >40% tiller mortality over the season (Figures 1 and 2), while conversely "Sandlover," the selection with the lowest initial tiller density, did not display any tiller mortality on average.

A trend was also noted for differences in decrease between selections when compared to their parent variety, appearing to often be due to lower initial tiller density in the selections at the beginning of the season. Redfearn et al. [48] examined tiller density in switchgrass plots in Iowa and Nebraska managed for forage and found G × E interactions to be present for both yield and tiller density, with "Cave-in-Rock" producing the lowest tiller number but highest forage yield. The authors noted that tiller density differed with location, even between selections from the same base population. In the present study at Sainte-Anne-de-Bellevue average tiller mortality over both years was 10% lower in "Blue Jacket II" than in "Sunburst" and 13% lower in "Cave-in-Rock II" than in the "Cave-in-Rock" base population. However,

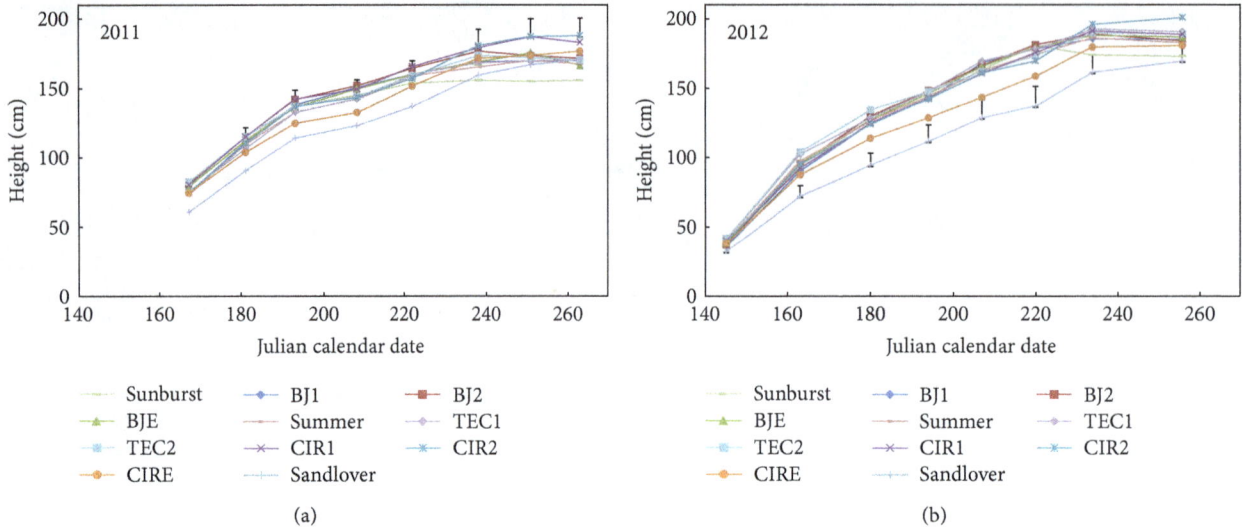

FIGURE 3: Height of 11 selections of switchgrass at Sainte-Anne-de-Bellevue, Quebec, in 2011 and 2012. Vertical bars represent the least significant difference (LSD, $P = 0.05$) at each sampling point throughout the season. BJ1, Blue Jacket; BJ2, Blue Jacket II; BJE, Blue Jacket Early; TEC1, Tecumseh; TEC2, Tecumseh II; CIR, Cave-in-Rock; CIR2, Cave-in-Rock II; CIRE, Cave-in-Rock Early.

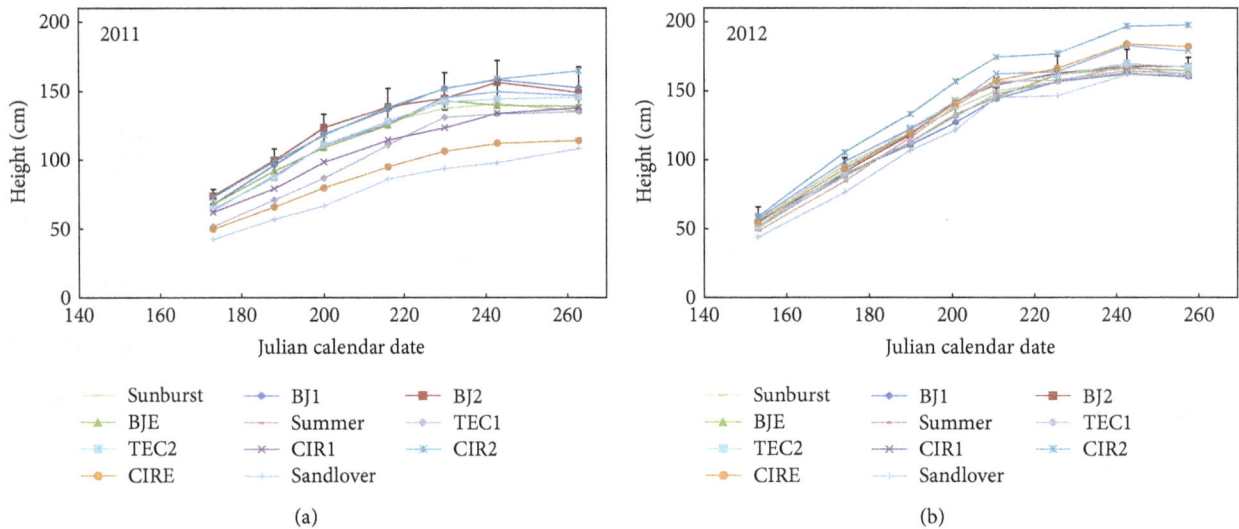

FIGURE 4: Height of 11 selections of switchgrass at Cookshire-Eaton, Quebec, in 2011 and 2012. Vertical bars represent the least significant difference (LSD, $P = 0.05$) at each sampling point throughout the season. BJ1, Blue Jacket; BJ2, Blue Jacket II; BJE, Blue Jacket Early; TEC1, Tecumseh; TEC2, Tecumseh II; CIR, Cave-in-Rock; CIR2, Cave-in-Rock II; CIRE, Cave-in-Rock Early.

there was an average increase in tiller loss in the "Summer" lineage, with mortality on average 16% higher in "Tecumseh II" than in "Summer." This was likely due to "Tecumseh II" establishing more quickly and having a higher initial density than "Summer" in 2011. In 2012 the "Summer" population was more fully established and there was no significant difference between the two. These trends suggest progress in modification of tillering among the selection lineages where populations were well established.

*3.3. Height.* Significant differences were observed between the heights of selections at all periods observed throughout

all site-years (Figures 3 and 4); however, trends varied by site. The trends observed were generally supported by previous studies which have indicated that later maturing varieties produce taller tillers due to their slower rates of leaf appearance [53]. "Cave-in-Rock II" was significantly taller than the other selections in its lineage at both sites in 2012, reaching an average of 201 cm in Sainte-Anne-de-Bellevue. Over all site-years this represented an average increase of 12 cm or 7% over the parent variety. The "Cave-in-Rock" selections were the tallest three recorded in 2012 at Cookshire-Eaton. In 2012 the relatively early maturing "Tecumseh" and "Tecumseh II" selections performed very well in Sainte-Anne-de-Bellevue,

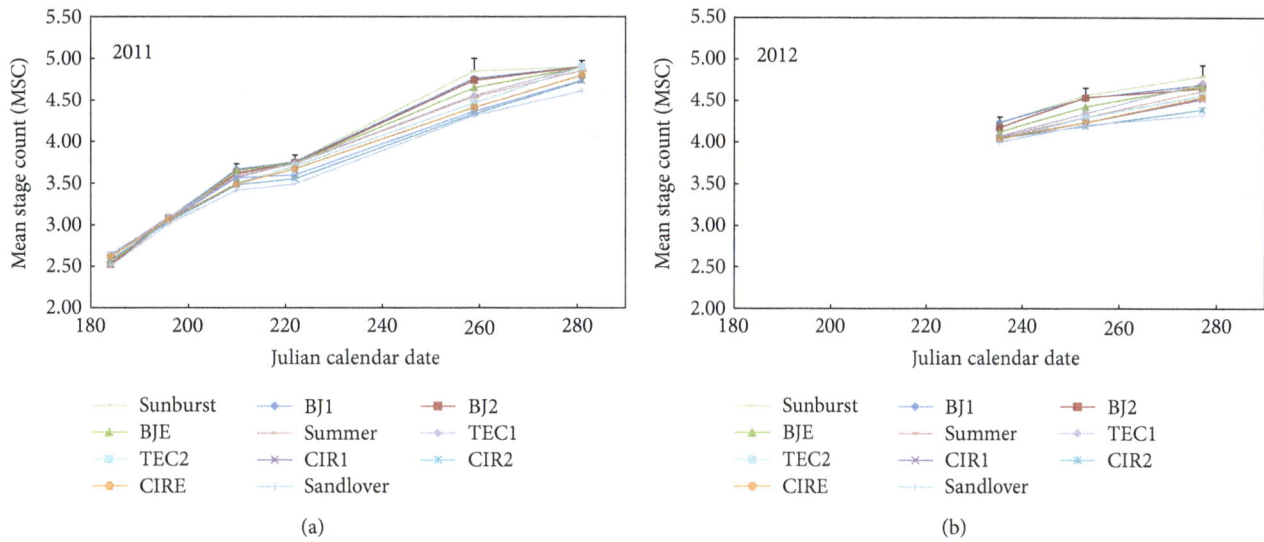

FIGURE 5: Mean stage count of 11 selections of switchgrass at Sainte-Anne-de-Bellevue, Quebec, in 2011 and 2012. Vertical bars represent the least significant difference (LSD, $P = 0.05$) at each sampling point throughout the season. BJ1, Blue Jacket; BJ2, Blue Jacket II; BJE, Blue Jacket Early; TEC1, Tecumseh; TEC2, Tecumseh II; CIR, Cave-in-Rock; CIR2, Cave-in-Rock II; CIRE, Cave-in-Rock Early.

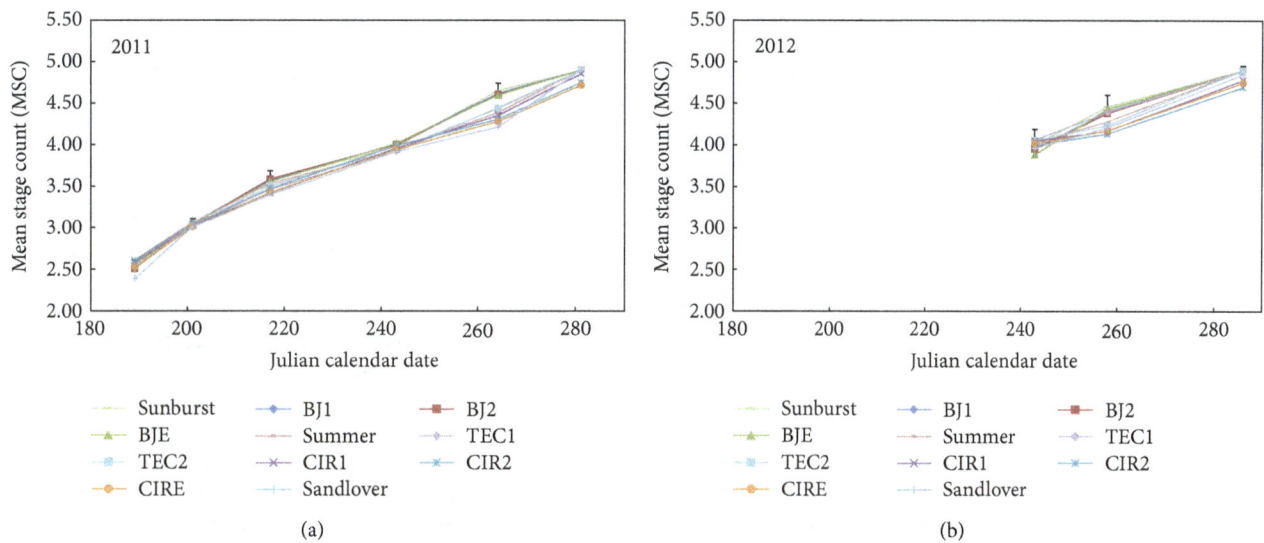

FIGURE 6: Mean stage count of 11 selections of switchgrass at Cookshire-Eaton, Quebec, in 2011 and 2012. Vertical bars represent the least significant difference (LSD, $P = 0.05$) at each sampling point throughout the season. BJ1, Blue Jacket; BJ2, Blue Jacket II; BJE, Blue Jacket Early; TEC1, Tecumseh; TEC2, Tecumseh II; CIR, Cave-in-Rock; CIR2, Cave-in-Rock II; CIRE, Cave-in-Rock Early.

attaining the second and third highest average heights; however, there was no significant increase in height over the parent variety "Summer."

No meaningful height differences were observed among big bluestem selections (data not shown); however, there was a significant increase in height in the second year at both sites. This result is consistent with the slower establishment in big bluestem. Average height increased from 144 to 192 cm (33%) at the Cookshire-Eaton site and from 194 to 216 cm (11%) at Sainte-Anne-de-Bellevue in 2011 and 2012, respectively.

*3.4. Maturity.* There were significant differences observed in phenological development at both sites for all measurement periods (Figures 5 and 6). The differences observed were almost exclusively between selections in different selection lineages and not within a lineage. The "Sunburst" selections displayed the earliest maturity, followed by the "Summer" selections, with the "Cave-in-Rock" and "Sandlover" selections displaying the latest maturity. These findings are supported by recent literature linking latitude of origin with maturity [46, 54]. The latitudes of origin of the selections in

TABLE 3: Moisture concentration (%) at fall harvest of 11 selections of switchgrass and 3 selections of big bluestem at two sites in Southern Quebec.

| Entry | Sainte-Anne-de-Bellevue | | Cookshire-Eaton | |
|---|---|---|---|---|
| | 2011 | 2012 | 2011 | 2012 |
| Switchgrass | | | | |
| Sunburst | 36.8 | 34.1 | 33.8 | 27.6 |
| Blue Jacket | 36.1 | 33.8 | 31.6 | 31.8 |
| Blue Jacket II | 35.5 | 35.4 | 32.7 | 30.1 |
| Blue Jacket Early | 38.1 | 35.2 | 30.1 | 28.0 |
| Summer | 39.7 | 31.9 | 26.1 | 24.4 |
| Tecumseh | 37.7 | 35.8 | 37.5 | 29.3 |
| Tecumseh II | 40.3 | 34.3 | 31.4 | 20.2 |
| Cave-in-Rock | 46.1 | 41.0 | 43.6 | 29.8 |
| Cave-in-Rock II | 45.8 | 44.5 | 44.9 | 39.6 |
| Cave-in-Rock Early | 44.1 | 41.8 | 35.5 | 31.6 |
| Sandlover | 49.1 | 44.5 | 38.7 | 34.8 |
| Switchgrass mean | 40.8 | 37.5 | 35.1 | 29.7 |
| Big Bluestem | | | | |
| Prairie view | 43.7 | 39.6 | 44.3 | 38.9 |
| Prairie view II | 46.8 | 41.0 | 35.6 | 30.9 |
| Prairie view Early | 43.8 | 39.0 | 44.0 | 33.7 |
| Big bluestem mean | 44.8 | 39.9 | 41.3 | 34.5 |
| Overall mean | 41.7 | 38.0 | 36.4 | 30.8 |
| P value | <0.0001 | <0.0001 | 0.0002 | 0.0023 |
| LSD (5%) | 3.3 | 3.2 | 7.9 | 7.6 |

this study rank in the same order as the maturity observed (Table 1). The moisture concentration of the selections at harvest also mirrored this pattern, with significant differences observed between selections related to maturity at harvest (Table 3). No significant moisture differences were observed between selections from a same parent variety.

3.5. *Effects of Harvest Date on Biomass Yield and Quality Variables.* Differences were observed at both sites for some selections between fall and spring harvest dates for yield, moisture concentration, and cellulose concentration (Tables 4 and 5). The Sainte-Anne-de-Bellevue spring harvest resulted in an overall 41% yield reduction compared to a fall harvest. In Cookshire-Eaton a spring harvest only reduced yield in the highest-yielding selection, "Cave-in-Rock II", which had an average loss of 56%. Reductions of up to 40% in biomass yield have also been reported in Pennsylvania, where this yield loss was attributed to the loss of panicles and leaves, the leaching of stem components such as minerals and cell solubles [27], and the translocation of mobile nutrients and N-rich compounds into underground perennial structures, a process which may continue into December [3]. These processes have also been shown to decrease the ash fraction in the spring [27, 28]. In the present study no differences were observed between ash concentrations of spring versus fall harvested material (Tables 4 and 5). This lack of change in ash concentration may have been due to soil contamination from the grasses overwintering on the site in a lodged state. Current best management practice by farmers in Eastern Canada is to

mow the grasses in late fall and to bale the biomass in spring. This system helps to reduce biomass losses by minimizing the breakage of the leaves and seed heads resulting from wind and snow exposure in material standing over winter [55]. Small plot studies may, therefore, overestimate yield losses compared to an on-farm spring harvest regime. In this study the yield and quality comparisons of the fall and spring material were only made on first production year (2nd year after seeding) material, which had not yet reached full establishment at the marginal site.

The decrease in moisture concentration in the spring biomass compared to the fall averaged 74% across selections and sites, falling from 45% to 11% moisture concentration (Table 4). The high fall moisture values for "Cave-in-Rock" selections are a result of its late maturity and high biomass production. Plant materials are difficult to dry in the short, cool, and high humidity days in Eastern Canada in the late fall [31]. The calorific value for combustion of herbaceous biomass declines linearly with rising moisture concentration, making this an important measure for certain applications [56]. Drying can also take place in covered storage, with fall-baled material at 16-17% moisture declining to 12–14% by the following spring [31].

The cell walls of grasses contain cellulose, hemicellulose, lignin, and other compounds including p-coumaric, phenolic, and ferulic acids [26, 57]. The increase in cellulose concentration for some selections at both sites (Tables 4 and 5) is likely related to increased stem fraction combined with an overall reduction of mineral and cell soluble concentrations

TABLE 4: Fall 2011 and spring 2012 values on dry matter basis for yield, moisture concentration, cell wall components, ash, and higher heating value (HHV) of three selections of switchgrass and three selections of big bluestem in Sainte-Anne-de-Bellevue, Quebec.

| Entry | | CIR | CIR2 | CIRE | PV | PV2 | PVE | P value | LSD |
|---|---|---|---|---|---|---|---|---|---|
| Moisture (%) | Fall | 46.1 | 45.8 | 44.1 | 43.7 | 46.8 | 43.8 | 0.1843 | 3.076 |
| | Spring | 11.2 | 14 | 10.6 | 10.7 | 11.5 | 10.9 | 0.326 | 3.411 |
| | % change[†] | −75.7 | −69.4 | −76 | −75.4 | −75.4 | −75.1 | | |
| | P value | 0.0001 | <0.0001 | 0.0002 | <0.0001 | 0.0002 | 0.0002 | | |
| | LSD | 4.209 | 2.812 | 4.925 | 3.737 | 4.706 | 4.2826 | | |
| Yield (Mg ha$^{-1}$) | Fall | 10 | 10.2 | 8.8 | 6 | 5.6 | 6.6 | <0.0001 | 1.202 |
| | Spring | 6.2 | 6.1 | 6.6 | 3.8 | 2.8 | 3 | <0.0001 | 1.518 |
| | % change[†] | −38.3 | −40.4 | −25.2 | −37.9 | −49.5 | −55.1 | | |
| | P value | 0.0052 | 0.0102 | 0.0493 | 0.053 | 0.0018 | 0.0034 | | |
| | LSD | 1.67 | 2.253 | 2.205 | 2.341 | 0.8374 | 1.3616 | | |
| Cellulose (g kg$^{-1}$) | Fall | 419 | 414 | 389 | 439 | 431 | 438 | 0.0056 | 25.05 |
| | Spring | 456 | 438 | 429 | 476 | 449 | 447 | 0.001 | 17.59 |
| | % change[†] | 8.8 | 5.6 | 10.4 | 8.3 | 4.2 | 2.1 | | |
| | P value | 0.0654 | 0.0004 | 0.0042 | 0.029 | 0.1361 | 0.0143 | | |
| | LSD | 41.2 | 4.227 | 16.29 | 29.42 | 28.491 | 5.676 | | |
| Hemicellulose (g kg$^{-1}$) | Fall | 315 | 321 | 324 | 316 | 303 | 300 | 0.0593 | 17.42 |
| | Spring | 324 | 336 | 340 | 291 | 313 | 307 | 0.0005 | 18.97 |
| | % change[†] | 2.9 | 4.7 | 4.9 | −7.9 | 3.3 | 2.3 | | |
| | P value | 0.457 | 0.2528 | 0.0061 | 0.056 | 0.209 | 0.1208 | | |
| | LSD | 34.56 | 35.1 | 7.173 | 26.219 | 20.403 | 10.329 | | |
| Lignin (g kg$^{-1}$) | Fall | 77 | 69 | 65.5 | 60.1 | 62.9 | 63.1 | 0.2727 | 14.94 |
| | Spring | 90.3 | 73.5 | 64.5 | 79.3 | 63.7 | 67.9 | <0.0001 | 8.647 |
| | % change[†] | 17.3 | 6.5 | −1.5 | 31.9 | 1.3 | 7.6 | | |
| | P value | 0.2283 | 0.3488 | 0.8434 | 0.0094 | 0.8919 | 0.2655 | | |
| | LSD | 28.093 | 12.857 | 14.786 | 9.9 | 16.149 | 11.246 | | |
| Ash (g kg$^{-1}$) | Fall | 45.2 | 48.9 | 47.4 | 48.2 | 49.1 | 46.3 | 0.9689 | 11.14 |
| | Spring | 50.4 | 53.2 | 50.8 | 48.3 | 50.3 | 55.5 | 0.9601 | 17.25 |
| | % change[†] | 11.5 | 8.8 | 7.2 | 0.2 | 2.4 | 19.9 | | |
| | P value | 0.5503 | 0.4563 | 0.2225 | 0.9562 | 0.7607 | 0.1186 | | |
| | LSD | 24.9 | 16.041 | 6.897 | 9.329 | 11.929 | 13.538 | | |
| HHV (MJ kg$^{-1}$) | Fall | 18.71 | 18.96 | 18.78 | 18.73 | 19 | 18.94 | 0.5896 | 0.4463 |
| | Spring | 18.96 | 18.99 | 18.73 | 18.85 | 18.88 | 18.92 | 0.761 | 0.399 |
| | % change[†] | 1.3 | 0.2 | −0.3 | 0.6 | −0.6 | −0.1 | | |
| | P value | 0.4213 | 0.8845 | 0.7915 | 0.4774 | 0.4568 | 0.9009 | | |
| | LSD | 0.887 | 0.5706 | 0.5966 | 0.4781 | 0.46 | 0.4408 | | |

[†]Percent change relative to fall value.
CIR, Cave-in-Rock; CIR2, Cave-in-Rock II; CIRE, Cave-in-Rock Early; PV, Prairie view; PV2, Prairie view II; PVE, Prairie view Early.

in overwintered materials. Whole plant big bluestem tended to have slightly higher cellulose and lower hemicellulose concentrations than switchgrass, again likely due to increased stem fraction in the big bluestem. Overwintering generally had a tendency to increase concentration of cellulose in both switchgrass and big bluestem. If big bluestem genetics can be improved to produce similar yields to switchgrass it could be a promising species where high cellulose content biomass feedstocks are required for markets such as bioethanol production and biofibre applications. For lignin there were no trends

evident between species or selections. Overall big bluestem appeared to have a higher cellulose to hemicellulose ratio than upland switchgrass. This could be of interest for specialty applications requiring high cellulose content feedstocks.

There were no differences in HHV observed between selections or between harvest dates, with Sainte-Anne-de-Bellevue and Cookshire-Eaton averaging 18.9 and 19.3 kJ g$^{-1}$, respectively, over both harvest times. A previous analysis of four varieties of fall-harvested switchgrass determined an average energy concentration of 18.8 ± 0.2 kJ g$^{-1}$ with no

TABLE 5: Fall of 2011 and spring of 2012 values on dry matter basis for yield, moisture concentration, cell wall components, ash, and higher heating value (HHV) of three selections of switchgrass and three selections of big bluestem in Cookshire-Eaton, Quebec.

| Entry | | CIR | CIR2 | CIRE | PV | PV2 | PVE | P value | LSD |
|---|---|---|---|---|---|---|---|---|---|
| Moisture (%) | Fall | 43.6 | 44.9 | 35.5 | 44.3 | 35.6 | 43.9 | 0.0962 | 8.8360 |
| | Spring | 8.5 | 8.8 | 7.3 | | | | | |
| | % change[†] | −80.4 | −80.3 | −79.5 | | | | | |
| | P value | 0.0007 | 0.001 | 0.0106 | | | | | |
| | LSD | 3.8578 | 4.8707 | 12.587 | | | | | |
| Yield (Mg ha$^{-1}$) | Fall | 3.5 | 5.9 | 2.5 | 2.1 | 1.9 | 1.6 | 0.0002 | 1.2830 |
| | Spring | 2.6 | 2.6 | 2.3 | 0.0 | 0.0 | 0.0 | | |
| | % change[†] | −26.2 | −56.3 | −8.5 | −100.0 | −100.0 | −100.0 | | |
| | P value | 0.2539 | 0.0188 | 0.6558 | | | | | |
| | LSD | 2.5201 | 1.9826 | 1.7471 | | | | | |
| Cellulose (g kg$^{-1}$) | Fall | 415 | 396 | 376 | 385 | 381 | 371 | 0.2719 | 40.852 |
| | Spring | 445 | 454 | 415 | | | | 0.001 | 10.307 |
| | % change[†] | 7.2 | 14.7 | 10.3 | | | | | |
| | P value | 0.3218 | 0.1236 | 0.0132 | | | | | |
| | LSD | 98.893 | 97.169 | 19.321 | | | | | |
| Hemicellulose (g kg$^{-1}$) | Fall | 352 | 317 | 339 | 316 | 301 | 290 | 0.0088 | 29.914 |
| | Spring | 345 | 304 | 333 | | | | 0.0001 | 6.325 |
| | % change[†] | −1.7 | −4.0 | −1.7 | | | | | |
| | P value | 0.3591 | 0.2353 | 0.2808 | | | | | |
| | LSD | 22.334 | 32.475 | 17.053 | | | | | |
| Lignin (g kg$^{-1}$) | Fall | 73 | 77 | 61 | 55 | 70 | 66 | 0.1155 | 16.628 |
| | Spring | 72 | 89 | 63 | | | | 0.0009 | 6.488 |
| | % change[†] | −0.5 | 16.2 | 3.4 | | | | | |
| | P value | 0.9687 | 0.3272 | 0.5228 | | | | | |
| | LSD | 34.118 | 41.443 | 11.684 | | | | | |
| Ash (g kg$^{-1}$) | Fall | 39 | 48 | 45 | 45 | 46 | 45 | 0.6675 | 11.274 |
| | Spring | 33 | 30 | 32 | | | | 0.0179 | 1.539 |
| | % change[†] | −17.5 | −38.0 | −30.2 | | | | | |
| | P value | 0.1626 | 0.0725 | 0.0625 | | | | | |
| | LSD | 13.665 | 22.343 | 15.498 | | | | | |
| HHV (MJ kg$^{-1}$) | Fall | 19.15 | 19.09 | 19.31 | 19.22 | 19.20 | 19.27 | 0.7519 | 3.452 |
| | Spring | 19.35 | 19.36 | 19.25 | | | | 0.6191 | 3.264 |
| | % change[†] | 1.0 | 1.4 | −0.3 | | | | | |
| | P value | 0.158 | 0.4399 | 0.2681 | | | | | |
| | LSD | 0.387 | 1.2138 | 0.1836 | | | | | |

[†]Percent change relative to fall value.
CIR, Cave-in-Rock; CIR2, Cave-in-Rock II; CIRE, Cave-in-Rock Early; PV, Prairie view; PV2, Prairie view II; PVE, Prairie view Early.

significant differences between varieties [58], while Madakadze et al. [52] reported a lower energy yield of 17.4 kJ g$^{-1}$ for "Cave-in-Rock" harvested in the fall in Southern Quebec.

Turn et al. [59] have indicated that the leaching process that often occurs in the winter increases HHV in herbaceous biomass through removal of inorganic constituents. However, in the present study harvest time had no effect on HHV. Experiences with spring harvested reed canary grass (*Phalaris arundinacea* L.) in Sweden demonstrated significant reductions in Cl, K, S, alkali, ash, and several other important quality factors but curiously also a decrease in HHV [28]. The authors also noted large variation in HHV and ash concentration related to soil type, with clay soils having much lower values than humus rich or sandy soils.

## 4. Conclusions

A two-year field study of elite selections of switchgrass and big bluestem found significant differences among selections for yield, tiller density, height, and maturity. Switchgrass was quicker to reach its optimum yield than big bluestem on both productive and marginal sites. Big bluestem was slower to establish and exhibited a higher cellulose content, which was attributed to its high stem content. On both productive and marginal sites switchgrass reached a similar optimum tiller density at the end of the growing season of 690 and 379 tillers m$^{-2}$, respectively. Tiller mortality throughout the season ranged from 25 to 38%, from 21 to 26%, and from 11 to 19% on the "Sunburst," "Cave-in-Rock," and "Summer" lineages, respectively, on the two sites. Two of the elite selections of switchgrass ("Blue Jacket II" and "Cave-in-Rock II") derived from lineages with high tiller mortality showed upwards trends in height, yield, and reduction of tiller mortality when compared to their parent materials. The present study is one of the most detailed to date on tiller mortality in upland switchgrass. Selection for reduced tillering appears to be a promising strategy for improving yield and quality of upland switchgrass biomass. The differences in maturity and moisture concentration observed at harvest in the fall appear to correspond to the germplasm latitude of origin, and earlier maturity typically was related to lower moisture concentration at harvest. Delayed spring harvest resulted in a large decrease in moisture concentration and a slight increase in cellulose concentration; however, yield losses approached 40%. Unlike in previous studies the ash concentration was not affected by harvest date, nor was the HHV.

The results of this study indicate that regional selection programmes have the potential to produce improved selections of upland switchgrass for bioenergy and biofibre markets. From a biomass quality standpoint, big bluestem also appears to be a species of significant potential; however, there is a need for plant breeders to focus on improving establishment to enhance its agronomic potential.

## Conflict of Interests

REAP-Canada is a nonprofit registered charity that has developed some of the lines evaluated in the present project; it could, in the future, receive seed royalties if some of these lines are developed commercially. Data management and analysis were independently managed by researchers from McGill University who have no financial interest in the lines evaluated. They did not receive any financial compensation from REAP-Canada to complete this project, which was funded by MAPAQ.

## Acknowledgments

This research was supported by a research grant from the Ministère de l'Agriculture, des Pêcheries, et de l'Alimentation du Québec (MAPAQ) through its Programme de Soutien à l'Innovation en Agroalimentaire. Philippe Seguin also acknowledges financial support from the Centre SEVE.

## References

[1] M. T. Panciera and G. A. Jung, "Switchgrass establishment by conservation tillage: planting date responses of two varieties," *Journal of Soil and Water Conservation*, vol. 39, pp. 68–70, 1984.

[2] M. A. Liebig, H. A. Johnson, J. D. Hanson, and A. B. Frank, "Soil carbon under switchgrass stands and cultivated cropland," *Biomass and Bioenergy*, vol. 28, no. 4, pp. 347–354, 2005.

[3] D. J. Parrish and J. H. Fike, "The biology and agronomy of switchgrass for biofuels," *Critical Reviews in Plant Sciences*, vol. 24, no. 5-6, pp. 423–459, 2005.

[4] R. Samson, S. Mani, R. Boddey et al., "The potential of C$_4$ perennial grasses for developing a global BIOHEAT industry," *Critical Reviews in Plant Sciences*, vol. 24, no. 5-6, pp. 461–495, 2005.

[5] H. Kludze, B. Deen, A. Weersink, R. van Acker, K. Janovicek, and A. de Laporte, "Impact of land classification on potential warm season grass biomass production in Ontario, Canada," *Canadian Journal of Plant Science*, vol. 93, no. 2, pp. 249–260, 2013.

[6] P. G. Jefferson and W. P. McCaughey, "Switchgrass (*Panicum virgatum* L.) cultivar adaptation, biomass production, and cellulose concentration as affected by latitude of origin," *ISRN Agronomy*, vol. 2012, Article ID 763046, 9 pages, 2012.

[7] K. P. Vogel, G. Sarath, J. Aaron, and R. B. Mitchell, "Switchgrass," in *Energy Crops*, N. G. Halford and A. Karp, Eds., pp. 341–380, Royal Society of Chemistry, Cambridge, UK, 2011.

[8] M. D. Casler, "Switchgrass breeding, genetics, and genomics," in *Switchgrass*, A. Monti, Ed., Green Energy and Technology, Springer, London, UK, 2012.

[9] P. G. Jefferson, W. P. Mccaughey, K. May, J. Woosaree, L. Mac-Farlane, and S. M. Wright, "Performance of American native grass cultivars in the Canadian prairie provinces," *Native Plants Journal*, vol. 3, pp. 24–33, 2002.

[10] A. Tubeileh, T. J. Rennie, A. Kerr, A. A. Saita, and C. Patanè, "Biomass production by warm-season grasses as affected by nitrogen application in Ontario," *Agronomy Journal*, vol. 106, pp. 416–422, 2013.

[11] I. C. Madakadze, T. Radiotis, J. Li, K. Goel, and D. L. Smith, "Kraft pulping characteristics and pulp properties of warm season grasses," *Bioresource Technology*, vol. 69, no. 1, pp. 75–85, 1999.

[12] A. Boe, R. Bortnem, and K. D. Kephart, "Quantitative description of the phytomers of big bluestem," *Crop Science*, vol. 40, no. 3, pp. 737–741, 2000.

[13] A. Boe and M. D. Casler, "Hierarchical analysis of switchgrass morphology," *Crop Science*, vol. 45, no. 6, pp. 2465–2472, 2005.

[14] H. G. Jung and K. P. Vogel, "Lignification of switchgrass (*Panicum virgatum*) and big bluestem (*Andropogon gerardii*) plant parts during maturation and its effect on fibre degradability," *Journal of the Science of Food and Agriculture*, vol. 59, pp. 169–176, 1992.

[15] W. Vermerris, *Genetic Improvement of Bioenergy Crops*, Springer, New York, NY, USA, 2008.

[16] B. A. Robertson, P. J. Doran, L. R. Loomis, J. R. Robertson, and D. W. Schemske, "Perennial biomass feedstocks enhance avian diversity," *GCB Bioenergy*, vol. 3, no. 3, pp. 235–246, 2011.

[17] M. A. Gardiner, J. K. Tuell, R. Isaacs, J. Gibbs, J. S. Ascher, and D. A. Landis, "Implications of three biofuel crops for beneficial arthropods in agricultural landscapes," *Bioenergy Research*, vol. 3, no. 1, pp. 6–19, 2010.

[18] C. J. Zilverberg, W. C. Johnson, V. Owens et al., "Biomass yield from planted mixtures and monocultures of native prairie vegetation across a heterogeneous farm landscape," *Agriculture, Ecosystems & Environment*, vol. 186, pp. 148–159, 2014.

[19] V. R. Mulkey, V. N. Owens, and D. K. Lee, "Management of warm-season grass mixtures for biomass production in South

Dakota USA," *Bioresource Technology*, vol. 99, no. 3, pp. 609–617, 2008.

[20] USDA-NRCS Elsberry Plant Materials Center, *Cave-In-Rock Switchgrass, (Panicum virgatum L.) Conservation Plant Release Brochure*, 1986, http://www.nrcs.usda.gov/Internet/FSE_PLANTMATERIALS/publications/mopmcrb11259.pdf.

[21] S. B. McLaughlin and L. A. Kszos, "Development of switchgrass (*Panicum virgatum*) as a bioenergy feedstock in the United States," *Biomass & Bioenergy*, vol. 28, no. 6, pp. 515–535, 2005.

[22] A. J. Smart, L. E. Moser, and K. P. Vogel, "Morphological characteristics of big bluestem and switchgrass plants divergently selected for seedling tiller number," *Crop Science*, vol. 44, no. 2, pp. 607–613, 2004.

[23] K. M. Zarrough, C. J. Nelson, and D. A. Sleper, "Interrelationships between rates of leaf appearance and tillering in selected tall fescue populations," *Crop Science*, vol. 24, pp. 565–569, 1984.

[24] R. J. Jones, C. J. Nelson, and D. A. Sleper, "Seedling selection for morphological characters associated with yield of tall fescue," *Crop Science*, vol. 19, pp. 631–634, 1979.

[25] A. J. Smart and L. E. Moser, "Switchgrass seedling development as affected by seed size," *Agronomy Journal*, vol. 91, no. 2, pp. 335–338, 1999.

[26] G. Sarath, B. Dien, A. J. Saathoff, K. P. Vogel, R. B. Mitchell, and H. Chen, "Ethanol yields and cell wall properties in divergently bred switchgrass genotypes," *Bioresource Technology*, vol. 102, no. 20, pp. 9579–9585, 2011.

[27] P. R. Adler, M. A. Sanderson, A. A. Boateng, P. J. Weimer, and H.-J. G. Jung, "Biomass yield and biofuel quality of switchgrass harvested in fall or spring," *Agronomy Journal*, vol. 98, no. 6, pp. 1518–1525, 2006.

[28] J. Burvall, "Influence of harvest time and soil type on fuel quality in reed canary grass (*Phalaris Arundinacea L.*)," *Biomass & Bioenergy*, vol. 12, no. 3, pp. 149–154, 1997.

[29] S. Clarke, P. Eng, and F. Preto, *Biomass Burn Characteristics*, OMAFRA, Ontario, Canada, 2013, http://www.omafra.gov.on.ca/english/engineer/facts/11-033.pdf.

[30] B. M. Jenkins, L. L. Baxter, and T. R. Miles Jr., "Combustion properties of biomass," *Fuel Processing Technology*, vol. 54, no. 1–3, pp. 17–46, 1998.

[31] R. Samson, *Switchgrass Production in Ontario: A Management Guide*, 2007, http://www.reap-canada.com/lirary/Bioenergy/2007%20SG%20production%20guide-FINAL.pdf.

[32] R. Samson, E. Delaquis, and G. MacInnis, *Enhancing the Commercial Viability of Switchgrass on Marginal Farmland through Plant Breeding*, 2014, http://www.reap-canada.com/online_library/feedstock_biomass/Report%20-%20Switchgrass%20breeding%20on%20marginal%20farmland%20-%20Samson,%20Delaquis,%20MacInnis%202013.pdf.

[33] G. W. Burton, "Recurrent restricted phenotypic selection increases forage yields of *Pensacola Bahiagrass*," *Crop Science*, vol. 14, no. 6, pp. 831–835, 1974.

[34] K. J. Moore, L. E. Moser, K. P. Vogel, S. S. Waller, B. E. Johnson, and J. F. Pedersen, "Describing and quantifying growth stages of perennial forage grasses," *Agronomy Journal*, vol. 83, pp. 1073–1077, 1991.

[35] Association of Official Analytical Chemists, *Official Methods for Analysis*, AOAC, Arlington, Va, USA, 15th edition, 1990.

[36] P. J. van Soest, J. B. Robertson, and B. A. Lewis, "Methods for dietary fiber, neutral detergent fiber, and nonstarch polysaccharides in relation to animal nutrition," *Journal of Dairy Science*, vol. 74, no. 10, pp. 3583–3597, 1991.

[37] SAS, SAS Institute Inc. Cary, North Carolina, USA. 2003.

[38] A. Boe and P. O. Johnson, "Deriving a large-seeded switchgrass population using air-column separation of parent seed," *Crop science*, vol. 27, pp. 147–148, 1987.

[39] R. Lemus, E. C. Brummer, K. J. Moore, N. E. Molstad, C. L. Burras, and M. F. Barker, "Biomass yield and quality of 20 switchgrass populations in southern Iowa, USA," *Biomass and Bioenergy*, vol. 23, no. 6, pp. 433–442, 2002.

[40] J. H. Fike, D. J. Parrish, D. D. Wolf et al., "Long-term yield potential of switchgrass-for-biofuel systems," *Biomass & Bioenergy*, vol. 30, no. 3, pp. 198–206, 2006.

[41] D. R. West and D. R. Kincer, "Yield of switchgrass as affected by seeding rates and dates," *Biomass and Bioenergy*, vol. 35, no. 9, pp. 4057–4059, 2011.

[42] A. A. Hopkins, K. P. Vogel, K. J. Moore, K. D. Johnson, and I. T. Carlson, "Genotype effects and genotype by environment interactions for traits of elite switchgrass populations," *Crop Science*, vol. 35, no. 1, pp. 125–132, 1995.

[43] S. D. Wullschleger, E. B. Davis, M. E. Borsuk, C. A. Gunderson, and L. R. Lynd, "Biomass production in switchgrass across the United States: database description and determinants of yield," *Agronomy Journal*, vol. 102, no. 4, pp. 1158–1168, 2010.

[44] R. Jannasch, P. Duxbury, and R. Samson, Development of bioenergy feedstocks: agronomy data from Eastern Canada, 2001, http://www.reap-canada.com/online_library/feedstock_biomass/8-Development%20of%20Bioenergy%20Feedstocks-Agronomy%20Data%20from%20Eastern%20Canada-Jannasch%20et%20al%202001.pdf.

[45] D. S. Zamora, G. J. Wyatt, K. G. Apostol, and U. Tschirner, "Biomass yield, energy values, and chemical composition of hybrid poplars in short rotation woody crop production and native perennial grasses in Minnesota, USA," *Biomass & Bioenergy*, vol. 49, pp. 222–230, 2013.

[46] M. D. Casler, K. P. Vogel, C. M. Taliaferro, and R. L. Wynia, "Latitudinal adaptation of switchgrass populations," *Crop Science*, vol. 44, no. 1, pp. 293–303, 2004.

[47] I. C. Madakadze, B. E. Coulman, A. R. Mcelroy, K. A. Stewart, and D. L. Smith, "Evaluation of selected warm-season grasses for biomass production in areas with a short growing season," *Bioresource Technology*, vol. 65, no. 1-2, pp. 1–12, 1998.

[48] D. D. Redfearn, K. J. Moore, K. P. Vogel, S. S. Waller, and R. B. Mitchell, "Canopy architecture and morphology of switchgrass populations differing in forage yield," *Agronomy Journal*, vol. 89, no. 2, pp. 262–269, 1997.

[49] D. M. Gustafson, A. Boe, and Y. Jin, "Genetic variation for *Puccinia emaculata* infection in switchgrass," *Crop Science*, vol. 43, no. 3, pp. 755–759, 2003.

[50] C. K. Ong, C. Marshall, and G. R. Saoar, "The physiology of tiller death in grasses. 2. Causes of tiller death in a grass sward," *Journal of the British Grassland Society*, vol. 33, pp. 205–211, 1978.

[51] R. B. Mitchell, L. E. Moser, K. J. Moore, and D. D. Redfearn, "Tiller demographics and leaf area index of four perennial pasture grasses," *Agronomy Journal*, vol. 90, no. 1, pp. 47–53, 1998.

[52] I. C. Madakadze, B. E. Coulman, P. Peterson, K. A. Stewart, R. Samson, and D. L. Smith, "Leaf area development, light interception, and yield among switchgrass populations in a short-season area," *Crop Science*, vol. 38, no. 3, pp. 827–834, 1998.

[53] I. Madakadze, B. E. Coulman, K. Stewart, P. Peterson, R. Samson, and D. L. Smith, "Phenology and tiller characteristics of big

bluestem and switchgrass cultivars in a short growing season area," *Agronomy Journal*, vol. 90, no. 4, pp. 489–495, 1998.

[54] A. Boe, "Variation between two switchgrass cultivars for components of vegetative and seed biomass," *Crop Science*, vol. 47, no. 2, pp. 636–642, 2007.

[55] R. Samson, C. Ho Lem, and S. Bailey-Stamler, "Optomization of switchgrass management for commercial fuel pellet production," 2008, http://www.reap-canada.com/online_library/ feedstock_biomass/Optimization%20of%20switchgrass%20-management%20for%20commercial%20fuel%20pellet%20pro-duction%20(Samson%20et%20al.,%202007).pdf.

[56] A. Prochnow, M. Heiermann, M. Plöchl, T. Amon, and P. J. Hobbs, "Bioenergy from permanent grassland—a review: 2. Combustion," *Bioresource Technology*, vol. 100, no. 21, pp. 4945–4954, 2009.

[57] J. Vogel, "Unique aspects of the grass cell wall," *Current Opinion in Plant Biology*, vol. 11, no. 3, pp. 301–307, 2008.

[58] Z. Hu, R. Sykes, M. F. Davis, E. C. Brummer, and A. J. Ragauskas, "Chemical profiles of switchgrass," *Bioresource Technology*, vol. 101, no. 9, pp. 3253–3257, 2010.

[59] S. Q. Turn, C. M. Kinoshita, and D. M. Ishimura, "Removal of inorganic constituents of biomass feedstocks by mechanical dewatering and leaching," *Biomass & Bioenergy*, vol. 12, no. 4, pp. 241–252, 1997.

# 5

# Fine Screening for Resistance to Cold-Induced Sweetening in Potato Hybrids Containing *Solanum raphanifolium* Germplasm

## Arfan Ali[1] and Shelley Jansky[2]

[1]*Centre of Excellence in Molecular Biology, University of the Punjab, Lahore, Pakistan*
[2]*USDA Agricultural Research Service and University of Wisconsin-Madison, 1575 Linden Drive, Madison, WI, USA*

Correspondence should be addressed to Shelley Jansky; shelley.jansky@ars.usda.gov

Academic Editor: Philippe Giordanengo

Potato is an indispensable part of human food. Many wild and cultivated potato relatives have been screened to find the best germplasm to improve productivity and quality, but only a small sample of the available biodiversity has been exploited. Most wild relatives are self-incompatible diploids. Genetic variability exists within and among populations, even within a species. Therefore, it is necessary to carry out fine screening to identify individuals carrying traits of interest. This study was carried out to quantify phenotypic variability for resistance to cold-induced sweetening, an important processing trait. Five families were evaluated for potato chip (crisp) color following cold storage of tubers harvested from four greenhouse trials and one field trial. The families were generated by crossing a single diploid clone to five plants from one accession of the wild potato relative *Solanum raphanifolium*. Analysis of variance revealed that resistance against cold-induced sweetening was dependent on family and trial. This study underscores the importance of fine screening to select individuals in potato accessions for use in potato improvement.

## 1. Introduction

In most potato producing regions worldwide, the crop is typically harvested during a narrow window of time, but it is consumed throughout the year. Consequently, most of the potato crop is stored before it is shipped to packing facilities and processing plants. Metabolic stability of potato tubers during this storage period is one of the prime trait targets for breeding programs worldwide [1–5]. Potatoes need to be cool-stored throughout the year to maintain a continuous supply to the industry, and storage at low temperature ($<8°C$) is beneficial because it reduces bacterial soft rots, decreases water and dry matter loss, and prevents sprouting without the need to add sprout inhibitors [6]. However, storage at low temperatures leads to an accumulation of the reducing sugars glucose and fructose, in a process known as cold-induced sweetening (CIS) [7]. Sugars that accumulate in nonphotosynthetic tissues are recruited from the starch degradation pathway [8, 9]. Tuber sugar content is affected by abundance and activity of carbohydrate metabolizing enzymes in source (leaf) and sink (tuber) tissues, and by the flux of sucrose

from source tissues [4]. When potatoes that have undergone CIS are processed into crisps or chips, high fry temperatures cause reducing sugars to react with free amino acids via the Maillard reaction, resulting in an unacceptable blackening of the product [10]. The Maillard reaction also leads to the formation of the probable carcinogen acrylamide [11].

Studies on carbohydrate metabolism in potato have shown that several enzymes contribute to low temperature sweetening. During low temperature storage, starch breakdown into sucrose is usually driven by UDP-glucose pyrophosphorylase and sucrose-6-phosphate synthase [5, 12]. Acid invertase converts sucrose into the reducing sugars glucose and fructose. A relationship between light chip color and a low level of UDP-glucose pyrophosphorylase activity has been demonstrated [4, 6, 12]. Similarly, low acid invertase activity is associated with light chip color [12–16]. However, QTL studies have found associations between a number of additional carbohydrate metabolism genes and CIS. More recently, association genetics studies have demonstrated that DNA polymorphisms in genes encoding invertase and starch phosphorylases are associated with potato chip color, starch content, and starch yield [17, 18]. Association analysis found that SNP2746 in the *StLapN* gene was strongly associated with chip quality [19]. These genetic studies support the working model that natural variation in tuber starch and sugar content is controlled by allelic variants of enzymes that function in starch and sugar metabolism [20].

Microarray hybridization experiments using a tomato gene chip hybridized with potato mRNA allowed the identification of known and novel genes that were differentially expressed during tuber cold storage in a potato clone. Transcript levels of known candidate genes, such as invertase, were correlated with sugar accumulation [21]. Comparative proteome analysis has previously proven successful in identifying new candidate genes controlling tuber quality traits [22, 23]. One approach to breeding for resistance to CIS is to reduce acid invertase activity. Silencing of the acid invertase gene has effectively reduced CIS, resulting in acceptable fry products [13]. Alternatively, resistance to CIS is found in wild relatives of potato.

Wild potato species are useful sources of genes for potato improvement [1, 7, 24, 25]. The wild diploid species *Solanum raphanifolium* is sexually compatible with diploid forms of cultivated potato. Its hybrid offspring have been shown to exhibit resistance to CIS [26]. In fact, acid invertase activity in germplasm carrying CIS resistance genes from *S. raphanifolium* is as low as that in clones in which the gene has been silenced [13]. However, *S. raphanifolium* and most diploid wild germplasm are self-incompatible and are maintained by intercrossing within populations. Consequently, genetic variation within and among accessions is common [24, 27–30]. It is important, therefore, to identify individuals within a population carrying genes of interest.

The purpose of this study was to carry out fine screening for resistance to CIS on a population of *S. raphanifolium*. Since tubers of wild species are typically very small, individuals in the population were crossed with a clone that would allow for the production of offspring with larger tubers suitable for CIS evaluation.

## 2. Materials and Methods

Crosses were made between clone hap-chc (HC) as a female and five plants of *S. raphanifolium* PI 310998 as a male. HC is a hybrid between *S. tuberosum* dihaploid (US-W730) and the potato wild relative *S. chacoense* PI 310998. In previous work, HC x PI 310998 hybrids have exhibited exceptional resistance to cold-induced sweetening [13, 26]. Seeds were sown on August 14, 2009, and transplanted to 48-well flats three weeks later, and then 16–48 plants per family were transplanted into 10 cm square pots on October 5. They were grown in a greenhouse at Arlington, WI, until maturity. Plants were harvested on January 28, 2010, and placed in 4°C storage on February 2. Two tubers of each plant in each family were chipped on June 3 (121 days in storage); two tubers were retained for clonal maintenance. This trial is designated A1. On September 30, 2010, one tuber per clone from the A1 trial was planted again at Arlington. Tubers were harvested from mature plants on January 28, 2011. They were stored at 4°C

until June 6 (129 days in storage), when they were chipped. This was trial A2.

Thirty-five clones in trial A1 produced at least eight tubers each, so four tubers were planted in the field at Rhinelander, WI, on May 6, 2010. Two replications of two plants each were included in a randomized complete block design. On September 10, each plant was harvested by hand and tubers were collected. The tubers were stored at 4.4°C until June 8, 2011, when two tubers per plot were chipped (271 days in storage). This was trial (R).

One tuber from each clone in the A2 trial was planted on September 15, 2011, at a greenhouse in Madison, WI. Mature plants were harvested on March 8, 2012, and stored at 6°C until April 9, when two tubers per clone were chipped (32 days in storage). This was trial M1. Each clone was again planted in the Madison greenhouse on April 16 and harvested on July 27, and tubers were stored at 6°C for 30 days. Two tubers per clone were then chipped. This was trial M2. All greenhouse trials used supplemental high intensity light to support plant growth and maintain an 18-hour photoperiod. Photoperiod was shortened to 12 hours to induce tuberization one month before harvest. Plants were grown in soil-less peat-based potting mix and fertilized weekly. The number of clones in each family is presented in Table 1. Storage conditions are summarized in Table 2.

Chip color was evaluated by taking a 1-2 mm slice from the center of a transverse tuber cut, rinsing it in tap water, and frying it in 190°C corn oil until bubbling ceased. Each chip was visually scored for color using a scale of 1 (light) to 10 (dark), at 0.5 intervals, based on the International Chip Color Institute (Cleveland, OH) color chart.

Analysis of variance was carried out using a general linear model in SAS (version 9.3; SAS Institute, Cary, NC). Means separation was carried out using Fisher's protected least significant difference (LSD) test at $P = 0.05$. Error variances were homogeneous, so no transformation of the data set was necessary.

## 3. Results and Discussion

Analysis of variance revealed highly significant effects of family, trial, and the family by trial interaction on chip color (Table 3). Consequently, subsequent evaluations considered each family in each trial, rather than averaging over trial or family. Not all clones tuberized in each trial, so the number of individuals evaluated varied from trial to trial. However, within each trial, a similar number of clones per family was evaluated. Many clones did not tuberize in the field, so the Rhinelander trial contained the smallest number of individuals.

Within each trial, ANOVA revealed a significant effect of family ($P < 0.05$), on chip score at all locations except M1 (Table 4). The effect of tuber sample and replication was not significant. Family 15 consistently produced light chips, while family 14 produced darker chips. All five families were generated by crossing the same female clone (HC) to plants from one wild species accession. HC is a heterozygous interspecific hybrid, so it may segregate for CIS alleles. However, the set of

TABLE 1: Number of clones evaluated for chip color in five families (F11–F18) across five trials (M1–R).

|     | M1 | M2 | A1 | A2 | R  |
|-----|----|----|----|----|----|
| F11 | 19 | 15 | 36 | 36 | 13 |
| F13 | 25 | 15 | 44 | 44 | 4  |
| F14 | 28 | 19 | 41 | 40 | 4  |
| F15 | 22 | 19 | 34 | 34 | 8  |
| F18 | 20 | 14 | 44 | 44 | 4  |

TABLE 2: Storage conditions for each of the five trials.

| Trial | Storage temperature | Days in storage |
|-------|---------------------|-----------------|
| M1 | 6°C | 32 |
| M2 | 6°C | 30 |
| A1 | 4°C | 121 |
| A2 | 4°C | 129 |
| R  | 4°C | 271 |

TABLE 3: ANOVA table for effect of family and trial on chip color.

| Source | Df | Sum of squares | $F$ value | $\Pr > F$ |
|--------|-----|---------|-------|----------|
| Family | 4 | 135.18 | 16.29 | <0.0001 |
| Trial | 4 | 63.91 | 7.70 | <0.0001 |
| Family * trial | 16 | 89.54 | 2.70 | 0.0004 |
| Error | 602 | 1249.12 | | |

TABLE 4: Mean chip score of five families in five cold storage trials. Chip score ratings were based on a scale of 1 (light) to 10 (dark).

|     | M1 | M2 | A1 | A2 | R |
|-----|------|---------|---------|---------|---------|
| F11 | 4.82 | 4.00 b[*] | 4.49 ab | 4.25 b | 6.77 ab |
| F13 | 4.38 | 4.43 b | 4.11 bc | 3.93 bc | 4.75 bc |
| F14 | 4.69 | 6.18 a | 4.76 a | 5.00 a | 8.00 a |
| F15 | 4.45 | 4.03 b | 3.53 c | 3.53 c | 4.38 c |
| F18 | 5.03 | 3.93 b | 4.11 bc | 4.14 bc | 4.00 c |

[*]Within a column, means followed by different letters are different at $p = 0.05$.

alleles it contributes to offspring is expected to be the same among the five families. Consequently, differences among families are likely due mainly to genetic variability among male parents from the same *S. raphanifolium* accession. Individuals from a population of an outcrossing species are likely to be heterozygous and heterogeneous. While PI 310998 is a good source of CIS resistance genes, fine screening of the population for individuals that produce a high proportion of resistant offspring is likely to be productive. For example, the use of the clone that produced family 15, rather than the one that produced family 14, would allow a breeder to make more progress toward breeding for reduced CIS. Fine screening of accessions has been reported for other traits in potato [31–33].

The darkest chips were produced from tubers grown at Rhinelander and stored for 271 days at 4°C (Table 4). This is an extreme storage environment for both temperature and duration. Long-term storage of potato tubers has significant,

TABLE 5: Number (percent) of clones in each family and trial with acceptable chip color scores (4.5 or less).

|     | M1 | M2 | A1 | A2 | R |
|-----|-----------|-------------|-------------|-------------|------------|
| F11 | 9 (47.36%) | 12 (80%) | 28 (77.7%) | 28 (77.7%) | 3 (23.07%) |
| F13 | 21 (84%) | 9 (60%) | 40 (93.1%) | 40 (90.9%) | 2 (50%) |
| F14 | 19 (71.4%) | 8 (42.10%) | 28 (68.29%) | 26 (65%) | 0 |
| F15 | 16 (72.7%) | 28 (84%) | 34 (100%) | 34 (100%) | 7 (87.5%) |
| F18 | 9 (45%) | 10 (71.4%) | 37 (84.09%) | 37 (84.09%) | 4 (100%) |

variety-dependent effects on sugar and amino acid concentrations [34]. However, in this study, nine clones produced chips with commercially acceptable color, illustrating the value of this germplasm for potato cultivar improvement.

Because potato cultivars are clones rather than populations, it is most important for breeders to identify parental combinations that will produce a large proportion of offspring carrying the trait of interest when introgressing wild germplasm into breeding lines. Then, after selecting clones expressing that trait, the breeder still has a large number of individuals with which to select for other important agronomic traits. The parents of families 13 and 15 consistently produced families in which at least 50% of the offspring had acceptable resistance to CIS (Table 5). These would be desirable parents to use in a breeding program to improve processing quality after cold storage in potato.

## Conflict of Interests

The authors have no competing interests.

## Acknowledgment

Funding for the senior author's salary was provided by the Higher Education Commission (HEC) of Pakistan.

## References

[1] J. E. Bradshaw and G. Ramsay, "Utilisation of the commonwealth potato collection in potato breeding," *Euphytica*, vol. 146, no. 1-2, pp. 9–19, 2005.

[2] M. F. B. Dale and J. E. Bradshaw, "Progress in improving processing attributes in potato," *Trends in Plant Science*, vol. 8, no. 7, pp. 310–312, 2003.

[3] C. N. Hirsch, C. D. Hirsch, K. Felcher et al., "Retrospective view of North American potato (*Solanum tuberosum* L.) breeding in the 20th and 21st centuries," *G3: Genes, Genomes, Genetics*, vol. 3, no. 6, pp. 1003–1013, 2013.

[4] C. M. Menéndez, E. Ritter, R. Schäfer-Pregl et al., "Cold sweetening in diploid potato: mapping quantitative trait loci and candidate genes," *Genetics*, vol. 162, no. 3, pp. 1423–1434, 2002.

[5] J. R. Sowokinos, "Biochemical and molecular control of cold-induced sweetening in potatoes," *American Journal of Potato Research*, vol. 78, no. 3, pp. 221–236, 2001.

[6] J. R. Sowokinos, "Allele and isozyme patterns of UDP-glucose pyrophosphorylase as a marker for cold-sweetening resistance

in potatoes," *American Journal of Potato Research*, vol. 78, no. 1, pp. 57–64, 2001.

[7] T. J. Herrman, S. L. Love, B. Shafii, and R. B. Dwelle, "Chipping performance of three processing potato cultfvars during long-term storage at two temperature regimes," *American Potato Journal*, vol. 73, no. 9, pp. 411–425, 1996.

[8] J. G. Malone, V. Mittova, R. G. Ratcliffe, and N. J. Kruger, "The response of carbohydrate metabolism in potato tubers to low temperature," *Plant and Cell Physiology*, vol. 47, no. 9, pp. 1309–1322, 2006.

[9] J. R. Sowokinos, V. Vigdorovich, and M. Abrahamsen, "Molecular cloning and sequence variation of UDP-glucose pyrophosphorylase cDNAs from potatoes sensitive and resistant to cold sweetening," *Journal of Plant Physiology*, vol. 161, no. 8, pp. 947–955, 2004.

[10] L. M. Benzing-Purdie, J. A. Ripmeester, and C. I. Ratcliffe, "Effects of temperature on Maillard reaction products," *Journal of Agricultural and Food Chemistry*, vol. 33, no. 1, pp. 31–33, 1985.

[11] D. S. Mottram, B. L. Wedzicha, and A. T. Dodson, "Food chemistry: acrylamide is formed in the Maillard reaction," *Nature*, vol. 419, no. 6906, pp. 448–449, 2002.

[12] M. J. McKenzie, J. R. Sowokinos, I. M. Shea, S. K. Gupta, R. R. Lindlauf, and J. A. D. Anderson, "Investigations on the role of acid invertase and UDP-glucose pyrophosphorylase in potato clones with varying resistance to cold-induced sweetening," *The American Journal of Potato Research*, vol. 82, no. 3, pp. 231–239, 2005.

[13] P. B. Bhaskar, L. Wu, J. S. Busse et al., "Suppression of the vacuolar invertase gene prevents cold-induced sweetening in potato," *Plant Physiology*, vol. 154, no. 2, pp. 939–948, 2010.

[14] L. Li, J. Strahwald, H.-R. Hofferbert et al., "DNA variation at the invertase locus invGE/GF is associated with tuber quality traits in populations of potato breeding clones," *Genetics*, vol. 170, no. 2, pp. 813–821, 2005.

[15] X. Liu, C. Zhang, Y. Ou et al., "Systematic analysis of potato acid invertase genes reveals that a cold-responsive member, *StvacINV1*, regulates cold-induced sweetening of tubers," *Molecular Genetics and Genomics*, vol. 286, no. 2, pp. 109–118, 2011.

[16] L. Wu, P. B. Bhaskar, J. S. Busse, R. Zhang, P. C. Bethke, and J. Jiang, "Developing cold-chipping potato varieties by silencing the vacuolar invertase gene," *Crop Science*, vol. 51, no. 3, pp. 981–990, 2011.

[17] A. M. Draffehn, S. Meller, L. Li, and C. Gebhardt, "Natural diversity of potato (*Solanum tuberosum*) invertases," *BMC Plant Biology*, vol. 10, no. 1, article 271, 2010.

[18] C. Urbany, T. Colby, B. Stich, L. Schmidt, J. Schmidt, and C. Gebhardt, "Analysis of natural variation of the potato tuber proteome reveals novel candidate genes for tuber bruising," *Journal of Proteome Research*, vol. 11, no. 2, pp. 703–716, 2012.

[19] M. Fischer, L. Schreiber, T. Colby et al., "Novel candidate genes influencing natural variation in potato tuber cold sweetening identified by comparative proteomics and association mapping," *BMC Plant Biology*, vol. 13, no. 1, article 113, 2013.

[20] A. Theocharis, C. Clément, and E. A. Barka, "Physiological and molecular changes in plants grown at low temperatures," *Planta*, vol. 235, no. 6, pp. 1091–1105, 2012.

[21] P. Bagnaresi, A. Moschella, O. Beretta, F. Vitulli, P. Ranalli, and P. Perata, "Heterologous microarray experiments allow the identification of the early events associated with potato tuber cold sweetening," *BMC Genomics*, vol. 9, no. 1, article 176, 2008.

[22] Y. Yang, X. Qiang, K. Owsiany, S. Zhang, T. W. Thannhauser, and L. Li, "Evaluation of different multidimensional LC-MS/MS pipelines for isobaric tags for relative and absolute quantitation (iTRAQ)-based proteomic analysis of potato tubers in response to cold storage," *Journal of Proteome Research*, vol. 10, no. 10, pp. 4647–4660, 2011.

[23] W. Hoehenwarter, J. T. van Dongen, S. Wienkoop et al., "A rapid approach for phenotype-screening and database independent detection of cSNP/protein polymorphism using mass accuracy precursor alignment," *Proteomics*, vol. 8, no. 20, pp. 4214–4225, 2008.

[24] S. H. Jansky, R. Simon, and D. M. Spooner, "A test of taxonomic predictivity: resistance to the Colorado potato beetle in wild relatives of cultivated potato," *Journal of Economic Entomology*, vol. 102, no. 1, pp. 422–431, 2009.

[25] R. Hanneman, "The potato germplasm resource," *American Potato Journal*, vol. 66, no. 10, pp. 655–667, 1989.

[26] A. J. Hamernik, R. E. Hanneman Jr., and S. H. Jansky, "Introgression of wild species germplasm with extreme resistance to cold sweetening into the cultivated potato," *Crop Science*, vol. 49, no. 2, pp. 529–542, 2009.

[27] X. K. Cai, D. M. Spooner, and S. H. Jansky, "A test of taxonomic and biogeographic predictivity: resistance to Potato virus y in wild relatives of the cultivated potato," *Phytopathology*, vol. 101, no. 9, pp. 1074–1080, 2011.

[28] Y. S. Chung, K. Holmquist, D. M. Spooner, and S. H. Jansky, "A test of taxonomic and biogeographic predictivity: resistance to soft rot in wild relatives of cultivated potato," *Phytopathology*, vol. 101, no. 2, pp. 205–212, 2011.

[29] S. Jansky, "Breeding for disease resistance in potato," *Plant Breeding Reviews*, vol. 19, pp. 69–155, 2000.

[30] S. H. Jansky, R. Simon, and D. M. Spooner, "A test of taxonomic predictivity," *Crop Science*, vol. 46, no. 6, pp. 2561–2570, 2006.

[31] J. B. Bamberg, C. Longtine, and E. Radcliffe, "Fine screening Solanum (potato) germplasm accessions for resistance to Colorado potato beetle," *American Journal of Potato Research*, vol. 73, pp. 211–223, 1996.

[32] J. B. Bamberg, J. P. Palta, L. A. Peterson, M. Martin, and A. R. Krueger, "Fine screening potato (*Solanum*) species germplasm for tuber calcium," *American Journal of Potato Research*, vol. 75, no. 4, pp. 181–186, 1998.

[33] D. S. Douches, J. B. Bamberg, W. Kirk et al., "Evaluation of wild Solanum species for resistance to the US-8 genotype of *Phytophthora infestans* utilizing a fine-screening technique," *American Journal of Potato Research*, vol. 78, no. 2, pp. 159–165, 2001.

[34] N. Muttucumaru, S. J. Powers, J. S. Elmore, A. Briddon, D. S. Mottram, and N. G. Halford, "Evidence for the complex relationship between free amino acid and sugar concentrations and acrylamide-forming potential in potato," *Annals of Applied Biology*, vol. 164, no. 2, pp. 286–300, 2014.

# 6

# Weed Control and Corn (*Zea mays*) Response to Planting Pattern and Herbicide Program with High Seeding Rates in North Carolina

**Mitchell K. Williams, Ronnie W. Heiniger, Wesley J. Everman, and David L. Jordan**

*Department of Crop Science, North Carolina State University, P.O. Box 7620, Raleigh, NC 27695-7620, USA*

Correspondence should be addressed to David L. Jordan; david_jordan@ncsu.edu

Academic Editor: Tibor Janda

Effective weed control in corn (*Zea mays* L.) is important to optimize yield. Concern over environmental impact of atrazine and selection for glyphosate resistance has increased the need to develop alternative strategies that use herbicides other than atrazine and glyphosate and appropriate cultural practices to control weeds. Research was conducted during 2011 and 2012 to determine weed and corn response to herbicide programs containing dicamba, glufosinate, and glyphosate applied postemergence alone or with atrazine in single- and twin-row planting patterns. Planting pattern had no effect on common ragweed (*Ambrosia artemisiifolia* L.) and Texas panicum (*Panicum texanum* L.) population and did not interact with herbicide program. Effective weed control hastened maturity in some but not all instances. Under weed-free conditions, corn grain yield was higher in 5 of 7 trials when planted in twin rows versus single rows at equivalent corn populations (141,000 plants $ha^{-1}$). These results suggest that while planting pattern may not impact weed control dramatically, planting corn in twin rows may be an effective alternative to single-row planting patterns because of increased yield under high corn populations.

## 1. Introduction

Effective weed management continues to be important in obtaining optimum corn yields [1]. Common ragweed is among the top ten most common weeds in corn in North Carolina [2]. Other troublesome broadleaf weeds include Palmer amaranth (*Amaranthus palmeri* S. Wats.), common lambsquarters (*Chenopodium album* L.), and sicklepod (*Senna obtusifolia* (L.) Irwin and Barneby) [2]. Texas panicum is considered one of the most troublesome weeds in corn production in the southeast US [3]. Several herbicides control Texas panicum in corn but not always completely [3–5].

Atrazine controls many broadleaf weeds and some grasses in corn and can be applied preplant incorporated (PPI), pre-emergence (PRE), or postemergence (POST) [6, 7]. Although used on many hectares in the US [8], atrazine is no longer used in Europe because of concern of endocrine disruption in amphibians [9]. Residues of atrazine have been found in groundwater and surface water [10] and use of atrazine continues to be reviewed by the EPA in the US [7].

Development of herbicide-resistant biotypes has increased challenges in managing weeds in corn and other agronomic crops [11]. To decrease the likelihood of development of herbicide-resistant biotypes, rotation of herbicide modes of action is essential and should be coupled with cultural practices that minimize reproduction of herbicide-resistant biotypes [11–14]. Although glyphosate can be very effective in glyphosate-tolerant corn to control common ragweed and Texas panicum, resistance of weeds to glyphosate has increased the need to use alternatives to glyphosate [14]. While biotypes of Texas panicum expressing resistance to glyphosate have not been reported, biotypes of common ragweed resistant to glyphosate have been documented in several geographic regions of the US [15]. Dicamba alone or in combination with other herbicides can be effective in controlling common ragweed and other broadleaf weeds [16, 17]. Dicamba and glufosinate applied to small common ragweed can control this weed effectively and can serve as alternatives to atrazine and glyphosate in corn [8, 17–20].

Higher plant populations and narrower row patterns enable corn to compete more effectively with weeds by expediting canopy closure and light interception [21–25]. However, excessive populations can negatively impact corn grain yield. Yield per plant decreased as plant population increased [21, 26]. Increasing plant populations in corn also decreased the number of leaves per plant and percent moisture in grain at harvest [27]. These changes reduced yield per plant because of less radiation interception per plant and consequently synthesized less assimilate for grain. High populations of corn also allow shading of the lower leaves of the corn plant, despite having a lower number of leaves per plant [28]. More frequent lodging was reported in corn planted in twin rows [29].

Research has shown varying results relative to weed suppression by higher corn populations or reduced row spacing of corn [25]. Increasing corn population from 33,000 to 133,000 plants ha$^{-1}$ reduced yellow nutsedge (*Cyperus esculentus* L.) growth [30]. Redroot pigweed (*Amaranthus retroflexus* L.) vegetative biomass was reduced by increased corn population [31]. However, response can be inconsistent due to hybrid selection, weed species, or environmental factors [26, 27, 32].

Historically, equipment has limited transition to rows less than 76 cm for corn, but recent equipment modification has made planting in narrow rows feasible. One alternative to narrow row planting is planting corn in twin row patterns. Conventional corn grain combine headers are compatible with this planting pattern, allowing the use of existing equipment for production [33]. Yield response to planting in single-row versus twin-row planting patterns has been variable in several crops. In soybean (*Glycine max* (L.) Merr.), twin rows planted 10–15 cm apart on 100 cm centers yielded more in twin rows compared to single rows [34]. Soybean yielded the same in the Mississippi Delta of the US when planted in twin-row and single-row patterns [35, 36]. Gozubenli et al. [37] reported a 4% yield increase in twin-row corn at high populations (60,000 plants ha$^{-1}$ to 135,000 plants ha$^{-1}$) compared with single-rows. Novacek et al. [38] reported no yield advantage of twin rows over single rows in Nebraska. Nelson and Smoot [39] reported no difference in corn yield in single and twin rows when planted on 76 cm centers. Yield of twin-row peanut (*Arachis hypogaea* L.) is often higher than yield of single-row peanut [40–43]. Response of cotton to twin-row plantings has been inconsistent [44, 45]. Although results for these crops are mixed relative to improvement in yield due to twin-row planting compared with singe-row planting, yield of corn, cotton, peanut, and soybean is generally not reduced when planted in twin-row planting patterns.

Drought is a major concern in many regions of the US with respect to corn production, especially on those in the coastal plain regions with sandy soils [46]. Stetzel et al. [47] reported that poor ear development has a profound impact on corn yields but may be attributed to factors other than drought stress. Arrested ears are shorter than normal ears and have a much lower kernel count; however, plants do not show signs of poor ear development until silk emergence

and grain fill. Husks are more slender in appearance, and silk numbers are lower compared to plants which exhibit normal ear development. Stetzel et al. [47] also suggested that arrested ear development may result from the application of a nonionic surfactant prior to full tassel of corn [48]. Arrested ear development is generally not uniform across fields, as normal ears can be found towards the outer edges of the field [47] suggesting that stress from crowding of plants may contribute to arrested ear syndrome. Corn growing at relatively high densities can result in reduced pollination caused by asynchrony between delayed silk emergence compared with tassel emergence [26, 49]. Although the number of days to tasseling was not affected across populations ranging from 60,000 to 135,000 plants ha$^{-1}$, tasseling period was affected differently by twin- and single-row planting patterns [26, 39, 49]. Impact of twin-row planting at higher than recommended seeding rates has not been evaluated in North Carolina with respect to ear development and grain yield of corn.

Developing weed management strategies in corn that use more than one herbicide mode of action is important in managing herbicide-resistant biotypes. Coupling effective and diverse herbicide modes of actions with cultural practices such as twin-row panting patterns could lead to more effective weed management in corn. Interactions of these variables have not been evaluated with corn grown in the coastal plain of North Carolina with glyphosate-resistant and glufosinate-resistant hybrids. Therefore, research was conducted to compare weed control and corn development and yield when diverse herbicide programs were used in single- and twin-row planting patterns.

## 2. Materials and Methods

*2.1. Locations, Soils, Tillage, and Irrigation.* Field experiments were conducted in North Carolina during 2011 in one field and during 2012 in two separate fields at the Peanut Belt Research Station located near Lewiston-Woodville (36.07N, −77.11W). The experiment was also conducted during 2012 in two separate fields at the Upper Coastal Plain Research Station located near Rocky Mount (35.84N, −77.67W) and in one field at the Tidewater Research Station located near Plymouth (35.86N, −76.74W). Soils at Rocky Mount, Lewiston-Woodville, and Plymouth were a Norfolk loamy sand (fine-loamy, kaolinitic, thermic *Typic Kandiudults*), a Goldsboro sandy loam (fine-loamy, siliceous, subactive, thermic *Aquic Paleudults*), and an Autryville loamy sand (loamy, siliceous, subactive, thermic *Aquic Paleudults*), respectively. Experiments were planted in fields with natural and relatively high populations of common ragweed and Texas panicum at Lewiston-Woodville. Weed population was too low and nonuniform in distribution at Rocky Mount and Plymouth to assess weed control.

Corn was planted after disking, field cultivation, and bedding/subsoiling in rows spaced 91 cm apart to establish a population of 141,000 plants ha$^{-1}$ during both years at Rocky Mount and Lewiston-Woodville. At Plymouth in 2012, corn density was 91,000 plants ha$^{-1}$. Corn was planted in single rows on 91 cm centers or in twin rows spaced 20 cm apart on 91 cm centers with in-row seeding rate adjusted to establish the same plant population for both planting patterns. Plot size was 2 rows by 9 m in length. Corn was planted on April 15, 2011 and April 16, 2012 at Rocky Mount and Lewiston-Woodville, respectively, and April 16, 2012 at Plymouth. Corn at Lewiston-Woodville and Rocky Mount was irrigated with overhead sprinklers delivering 20 mm each week. Corn was not irrigated at Plymouth.

*2.2. Herbicide Treatments.* During 2011 only, treatments consisted of 2 levels of corn hybrid expressing either glufosinate- or glyphosate-resistance; 2 levels of planting pattern consisting of single or twin rows; and 6 levels of POST herbicide. Herbicide treatments included no herbicide, dicamba, glufosinate or glyphosate (depending on hybrid), atrazine, dicamba plus atrazine, and glufosinate or glyphosate plus atrazine (depending on hybrid). The glufosinate-tolerant hybrid was P1516BVT (DuPont Crop Protection, Wilmington, DE). The glyphosate-tolerant hybrid was P1516HR (DuPont Crop Protection, Wilmington, DE). In 2012, a corn hybrid 1615HR (DuPont Crop Protection, Wilmington, DE) expressing tolerance to both glufosinate and glyphosate was planted. S-metolachlor (Dual II Magnum herbicide, Syngenta Crop Protection, Greensboro, NC) was applied PRE to all plots at all locations at $1700 \, g \, ai \, ha^{-1}$. Dicamba (Clarity herbicide, BASF Crop Protection, Research Triangle Park, NC), glufosinate-ammonium (Liberty herbicide, Bayer CropScience, Research Triangle Park, NC), potassium salt of glyphosate (Roundup PowerMAX herbicide, Monsanto Company, St. Louis, MO), and atrazine (Atrazine 4L herbicide, Helena Chemical Company, Collierville, TN) were applied at $140 \, g \, ae \, ha^{-1}$, $560 \, g \, ae \, ha^{-1}$, $840 \, g \, ae \, ha^{-1}$, and $1100 \, g \, ai \, ha^{-1}$, respectively. Crop oil concentrate (Crop Oil Concentrate, Helena Chemical Company, Collierville, TN) was included with all treatments containing atrazine. Adjuvant was not included with dicamba, glufosinate, and glyphosate when these herbicides were applied alone.

Herbicides were applied with a $CO_2$-pressurized backpack sprayer calibrated to deliver $145 \, L \, ha^{-1}$ using 8002 regular flat-fan nozzles (Teejet Corporation, Wheaton, IL) at 275 kPa. All POST treatments were applied on the same day when corn was 25–36 cm in height and weeds were less than 10 cm in height.

*2.3. Weed Density.* Density of common ragweed was recorded 6 weeks after treatment (WAT) at Lewiston-Woodville during 2011 in a $1 \, m^2$ section of each plot. Density of both common ragweed and Texas panicum was recorded at this location during 2012. Texas panicum population was inconsistent during 2011 across the experimental area at Lewiston-Woodville and at Rocky Mount and Plymouth for all weeds present.

*2.4. Corn Measurements.* The number of days from planting to silk emergence was determined for corn during both years at all locations. Emergence of silks from all plants was recorded in a 3 m section of each row for all plots. Once a corn plant reached silk emergence, the stalk was marked with a felt marker directly above the ear to prevent counting a plant more than once until pollination ended.

Corn ears were removed by hand, shelled, and dried to $140 \, g \, kg^{-1}$ moisture. Prior to shelling, ear type categories of incomplete, complete, blunt, or nubbin were determined using illustrations [50]. Once each ear was classified, all related ears were bulked into ear type and grain yield was determined for each plot.

The experimental design during 2011 was a split-plot with hybrid serving as whole plot units and combinations of planting patterns and herbicide treatments serving as subplot units. In 2012, the experimental design was a randomized complete block because the hybrid used was tolerant to both glufosinate and glyphosate. Treatments were replicated four times.

*2.5. Statistical Analyses.* Data for common ragweed and Texas panicum population, silking and ear type, and grain yield were subjected to analysis of variance for a two (hybrid) by two (planting pattern) by six (POST herbicide treatment) factorial arrangement of treatments during 2011, using the PROC GLM procedure in SAS (SAS v9.2, SAS Institute Inc., Cary, NC). In 2012, data were subjected to analysis of variance for a two (planting pattern) by six (POST herbicide treatment) factorial arrangement of treatments. Means of significant main effects and interactions were separated using Fisher's protected LSD test at $P \leq 0.05$. Pearson correlation coefficients were determined for all parameters at $P \leq 0.05$ (SAS v9.2, SAS Institute, Cary, NC).

At Rocky Mount and Plymouth, weed populations were very low and nonuniform; therefore herbicide program was not considered in the statistical analysis. Data from these locations and for the most effective herbicide treatment in controlling common ragweed and Texas panicum at Lewiston-Woodville during both years were combined to determine the effect of twin-row and single-row planting patterns on corn yield under weed-free conditions. Silk emergence, ear type, and yield data were subjected to ANOVA using the PROC GLM procedure in SAS for a 7 (trial) by 2 (planting pattern) factorial arrangement of treatments. Significant main effects and interactions were separated using Fisher's protected LSD test at $P \leq 0.05$.

# 3. Results and Discussion

*3.1. Weed Density.* During 2011, with the exception of the main effect of POST herbicide program ($P \leq 0.0001$), no other main effects or interactions ($P \geq 0.1937$) were observed for common ragweed density. Application of a POST herbicide decreased density of common ragweed regardless of POST herbicide treatment compared with S-metolachlor alone (Table 1). Lack of an interaction of hybrid and herbicide program suggests that both glufosinate and glyphosate are equally effective in controlling common ragweed when applied using the appropriate herbicide-tolerant hybrid [17].

TABLE 1: Influence of herbicide program on common ragweed density 6 weeks after treatment (WAT), corn grain yield, and days from planting to silking during 2011[a].

| Herbicides applied postemergence[b] | Herbicide rate | Common ragweed | Corn | |
|---|---|---|---|---|
| | | Density 6 WAT | Grain yield | Time to silking |
| | g ha$^{-1}$ | Number m$^{-2}$ | kg ha$^{-1}$ | d |
| No POST herbicide | — | 19$^a$ | 3290$^b$ | 70$^a$ |
| Dicamba | 140 | 0$^b$ | 4140$^a$ | 68$^c$ |
| Glyphosate or glufosinate | 560 or 840 | 1$^b$ | 4140$^a$ | 68$^c$ |
| Atrazine | 1100 | 2$^b$ | 3910$^{ab}$ | 69$^b$ |
| Atrazine plus dicamba | 1100 + 140 | 0$^b$ | 4180$^a$ | 69$^b$ |
| Atrazine plus dicamba plus glyphosate or glufosinate | 1100 + 560 or 840 | 0$^b$ | 4710$^a$ | 69$^b$ |

[a]Means within a column followed by the same letter are not significantly different at $P \leq 0.05$. Data are pooled over hybrid and planting patterns.
[b]Glufosinate and glyphosate applied to the appropriate herbicide-tolerant hybrid. S-metolachlor was applied PRE over the entire test area immediately after planting.

TABLE 2: Common ragweed and Texas panicum densities 6 weeks after treatment as influenced by herbicide program during 2012 at Lewiston-Woodville.

| Herbicides applied postemergence[ab] | Herbicide rate | Common ragweed[c] | Texas panicum[d] | |
|---|---|---|---|---|
| | | | Field 1 | Field 2 |
| | g ha$^{-1}$ | Number in m$^2$ | | |
| No POST herbicide | — | 20.1$^a$ | 10.5$^b$ | 22.1$^{bc}$ |
| Dicamba | 140 | 0$^b$ | 21$^a$ | 42.5$^a$ |
| Glufosinate | 560 | 1.1$^b$ | 17.3$^{ab}$ | 27.5$^b$ |
| Glyphosate | 840 | 3.3$^b$ | 12.8$^{ab}$ | 17.2$^c$ |
| Atrazine | 1100 | 0.3$^b$ | 18$^{ab}$ | 41.7$^a$ |
| Atrazine plus dicamba | 1100 + 140 | 0.1$^b$ | 13.3$^{ab}$ | 44.1$^a$ |
| Atrazine plus glufosinate | 1100 + 560 | 1.1$^b$ | 15.4$^{ab}$ | 19.1$^{bc}$ |
| Atrazine plus glyphosate | 1100 + 840 | 0.5$^b$ | 13.8$^{ab}$ | 23.5$^{bc}$ |

[a]S-metolachlor applied PRE to all plots immediately after planting.
[b]The corn hybrid was tolerant of both glufosinate and glyphosate.
[c]Means followed by the same letter are not significantly different at $P \leq 0.05$. Data are pooled over fields and planting patterns.
[d]Means followed by the same letter are not significantly different at $P \leq 0.05$. Data are pooled over planting patterns.

During 2012, the interaction of trial by planting pattern by herbicide program was not significant for common ragweed density ($P = 0.7028$) or Texas panicum density ($P = 0.2004$). The main effect of herbicide was significant for both common ragweed and Texas panicum density ($P \leq 0.0001$). The interaction of trial by herbicide was not significant for common ragweed density ($P = 0.6448$) but was significant for Texas panicum density ($P \leq 0.0001$). Density of both weed species was not influenced by the main effect of planting pattern ($P \geq 0.6728$) or the interaction of trial by planting pattern ($P \geq 0.4368$). The interaction of planting pattern and POST herbicide program was not significant for common ragweed density ($P = 0.7909$) or Texas panicum density ($P = 0.2898$).

When pooled over fields during 2012, common ragweed density decreased from 20.1 plants m$^{-2}$ to 3.3 plants m$^{-2}$ or less 6 WAT (Table 2) corresponding to a reduction in weed density of at least 84%. Other research [17–20] has demonstrated that atrazine, dicamba, glufosinate, and glyphosate control common ragweed. Differences in Texas panicum density between fields were associated with fewer differences observed between treatments in field 1 compared with more

differences among treatments in field 2 (Table 2). A higher density in the S-metolachlor control was noted in field 2 compared with field 1 (22.1 versus 10.5 plant m$^{-2}$) and may have contributed to the ability of treatment means to separate statistically. In field 1, density of Texas panicum was higher when dicamba was applied compared with S-metolachlor alone (Table 2). There was no difference in density when comparing all other POST herbicide treatments with S-metolachlor in this field. In contrast, in field 2 density of Texas panicum was similar to the S-metolachlor control for all treatments except atrazine or dicamba alone or when these herbicides were co-applied (Table 2). Higher density of Texas panicum following dicamba or atrazine could be explained partially by ineffectiveness of these herbicides in controlling Texas panicum [17]. However, differences in density of weeds may have been associated with weeds emerging after herbicides had their full impact and may not have been related directly to herbicide efficacy. This can be explained by comparing common ragweed density to Texas panicum density. The higher control of common ragweed by herbicides most likely led to a higher density of Texas panicum because there was very little interference of Texas panicum growth

TABLE 3: Days from planting to initial silk emergence and silking duration as influenced by herbicide program during 2012 at Lewiston-Woodville.

| Herbicides applied postemergence[a] | Herbicide rate | Silking[b] | | | | Grain yield[c] |
|---|---|---|---|---|---|---|
| | | Time from planting to initial silking | | Time from silk initiation to final silking | | |
| | | Field 1 | Field 2 | Field 1 | Field 2 | |
| | g ha$^{-1}$ | d | | | | kg ha$^{-1}$ |
| No POST herbicide | — | 66$^a$ | 68$^a$ | 12$^a$ | 11$^d$ | 3570$^c$ |
| Dicamba | 140 | 66$^a$ | 64$^{bc}$ | 11$^a$ | 15$^{ab}$ | 4050$^{bc}$ |
| Glufosinate | 560 | 66$^a$ | 63$^c$ | 11$^a$ | 15$^{ab}$ | 4950$^a$ |
| Glyphosate | 840 | 66$^a$ | 63$^c$ | 11$^a$ | 15$^{ab}$ | 4530$^{ab}$ |
| Atrazine | 1100 | 66$^a$ | 65$^b$ | 11$^a$ | 14$^{bc}$ | 3620$^c$ |
| Atrazine plus dicamba | 1100 + 140 | 66$^a$ | 65$^b$ | 11$^a$ | 13$^c$ | 4220$^b$ |
| Atrazine plus glufosinate | 1100 + 560 | 66$^a$ | 63$^c$ | 12$^a$ | 16$^a$ | 3660$^c$ |
| Atrazine plus glyphosate | 1100 + 840 | 66$^a$ | 64$^{bc}$ | 12$^a$ | 15$^{ab}$ | 4290$^b$ |

[a] S-metolachlor applied PRE to all plots immediately after planting.
[b] Means within a field followed by the same letter are not significantly different at $P \leq 0.05$. Data are pooled over planting patterns.
[c] Means followed by the same letter are not significantly different at $P \leq 0.05$. Data are pooled over planting patterns.

by common ragweed. These data also suggest that in fields with common ragweed and Texas panicum, additional POST herbicides most likely will be required to obtain season-long weed control in corn.

*3.2. Corn Response.* The interaction of hybrid, herbicide program, and planting pattern was not significant for any of the measurements associated with corn growth during 2011 ($P = 0.1836$). Corn stand and height were affected by main effect of hybrid and both main effects of hybrid and POST herbicides, respectively ($P \leq 0.0223$). Grain yield was affected only by herbicide program ($P = 0.0368$) but not by hybrid ($P = 0.6721$), planting pattern ($P = 0.6861$), or the interaction of these treatment factors ($P = 0.0589$). With the exception of time to silking, herbicide treatments did not affect ear type or duration of silking ($P \geq 0.1960$). These data suggest that twin row planting patterns offer no advantage over single rows when common ragweed is the only weed present. Corn grain yield was higher when POST herbicides were applied compared with S-metolachlor except when atrazine was applied alone (Table 1). Fewer days to silking were noted when POST herbicides were applied, with dicamba, glufosinate, and glyphosate having the fewest days to first silking (Table 1).

During 2012, the interaction of trial by herbicide program was noted for days to first silking ($P \leq 0.0001$) and duration of silking ($P \leq 0.0001$). Corn grain yield was affected by the main effect of herbicide program ($P \leq 0.0001$) and the interaction of trial by planting pattern ($P = 0.0476$). Herbicide program, planting pattern, and the interaction of planting pattern and herbicide program did not affect ear type ($P \geq 0.1062$).

The interaction of trial by herbicide program occurred because there was no difference in silk emergence among POST herbicide programs in field 1 compared with differences up to 5 days in silk emergence among treatments in field 2 (Table 3). Time to silking decreased when POST herbicides were applied when compared with S-metolachlor alone

TABLE 4: Influence of planting pattern on corn grain yield during 2012 at Lewiston-Woodville.

| Planting pattern[a] | Corn grain yield[b] | |
|---|---|---|
| | Field 1 | Field 2 |
| | kg ha$^{-1}$ | |
| Single row | 4830$^b$ | 3150$^a$ |
| Twin row | 5400$^a$ | 3120$^a$ |

[a] Single rows consisted of planting corn on 91 cm centers. Twin rows consisted of planting corn in rows spaced 20 cm apart on 91 cm centers. In-row seeding rate adjusted to establish the same plant population for both planting patterns.
[b] Means within a field followed by the same letter are not significantly different at $P \leq 0.05$. Data are pooled over herbicide programs.

(Table 3). Similarly, applying POST herbicides increased the duration of silking compared with S-metolachlor alone (Table 3). The difference in silk emergence between fields 1 and 2 could not be explained.

When pooled over fields during 2012, corn grain yield was similar when dicamba and atrazine were applied alone and when atrazine was applied with glufosinate, and yield following these herbicide treatments did not exceed that of S-metolachlor alone (Table 3). The highest yields were noted when glufosinate and glyphosate were applied alone. Grain yield was higher in one of two fields when corn was planted in twin rows compared with single rows (Table 4). Although inconsistent, yield of corn can be higher when planted in twin rows compared to planting in single rows [33, 38, 39].

*3.3. Correlations of Weed and Corn Response.* When pooled over trials where both weed species were present, correlations of common ragweed density and Texas panicum density with corn stand and height were not significant (Table 5). However, density of these weeds was negatively correlated with grain yield. In contrast, common ragweed density was positively correlated with days to first silk while Texas panicum density was negatively correlated with days to first

TABLE 5: Pearson correlations of common ragweed and Texas panicum density with corn growth and yield response at Lewiston-Woodville.

| Source | Common ragweed | | Texas panicum | |
|---|---|---|---|---|
| | $P > F$ | $R^2$ | $P > F$ | $R^2$ |
| Stand | 0.0730 | −0.16 | 0.1628 | 0.12 |
| Height | 0.0757 | −0.16 | 0.4520 | 0.07 |
| Yield | 0.0141 | −0.23 | <0.0001 | −0.43 |
| Days to first silk | 0.0008 | 0.30 | 0.0147 | −0.22 |
| Silk duration | 0.3024 | −0.09 | 0.0046 | 0.25 |

silk. Silk duration was correlated with Texas panicum density but not common ragweed density. A higher density of both common ragweed and Texas panicum would be expected to decrease grain yield and explain the negative correlation. The significant but differential response of days to first silking is more difficult to explain when comparing the weed species. A delay in reaching silking due to interference by common ragweed is reasonable in that weeds can delay early season growth and development which may adversely affect grain yield. A negative correlation between common ragweed and Texas panicum density and days to first silking implies that at higher Texas panicum densities the number of days to silking would be less than the number of days required with fewer weeds. These differences may be due in part to when weeds emerged in corn and how POST herbicides impacted densities later in the season. Common ragweed density and Texas panicum density were negatively correlated ($P$ = 0.1178, $R$ = −0.14, data not shown in tables). In these fields, common ragweed emerged before Texas panicum and would have resulted in less Texas panicum present when herbicides were not applied to control common ragweed. When common ragweed was controlled by herbicides, more Texas panicum would have been present because of little interference from common ragweed. Silk duration was not affected by common ragweed but was affected by Texas panicum (Table 5). The positive relationship of weed population with days to silking most likely resulted from greater interference with corn and accompanying stress that delayed crop development. Blunt weight ($P \le 0.0001$, $R$ = 0.47) and complete weight ($P \le 0.0001$, $R$ = 0.45) were positively correlated with corn grain yield while duration of silking and grain yield were negatively correlated ($P \le 0.0001$, $R$ = −0.44) (data not shown in tables).

*3.4. Corn Response to Planting Pattern in Absence of Weeds.* When pooled over fields, locations (Lewiston-Woodville, Plymouth, and Rocky Mount), years (2011 and 2012), days to silking, silk duration, and corn stand were not affected by planting pattern or the interaction of trial by planting pattern ($P$ = 0.1669 to 0.5299) (data not shown in tables). In contrast, two distinct groups of trials emerged when considering the impact of plant population on corn grain yield. For 5 of the trials, corn grain yield increased from 5830 kg ha$^{-1}$ in single rows to 7180 kg ha$^{-1}$ in twin rows ($P$ = 0.0007, data not shown in tables). In the remaining two trials grain yield was not affected by planting pattern ($P$ = 0.2442) with yield ranging from 3870 to 4380 kg ha$^{-1}$ (data not shown in tables). The higher yield of corn in twin rows occurred in trials where higher yield potential

was observed (7180 kg ha$^{-1}$ versus 4380 kg ha$^{-1}$ for maximum yield). These data suggest that a positive response to twin-row planting may be possible when environmental and edaphic conditions favor higher yields compared with conditions that are less favorable. The positive response to twin rows also may have been a function of higher than recommended plant populations in single rows that lowered yield potential in this planting pattern. Competition of plants for water and other essential components that contribute to yield may have been more limited in this planting pattern compared with the twin rows because of plant arrangement. Less competition would have occurred between individual plants in twin rows compared with single rows when total plant population was similar for both planting patterns. While the experimental procedure did not allow comparison of twin rows at higher plant populations with more modest populations generally recommended for single-row planting patterns for corn, results do suggest that twin rows are a planting pattern with potential to increase yield by minimizing interplant competition often observed at higher plant populations.

## 4. Conclusion

While broad conclusions from this experiment are limited due to the number of years, experimental procedures, and locations, some reasonable observations can be made. Postemergence herbicides reduced weed populations compared with populations for S-metolachlor alone during both years regardless of corn hybrid. Weed density was not affected by planting pattern; however, in absence of weeds, yield was higher in 5 of 7 trials when corn was planted in twin rows rather than single rows, suggesting that twin-row planting patterns may be an effective alternative to single-row planting patterns in high-yielding environments when total plant populations are similar and corn is seeded at relatively high rates. Only three genotypes were used in the present experiment. Additional research with a broader group of corn hybrids is needed to determine the value of twin-row patterns on wide-scale adoption in corn. However, the value of twin-row planting in corn has not been clearly determined in North Carolina and surrounding states. Results from these experiments will assist growers and their advisors in making decisions on adoption of twin-row planting patterns in corn.

## Conflict of Interests

None of the authors has a conflict of interests in terms of the products mentioned in the paper.

## Acknowledgments

The North Carolina Corn Growers Association provided financial support for this research. Appreciation is expressed to staff at the Peanut Belt Research Station, Tidewater Research Station, and Upper Coastal Plain Research Station for technical assistance. Appreciation is addressed to Dewayne Johnson, Leah Boerema, Matthew Barrow, Joseph Oakes, Jamie Hinton, and Chelsea McPherson for technical assistance.

## References

[1] G. W. Bird, K. M. Maredia, D. Dakouo, and D. Mota-Sanchez, "Role of integrated pest management and sustainable development," in *Integrated Pest Management in the Global Arena*, pp. 73–85, CABI Publishing, Wallingford, UK, 2003.

[2] T. M. Webster, "Weed survey-southern states," *Proceedings Southern Weed Science Society*, vol. 53, pp. 247–252, 2000.

[3] E. P. Prostko, T. L. Grey, and J. W. Davis, "Texas panicum (*Panicum texanum*) control in irrigated field corn (*Zea mays*) with foramsulfuron, glyphosate, nicosulfuron, and pendimethalin," *Weed Technology*, vol. 20, no. 4, pp. 961–964, 2006.

[4] W. C. Johnson III and B. G. Mullinix, "Efficacy and economic analysis of Texas panicum (*Panicum texanum*) management systems in corn (*Zea mays*)," *Weed Technology*, vol. 4, no. 4, pp. 754–758, 1990.

[5] W. E. Thomas, I. C. Burke, and J. W. Wilcut, "Weed management in glyphosate-resistant corn with glyphosate, halosulfuron, and mesotrione," *Weed Technology*, vol. 18, no. 3, pp. 826–834, 2004.

[6] M. R. Obermeier and G. Kapusta, "Postemergence broadleaf weed control in corn (*Zea mays*) with CGA-152005," *Weed Technology*, vol. 10, no. 4, pp. 689–698, 1996.

[7] H. M. LeBaron, J. E. McFarland, and O. C. Burnside, *The Trizaine Herbicides: A Milestone in the Development of Weed Control Technology*, The Triazine Herbicides, 50 Years Revolutionizing Agriculture, 2008.

[8] F. Ackerman, "The economics of atrazine," *International Journal of Occupational and Environmental Health*, vol. 13, no. 4, pp. 441–449, 2007.

[9] T. B. Hayes, K. Haston, M. Tsui, A. Hoang, C. Haeffele, and A. Vonk, "Atrazine-induced hermaphroditism at 0.1 ppb in American leopard frogs (*Rana pipiens*): laboratory and field evidence," *Environmental Health Perspectives*, vol. 111, no. 4, pp. 568–575, 2003.

[10] B. T. Croll, "Pesticides in surface waters and groundwaters," *Journal of the Institution of Water and Environmental Management*, vol. 5, no. 4, pp. 389–395, 1991.

[11] J. S. Holt and H. M. LeBaron, "Significance and distribution of herbicide resistance," *Weed Technology*, vol. 4, no. 1, pp. 141–149, 1990.

[12] J. Gressel and L. A. Segel, "Modelling the effectiveness of herbicide rotations and mixtures as strategies to delay or preclude resistance," *Weed Technology*, vol. 4, pp. 186–198, 1990.

[13] G. R. Stephenson, M. D. Dykstra, R. D. McLaren, and A. S. Hamill, "Agronomic practices influencing triazine-resistant weed distribution in Ontario," *Weed Technology*, vol. 4, pp. 199–207, 1990.

[14] V. K. Nandula, K. N. Reddy, S. O. Duke, and D. H. Poston, "Glyphosate-resistant weeds: current status and future outlook," *Outlooks on Pest Management*, vol. 16, no. 4, pp. 183–187, 2005.

[15] I. Heap, "The International Survey of Herbicide Resistant Weeds," 2013, http://www.weedscience.org.

[16] N. Soltani, C. Shropshire, and P. H. Sikkema, "Giant ragweed (*Ambrosia trifida* L.) control in corn," *Canadian Journal of Plant Science*, vol. 91, no. 3, pp. 577–581, 2011.

[17] W. Everman, "Weed control in corn," in *North Carolina Agricultural Chemicals Manual*, vol. AG-1, pp. 224–230, 2014.

[18] S. D. Askew and J. W. Wilcut, "Cost and weed management with herbicide programs in glyphosate-resistant cotton (*Gossypium hirsutum*)," *Weed Technology*, vol. 13, no. 2, pp. 308–313, 1999.

[19] J. L. Corbett, S. D. Askew, W. E. Thomas, and J. W. Wilcut, "Weed efficacy evaluations for bromoxynil, glufosinate, glyphosate, pyrithiobac, and sulfosate," *Weed Technology*, vol. 18, no. 2, pp. 443–453, 2004.

[20] A. S. Culpepper, A. C. York, R. B. Batts, and K. M. Jennings, "Weed management in glufosinate- and glyphosate-resistant soybean (*Glycine max*)," *Weed Technology*, vol. 14, no. 1, pp. 77–88, 2000.

[21] W. G. Duncan, "The relationship between corn population and yield," *Agronomy Journal*, vol. 50, pp. 82–84, 1958.

[22] W. G. Duncan, "A theory to explain the relationship between corn population and grain yield," *Crop Science*, vol. 24, pp. 1141–1145, 1984.

[23] P. H. Sikkema, R. E. Nurse, T. Welacky, and A. S. Hamill, "Reduced herbicide rates provide acceptable weed control regardless of corn planting strategy in Ontario field corn," *Canadian Journal of Plant Science*, vol. 88, no. 2, pp. 373–378, 2008.

[24] J. R. Teasdale, "Influence of narrow row/high population corn (*Zea mays*) on weed control and light transmittance," *Weed Technology*, vol. 9, no. 1, pp. 113–118, 1995.

[25] J. R. Teasdale, "Influence of corn (*Zea mays*) population and row spacing on corn and velvetleaf (*Abutilon theophrasti*) yield," *Weed Science*, vol. 46, no. 4, pp. 447–453, 1998.

[26] A. M. Hashemi, S. J. Herbert, and D. H. Putnam, "Yield response of corn to crowding stress," *Agronomy Journal*, vol. 97, no. 3, pp. 839–846, 2005.

[27] E. E. N. A. Bonaparte and R. I. Brawn, "Effects of plant density and planting date on leaf number and some developmental events in corn," *Canadian Journal of Plant Science*, vol. 56, pp. 691–698, 1976.

[28] M. J. Ottman and L. F. Welch, "Planting patterns and radiation interception, plant nutrient concentration, and yield in corn," *Agronomy Journal*, vol. 81, pp. 167–174, 1989.

[29] D. M. Lambert and J. Lowenberg-DeBoer, "Economic analysis of row spacing for corn and soybean," *Agronomy Journal*, vol. 95, no. 3, pp. 564–573, 2003.

[30] Z. Ghafar and A. K. Watson, "Effect of corn population on the growth of yellow nutsedge," *Weed Science*, vol. 31, pp. 588–592, 1983.

[31] S. M. McLachlan, M. Tollenaar, C. J. Swanton, and S. F. Weise, "Effect of corn-induced shading on dry matter accumulation, distribution, and architecture of redroot pigweed," *Weed Science*, vol. 41, no. 4, pp. 568–573, 1993.

[32] G. A. Johnson, T. R. Hoverstad, and R. E. Greenwald, "Integrated weed management using narrow corn row spacing, herbicides, and cultivation," *Agronomy Journal*, vol. 90, no. 1, pp. 40–46, 1998.

[33] H. A. Bruns, M. W. Ebelhar, and H. K. Abbas, "Comparing single-row and twin-row corn production in the Mid South," *Crop Management*, vol. 11, no. 1, 2012.

[34] W. J. Grichar, "Row spacing, plant populations, and cultivar effects on soybean production along the Texas Gulf Coast," *Crop Management*, vol. 6, no. 1, 2007.

[35] H. A. Bruns, "Comparisons of single-row and twin-row soybean production in the Mid-South," *Agronomy Journal*, vol. 103, no. 3, pp. 702–708, 2011.

[36] H. A. Bruns, "Planting date, rate, and twin-row vs. single-row soybean in the mid-south," *Agronomy Journal*, vol. 103, no. 5, pp. 1308–1313, 2011.

[37] H. Gozubenli, M. Kilinc, O. Sener, and O. Konuskan, "Effects of single and twin row planting on yield and yield components in maize," *Asian Journal of Plant Sciences*, vol. 3, pp. 203–206, 2004.

[38] M. J. Novacek, S. C. Mason, T. D. Galusha, and M. Yaseen, "Twin rows minimally impact irrigated maize yield, morphology, and lodging," *Agronomy Journal*, vol. 105, no. 1, pp. 268–276, 2013.

[39] K. A. Nelson and R. L. Smoot, "Twin- and single-row corn production in northeast Missouri," *Crop Management*, 2009.

[40] E. W. Hauser and G. A. Buchanan, "Influence of row spacing, seeding rates, and herbicide systems on the competitiveness and yield of peanuts," *Peanut Science*, vol. 8, pp. 78–81, 1981.

[41] E. W. Hauser and G. A. Buchanan, "Production of peanuts as affected by weed competition and row spacing," *Alabama Agricultural Experiment Station Bulletin*, vol. 538, p. 35, 1982.

[42] G. Wehtje, R. H. Walker, M. G. Patterson, and J. A. McGuire, "Influence of twin rows on yield and weed control in peanuts," *Peanut Science*, vol. 11, pp. 88–91, 1984.

[43] J. E. Lanier, D. L. Jordan, J. F. Spears et al., "Peanut response to planting pattern, row spacing, and irrigation," *Agronomy Journal*, vol. 96, no. 4, pp. 1066–1072, 2004.

[44] K. N. Reddy, I. C. Burke, J. C. Boykin, and J. Ray Williford, "Narrow-row cotton production under irrigated and non-irrigated environment: plant population and lint yield," *Journal of Cotton Science*, vol. 13, no. 2, pp. 48–55, 2009.

[45] K. N. Reddy and J. C. Boykin, "Weed control and yield comparisons of twin- and single-row glyphosate-resistant cotton production systems," *Weed Technology*, vol. 24, no. 2, pp. 95–101, 2010.

[46] E. J. Sadler, P. J. Bauer, and W. J. Busscher, "Site-specific analysis of a droughted corn crop: I. Growth and grain yield," *Agronomy Journal*, vol. 92, no. 3, pp. 395–402, 2000.

[47] N. Stetzel, K. Wise, B. Nielsen, and C. Gerber, *Arrested Ear Development in Hybrid Corn*, Purdue Extension Publication, BP-85-W, West Lafayette, Ind, USA, 2011.

[48] L. J. Abendroth, R. W. Elmore, M. J. Boyer, and S. K. Marlay, *Corn Growth and Development*, Iowa State University Cooperative Extension Publication, Ames, Iowa, USA, 2011.

[49] M. E. Otegui, "Kernel set and flower synchrony within the ear of maize: II. Plant population effects," *Crop Science*, vol. 37, no. 2, pp. 448–455, 1997.

[50] D. Mueller and R. Pope, *Corn Field Guide*, Iowa State University Cooperative Extension Service, 2009.

# 7

# Effect of Irrigation Regimes and Nitrogen Levels on the Growth and Yield of Wheat

**S. M. Shirazi,[1] Zulkifli Yusop,[1] N. H. Zardari,[1] and Z. Ismail[2]**

[1] Institute of Environmental and Water Resource Management (IPASA), Universiti Teknologi Malaysia (UTM),
  81310 Skudai, Johor, Malaysia
[2] Department of Civil Engineering, Faculty of Engineering, University of Malaya, 50603 Kuala Lumpur, Malaysia

Correspondence should be addressed to S. M. Shirazi; smshirazi@gmail.com

Academic Editor: Ayman Suleiman

A field experiment was carried out to evaluate the effect of irrigation regimes and nitrogen levels on the growth and yield of wheat cv. Kanchan (*Triticum aestivum* L.). The experiment includes two factors such as four irrigation regimes and four nitrogen levels. Three farmer's fields were selected for experimentation as replication. Yield and yield contributing factors were significantly affected by irrigation regimes and different doses of nitrogen. Maximum grain yield of 2.27 t ha$^{-1}$ by the application of 200 mm irrigation treatment. Interaction between 200 mm irrigation and 120 kg N ha$^{-1}$ was the best combination treatment.

## 1. Introduction

Wheat is the second most important cereal crop and covers 0.64 million ha of land with an annual production of 1.2 million tons in Bangladesh [1]. Area under wheat is increasing because wheat requires less amount of irrigation compared to other cereal crops like rice. Wheat is grown in the driest months of the year when rainfall is scarce. Water balance analysis at Mymensingh region in Bangladesh depicts that during January to April and November to December potential evapotranspiration (PET) is higher than precipitation (Figure 1). Irrigation is necessary in order to grow crops during this period because of insufficient amount of rain water and high atmospheric evaporative demand by crops (Figure 2). Nitrogen (N) is a key element for plant nutrition. Applying N and phosphorus (P) fertilizers and other management practices increased the yield of wheat but in some cases these show adverse effects due to severely limiting irrigation [2, 3]. Nitrogen use efficiency can be increased by combining fertilizer, soil, water, and management. Two main approaches can be undertaken: increasing the use of N during crop growing season and decreasing the losses of N by applying optimum doses [4]. Proper growth and development of wheat needs favorable soil moisture in the root zone. Extractable water capacity of soil has significant influence on wheat grain yield and water productivity response to irrigation [5]. The moisture content in the soil gradually decreases with time in dry season and simultaneously soil moisture tension increases. Excessive irrigation increases evapotranspiration and decreases water use efficiency and may also reduce grain yield [6]. Limited irrigation is an important constraint for wheat production in rainfed, tropical, arid, and semi-arid regions. In scarce rainfall conditions and less irrigation, the best management option for N and irrigation levels is to maintain the maize-wheat cropping sequence [7] and limited irrigation water is combined with N fertilizer [8] to get maximum productivity of wheat. Supplemental irrigation significantly increased the yield of wheat with respect to rainfed treatment. According to Karam et al. [9], about 50% of soil water deficit as supplemental irrigation and 150 kg N ha$^{-1}$ was the optimum combination for maximum grain yield of wheat. Level of fertilization and irrigation is very important to increase the fertilizer efficiency and decrease the loss of water. At a very high tension (about 15 bar), plants cannot absorb water from the soil through root zone and as a result it has an ultimate impact on crop yield. Proper timing and frequency

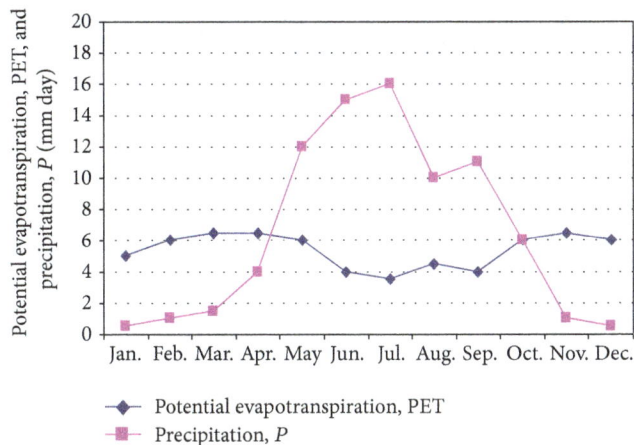

FIGURE 1: Hydrological balance of PET and precipitation at Mymensingh [28].

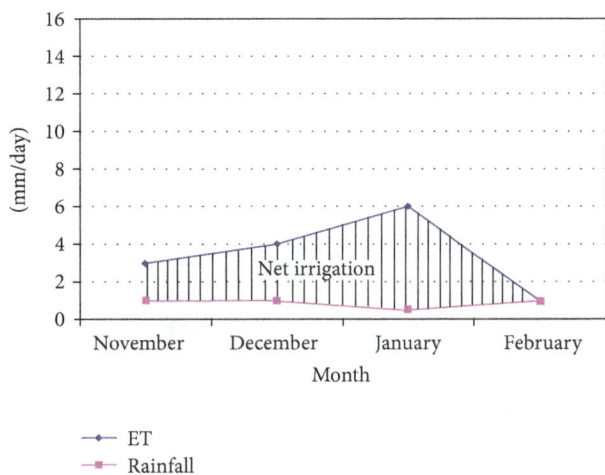

FIGURE 2: Net irrigation requirement of wheat [28].

of supplementary irrigation in relation to crop yield are crucial in irrigation scheduling for the most effective use of available water in optimizing wheat production [10–19]. Lack of irrigation at heading and grain formulation and during maturation significantly reduced the yield of wheat grain [20–26]. Grain yield increased with increase in the frequency of irrigation [27]. Improper scheduling of irrigation results not only in wastage of water but also in decreasing crop growth and yield [28–30]. Nitrogen for wheat production is equally important to realize the yield potential. Among the different elements of Bangladesh soil, N is the key input for achieving higher yield of wheat. Although application of nitrogenous fertilizer could increase yield to a certain level, it has adverse effect when the optimum level is exceeded [31]. Irrigation water dissolved the fertilizers and was made available to the crop for proper growth and development. Therefore, an attempt has been made to evaluate the effect of irrigation regimes and N levels and the best combination on the yield and yield contributing parameters of wheat.

## 2. Materials and Methods

The experiment was conducted at Mymensingh region in Bangladesh using three farmer's fields to evaluate the effect of irrigation regimes and N level on the performance of wheat cv. Kanchan. The experiment included two factors (1) four irrigation regimes ((no irrigation), 100 mm at 30 days after sowing (DAS), 200 mm (100 mm at 30 DAS + 100 mm at 45 DAS), and 300 mm (100 mm at 30 DAS + 100 mm at 45 DAS + 100 mm at 60 DAS)) and (2) four N levels (00, 80, 100, and 120 kg N ha$^{-1}$). The experiment was laid out in a split plot design with irrigation as main plot and N as subplot. The soil texture is loam. During land preparation the soil pH ranged from 6.02 to 7.62 and organic carbon varied from 0.51 to 1.63%. The land was prepared by the farmers with country plough followed by laddering. Three farmer's fields were selected for experimentation as replication. The unit plot size was 2.5 m × 4 m (10 m$^2$). Triple superphosphate (TSP), muriate of potash (MP), and gypsum were applied at the rates of 80 kg P$_2$O$_5$, 60 kg K$_2$O, and 30 kg S ha$^{-1}$, respectively. One-third of the N along with the complete doze of TSP, MP, and S were applied at the time of final land preparation. The remaining two-thirds urea was top-dressed in two equal splits, namely, at 30 and 45 DAS. Data on grain and straw yield, plant height, number of ears per plant, length of ear, number of grains per ear, and 1000 grain weight were recorded. Data were analyzed following analysis of variance technique with the computer package M-stat and mean comparison was performed using Duncan's Multiple Range Test [32].

## 3. Results and Discussion

Effect of irrigation regimes and N levels on the yield and yield contributing characters of wheat are presented in Table 1. The yield and yield parameters were significantly affected due to application of nitrogen and irrigation water. For higher yield, it is better to apply nitrogen at the stem elongation and heading unless it is not irrigated with minimum amount of water [33]. Maximum plant height was recorded in 300 mm irrigation treatment and shortest in the control. Availability of well distributed soil moisture at different growth stages due to irrigation probably enhanced the growth of plant. Due to application of irrigation water, ears per plant were significantly increased and followed similar pattern as in number of tillers per plant. Similar trend was found in the case of ear length. Plant height and ears per plant increased with the increasing rate of N doses but were not statistically different. Maximum number of tiller per plant was produced in 200 mm irrigation treatment which was statistically similar with other irrigation treatments except for the control. The effective tillers per plant were statistically different at 5% level of significance but identical among the nitrogen treatments. The maximum number of effective tillers per plant was obtained by applying 100 kg N ha$^{-1}$ and the lowest in control. The result revealed that the applications of 80 to 120 kg N ha$^{-1}$ treatment were statistically identical in respect of tillers per plant. Islam [34] reported that the maximum number of tillers per plant was obtained with 120 kg N ha$^{-1}$ and the minimum in control. Ear length showed significant

TABLE 1: Effect of irrigation and nitrogen on the yield and yield contributing characters of wheat.

| Treatments | Plant height (cm) | Tillers plant$^{-1}$ (no.) | Ears plant$^{-1}$ (no.) | Ear length (cm) | Grains ear$^{-1}$ (no.) | Grain yield (t ha$^{-1}$) | Straw yield (t ha$^{-1}$) | 1000 grain weight (g) |
|---|---|---|---|---|---|---|---|---|
| Irrigation | | | | | | | | |
| $I_0$ (00 mm) | 62.75$^b$ | 2.49$^b$ | 2.18$^b$ | 7.35$^b$ | 35.75$^b$ | 1.74$^d$ | 4.12$^d$ | 42.83$^b$ |
| $I_1$ (100 mm) | 65.62$^{ab}$ | 2.85$^a$ | 2.53$^a$ | 7.78$^a$ | 36.82$^{ab}$ | 1.99$^c$ | 4.38$^c$ | 43.50$^{ab}$ |
| $I_2$ (200 mm) | 65.87$^{ab}$ | 2.90$^a$ | 2.70$^a$ | 8.04$^a$ | 38.23$^a$ | 2.27$^a$ | 5.12$^a$ | 44.57$^a$ |
| $I_3$ (300 mm) | 67.80$^a$ | 2.81$^a$ | 2.59$^a$ | 7.95$^a$ | 38.10$^a$ | 2.14$^b$ | 4.71$^b$ | 44.50$^a$ |
| P | <0.05* | <0.01** | <0.01** | <0.01** | <0.01** | <0.01** | <0.01** | <0.01** |
| Nitrogen | | | | | | | | |
| $N_0$ (00 kg/h) | 63.92 | 2.63$^b$ | 2.40 | 7.02$^b$ | 34.15$^b$ | 1.84$^c$ | 3.96$^c$ | 42.58$^c$ |
| $N_1$ (80 kg/ha) | 65.62 | 2.77$^a$ | 2.48 | 7.98$^a$ | 37.37$^a$ | 1.97$^b$ | 4.53$^b$ | 43.50$^{bc}$ |
| $N_2$ (100 kg/ha) | 64.65 | 2.83$^a$ | 2.53 | 7.88$^a$ | 38.92$^a$ | 2.15$^a$ | 4.87$^a$ | 44.58$^{ab}$ |
| $N_3$ (120 kg/ha) | 67.75 | 2.82$^a$ | 2.59 | 8.25$^a$ | 38.47$^a$ | 2.18$^a$ | 4.97$^a$ | 44.83$^a$ |
| P | ns | <0.05** | ns | <0.01** | <0.01** | <0.01** | <0.01** | <0.01** |

In a column, having common letters (s) do not differ significantly but dissimilar letter differ significantly. ns, *, **, mean non-significant, significant at $P \leq$ 0.05 and $P \leq 0.01$, respectively.

difference by the use of different levels of nitrogen. The longest ear was observed from application of 120 kg N ha$^{-1}$ and the shortest in control. S. P. Singh and H. B. Singh [35] stated that ear length increased with increasing of nitrogen levels (0 to 120 kg N ha$^{-1}$). Grains per ear increased with increasing rate of nitrogen up to 100 kg N ha$^{-1}$ and then decreased. The number of grains per ear was statistically similar due to application of 80 to 120 kg N ha$^{-1}$ and significantly superior over control. Similarly results were obtained by Patel et al. [36]. Influence of irrigation on grain yield was statistically significant at 1% level of significance. Maximum grain yield was obtained in 200 mm irrigation treatment and minimum in control. These results are in conformity with the findings of Islam [34]. He reported that grain yield of 2.10 t ha$^{-1}$ was obtained in 120 mm irrigation and 1.70 t ha$^{-1}$ in control. The grain yields were significantly influenced by different levels of nitrogen. The grain yield was maximum due to application of 120 kg N ha$^{-1}$ and statistically similar to that of 100 kg N ha$^{-1}$ and both were significantly superior to 80 kg N ha$^{-1}$. These results are in agreement with the findings of Patel and Upadhyay [37]. Different irrigation regimes were found to have significant effect on the straw yield. Straw yield exhibited the tendency of increasing with the influence of irrigation levels. This might be due to the luxuriant vegetative growth in terms of plant height and number of tillers per plant. The maximum straw yield was obtained with 200 mm irrigation and minimum in control. Islam [34] reported that straw yield of 2.74 t ha$^{-1}$ was found with the application of 180 mm irrigation water. Straw yield followed similar pattern to grain yield. This might be due to the exuberant vegetative growth noted in case of higher doses of nitrogen. Higher straw yield was obtained in 120 kg N ha$^{-1}$ treatment and statistically superior to 0 to 80 kg N ha$^{-1}$. A significant variation was recorded for 1000 grain weight owing to difference in irrigation levels. 1000 grain weight was statistically similar in all irrigation treatments but the control

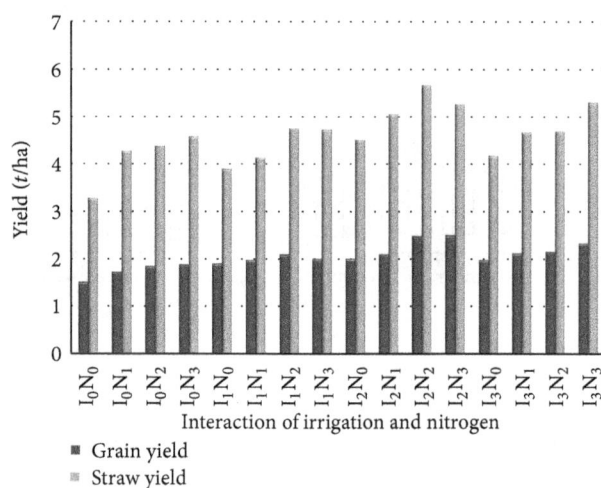

FIGURE 3: Combined effect of irrigation and nitrogen on the performance of wheat.

produced the lowest weight. S. P. Singh and H. B. Singh [35] reported that 1000 grain weight varied significantly with variable number of irrigation. Different N levels had significant influence on 1000 grain weight. Maximum weight of 1000 grain was found with 120 kg N ha$^{-1}$ and minimum weight was observed with control.

Interaction between irrigation and nitrogen did not show significant effect on plant height, tillers plant$^{-1}$, grains ear$^{-1}$, and 1000 grain weight. The interaction effect of irrigation and nitrogen application on ear length was statistically significant. The longest ear was observed due to 200 mm irrigation with 80 kg N ha$^{-1}$. Maximum grain yield was obtained due to application of 200 mm irrigation water and 120 kg N ha$^{-1}$ (Figure 3) which was statistically similar to the 200 mm irrigation and 100 kg N ha$^{-1}$ treatment. Minimum grain yield was obtained in the combination of no irrigation and no

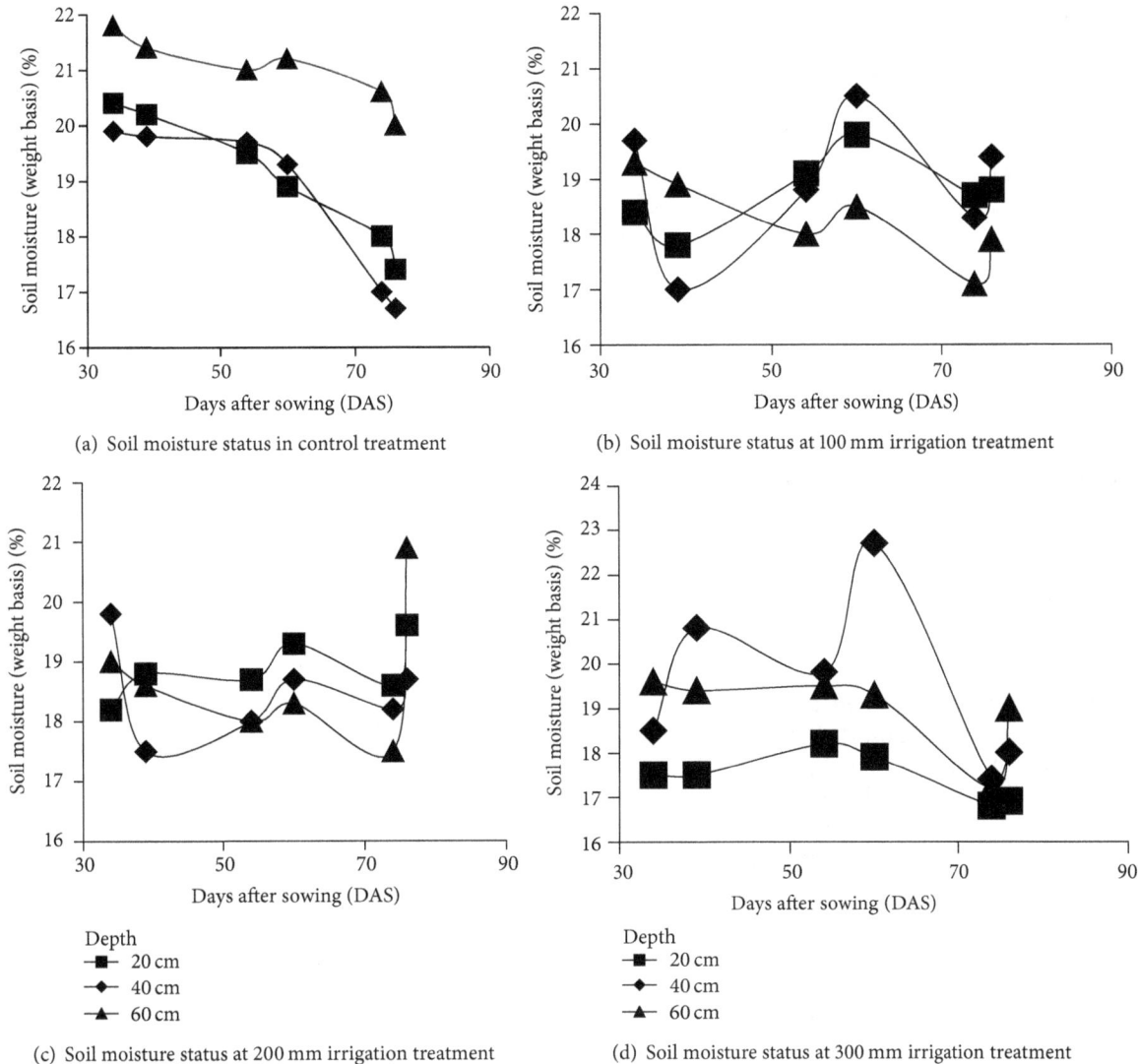

(a) Soil moisture status in control treatment

(b) Soil moisture status at 100 mm irrigation treatment

(c) Soil moisture status at 200 mm irrigation treatment

(d) Soil moisture status at 300 mm irrigation treatment

FIGURE 4: Soil moisture status at different depth of soil in irrigation treatments.

nitrogen treatment. These findings are corroborated with the observations by Kumar et al. [38]. Straw yields varied significantly at 1% level under different irrigation regimes and doses of nitrogen. It can be seen that straw yield increased with the increasing depth of irrigation and the highest rates of nitrogen fertilizer. Maximum straw yield was recorded by the application of 200 mm irrigation with 100 kg N ha$^{-1}$ and the minimum straw yield was produced in control. Eventually, it may be concluded that a combination of 200 mm irrigation and 120 kg N ha$^{-1}$ is the best in respect of yield and yield contributing characters.

The moisture status of the experimental plots during growing period of wheat is shown in Figure 4. Percentage of soil moisture gradually decreased through the soil profile in control (Figure 4(a)). Moisture status at 20 and 40 cm depths of soil decreased days after sowing and then gradually increased at every time of irrigation applied (Figures 4(b)–4(d)). It might be due to elapsed time of percolation through the soil profile. Soil moisture status at 60 cm depths

followed similar trend but moisture status at 20 cm depth fluctuated at prior and following irrigation, respectively. It might be due to infiltration, evaporation, and water holding capacity of the soil.

The N contents in plots applied with 80, 100, and 120 kg N/ha treatments increased gradually up to 60 days after sowing (Figure 5). It may be due to the fact that one-third of the nitrogen was applied at sowing and the rest in two equal splits at 35 and 65 days after sowing (DAS). It can be seen that the nitrogen content decreased rapidly from 90 to 110 DAS because nitrogen was not applied at that period and also nitrogen uptake by plants may be higher due to flowering and grain formation.

## 4. Conclusions

The effects of combinations of irrigation regimes and nitrogen rates on yield and yield contributing parameters were studied at field level. Different doses of nitrogen significantly

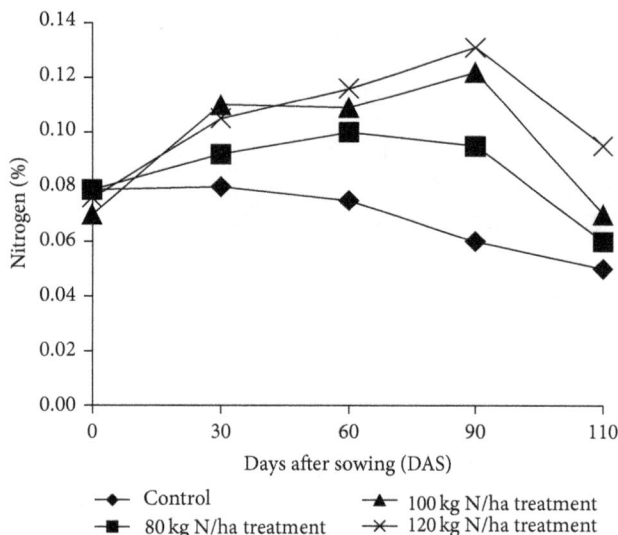

FIGURE 5: Nitrogen status of soil with different rates of N-application at different irrigation intervals (days).

influenced the grain yield and yield parameters. For the highest grain yield, nitrogen doses of $100\,kg\,N\,ha^{-1}$ were the best treatment when considering nitrogen fertilizer only. Irrigation regimes also have significant effect on yield and growth parameters of wheat. The combination of 200 mm irrigation and $120\,kg\,N\,ha^{-1}$ is the best treatment for optimal production of wheat.

## Conflict of Interests

The authors declare that there is no conflict of interests regarding the publication of this paper.

## Acknowledgment

Financial support by the research Grant no. PY/2014/01708 of Universiti Teknologi Malaysia (UTM) is gratefully acknowledged.

## References

[1] BBS., *Statistical Pocket Book of Bangladesh Bureau of Statistics*, Statistic Division, Ministry of Planning, Government of the Republic of Bangladesh, 2007.

[2] K. El Mejahed and L. Aouragh, "Green manure and n fertilizer in soil quality and profitability of wheat based system in semiarid morocco using nuclear techniques," in *Management of Nutrients, Water in Rainfed Arid, Semi-Arid Areas for Increasing Crop, Production*, IAEA-TECDOC-1468, pp. 89–106, 2005.

[3] M. M. Rusan, A. Battikhi, and S. Zuraiqi, "Enhancement of nitrogen and water use efficiency by optimizing the combination of soil, crop and nitrogen management," in *Management of Nutrients and Water in Rainfed Arid and Semi-arid Areas for Increasing Crop Production*, IAEA-TECDOC-1468, pp. 155–177, 2005.

[4] Z. Cui, X. Chen, and F. Zhang, "Current Nitrogen management status and measures to improve the intensive wheat-maize system in China," *Ambio*, vol. 39, no. 6, pp. 376–384, 2010.

[5] V. K. Arora, H. Singh, and B. Singh, "Analyzing wheat productivity responses to climatic, irrigation and fertilizer-nitrogen regimes in a semi-arid sub-tropical environment using the CERES-Wheat model," *Agricultural Water Management*, vol. 94, no. 1–3, pp. 22–30, 2007.

[6] H. Sun, C. Liu, X. Zhang, Y. Shen, and Y. Zhang, "Effects of irrigation on water balance, yield and WUE of winter wheat in the North China Plain," *Agricultural Water Management*, vol. 85, no. 1-2, pp. 211–218, 2006.

[7] S. Lenka and A. K. Singh, "Simulating interactive effect of irrigation and nitrogen on crop yield and water productivity in maize-wheat cropping system," *Current Science*, vol. 101, no. 11, pp. 1451–1461, 2011.

[8] M. Montazar and M. Mohseni, "Influence of supplemental irrigation and applied nitrogen on wheat water productivity and yields," *Journal of Agricultural Science*, vol. 3, pp. 78–90, 2011.

[9] F. Karam, R. Kabalan, J. Breidi, Y. Rouphael, and T. Oweis, "Yield and water-production functions of two durum wheat cultivars grown under different irrigation and nitrogen regimes," *Agricultural Water Management*, vol. 96, no. 4, pp. 603–615, 2009.

[10] D. Shimshi and U. Kafkafi, "The Effect of supplemental irrigation and nitrogen fertilisation on wheat (*Triticum aestivum* L.)," *Irrigation Science*, vol. 1, no. 1, pp. 27–38, 1978.

[11] R. H. Moll, E. J. Kamprath, and W. A. Jackson, "Analysis and interpretation of factors which contribute to efficiency of nitrogen utilization," *Agron Journal*, vol. 74, pp. 562–565, 1982.

[12] M. A. Quayum and F. Kamal, "Effect of irrigation at different growth stages of wheat and grain yield," *Bangladesh Journal of Agricultural Sciences*, vol. 11, pp. 47–55, 1986.

[13] K. D. Shing and B. M. Sharma, "Soil-test based specification for efficient use of fertilizer and targeted yield of wheat (*Triticum aestivum*) in Typic Ustochrept soil of Delhi Region," *Indian Journal of Agricultural Sciences*, vol. 60, pp. 36–40, 1990.

[14] C. Cabeza, A. Kin, and J. F. Ledent, "Effect of water shortage on main shoot development and tillering of common and spelt wheat," *Journal of Agronomy and Crop Science*, vol. 170, no. 4, pp. 243–250, 1993.

[15] G. Hussain and A. A. Al-Jaloud, "Effect of irrigation and nitrogen on water use efficiency of wheat in Saudi Arabia," *Agricultural Water Management*, vol. 27, no. 2, pp. 143–153, 1995.

[16] H. Zhang and T. Oweis, "Water-yield relations and optimal irrigation scheduling of wheat in the Mediterranean region," *Agricultural Water Management*, vol. 38, no. 3, pp. 195–211, 1999.

[17] G. Hussain, A. A. Al-Jaloud, and S. Karimulla, "Effect of treated effluent irrigation and nitrogen on yield and nitrogen use efficiency of wheat," *Agricultural Water Management*, vol. 30, no. 2, pp. 175–184, 1996.

[18] A. R. Tavakkoli and T. Y. Oweis, "The role of supplemental irrigation and nitrogen in producing bread wheat in the highlands of Iran," *Agricultural Water Management*, vol. 65, no. 3, pp. 225–236, 2004.

[19] S. M. Shirazi, S. Akib, F. A. Salman, U. J. Alengaram, and M. Jameel, "Agro-ecological aspects of groundwater utilization: a case study," *Scientific Research and Essays*, vol. 5, no. 18, pp. 2786–2795, 2010.

[20] J. L. Cooper, "The effect of nitrogen fertilizer and irrigation frequency on a semi-dwarf wheat in south-east Australia. 1. Growth and yield," *Australian Journal of Experimental Agriculture and Animal Husbandry*, vol. 20, pp. 359–364, 1980.

[21] P. R. Gajri, S. S. Prihar, and V. K. Arora, "Interdependence of nitrogen and irrigation effects on growth and input-use efficiencies in wheat," *Field Crops Research*, vol. 31, no. 1-2, pp. 71–86, 1993.

[22] T. Kätterer, A. Hansson, and O. Andrén, "Wheat root biomass and nitrogen dynamics—effects of daily irrigation and fertilization," *Plant and Soil*, vol. 151, no. 1, pp. 21–30, 1993.

[23] J. R. Frederick and J. J. Camberato, "Water and nitrogen effects on winter wheat in the southeastern coastal plain: I. Grain yield and Kernel traits," *Agronomy Journal*, vol. 87, no. 3, pp. 521–526, 1995.

[24] S. Abderrazak, K. M. Ezzarouk, and M. Amuslim, "Effects of water stress and nitrogen fertilizer rate on nitrogen uptake, water use and wheat yield," *Al Awamia*, no. 89, pp. 49–75, 1995.

[25] A. K. Singh and G. L. Jain, "Effect of sowing time, irrigation and nitrogen on grain yield and quality of durum wheat (*Triticum durum*)," *Indian Journal of Agricultural Sciences*, vol. 70, no. 8, pp. 532–533, 2000.

[26] Z. Z. Xu, Z. W. Yu, D. Wang, and Y. L. Zhang, "Nitrogen accumulation and translocation for winter wheat under different irrigation regimes," *Journal of Agronomy and Crop Science*, vol. 191, no. 6, pp. 439–449, 2005.

[27] B. D. Sharma, S. S. Cheema, and S. Kar, "Water and nitrogen uptake of wheat as related to nitrogen application rate and irrigation water regime," *Fertilizer News*, vol. 35, pp. 31–35, 1990.

[28] S. M. Shirazi, Z. Ismail, S. Akib, M. Sholichin, and M. A. Islam, "Climatic parameters and net irrigation requirement of crops," *International Journal of Physical Sciences*, vol. 6, no. 1, pp. 15–26, 2011.

[29] S. M. Shirazi, M. Sholichin, M. Jameel, S. Akib, and M. Azizi, "Effects of different irrigation regimes and nitrogenous fertilizer on yield and growth parameters of maize," *International Journal of Physical Sciences*, vol. 6, no. 4, pp. 677–683, 2011.

[30] S. M. Shirazi, M. A. Islam, Z. Ismail, M. Jameel, U. J. Alengaram, and A. Mahrez, "Arsenic contamination of aquifers: a detailed investigation on irrigation and portability," *Scientific Research and Essays*, vol. 6, no. 5, pp. 1089–1100, 2011.

[31] N. Kataria and K. Bassi, "Effect of organic mulch and nitrogen on early-sown wheat (*Triticum aestivum*) under rainfed conditions," *Indian Journal of Agronomy*, vol. 42, no. 1, pp. 94–97, 1997.

[32] K. A. Gomez and A. A. Gomez, *Statistical Procedure for Agricultureal Research*, John Willey & Sons, New York, NY, USA, 2nd edition, 1984.

[33] C. G. Abourached, S. K. Yau, M. N. Nimah, and I. I. Bashour, "Deficit irrigation and split N fertilization on wheat and barley yields in a semi-arid mediterranean area," *The Open Agriculture Journal*, vol. 2, pp. 28–34, 2008.

[34] M. S. Islam, *Growth and yield response of wheat to irrigation and nitrogen application [M.S. Thesis]*, Department of Soil Science, Bangladesh Agricultural University, Mymensingh, Bangladesh, 1997.

[35] S. P. Singh and H. B. Singh, "Effect of irrigation time and nitrogen level on wheat (*Triticum aestivum*) under late-sown condition of Western Uttar Pradesh," *Indian Journal of Agronomy*, vol. 36, pp. 41–42, 1991.

[36] N. M. Patel, S. G. Sadaria, B. B. Kaneria, and V. D. Khanpura, "Effect of irrigation, potassium and zinc on growth and yield of wheat (*Triticum aestivum*)," *Indian Journal of Agronomy*, vol. 40, pp. 290–292, 1995.

[37] R. M. Patel and P. N. Upadhyay, "Response of wheat (*Triticum aestivum*) to irrigation under varying levels of nitrogen and phosphorus," *Indian Journal of Agronomy*, vol. 38, pp. 113–115, 1993.

[38] A. Kumar, D. K. Sharma, and H. C. Sharma, "Response of wheat (Triticum aestivum) to irrigation and nitrogen in sodic soils," *Indian Journal of Agronomy*, vol. 40, no. 1, pp. 338–342, 1995.

# 8

# Isolation and Molecular Characterization of Potential Plant Growth Promoting *Bacillus cereus* GGBSTD1 and *Pseudomonas* spp. GGBSTD3 from Vermisources

**Balayogan Sivasankari and Marimuthu Anandharaj**

*Department of Biology, Gandhigram Rural Institute-Deemed University, Gandhigram, Dindigul, Tamil Nadu 624 302, India*

Correspondence should be addressed to Balayogan Sivasankari; balayogansivasankari@gmail.com

Academic Editor: Tibor Janda

Vermicompost was prepared from leaf materials of *Gliricidia sepium* + *Cassia auriculata* + *Leucaena leucocephala* with cow dung (1:1:2) using *Eudrilus eugeniae* (Kinberg) and *Eisenia fetida* for 60 days. Nineteen bacterial strains which have the capability to fix nitrogen, solubilize inorganic phosphate, and produce phytohormones were isolated from vermicompost, vermisources, and earthworm (fore, mid, and hind) guts and tested for plant growth studies. Among the bacterial strains only five strains had both activities; among the five *Bacillus* spp. showed more nitrogen fixing activity and *Pseudomonas* spp. showed more phosphate solubilizing activity. Hence these bacterial strains were selected for further molecular analysis and identified *Bacillus cereus* GGBSTD1 and *Pseudomonas* spp. GGBSTD3. Plant growth studies use these two organisms separately and as consortium (*Bacillus cereus* + *Pseudomonas* spp.) in (1:1) ratio at different concentrations using *Vigna unguiculata* (L.) Walp. at different day intervals. The germination percent, shoot length, root length, leaf area, chlorophyll a content of the leaves, chlorophyll b content of the leaves, total chlorophyll content of the leaves, fresh weight of the whole plant, and dry weight of the whole plant were significantly enhanced by the consortium (*Bacillus cereus* + *Pseudomonas* spp.) of two organisms at 5 mL concentrations on the 15th day compared to others.

## 1. Introduction

Nowadays, under the modern agricultural practices, chemical fertilizers are used to boost the crop production. But the application of chemical fertilizers affects the total productivity of the crops and in the long run the soil becomes sterile and unfit for cultivation practices. Hence, in order to enhance the fertility status of the soil, the natural way of feeding the soil with different types of organic inputs (composts, vermicomposts, biofertilizers, farmyard manure, etc.) has been developed so as to ensure sustained productivity [1]. However, a better understanding of nutrient cycling and the factors governing their decomposition in soil is imperative for implementing sustainable management practices. Nutrient cycling in soil involves chemical, biochemical, and physicochemical reactions, with the biochemical reactions being catalyzed by soil enzymes associated with viable cells of microbial origin and plant roots. Therefore, any factor that affects soil microbial population will necessarily alter soil enzyme activity [2]. Sustainability of agricultural systems has become an important issue all over the world. Many of the issues of sustainability are related to soil quality and its change with time. It is well known that intensive cultivation has led to a rapid decline in organic matter and nutrient levels besides affecting soil physical properties. Conversely, management practices with organic materials influence agricultural sustainability by improving physical, chemical, and biological properties of soils [3].

Recently, there is increasing interest in the potential of vermicomposts, as plant growth media and as soil amendments. Vermicomposts are finely divided peat-like materials with high porosity, aeration, drainage, water-holding capacity, and microbial activity, which make them excellent soil amendments or conditioner. It is a sustainable source of macro- and micronutrients, which enlivens the soil through partial substitution of the horticultural container media. Enhancement in plant growth after substitution of soils or greenhouse container media with composts is attributed to modifications in soil structure, change in water availability, increased availability of macro- and micronutrients, stimulation of microbial activity, and augmentation of the activities of critical substances by microorganisms through interactions with earthworms [4]. The vermicompost is rich in available nutrients, soil beneficial microbes, and plant growth promoting substances.

Research during the past few decades has led to the identification of certain biological organisms and their products that could potentially be used as fertilizer sources. This strategy of fertilizing the soil with biological sources has been widely accepted and recognized as a viable alternative to the application of chemical fertilizers. Nitrogen (N) fixing and phosphate (P) solubilizing bacteria may be important for plant nutrition by increasing N and P uptake by the plants and playing a significant role as plant growth promoting bacteria in the biofertilization of crops. Increasing and extending the role of biofertilizers could reduce adverse environmental effects caused by chemical fertilizers. Nitrogen is one of the basic requirements for the growth, productivity, and yield of plants. On a worldwide basis it is estimated that about 175 million tons of nitrogen per year is added to soil through biological nitrogen fixation. Meanwhile superphosphate fertilizer is expensive and is in short supply and hence biofertilizers can bridge the gap. Soluble fertilizer phosphorus (P) is the world's second largest bulk agricultural chemical under use and as such is absolutely essential for food production. However, most soil phosphorus, approximately 95–99 percent, is present in the form of insoluble phosphates and hence cannot be utilized by plants [5]. To increase the availability of phosphorus to plants, large amounts of fertilizers are used on a regular basis. But, after application, a large proportion of the fertilizer phosphorus is quickly transferred to the insoluble form [6]. Therefore, a very little percentage of the applied phosphorus only is used, making continuous application necessary. It has been reported that many soil fungi and bacteria can solubilize inorganic phosphates [7]. The *Bacillus* species offer several advantages over the other genera because of their capacity to produce spores in unfavorable environmental conditions. This characteristic facilitates the conversion of spore suspensions to powder formulations without killing bacteria [8]. Vermicompost is rich in NPK, plant growth promoters, and other nutrients and when applied, the soil will be enriched with a variety of nutrients that will become available for the indigenous microflora.

The PGPR have been known to directly enhance plant growth by a variety of mechanisms, namely, fixation of atmospheric nitrogen that is transferred to the plant, production of siderophores that chelate iron and make it available to the plant root, solubilization of minerals such as phosphorus, and synthesis of phytohormones [9]. Plant growth promoting bacteria (PGPB) stimulate plant growth by nitrogen fixation [10], solubilization of nutrients [11], and production of growth hormones and 1-aminocyclopropane-1-carboxylate (ACC).

The present paper describes the potential of PGPR isolates from vermisources and their plant growth stimulating activity in pot trials under ambient conditions. Further studies will clarify the potential use of these bacteria used as biofertilizers.

The main objective of this study was to isolate, identify, and characterize the potential plant growth promoting rhizobacteria from vermisources, as well as evaluating their plant growth potential under controlled environment and characterizing the isolated bacteria using various molecular techniques. These bacteria can be considered promising candidates for application in sustainable agricultural management.

## 2. Materials and Methods

*2.1. Vermicompost Preparation.* For the present study epigeic earthworms, *Eudrilus eugeniae* (Kinberg) and *Eisenia fetida* (Savigny), were collected from the breeding stock of the Department of Biology, Gandhigram Rural Institute-Deemed University, Gandhigram, Tamil Nadu, India, and leaf materials of *Gliricidia sepium* Jacq, *Leucaena leucocephala* (Lam.) De Wit, and *Cassia auriculata* Linn. were collected from Gandhigram campus. The leaf materials were separately subjected to predigestion for 15 days by sprinkling water on the heap and covering it with gunny bag and turning it periodically in order to release out the initial heat produced during decomposition of organic material. The changes in temperature were observed every three days up to 15 days. The vermibeds were prepared in plastic containers of $45 \times 35 \times 15$ cm size and the substrate was moistened to hold 60–80 percent moisture and kept for 24 hours stabilization. 20 numbers of healthy clitellate *E. eugeniae* and 30 numbers of *E. fetida* were separately introduced in the vermibeds. The vermicomposting trials were carried out in the rearing room with the relative humidity and the temperature of 75–85 percent and 26–28°C, respectively. The substrate was turned (mixed) once in a week and maintained up to 60 days. The experiment was carried out with three replicates for each substrate with proper control [12].

*2.2. Microbial Study.* The total microbial counts in terms of colony forming units (CFU) of bacteria in the vermicomposts, vermicasts, and earthworm (fore, mid, and hind) guts were determined every 15 days (0, 15, 30, 45, and 60 d) using standard plate count method. From the total colony forming units, only those bacterial colonies (19 bacterial colonies) which showed predominant growth were restreaked onto appropriate agar medium to obtain pure cultures and subjected to characterization and identification.

*2.3. Screening for Biofertilization Activities*

*2.3.1. In Vitro Nitrogen Fixing Activity.* The bacterial isolates were screened for nitrogen fixing ability using nitrogen-free liquid medium [13]. Nitrogen-free medium was prepared and sterilized. The strains were inoculated separately on selective nitrogen-free liquid medium. All the test organisms were incubated at 37°C $\pm$ 2°C for 7 days. The nitrogen fixing activities were observed on the basis of the formation of turbidity in the flasks. Among all the nineteen bacterial isolates tested, only five bacterial isolates formed significantly higher rate of turbidity in the nitrogen-free liquid medium and selected for further molecular and plant growth studies.

### 2.3.2. In Vitro Phosphate Solubilization.

All bacterial isolates were screened for inorganic phosphate solubilization using Pikovskaya's agar medium, the selective medium for phosphate solubilization test [14]. Pikovskaya's agar medium was prepared and sterilized and the bacterial strains were inoculated separately on the selective medium containing 0.5 percent of (w/v) tricalcium phosphate ($Ca_3PO_4$) as complex insoluble phosphate source. All the test plates of the bacteria were incubated at $37°C \pm 2°C$ for 10 days. After incubation for up to 7 days at $30°C$, formation of yellow halos and/or clearing zones was evaluated. The results were expressed as solubilization index (SI) and they were measured using the following formula [15]:

$$SI = \frac{\text{Colony diameter} + \text{halozone diameter}}{\text{Colony diameter}}. \quad (1)$$

Five bacterial strains which showed phosphate solubilizing activity were identified and grouped as phosphate solubilizing bacteria (PSB). The bacterial strain which showed highest SI was selected for further molecular and plant growth studies.

### 2.4. Production of Phytohormones (IAA).

The bacterial cultures were inoculated in nutrient broth with tryptophan ($5 \mu g/mL$) and incubated at $28 \pm 2°C$ for 5 days. After incubation cultures were centrifuged at 3000 rpm for 30 min. Two milliliters of the supernatant was mixed with 2 drops of orthophosphoric acid and 4 mL of Salkowski's reagent (50 mL of 35% perchloric acid + 1 mL 0.5 $FeCl_3$) and incubated in the dark for 25 minutes. Development of pink colour indicates indole-3-acetic acid (IAA) production. The optical density was measured at 530 nm using Spectronic 200 (India). The quantity of IAA production was estimated using standard IAA graph and expressed as micrograms per milliliter [16, 17].

### 2.5. Siderophore Production.

A qualitative assay of siderophore production was conducted in Chrome Azurol S (CAS) agar medium [18]. CAS agar plates were prepared and spot-inoculated with test organism and incubated at $30°C$ for 3–5 days. Change of blue color of the medium surrounding the bacterial growth to fluorescent yellow indicated the production of siderophore. Bacillus subtilis was chosen as a positive control [19, 20].

### 2.6. 16S rDNA Gene Amplification and Sequencing.

Genomic DNA from each isolate was extracted using the modified method of Smoker and Barnum [21]. The extracted DNA was dissolved in $20 \mu L$ TE buffer and used as the template for the PCR reactions. PCR amplifications were performed in a total volume of $50 \mu L$ by mixing 20 ng of the template DNA with 2.5 mM concentrations of each deoxynucleotide triphosphate and $1 \mu m$ of each universal primer of Bac8uf (5′-AGAGTTTGATCCTGGCTCAG-3′) and Univ1492r (5′-CTACGGCTACCTTGTTACGA-3′) described by Edwars et al. [22]. The thermocycling profile was carried out with an initial denaturation at $95°C$ (4 min) followed by 30 cycles of denaturation at $95°C$ (1 min), annealing at $56°C$ (30 s), extension at $72°C$ (1 min), and a final extension at $72°C$ (10 min) in an Eppendorf Gradient thermocycler. The PCR amplified rDNA were purified by using the Quick PCR purification kit (Bangalore Genie, India). 16S rRNA gene sequence of the isolate was compared with 16S rRNA gene sequences available by the BLAST search in the NCBI, GenBank database (http://www.ncbi.nlm.nih.gov). The analysis of alignment and homology of the partial nucleotide sequence of Bacillus sp. was carried out by the basic local alignment search tool (BLAST). The PCR products were stored at $4°C$. Aliquots of the PCR products were separated by 2% agarose gel electrophoresis in TAE buffer (pH 8.0). A 100 bp DNA marker (Sigma, Bangalore) was used as a reference. Gels were stained with ethidium bromide (11 g/mL) and visualized under UV light.

### 2.7. Nucleotide Sequence Accession.

The sequence obtained in this study was deposited in the GenBank (USA) nucleotide sequence database under the accession number GQ413962 (B. cereus) and HM753262 (Pseudomonas spp.).

### 2.8. Phylogenetic Analysis.

The 16S rRNA gene sequences were obtained in the present study and the taxonomically related Bacillus spp. were retrieved from the National Center from Biotechnology Information (NCBI) database (http://www.ncbi.nlm.nih.gov). All the sequences were aligned using multiple sequence alignment program CLUSTAL W developed by Higgins et al. [23]. The pairwise evolutionary distances were computed using the method of Kimura [24]. The multiple distance matrix obtained was then used to construct phylogenetic trees using neighbour joining method of Saitou and Nei [25]. Phylogenetic trees were constructed by using Mega 5 software (Molecular Evolutionary Genetic Analysis).

### 2.9. Prediction of RNA Secondary Structure.

The secondary structure of 16S rDNA of Bacillus spp. was predicted using the bioinformatics Genebee tool.

### 2.10. Restriction Site Analysis in 16S rDNA.

The restriction sites in DNA of Bacillus spp. were analyzed using the NEB Cutter program version 2.0 tools available online at http://tools.neb.com/NEBcutter2/index.php.

### 2.11. Fermentation Process.

For the preparation of inoculum, 50 mL of nutrient broth medium (HiMedia, Mumbai, India) was inoculated with appropriate overnight bacterial culture and incubated at $37°C$ for 24 h at 120 rpm ($OD_{540}$ = 0.45; $10^7$-$10^8$ CFU $mL^{-1}$). The cell-free supernatant was obtained by centrifuging the fermented broth at

10,000 rpm for 10 min. The supernatant was analyzed for plant growth promotion at various concentrations.

### 2.12. Evaluation of Plant Growth Promoting Activity under Controlled Environment.

The nitrogen fixing bacteria (Bacillus spp.) and the phosphate solubilizing bacteria (Pseudomonas spp.) were separately tested for plant growth promoting activity and also as consortium (1:1) at 5 different concentrations, that is, 1, 2, 3, 4, and 5 mL at three different day intervals (5, 10, and 15 days) with proper control through Exp1, Exp2, Exp3, and Exp0, respectively, using Vigna unguiculata (L.) Walp. The seeds (30 seeds in each pot) were sown in plastic pots containing sterilized vermiculite and arranged in a completely randomized factorial design. The seedlings were grown in a greenhouse at a temperature of 28–32°C and 85% relative humidity. The pots were watered to 50% water-holding capacity and were maintained at this moisture content by watering to weight every day and plant growth was observed for 15 days. Parameters such as germination percent, shoot length, root length, leaf area, chlorophyll content of the leaves, fresh weight of the whole plant, and dry weight of the whole plant were measured using standard procedure.

### 2.13. Statistical Analysis.

The following statistical tools were used for the analyses and interpretation of the data. The experimental results are presented in the form of tables and graphs using Microsoft Excel 2007. Data obtained from the different treatments were statistically analyzed using the one-way ANOVA and the mean values were compared by the Duncan's multiple range test for multiple comparisons. Differences were considered significant at the 0.05 level using Origin software (Version 8.5.0) 2010, Origin Lab Corporation.

## 3. Results

### 3.1. Isolation and Characterization of Bacterial Isolates.

A total of nineteen bacterial strains were isolated from vermicompost, vermicast, and earthworm (fore, mid, and hind) guts (Table 1). The morphological and biochemical characteristics of bacterial isolates were elucidated in Table 1. All the isolates were tested for plant growth promoting activity. Among the 19 bacterial isolates only five bacterial strains showed plant growth promoting activity. Based on colony morphology, microscopic observations, and cultural, biochemical, and physiological properties, the bacterium was given the name Bacillus spp., Alcaligenes spp., Erwinia spp., Serratia spp., or Pseudomonas spp. These five bacterial strains were able to grow in the nitrogen-free medium and they were also identified as phosphate solubilizing microorganisms on the basis of their solubilization index (Table 2). Bacillus spp. have the highest SI (38.55), followed by Pseudomonas spp. (35.22) and Serratia spp. (30.45). All the isolates produced a significant amount of IAA from the medium in the range of 11.3 to 24.2 μg/mL; Bacillus spp. showed higher IAA production (24.2 μg/mL) and Serratia spp. showed lower IAA production (11.3 μg/mL). All the strains (except Alcaligenes spp.) were positive for siderophore production, showing a yellow zone on the CAS agar medium plate (Table 2).

### 3.2. Molecular Studies.

The partial 16S rRNA sequences carried out in the present study for Bacillus spp. and Pseudomonas spp. covered a stretch of approximately 1500 nucleotides for each. About half of the sequences found in the clone library showed only slight relationship to other known sequences, while the other half were highly similar (approximately 95 percent sequence identity) to other database entries for Bacillus spp. and Pseudomonas spp. Less than 0.5 percent of all nucleotides was found to be unique within the conserved regions of the cloned sequence and could almost always be related to reading errors in ambiguous regions of the sequencing gel. On the basis of phylogenetic analysis of 16S rDNA partial sequences, Bacillus spp. GGBSTD1 is identified as Bacillus cereus and Pseudomonas spp. GGBSTD3 is identified as Pseudomonas spp. Phylogenetic analyses of the strains based on the neighbor joining method were represented in Figures 1 and 2. Among eubacteria, the mean guanine and cytosine (G + C) content of genomic DNA varies from approximately 25% to 75%. That the bacterial genomic G + C content is somehow related to phylogeny has been suggested [26, 27]. The phylogenetic tree of eubacterial 16S rRNA clearly indicates this relationship. Gram-positive bacteria Bacillus subtilis have genomic G + C, 42%. Gram-negative bacteria with intermediate G + C content, such as Escherichia coli (50%), Serratia marcescens (58%), Salmonella typhimurium (51%), and Pseudomonas fluorescens (60%), belong to the common Gram-negative bacteria [28]. Bacillus cereus have the G + C content 53% and A + T content 47%, similarly Pseudomonas spp. having G + C content 53% and A + T content 47%. Most functional RNA molecules have characteristic secondary structures that are highly conserved in evolution. In this study RNA secondary structures of Bacillus cereus and the Pseudomonas spp. were predicted, Bacillus cereus have free energy of −294.1 kkal/mol, and Pseudomonas spp. have free energy of −299.8 kkal/mol.

### 3.3. Evaluation of Biofertilization Abilities under Controlled Conditions.

Germination percentage of Vigna unguiculata seeds sown in vermiculite supplemented with Bacillus cereus (Exp1), with Pseudomonas spp. (Exp2), and with consortium (Exp3) (1:1 ratio) at five different concentrations (1, 2, 3, 4, and 5 mL) and in the control (Exp0) is shown in Figure 3. The highest germination percentage was observed in Exp3 (consortium) at 5 mL concentration compared to others. The increase in germination percentage and in other growth parameters can be attributed to the combined effect of the nitrogen fixing and phosphate solubilizing microbes which are able to fix atmospheric nitrogen and solubilize P and also produce growth promoting substances. These organisms have been isolated from vermisources and this observation indicates the reason for promotion and enhancement of crop growth and yield when vermicompost is applied. It is a proof that these symbiotic microbes present in vermisources benefit the soil fertility. Observation on growth parameters such as shoot length, root length, and leaf area of V. unguiculata

TABLE 1: Predominant bacterial strains isolated from vermisources and their biochemical characteristics. T1-control—*G. sepium* + *L. leucocephala* + cow dung in (1:1:2) (worm free), T2—*G. sepium* + *L. leucocephala* + cow dung in (1:1:2) + *E. eugeniae*, T3—*G. sepium* + *L. leucocephala* + cow dung in (1:1:2) + *E. fetida*, T4-control (2)—*C. auriculata* + *L. leucocephala* + cow dung in (1:1:2) (worm free), T5—*C. auriculata* + *L. leucocephala* + cow dung in (1:1:2) + *E. eugeniae*, T6—*C. auriculata* + *L. leucocephala* + cow dung in (1:1:2) + *E. fetida*.

| S. number | Sample source | Gram's reaction[a] | Cell shape | Motility[b] | Indole test | Methyl Red test | Voges Proskauer test | Citrate Utilization test | Catalase test | Urease test | Gelatin hydrolysis test | Nitrate reductase test | Starch hydrolysis test | Casein hydrolysis test | Glucose utilization test | Name of the strain |
|---|---|---|---|---|---|---|---|---|---|---|---|---|---|---|---|---|
| 1. | Vermicompost (T2) | + | Cocci | − | − | + | − | − | + | − | + | + | − | − | − | *Staphylococcus* spp. |
| 2. | Vermicompost (T3) | + | Rod | + | − | − | − | − | + | − | − | + | − | − | + | *Corynebacterium* spp. |
| 3. | Vermicompost (T5) | + | Rod | − | + | + | + | + | + | − | + | + | + | − | + | *Bacillus* spp. |
| 4. | Vermicompost (T6) | + | Cocci | − | + | + | − | − | + | − | + | − | − | − | + | *Micrococcus* spp. |
| 5. | Vermicast (T3) | + | Cocci | − | + | + | − | − | − | + | + | + | + | − | − | *Klebsiella* spp. |
| 6. | Vermicast (T5) | + | Rod | − | + | − | − | + | − | + | − | + | − | − | + | *Streptococcus* spp. |
| 7. | Vermicast (T6) | + | Cocci | − | + | − | + | + | − | + | + | − | − | − | + | *Enterobacter* spp. |
| 8. | Earthworm hind gut (T2) | + | Cocci | − | − | + | + | − | + | − | − | − | + | + | + | *Tricoccus* spp. |
| 9. | Earthworm hind gut (T3) | + | Cocci | − | − | + | − | + | + | + | − | + | − | − | − | *Yersinia* spp. |
| 10. | Earthworm hind gut (T5) | + | Cocci | − | − | − | − | − | + | + | + | + | − | − | − | *Micrococcus* spp. |
| 11. | Vermicompost (T1) | − | Rod | − | + | + | − | − | − | − | + | + | − | − | + | *Escherichia* spp. |
| 12. | Vermicompost (T2) | − | Rod | − | − | + | − | − | − | + | + | + | − | − | + | *Proteus* spp. |
| 13. | Vermicompost (T4) | − | Rod | − | − | − | + | − | + | − | − | + | + | + | + | *Paracoccus* spp. |
| 14. | Vermicompost (T5) | − | Rod | + | + | − | − | − | + | − | + | + | − | − | − | *Pseudomonas* spp. |
| 15. | Vermicast (T2) | − | Rod | + | − | + | − | + | + | − | − | + | − | − | − | *Citrobacter* spp. |
| 16. | Vermicast (T3) | − | Cocci | − | + | + | − | − | + | − | + | − | − | − | − | *Erwinia* spp. |
| 17. | Vermicast (T5) | − | Rod | + | + | + | + | + | + | − | + | − | − | − | − | *Alcaligenes* spp. |
| 18. | Vermicast (T6) | − | Cocci | − | − | + | + | + | + | − | + | + | − | − | − | *Serratia* spp. |
| 19. | Earthworm mid gut (T2) | − | Rod | + | − | − | − | + | − | − | − | − | − | − | − | *Pediococcus* spp. |

<sub></sub>

[a] Gram's staining: (+) positive, (−) negative.

[b] Motility: (+) motile, (−) nonmotile.

[c] Biochemical characteristics: (+) positive, (−) negative.

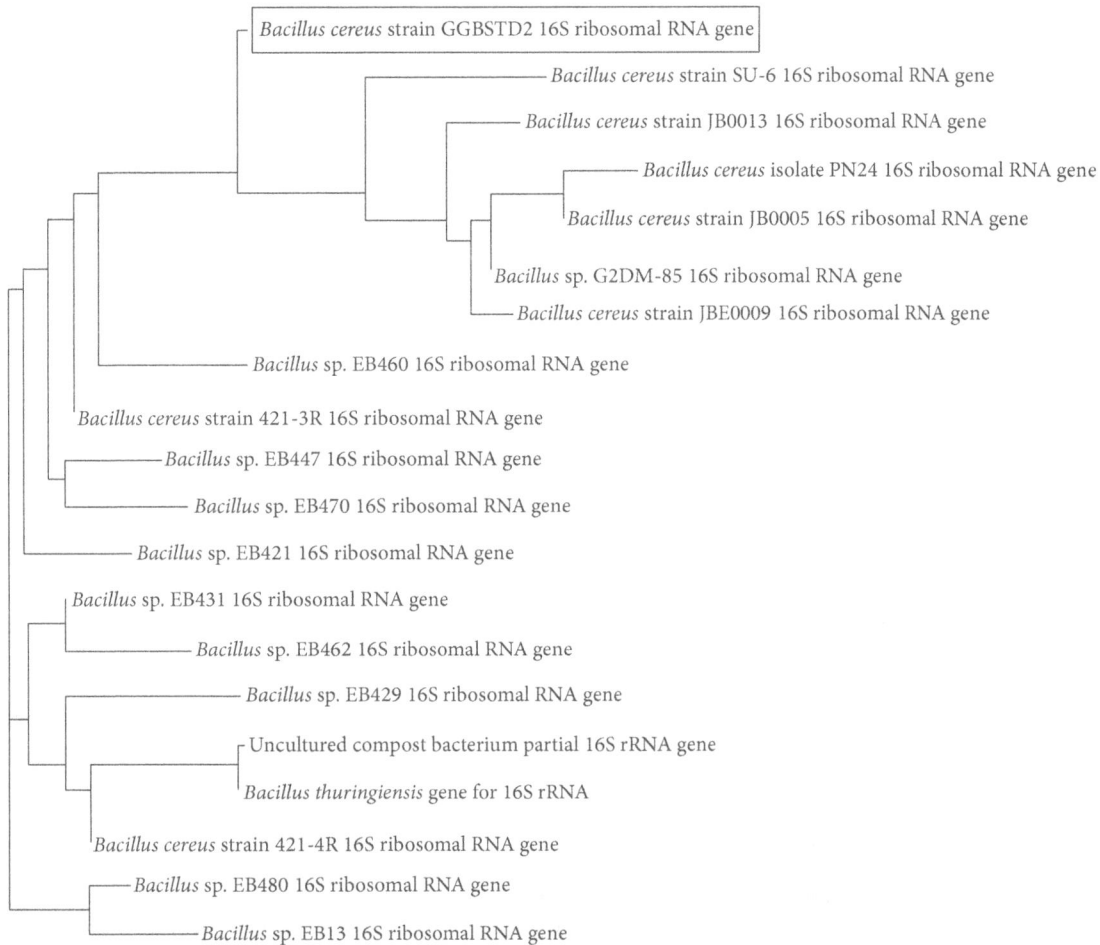

FIGURE 1: Neighbor joining phylogenetic tree of full 16S rRNA gene sequences showing the relationship among *Bacillus cereus* GGBSTD1 and representatives of some related taxa. The sequence data for several closely related *Bacillus* cultures were recovered from GenBank and included in the tree.

TABLE 2: The phosphate solubilizing activity, phytohormone (IAA) production, and siderophores production of the bacterial isolates.

| Name of the organism | Phosphate solubilization index (SI) | IAA production ($\mu$g/mL)[a] | Siderophores production[b] |
| --- | --- | --- | --- |
| *Bacillus* spp. | 38.55 | 24.2 | +++ |
| *Alcaligenes* spp. | 29.30 | 14.7 | − |
| *Erwinia* spp. | 28.90 | 11.3 | + |
| *Serratia* spp. | 30.45 | 16.8 | ++ |
| *Pseudomonas* spp. | 35.22 | 19.6 | +++ |

[a]Indole acetic acid production in culture media supplemented with tryptophan (5 $\mu$g/mL) after 5 days.

[b]*In vitro* siderophores production: − represents the absence of siderophores production and +++ represents >10 mm wide yellow-orange zone.

is given in Tables 3 and 4, respectively. The shoot length (11.47 ± 0.06 cm), root length (10.27 ± 0.06 cm), and leaf area (9.05 ± 0.26 cm$^2$) of the plant get significantly ($P <$ 0.001) increased in Exp3 (consortium of *Bacillus cereus* + *Pseudomonas* spp. [1:1]) at 5 mL concentration on the 15th day compared to other experiments. The chlorophyll a, b and total chlorophyll content of the leaves on d 5, d 10, and d 15 are given in Table 5. The chlorophyll a (0.10 ± 0.09 mg/g of fresh leaf), chlorophyll b (10.85 ± 0.16 mg/g of fresh leaf), and total chlorophyll (10.95 ± 0.07 mg/g of fresh leaf) content

of leaves were significantly ($P <$ 0.001) higher in Exp3 (consortium of *Bacillus cereus* + *Pseudomonas* spp. (1:1)) at 5 mL concentration on the 15th day compared to other experiments. Observations on the fresh weight of the whole plant are given in Table 6. The fresh weight of the whole plant (0.92 ± 0.00 g) was significantly ($P <$ 0.001) higher in Exp3 (consortium of *Bacillus cereus* + *Pseudomonas* spp. (1:1)) at 5 mL concentration on the 15th day compared to other experiments. Observations on the dry weight of the whole plant are given in Table 6. The dry weight of the whole

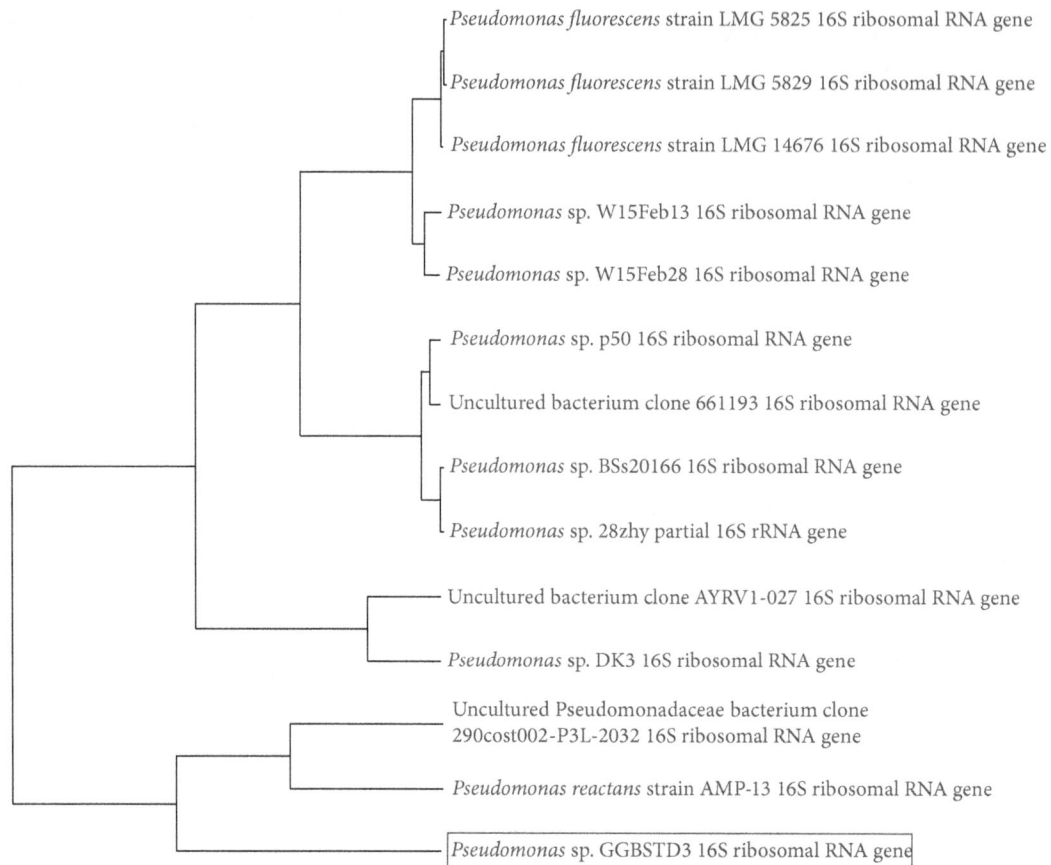

FIGURE 2: Neighbor joining phylogenetic tree of full 16S rRNA gene sequences showing the relationship among *Pseudomonas* spp. GGBSTD3 and representatives of some related taxa. The sequence data for several closely related *Pseudomonas* cultures were recovered from GenBank and included in the tree.

plant ($0.09 \pm 0.00$ g) was significantly ($P < 0.001$) higher in Exp3 (consortium of *Bacillus cereus* + *Pseudomonas* spp. (1 : 1)) at 5 mL concentration on the 15th day compared to other experiments.

## 4. Discussion

Inoculation of these newly identified nitrogen fixing and phosphate solubilizing microbes in the plant growth medium enhanced various physiological activities culminating in higher biomass production in *V. unguiculata*. Since this is a nonhazardous way of fertilization of crop plants, it is very relevant to a developing country like India. This technique can save the farmers of India from the problem posed by high cost of fertilizers by offering a comparatively inexpensive alternative (biofertilizers), while additionally preventing the degradation of the soil and thereby ensuring sustainable agriculture. Bess (1999) reported that the content in composting needs to be determined through the concentration of six functional groups of microorganisms such as aerobic bacteria, anaerobic bacteria, fungi, actinomycetes, *Pseudomonas*, and nitrogen fixing bacteria [29]. Now there are ways to evaluate the concentrations of these organisms in the finished compost and that can serve as an interpretation guide to

determine the quality of the compost as an inoculant of soil microorganisms. Grafff (1981) and Atlavinyte have reported a higher P content in the cast than in the surrounding soil [30]. Satchell (1983) has linked the phosphatase activity in the gut of worms to the availability of bound P in soils. The bacterial strains which have the capability to fix nitrogen only can grow in the nitrogen-free medium [31]. Hence the nitrogen fixing ability was tested in the present study based on the growth of bacterial strain in nitrogen- free medium. Five bacteria strains showed more nitrogen fixation; among all the five bacterial strains tested only one strain (i.e., *Bacillus* spp.) showed significantly more turbidity compared to the other strains. The ultimate source of the nitrogen used by plants is $N_2$ gas, which constitutes 78% of the Earth's atmosphere. Unfortunately higher plants cannot metabolize $N_2$ directly into protein. $N_2$ gas must be converted to a plant available form [32] isolated many species of nitrogen fixing *Bacillus* (*B. megaterium*, *B. cereus*, *B. subtilis*, *B. licheniformis*, and *B. azotoformans*) from the rhizospheric zone of rice fields in Central China. The widely studied *Bacillus* genus represents one of the most diverse genera in the bacilli group. Numerous *Bacillus* and *Paenibacillus* strains express plant growth promoting (PGP) activities and a number of these strains have already been commercially developed as biological fungicides, insecticides, and nematicides or generic plant

TABLE 3: Shoot length and Root length of *Vigna unguiculata* grown on vermiculite supplemented with *Bacillus cereus*, with *Pseudomonas* spp., and with their consortium (1 : 1) at five different concentrations (1, 2, 3, 4, and 5 mL) on the 5th, the 10th, and the 15th day and in the control.

| Experiments[a] | Quantity of bacterial cultures added | Shoot length (cm)[b] | | | Root length (cm)[b] | | |
|---|---|---|---|---|---|---|---|
| | | 5th day | 10th day | 15th day | 5th day | 10th day | 15th day |
| Exp0 | 1 mL | 2.97 (ab) | 3.97 (a) | 4.97 (ab) | 1.97 (de) | 2.83 (ab) | 3.97 (bcd) |
| | 2 mL | 3.03 (bcd) | 4.03 (ab) | 5.00 (ab) | 2.00 (abc) | 2.87 (de) | 4.00 (a) |
| | 3 mL | 3.10 (abc) | 4.07 (a) | 5.03 (ab) | 2.03 (de) | 2.90 (abc) | 4.03 (ab) |
| | 4 mL | 3.10 (bcd) | 4.10 (bcd) | 5.07 (bcd) | 2.07 (a) | 2.97 (de) | 4.10 (abc) |
| | 5 mL | 3.13 (a) | 4.13 (ab) | 5.10 (ab) | 2.10 (bcd) | 3.00 (ab) | 4.13 (de) |
| Exp1 | 1 mL | 3.40 (ab) | 4.40 (abc) | 5.40 (ab) | 2.33 (abc) | 3.27 (bcd) | 4.33 (f) |
| | 2 mL | 4.50 (ab) | 5.40 (abc) | 6.47 (ab) | 3.33 (ab) | 4.30 (a) | 5.40 (abc) |
| | 3 mL | 5.50 (de) | 6.30 (ab) | 7.50 (a) | 4.40 (de) | 5.50 (abc) | 6.20 (a) |
| | 4 mL | 6.53 (ab) | 7.30 (a) | 8.33 (ab) | 5.50 (bcd) | 6.53 (de) | 7.50 (abc) |
| | 5 mL | 7.40 (ab) | 8.40 (abc) | 9.23 (f) | 6.40 (abc) | 7.53 (ab) | 8.27 (ab) |
| Exp2 | 1 mL | 2.77 (bcd) | 3.57 (f) | 4.40 (de) | 1.60 (de) | 2.17 (abc) | 3.50 (f) |
| | 2 mL | 3.43 (abc) | 4.47 (ab) | 5.53 (ab) | 2.53 (abc) | 3.40 (bcd) | 4.27 (bcd) |
| | 3 mL | 4.37 (de) | 5.43 (abc) | 6.33 (abc) | 3.40 (ab) | 4.43 (f) | 5.33 (ab) |
| | 4 mL | 5.47 (bcd) | 6.47 (de) | 7.33 (a) | 4.20 (bcd) | 5.50 (ab) | 6.33 (ab) |
| | 5 mL | 6.40 (f) | 7.33 (bcd) | 8.33 (bcd) | 5.33 (ab) | 6.30 (ab) | 7.30 (bcd) |
| Exp3 | 1 mL | 5.30 (bcd) | 6.47 (ab) | 7.30 (ab) | 4.33 (de) | 5.50 (abc) | 6.20 (ab) |
| | 2 mL | 6.43 (ab) | 7.40 (bcd) | 8.27 (f) | 5.47 (a) | 6.30 (de) | 7.33 (ab) |
| | 3 mL | 7.37 (bcd) | 8.40 (bcd) | 9.27 (ab) | 6.40 (ab) | 7.40 (bcd) | 8.40 (bcd) |
| | 4 mL | 8.47 (abc) | 9.30 (a) | 10.43 (abc) | 7.40 (f) | 8.33 (a) | 9.37 (abc) |
| | 5 mL | 9.57 (ab) | 10.30 (f) | 11.47 (bcd) | 8.33 (abc) | 9.40 (ab) | 10.27 (bcd) |

[a] Exp0: nutrient solution (control), Exp1: *Bacillus cereus*, Exp2: *Pseudomonas* spp., Exp3: consortium of *Bacillus cereus* + *Pseudomonas* spp. (1 : 1).
[b] Mean of two repeated experiments (30 plants). Different letters within parenthesis indicate significant difference between treatments for each growth parameter using Duncan's multiple range test ($P = 0.05$).

growth promoters. The use of these strains in agriculture has recently been reviewed [33]. Those strains besides having several PGP properties can also fix nitrogen and also do phosphate solubilization. The symbiotic fixation of nitrogen through inoculation of legume crops with effective rhizobia is well known [34]. Asymbiotic nitrogen fixing bacteria, which live in the rhizosphere and/or endophytically, often increase yields of crops. Many bacterial species have nitrogen fixing properties, including *Bacillus* spp., *Azotobacter* spp., *Azospirillum* spp., *Beijerinckia* spp., and *Pseudomonas* spp. [35]. Nitrogen fixing bacteria have been used in foliar applications in mulberry. The solubilization index (SI) was formed due to solubilization of insoluble phosphates by organic acid secretion [36]. El-Komy (2005) indicated that *Pseudomonas fluorescence* and *Bacillus megaterium* strains were the most powerful phosphate solubilizers on PVK plates as well as on Pikovskaya's broth [37]. The P concentration in Pikovskaya's broth increased gradually, achieving a peak on the sixth day and declined slowly during the late days. In this study, pH values decreased gradually in PVK broth during the early incubation days and no revival was observed during later days for all the tested bacterial strains. Several bacteria, particularly those belonging to the genus *Bacillus* spp., convert insoluble phosphate into soluble forms by secreting organic acids such as formic acid, acetic acid, propionic acid, citric acid, fumaric acid, gluconic acid, glyoxylic acid, ketobutyric

acid, malonic acid, succinic acid, and tartaric acid. These acids lower the pH and bring about the dissolution of bound forms of phosphate. Some of the hydroxyl acids may chelate with calcium and iron resulting in effective solubilization and utilization of phosphates [38].

Beneduzi et al. (2008) had isolated the plant growth promoting strain SVPR30 and identified by 16S rRNA gene sequence as *Bacillus* spp. and they had tested it for plant growth through *in vivo* experiments [39]. This strain was characterized as a high IAA producer, able to solubilize phosphate and also fix a considerably high amount of nitrogen. The inoculation of rice with *Bacillus* spp. SVPR30 strain showed a significant increase in the root and shoot parts when compared with the controls within 15 and 30 days after sprouting. Numerous *Bacillus* and *Paenibacillus* strains expressed plant growth promoting (PGP) activities and a number of these strains have been commercially developed as generic plant growth promoters. The use of these strains in agriculture has recently been reviewed [33].

Various reports indicate that coinoculation of beneficial organisms generally increased plant growth relative to single inoculation with a sole beneficial organism [40]. Most of the effects of the individual microorganisms coinoculation are additive, although a synergistic effect has been reported in some cases. The average root to shoot ratio was higher in *B. subtilis* treated yam minisetts in comparison to those

TABLE 4: Leaf area of *Vigna unguiculata* grown on vermiculite supplemented with *Bacillus cereus*, with *Pseudomonas* spp., and with their consortium (1:1) at five different concentrations (1, 2, 3, 4, and 5 mL) on the 5th, the 10th, and the 15th days and in the control.

| Experiments[a] | Quantity of bacterial cultures added | Leaf area $(cm^2)$[b] | | |
|---|---|---|---|---|
| | | 5th day | 10th day | 15th day |
| Exp0 | 1 mL | 0.27 (ab) | 0.94 (abc) | 4.07 (b) |
| | 2 mL | 0.35 (ab) | 1.02 (ab) | 4.19 (abc) |
| | 3 mL | 0.40 (abc) | 1.25 (ab) | 4.71 (bcd) |
| | 4 mL | 0.48 (ab) | 1.38 (abc) | 5.10 (ab) |
| | 5 mL | 0.54 (abc) | 1.61 (b) | 5.38 (b) |
| Exp1 | 1 mL | 0.18 (bcd) | 1.23 (b) | 6.01 (abc) |
| | 2 mL | 0.25 (ab) | 1.38 (abc) | 6.32 (ab) |
| | 3 mL | 0.33 (abc) | 1.50 (ab) | 6.84 (b) |
| | 4 mL | 0.38 (ab) | 1.68 (ab) | 6.86 (abc) |
| | 5 mL | 0.44 (abc) | 1.91 (bcd) | 7.33 (ab) |
| Exp2 | 1 mL | 0.11 (b) | 1.07 (ab) | 5.34 (b) |
| | 2 mL | 0.14 (ab) | 1.19 (ab) | 5.59 (ab) |
| | 3 mL | 0.17 (abc) | 1.27 (ab) | 5.79 (ab) |
| | 4 mL | 0.21 (ab) | 1.53 (abc) | 6.31 (abc) |
| | 5 mL | 0.26 (ab) | 1.66 (b) | 6.67 (b) |
| Exp3 | 1 mL | 0.26 (bcd) | 2.97 (ab) | 7.53 (bcd) |
| | 2 mL | 0.33 (b) | 3.25 (ab) | 7.87 (ab) |
| | 3 mL | 0.40 (ab) | 3.50 (bcd) | 8.11 (bcd) |
| | 4 mL | 0.44 (b) | 3.84 (b) | 8.61 (b) |
| | 5mL | 0.62 (ab) | 4.00 (bcd) | 9.05 (ab) |

[a]Exp0: nutrient solution (control), Exp1: *Bacillus cereus*, Exp2: *Pseudomonas* spp., Exp3: consortium of *Bacillus cereus* + *Pseudomonas* spp. (1:1).
[b]Mean of two repeated experiments (30 plants). Different letters within parenthesis indicate significant difference between treatments for each growth parameter using Duncan's multiple range test ($P = 0.05$).

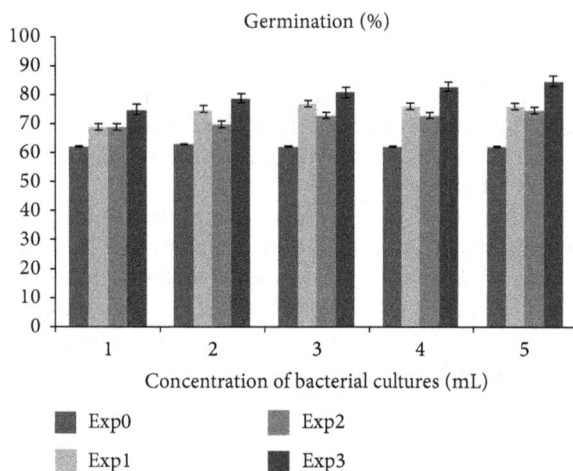

FIGURE 3: Germination percentage of *Vigna unguiculata* seeds sown in vermiculite supplemented with *Bacillus cereus*, with *Pseudomonas* spp., and with their consortium at five different concentrations (1, 2, 3, and 5 mL) and in the control. Exp0: nutrient solution (control), Exp1: *Bacillus cereus*, Exp2: *Pseudomonas* spp., Exp3: consortium of *Bacillus cereus* + *Pseudomonas* spp. (1:1).

not treated with the bacterial culture [41]. In earlier reports, root elongation was found to occur in *Sesbania aculeata* by inoculation with *Azotobacter* spp. and *Pseudomonas* spp. in

*Vigna radiata* by *Pseudomonas putida* [16] and in *Pennisetum americanum* by *Azospirillum brasilense* [42]. Esitken et al. (2006) had found out that *Bacillus* OSU-142 and *Pseudomonas* BA-8 alone or in combination had a great potential to increase the growth, yield, and nutrition of sweet cherry plant [43]. *Bacillus* M3, *Bacillus* OSU-142, and *Micobacterium* FS01 separately or in combination were found to have great potential for use as plant growth promoting rhizobacteria to increase production in apples and in many other crops [44]. These phosphate solubilizing microbes consist predominantly of fungal, bacterial, and actinomycetes species, collectively called phosphate solubilizing microorganisms—PSM [45]. There are several microorganisms which can also solubilize the cheaper sources of phosphorous such as rock phosphate. Bacteria such as *Bacillus* are widely used in plant production system and are important phosphorus solubilizing microorganisms, resulting in improved growth, yield of crops, and metabolic activities, especially in synthesis of protein [34, 46].

## 5. Conclusions

This study illustrates the isolation of plant growth promoting bacteria from vermisources, screening under *in vitro* conditions for multiple PGPR traits and their evaluation under controlled conditions in a pot experiment. The findings of this study indicated that the consortium of *B. cereus* and

TABLE 5: Chlorophyll content of leaves of *Vigna unguiculata* grown on vermiculite supplemented with *Bacillus cereus*, with *Pseudomonas* spp., and with their consortium (1:1) at five different concentrations (1, 2, 3, 4, and 5 mL) on the 5th day, the 10th day, and the 15th day and in the control.

| Experiments[a] | Quantity of bacterial Cultures added | Chlorophyll content (mg/g of fresh leaf)[b] | | | | | | | | |
|---|---|---|---|---|---|---|---|---|---|---|
| | | 5th day | | | 10th day | | | 15th day | | |
| | | Chlorophyll a | Chlorophyll b | Total chlorophyll | Chlorophyll a | Chlorophyll b | Total chlorophyll | Chlorophyll a | Chlorophyll b | Total chlorophyll |
| Exp0 | 1 mL | 0.21 (ab) | 2.85 (abc) | 3.07 (a) | 0.20 (abc) | 4.81 (a) | 5.01 (ab) | 0.15 (bcd) | 6.12 (abc) | 6.27 (bcd) |
| | 2 mL | 0.22 (abc) | 2.93 (ab) | 3.15 (abc) | 0.21 (ab) | 4.89 (ab) | 5.10 (bcd) | 0.12 (abc) | 6.18 (ab) | 6.30 (ab) |
| | 3 mL | 0.23 (abc) | 3.01 (ab) | 3.15 (abc) | 0.22 (bcd) | 4.97 (bcd) | 5.18 (a) | 0.14 (a) | 6.33 (cde) | 6.47 (bcd) |
| | 4 mL | 0.24 (ab) | 3.08 (f) | 3.32 (a) | 0.22 (a) | 5.04 (b) | 5.27 (abc) | 0.15 (bcd) | 6.41 (cde) | 6.56 (a) |
| | 5 mL | 0.25 (f) | 3.16 (abc) | 3.41 (bcd) | 0.19 (ab) | 5.11 (ab) | 5.30 (abc) | 0.11 (cde) | 6.45 (ab) | 6.50 (abc) |
| Exp1 | 1 mL | 0.35 (b) | 4.00 (cde) | 4.35 (abc) | 0.11 (b) | 5.81 (cde) | 5.93 (cde) | 0.10 (abc) | 7.04 (abc) | 7.15 (f) |
| | 2 mL | 0.26 (abc) | 4.62 (bcd) | 4.88 (f) | 0.11 (abc) | 6.47 (b) | 6.58 (abc) | 0.06 (ab) | 7.79 (bcd) | 7.85 (a) |
| | 3 mL | 0.21 (bcd) | 5.26 (abc) | 5.47 (ab) | 0.08 (bcd) | 7.18 (ab) | 7.26 (bcd) | 0.06 (f) | 8.47 (f) | 8.53 (ab) |
| | 4 mL | 0.08 (f) | 5.87 (cde) | 5.95 (b) | 0.11 (bcd) | 7.78 (abc) | 7.89 (abc) | 0.04 (bcd) | 9.33 (b) | 9.37 (bcd) |
| | 5 mL | 0.26 (abc) | 6.41 (bcd) | 6.67 (abc) | 0.25 (cde) | 8.34 (cde) | 8.59 (f) | 0.07 (abc) | 9.76 (cde) | 9.83 (abc) |
| Exp2 | 1 mL | 0.30 (abc) | 3.54 (ab) | 3.84 (b) | 0.12 (a) | 5.45 (abc) | 5.56 (ab) | 0.07 (bcd) | 6.81 (abc) | 6.88 (bcd) |
| | 2 mL | 0.29 (bcd) | 4.20 (f) | 4.49 (abc) | 0.11 (bcd) | 6.10 (ab) | 6.21 (abc) | 0.07 (abc) | 7.44 (f) | 7.52 (bcd) |
| | 3 mL | 0.22 (ab) | 5.04 (abc) | 5.27 (bcd) | 0.15 (a) | 6.77 (cde) | 6.92 (bcd) | 0.06 (bcd) | 8.12 (ab) | 8.17 (bcd) |
| | 4 mL | 0.10 (f) | 5.66 (a) | 5.76 (cde) | 0.18 (abc) | 7.44 (f) | 7.62 (ab) | 0.05 (ab) | 8.85 (a) | 8.90 (abc) |
| | 5 mL | 0.06 (bcd) | 6.22 (abc) | 6.28 (abc) | 0.18 (b) | 8.10 (abc) | 8.28 (a) | 0.04 (cde) | 9.48 (a) | 9.52 (f) |
| Exp3 | 1 mL | 0.27 (ab) | 4.70 (bcd) | 4.96 (cde) | 0.09 (ab) | 6.68 (b) | 6.77 (cde) | 0.10 (a) | 8.14 (ab) | 8.25 (a) |
| | 2 mL | 0.20 (ab) | 5.48 (a) | 5.68 (f) | 0.03 (bcd) | 7.46 (bcd) | 7.49 (ab) | 0.09 (ab) | 8.87 (f) | 8.97 (ab) |
| | 3 mL | 0.05 (bcd) | 5.84 (abc) | 5.89 (ab) | 0.08 (bcd) | 8.04 (abc) | 8.12 (cde) | 0.10 (ab) | 9.53 (abc) | 9.63 (bcd) |
| | 4 mL | 0.23 (f) | 6.63 (ab) | 6.86 (cde) | 0.07 (ab) | 8.70 (b) | 8.77 (f) | 0.10 (abc) | 10.19 (bcd) | 10.29 (f) |
| | 5 mL | 0.22 (ab) | 7.44 (cde) | 7.66 (a) | 0.09 (a) | 9.32 (abc) | 9.40 (ab) | 0.10 (a) | 10.85 (ab) | 10.95 (abc) |

[a]Exp0: nutrient solution (control), Exp1: *Bacillus cereus*, Exp2: *Pseudomonas* spp., Exp3: consortium of *Bacillus cereus* + *Pseudomonas* spp. (1:1).
[b]Mean of two repeated experiments (30 plants). Different letters within parenthesis indicate significant difference between treatments for each growth parameter using Duncan's multiple range test ($P = 0.05$).

TABLE 6: Fresh weight and dry weight of whole plant of *Vigna unguiculata* grown on vermiculite supplemented with *Bacillus cereus*, with *Pseudomonas* spp., and with their consortium (1:1) at five different concentrations (1, 2, 3, 4, and 5 mL) on the 5th, the 10th, and the 15 days and in the control.

| Experiments[a] | Quantity of bacterial cultures added | Fresh weight of the whole plant (g)[b] | | | Dry weight of the whole plant (g)[b] | | |
|---|---|---|---|---|---|---|---|
| | | 5th day | 10th day | 15th day | 5th day | 10th day | 15th day |
| | 1 mL | 0.58 (bcd) | 0.61 (abc) | 0.64 (ab) | 0.06 (a) | 0.06 (abc) | 0.06 (ab) |
| | 2 mL | 0.58 (ab) | 0.61 (a) | 0.64 (abc) | 0.06 (bcd) | 0.06 (ab) | 0.06 (ab) |
| Exp0 | 3 mL | 0.58 (abc) | 0.61 (ab) | 0.64 (bcd) | 0.06 (abc) | 0.06 (bcd) | 0.06 (b) |
| | 4 mL | 0.41 (abc) | 0.61 (ab) | 0.64 (ab) | 0.06 (abc) | 0.06 (ab) | 0.06 (abc) |
| | 5 mL | 0.58 (abc) | 0.61 (abc) | 0.64 (bcd) | 0.06 (a) | 0.06 (bcd) | 0.06 (def) |
| | 1 mL | 0.62 (abc) | 0.65 (abc) | 0.68 (ab) | 0.06 (bcd) | 0.06 (ab) | 0.07 (bcd) |
| | 2 mL | 0.65 (bcd) | 0.68 (bcd) | 0.71 (def) | 0.06 (abc) | 0.07 (a) | 0.07 (ab) |
| Exp1 | 3 mL | 0.68 (abc) | 0.71 (def) | 0.74 (bcd) | 0.07 (abc) | 0.07 (ab) | 0.07 (abc) |
| | 4 mL | 0.71 (ab) | 0.74 (abc) | 0.77 (abc) | 0.07 (bcd) | 0.07 (b) | 0.08 (bcd) |
| | 5 mL | 0.74 (abc) | 0.77 (def) | 0.80 (ab) | 0.07 (def) | 0.08 (a) | 0.08 (b) |
| | 1 mL | 0.56 (ab) | 0.59 (bcd) | 0.62 (ab) | 0.06 (ab) | 0.06 (ab) | 0.06 (abc) |
| | 2 mL | 0.59 (abc) | 0.62 (abc) | 0.65 (def) | 0.06 (ab) | 0.06 (abc) | 0.06 (bcd) |
| Exp2 | 3 mL | 0.62 (bcd) | 0.65 (a) | 0.68 (bcd) | 0.06 (abc) | 0.06 (bcd) | 0.07 (abc) |
| | 4 mL | 0.65 (abc) | 0.68 (ab) | 0.71 (b) | 0.06 (ab) | 0.07 (abc) | 0.07 (b) |
| | 5 mL | 0.68 (abc) | 0.71 (b) | 0.74 (a) | 0.07 (def) | 0.07 (abc) | 0.07 (bcd) |
| | 1 mL | 0.74 (ab) | 0.77 (ab) | 0.80 (abc) | 0.07 (a) | 0.08 (bcd) | 0.08 (ab) |
| | 2 mL | 0.77 (bcd) | 0.80 (a) | 0.83 (bcd) | 0.08 (a) | 0.08 (bcd) | 0.08 (abc) |
| Exp3 | 3 mL | 0.80 (bcd) | 0.83 (bcd) | 0.86 (abc) | 0.08 (ab) | 0.08 (b) | 0.09 (b) |
| | 4 mL | 0.83 (bcd) | 0.86 (abc) | 0.89 (ab) | 0.08 (abc) | 0.09 (ab) | 0.09 (bcd) |
| | 5 mL | 0.86 (ab) | 0.89 (b) | 0.92 (def) | 0.09 (bcd) | 0.09 (bcd | 0.09 (b) |

[a] Exp0: nutrient solution (control), Exp1: *Bacillus cereus*, Exp2: *Pseudomonas* spp., Exp3: consortium of *Bacillus cereus* + *Pseudomonas* spp. (1:1).
[b] Mean of two repeated experiments (30 plants). Different letters within parenthesis indicate significant difference between treatments for each growth parameter using Duncan's multiple range test ($P = 0.05$).

*Pseudomonas* spp. acts as a potential biofertilizer, due to its various plant growth promotion abilities such as solubilization of phosphate and production of IAA and siderophore; it can be developed as potential microbial inoculant for various crops.

## Conflict of Interests

The authors declared that there is no conflict of interests regarding this paper.

## Acknowledgments

The authors are thankful to the Department of Biology, Gandhigram Rural Institute—Deemed University, Gandhigram. This research was funded by University Grants Commission under UGC Research Fellowship in Science for Meritorious Students (BSR) Grant no. F.4-1/2008.

## References

[1] U. S. Bhawalkar, *Vermiculture Ecotechnology*, Earthworm Research Institute, Pune, India, 2nd edition, 1996.

[2] S. Kiss, D. M. Bularda, and D. Radulescu, "Biological Significance of Enzymes Accumulated in Soil," *Advances in Agronomy*, vol. 27, pp. 25–87, 1975.

[3] D. L. Karlen, M. J. Mausbach, J. W. Doran, R. G. Cline, R. F. Harris, and G. E. Schuman, "Soil quality: a concept, definition, and framework for evaluation," *Soil Science Society of America Journal*, vol. 61, no. 1, pp. 4–10, 1997.

[4] R. M. Atiyeh, C. A. Edwards, S. Subler, and J. Metzger, "Earthworm processed organic wastes as components of horticultural potting media for growing manifold and vegetable seedlings," *Compost Science and Utilization*, vol. 8, no. 3, pp. 215–223, 2000.

[5] N. Vassilev and M. Vassileva, "Biotechnological solubilization of rock phosphate on media containing agro-industrial wastes," *Applied Microbiology and Biotechnology*, vol. 61, no. 5-6, pp. 435–440, 2003.

[6] S. A. Omar, "The role of rock-phosphate-solubilizing fungi and vesicular-arbusular-mycorrhiza (VAM) in growth of wheat plants fertilized with rock phosphate," *World Journal of Microbiology and Biotechnology*, vol. 14, no. 2, pp. 211–218, 1998.

[7] J. A. J. Paul and T. Daniel, "Standardization of sampling method for physical characterization of municipal solid waste," *Asian Journal of Water, Environment and Pollution*, vol. 5, no. 1, pp. 95–98, 2007.

[8] R. Lalloo, D. Maharajh, J. Görgens, and N. Gardiner, "A downstream process for production of a viable and stable *Bacillus cereus* aquaculture biological agent," *Applied Microbiology and Biotechnology*, vol. 86, no. 2, pp. 499–508, 2010.

[9] B. Sivasankari, N. Kavikumar, and M. Anandharaj, "Indole-3-acetic acid production and enhanced Plant growth promotion by Indigenous bacterial species," *Journal of Current Research in Science*, vol. 1, no. 5, pp. 331–335, 2013.

[10] J. Han, L. Sun, X. Dong et al., "Characterization of a novel plant growth-promoting bacteria strain *Delftia tsuruhatensis* HR4 both as a diazotroph and a potential biocontrol agent against various plant pathogens," *Systematic and Applied Microbiology*, vol. 28, no. 1, pp. 66–76, 2005.

[11] H. Rodríguez and R. Fraga, "Phosphate solubilizing bacteria and their role in plant growth promotion," *Biotechnology Advances*, vol. 17, no. 4-5, pp. 319–339, 1999.

[12] T. Daniel and N. Karmegam, "Bio-conversion of selected leaf litters using an African epigeic earthworm, *Eudrilus eugeniae*," *Ecology, Environment and Conservation*, vol. 5, no. 3, pp. 271–275, 1999.

[13] J. Dobereiner, I. E. Marriel, and M. Nery, "Ecological distribution of *Spirillum lipoferum*, Beijerinck," *Canadian Journal of Microbiology*, vol. 22, no. 10, pp. 1464–1473, 1976.

[14] R. I. Pikovskaya, "Mobilization of phosphorus in soil in connection with vital activity of some microbial species," *Microbiology*, vol. 17, pp. 362–370, 1948.

[15] M. Edi Premono, A. M. Moawad, and P. L. G. Vlek, "Effect of phosphate-solubilizing *Pseudomonas putida* on the growth of maize and its survival in the rhizosphere," *Indonesian Journal of Crop Science*, vol. 11, pp. 13–23, 1996.

[16] C. L. Patten and B. R. Glick, "Bacterial biosynthesis of indole-3-acetic acid," *Canadian Journal of Microbiology*, vol. 42, no. 3, pp. 207–220, 1996.

[17] N. Amaresan, V. Jayakumar, K. Kumar, and N. Thajuddin, "Isolation and characterization of plant growth promoting endophytic bacteria and their effect on tomato (*Lycopersicon esculentum*) and chilli (*Capsicum annuum*) seedling growth," *Annals of Microbiology*, vol. 62, no. 2, pp. 805–810, 2012.

[18] B. Schwyn and J. B. Neilands, "Universal chemical assay for the detection and determination of siderophores," *Analytical Biochemistry*, vol. 160, no. 1, pp. 47–56, 1987.

[19] P. Rahi, P. Vyas, S. Sharma, and A. Gulati, "Plant growth promoting potential of the fungus *Discosia* sp. FIHB 571 from tea rhizosphere tested on chickpea, maize and pea," *Indian Journal of Microbiology*, vol. 49, no. 2, pp. 128–133, 2009.

[20] K. Kumar, N. Amaresan, S. Bhagat, K. Madhuri, and R. C. Srivastava, "Isolation and characterization of rhizobacteria associated with coastal agricultural ecosystem of rhizosphere soils of cultivated vegetable crops," *World Journal of Microbiology and Biotechnology*, vol. 27, no. 7, pp. 1625–1632, 2011.

[21] J. A. Smoker and S. R. Barnum, "Rapid small scale DNA isolation from filamentous cyanobacteria," *FEMS Microbiology Letters*, vol. 56, no. 1, pp. 119–122, 1988.

[22] U. Edwars, T. Rogall, H. Blocker, M. Emde, and E. C. Bottger, "Isolation and direct complete nucleotide determination of entire genes. Characterization of a gene coding for 16S ribosomal RNA," *Nucleic Acids Research*, vol. 17, no. 19, pp. 7843–7853, 1989.

[23] D. G. Higgins, A. J. Bleasby, and R. Fuchs, "CLUSTAL V: improved software for multiple sequence alignment," *Computer Applications in the Biosciences*, vol. 8, no. 2, pp. 189–191, 1992.

[24] M. Kimura, "A simple method for estimating evolutionary rates of base substitutions through comparative studies of nucleotide sequences," *Journal of Molecular Evolution*, vol. 16, no. 2, pp. 111–120, 1980.

[25] N. Saitou and M. Nei, "The neighbor-joining method: a new method for reconstructing phylogenetic trees," *Molecular Biology and Evolution*, vol. 4, no. 4, pp. 406–425, 1987.

[26] E. Barbu, K. Y. Lee, and R. Wahl, "Content of purine and pyrimidine base in desoxyribonucleic acid of bacteria," *Annales de l'Institut Pasteur*, vol. 91, no. 2, pp. 212–224, 1956.

[27] N. Sueoka, "Variation and heterogeneity of base composition of deoxyribonucleic acids: a compilation of old and new data," *Journal of Molecular Biology*, vol. 3, no. 1, pp. 31–40, 1961.

[28] H. Hori and S. Osawa, "Evolutionary change in 5S rRNA secondary structure and a phylogenic tree of 352 5S rRNA species," *BioSystems*, vol. 19, no. 3, pp. 163–172, 1986.

[29] V. Bess, "Evaluating microbiology of compost: microbial content of composting is helping the producers and growers to understand its role as a soil inoculant and plant," *BioCycle Magazine*, p. 62, 1999.

[30] O. Grafff, "Preliminary experiments of vermin composting of different waste materials using *Eudrilus eugeniae* Kinberg," in *Workshop on the Role of Earthworms in the Stabilization of Organic Residues*, M. Happelhof, Ed., pp. 178–191, Beech Leaf Press, Kalamazoo, Mich, USA, 1981.

[31] J. E. Satchell, *Earthworm Ecology from Darwin to Vermiculture*, Chapman and Hall, 1983.

[32] G. H. Xie, M. Y. Cai, G. C. Tao, and Y. Steinberger, "Cultivable heterotrophic N2-fixing bacterial diversity in rice fields in the Yangtze River Plain," *Biology and Fertility of Soils*, vol. 37, no. 1, pp. 29–38, 2003.

[33] B. B. McSpadden Gardener, "Ecology of *Bacillus* and *Paenibacillus* spp. in agricultural systems," *Phytopathology*, vol. 94, no. 11, pp. 1252–1258, 2004.

[34] J. Dobereiner, "Biological nitrogen fixation in the tropics: social and economic contributions," *Soil Biology and Biochemistry*, vol. 29, no. 5-6, pp. 771–774, 1997.

[35] V. M. Reis, F. L. Olivares, and J. Dobereiner, "Improved methodology for isolation of *Acetobacter diazotrophicus* and confirmation of its endophytic habitat," *World Journal of Microbiology and Biotechnology*, vol. 10, no. 4, pp. 401–405, 1994.

[36] S. Khalil, N. Ayub, S. Alam, and F. Latif, "Organic acids production and phosphate solubilization by phosphate solubilizing microorganisms (PSM) under in vitro conditions," *Pakistan Journal of Biological Sciences*, vol. 7, no. 2, pp. 187–196, 2004.

[37] H. M. A. El-Komy, "Coimmobilization of *Azospirillum lipoferum* and *Bacillus megaterium* for successful phosphorus and nitrogen nutrition of wheat plants," *Food Technology and Biotechnology*, vol. 43, no. 1, pp. 19–27, 2005.

[38] N. B. Paul and W. V. B. Sundara Rao, "Phosphate-dissolving bacteria in the rhizosphere of some cultivated legumes," *Plant and Soil*, vol. 35, no. 1, pp. 127–132, 1971.

[39] A. Beneduzi, D. Peres, L. K. Vargas, M. H. Bodanese-Zanettini, and L. M. P. Passaglia, "Evaluation of genetic diversity and plant growth promoting activities of nitrogen-fixing bacilli isolated from rice fields in South Brazil," *Applied Soil Ecology*, vol. 39, no. 3, pp. 311–320, 2008.

[40] M. P. Raimam, U. Albino, M. F. Cruz et al., "Interaction among free-living N-fixing bacteria isolated from *Drosera villosa* var. villosa and AM fungi (*Glomus clarum*) in rice (*Oryza sativa*)," *Applied Soil Ecology*, vol. 35, no. 1, pp. 25–34, 2007.

[41] M. R. Swain, S. K. Naskar, and R. C. Ray, "Indole-3-acetic acid production and effect on sprouting of yam (*Dioscorea rotundata* L.) minisetts by *Bacillus subtilis* isolated from culturable cow-dung microflora," *Polish Journal of Microbiology*, vol. 56, no. 2, pp. 103–110, 2007.

[42] T. M. Tien, M. H. Garkins, and D. H. Hubbell, "Plant growth substances produced by *Azospirillum brasilense* and their effect

on the growth of pearl Millet *Pennisetum americanum* (L.)," *Applied and Environmental Microbiology*, vol. 37, no. 5, pp. 1016–1024, 1979.

[43] A. Esitken, L. Pirlak, M. Turan, and F. Sahin, "Effects of floral and foliar application of plant growth promoting rhizobacteria (PGPR) on yield, growth and nutrition of sweet cherry," *Scientia Horticulturae*, vol. 110, no. 4, pp. 324–327, 2006.

[44] H. Karlidag, A. Esitken, M. Turan, and F. Sahin, "Effects of root inoculation of plant growth promoting rhizobacteria (PGPR) on yield, growth and nutrient element contents of leaves of apple," *Scientia Horticulturae*, vol. 114, no. 1, pp. 16–20, 2007.

[45] C. S. Nautiyal, S. Bhadauria, P. Kumar, H. Lal, R. Mondal, and D. Verma, "Stress induced phosphate solubilization in bacteria isolated from alkaline soils," *FEMS Microbiology Letters*, vol. 182, no. 2, pp. 291–296, 2000.

[46] A. K. Halder, A. K. Mishra, P. Bhattacharya, and P. K. Chakrabarthy, "Solubilization of inorganic phosphates by *Bradyrhizobium*," *Indian Journal of Experimental Biology*, vol. 29, pp. 28–31, 1991.

# 9

# Managing Soil Biota-Mediated Decomposition and Nutrient Mineralization in Sustainable Agroecosystems

**Joann K. Whalen**

*Department of Natural Resource Sciences, McGill University, Macdonald Campus, 21111 Lakeshore Road, Ste-Anne-de-Bellevue, QC, Canada H9X 3V9*

Correspondence should be addressed to Joann K. Whalen; joann.whalen@mcgill.ca

Academic Editor: Elke Bloem

Transformation of organic residues into plant-available nutrients occurs through decomposition and mineralization and is mediated by saprophytic microorganisms and fauna. Of particular interest is the recycling of the essential plant elements—N, P, and S—contained in organic residues. If organic residues can supply sufficient nutrients during crop growth, a reduction in fertilizer use is possible. The challenge is synchronizing nutrient release from organic residues with crop nutrient demands throughout the growing season. This paper presents a conceptual model describing the pattern of nutrient release from organic residues in relation to crop nutrient uptake. Next, it explores experimental approaches to measure the physical, chemical, and biological barriers to decomposition and nutrient mineralization. Methods are proposed to determine the rates of decomposition and nutrient release from organic residues. Practically, this information can be used by agricultural producers to determine if plant-available nutrient supply is sufficient to meet crop demands at key growth stages or whether additional fertilizer is needed. Finally, agronomic practices that control the rate of soil biota-mediated decomposition and mineralization, as well as those that facilitate uptake of plant-available nutrients, are identified. Increasing reliance on soil biological activity could benefit crop nutrition and health in sustainable agroecosystems.

## 1. Introduction

Agricultural scientists and practitioners face major challenges in the 21st century. They must produce enough nutritious food to feed a growing world population, which is increasing at rate of 1.14% per year and expected to reach 9.5 billion by 2050 [1]. Agroecosystems are also under pressure to produce greater quantities of fiber for biofuel production. For example, the European Union directive on renewable energy set a target of 10% biofuel, while legislation in the United States will require 20% renewables in fuel by 2022. It is critical that these targets be achieved without compromising food production, which is possible with careful agricultural management. The practices of multiple cropping and using by-products of biofuel production as animal feed resulted in an increase of 19 million ha of net harvested area from 2000 to 2010 in the United States, Brazil, Indonesia, Malaysia, China, Mozambique, South Africa, and 27 European Union member states [2]. By 2010, these countries produced 86 billion L

of ethanol and 15 billion L of biodiesel and also had a net gain in land available to produce food for human and animal consumption [2].

Boosting agricultural production to a level that meets concurrent demands for food and biofuel must be done in a sustainable manner, such that societal objectives and economic prosperity can be achieved without environmental damage. There is concern that agricultural activities have disturbed the natural N and P cycles to the point that they exceed their planetary boundaries [3]. However, initial estimates of the planetary boundaries appear to be low, leading de Vries et al. [4] to revise the N limits after considering N requirements to feed the world's population while avoiding adverse impacts of reactive N in water, air, and soils. Still, sustainable agroecosystems need to be managed to recycle nutrients efficiently, thus reducing reliance on external nutrient and energy inputs. They should also resist abiotic and biotic stresses, allowing them to tolerate periodic disturbances and extreme conditions, such as a short-term

TABLE 1: Attributes of sustainable agroecosystems, based on the ecosystem services concept (adapted from Whalen and Sampedro [17]).

| Supporting services | Provisioning services | Regulating services | Cultural services |
| --- | --- | --- | --- |
| Sustain energy and material flows<br>Support growth of high-quality agricultural crops<br>Provide habitat for diverse soil fauna | Produce food for people and animals<br>Produce fiber and biofuel<br>Biological products: novel organisms and compounds<br>Generate clean water | Filtration system for percolating water<br>Carbon sequestration and greenhouse gas mitigation<br>Decomposition and nutrient cycling<br>Remediation and attenuation of wastes and pollutants<br>Pest and disease control | Rural communities: past, present, and future<br>Agrotourism<br>Science and education |

flooding event or disease outbreak. These attributes also convey resilience so the agroecosystem will continue to be functional and productive if there is a permanent change in the abiotic and biotic stresses that it is subjected to. Adaptive management has a role in supporting agricultural resilience, permitting sustainable agroecosystems to provide ecosystem services to mankind (Table 1).

Soils feature prominently in sustainable agroecosystems because they support tremendous biodiversity that is linked to the health and productivity of crops. Among the millions of species that are permanent soil dwellers include single-celled prokaryotes (Bacteria and Archaea) and eukaryotes (Fungi, Protists, multicellular insects, and annelids). These organisms are sustained by energy inputs from plants, in the form of root exudates and secretions from living plants as well as residues from nonliving plants and other organic wastes. Interactions between plants and soil biota, both symbiotic and free-living inhabitants of the rhizosphere, are emerging as a hot research topic due to the fact that naturally occurring soil biota protect plants against pathogens. For example, vesicular arbuscular mycorrhizal fungi prevent plant diseases through physical and biochemical mechanisms [5], including upregulation of jasmonate induced plant defenses [6, 7]. Systemic resistance, induced by signal molecules released from plant growth promoting rhizobacteria and/or elicitors emitted by nonvirulent pathogens, is also stimulated by auxins released from earthworm activities [8].

Productive crops require nutrients to sustain high yields, and thus plants benefit when soil biota transform organically bound N, P, and S into soluble $NH_4^+$, $NO_3^-$, $H_2PO_4^-$, $HPO_4^{2-}$, and $SO_4^{2-}$, the ionic forms absorbed by root cells. It is known that plants assimilate amino acids, either directly through their roots or via arbuscular mycorrhiza [9], although microorganisms also acquire amino acids from the soil pore water and there is evidence for preferential retention of $^{15}$N-labeled amino acids in soil microbial biomass in low productivity ecosystems (including low input agroecosystems, e.g., [10]). Still, plant N uptake of $NH_4^+$, glutamate-N, and glycine-N was less than 10% of the microbial N uptake from these N forms in a controlled study with *Festuca gigantea* L. under constant soil moisture (60% water-filled pore space) or with wet-dry cycles [11]. In addition, plants bred for high production agroecosystems tend to be less efficient at assimilating organic N forms than $NH_4^+$ and $NO_3^-$ than their wild relatives [12]. The assumption

in this review is that N mineralization yielding $NH_4^+$, which could be taken up by plants or further transformed via ammonia oxidation and nitrification to produce $NO_3^-$, another plant-available N ion, would be of greater relevance in agroecosystems than uptake of amino acids. Similarly, S mineralization that produces $SO_4^{2-}$ was presumed to be of greater importance for plant S nutrition than uptake of the S-containing amino acids, cysteine and methionine [13]. Plants derive nutrition from organic P compounds by secreting extracellular phytases from the root, which facilitates the solubilization and subsequent uptake of $H_2PO_4^-$ [14]. Further information on the contribution of organic N, P, and S to plant nutrition is detailed in the review paper of Paungfoo-Lonhienne et al. [15].

Given that most agricultural crops will assimilate soluble ionic forms of N, P, and S, it is important to understand how all soil biota, from microorganisms to macrofauna, are involved in the decomposition of organic compounds and the mineralization of N, P, and S in the soil-plant system. We can potentially reduce fertilizer inputs if we know how much plant-available N, P, and S will be liberated from organic residues, but will plant-available nutrients be released at the right time during the growing season to satisfy plant nutrient demands? Knowing this will allow us to precisely select the application time and amount of supplemental fertilizers needed to achieve yield goals. Such judicious use of fertilizer is beneficial from an economic perspective, because it will improve nutrient use efficiency, and for the environment since excessive nutrient inputs are susceptible to be transported into waterways and the atmosphere.

Biotic conversion of organically bound nutrients to soluble ions follows a predictable sequence of physical transformations and biochemical reactions, which can be modulated by environmental conditions (e.g., temperature, moisture, and soil physicochemical properties) and agronomic factors (e.g., litter inputs, tillage, and manuring) that affect the activity of the soil biota, as reviewed by Whalen et al. [16]. Fundamentally, it can be viewed as a two-step process that proceeds as follows.

(1) Decomposition: fragmentation and physical disruption reduces the particle size of the organic residue, which increases the surface area for microbial colonization and hydrolysis by extracellular enzymes. After the chemically complex polymers in the residue

are degraded into monomeric compounds (e.g., amino acids) or ions (e.g., $H_2PO_4^-$ and $HPO_4^{2-}$) through these extracellular reactions, they can be absorbed into microbial cells for the second step.

(2) Mineralization: absorbed monomeric compounds are acted upon by intracellular enzymes, releasing energy and precursors for microbial metabolism (e.g., $NH_4^+$ and $SO_4^{2-}$ destined for protein synthesis). Ionic forms of N, P, and S are also absorbed by microbial cells. When microbial requirements are met or when microbial cells are lysed, excess ions can be released into the soil pore water, which is accessible to plant roots.

The purpose of this review is to describe how soil biota mediate the processes of decomposition and nutrient mineralization in sustainable agroecosystems. Next, I discuss how this knowledge can be used to synchronize nutrient release from organic residues and nutrient uptake by agricultural crops, followed by a description of experimental approaches to evaluate the decomposition and nutrient mineralization processes. This approach could permit agricultural managers to increase their reliance on soil biota for nutrient recycling, applying fertilizers strategically to boost the soil nutrient supply at critical growth stages when inherent nutrient reserves are insufficient to meet crop yield goals.

## 2. Decomposition and Nutrient Mineralization: A Two-Step Process

*2.1. Soil Biota Involved in the Decomposition Process.* Organic residue inputs in agroecosystems originate from many sources. Virtually all agroecosystems will have an input of non-harvested crop components—roots, above-ground residues (mostly leaves, stems, and husks) of grain crops, the above-ground biomass of cover crops, green manure crops, and fallow crops—and may also receive organic materials like forestry residues, animal wastes, compost from various feedstocks, and biosolids from municipal water treatment facilities. These wastes are characterized by large particle size and a variable degree of predecomposition at the time they are land applied.

Heterogeneity is the major challenge to describe biologically-mediated decomposition and it arises from a number of factors. First, the variability in physical size, chemical composition, and degree of predecomposition among organic residues makes it difficult to apply them uniformly in agricultural fields. Even if residue from a single crop (e.g., maize) is considered, the amount of organic residue left in the field, its chemical composition, and decomposability is affected by the above-ground and root biomass produced in the previous growing season, the cultivar grown, and genetic modification [22]. Once in the soil, the residue breakdown is mediated by soil organisms that exhibit an aggregated and uneven distribution throughout the soil profile (vertically) and across soil microenvironments (horizontally) [17]. Further, the action of the soil organisms is strongly controlled by oscillations in soil moisture, which is linked to rainfall,

irrigation, and snowmelt patterns as well as evapotranspiration, transpiration, and drainage of water through the soil profile. Wetting-drying cycles are a powerful modulator of soil biota-mediated decomposition and N mineralization [23, 24]. Other important abiotic controls on soil biotic activity are soil texture, soil pH, soil organic matter, and nutrient levels [25, 26]. Variability in these parameters arises during soil pedogenesis and from historical agricultural management. Although small-scale heterogeneity and large-scale gradients that affect the pattern of biologically mediated decomposition and nutrient mineralization need to be considered, this review takes a simplistic view of decomposition and assumes the researcher will account for extraneous, moderator, and mediator variables that produce site-specific conditions affecting decomposition. The conceptual model of decomposition illustrated in Figure 1 is based on senescent crop residue that is left on the soil surface after harvesting a grain crop.

Once a crop reaches physiological maturity, fungi and other microorganisms begin to colonize the senescing leaf and stem tissues of the standing crop, thus initializing the decomposition process before the unharvested residues hit the ground. Still, the senescent crop residues are mostly intact and must be fragmented into smaller particles so that plant cells are physically disrupted and permit microbial and enzymatic access to polymeric compounds. Saprophagous soil macrofauna and mesofauna like earthworms, millipedes, and collembola consume organic residues, which physically fragments the material and mixes it with soil particles containing microorganisms that are ingested by these organisms. For example, pure leaf litter passing through the gut of *Lumbricus terrestris* was reduced in size to $0.23\,mm^2$ and further ground to less than $0.001\,mm^2$ when sand and litter were consumed together [27]. Litter comminution by earthworms partially or completely buries organic residues, accounting for redistribution of 19 to 24% of the organic matter in crop residues and the top 15 cm of soil each year [28]. These activities result in particle size reduction, which stimulates the soil micro- and mesofauna to further degrade the materials and enhances microbial colonization of the residue. This is supported by the greater abundance and larger biomass of decomposer organisms (bacteria, fungi, protozoa, nematodes, and microarthropods) in buried litter than surface litter [29]. As noted by Beare et al. [29], buried sorghum litter decays more than twice as quickly (1.4–1.7% mass loss per day) as litter left on the soil surface (0.5–0.7% mass loss per day). A similar effect is achieved when residue size is reduced mechanically by tillage in agroecosystems.

After physical barriers to decomposition are overcome by reducing the particle size, the next barrier to decomposition is the chemistry of the polymeric compounds contained in the crop residue. The chemical recalcitrance of plant residues to decomposition and the role of lignin in slowing the decomposition process were reviewed extensively by Gul and Whalen [30] and Gul et al. [31]. Briefly, extracellular enzymes of microbial origin are responsible for cleaving monomeric units from complex polymers like cellulose, hemicellulose,

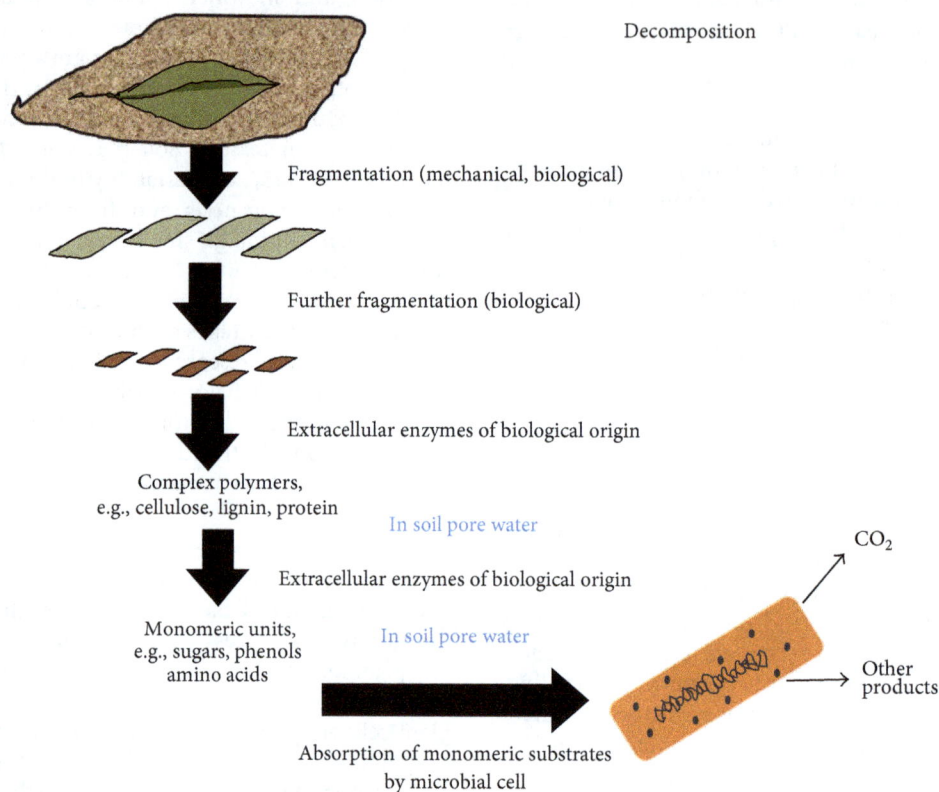

FIGURE 1: Decomposition of senescent crop residue on the soil surface of an agroecosystem. The leaf is fragmented biologically, through the action of soil meso- and macrofauna, and mechanically by tillage or agricultural machinery that passes over the soil surface. Physical fragmentation continues as microorganisms colonize the surface of leaf particles and those particles are ingested by soil fauna. As secondary cell walls are penetrated and cell membranes rupture, complex organic molecules like cellulose, lignin and proteins are released into the soil pore water. Extracellular enzymes produced by microorganisms and plants hydrolyse complex compounds into simple monomeric units like sugars, phenols, and amino acids. Monomeric substrates are then absorbed through the cell membranes of prokaryotes (shown) and eukaryotes (not shown), where they undergo intracellular enzymatic hydrolysis, leading to the production of carbon dioxide ($CO_2$) and other metabolic byproducts.

lignin, and proteins. Rates of enzymatic activity are controlled by access to substrates, where access is governed by the amount and chemistry of lignin and lignocellulosic compounds deposited in the secondary cell wall of plants. Water is required for all hydrolytic reactions, which directly affects the functioning of extracellular enzymes as well as the concentration and diffusion of substrates to the active site of the enzyme [32]. For example, low soil moisture (−4 to −1 MPa soil water potential) reduced $\beta$-glucosidase activity, such that soils under "drought" had 46% less $\beta$-glucosidase activity than those in the ambient moisture treatment in an agricultural field with permanent vegetation [33]. To retain its function, an extracellular enzyme present in soil pore water or attached to a soil organomineral surface must be able to catalyze reactions (i.e., substrates can bind to the active site, the binding affinity of the protein-ligand at the active site is maintained by the tertiary structure, and no inhibitory compounds are bound to the enzyme). Factors affecting extracellular enzyme production and activity include microbial species present, microbial requirements for energy and nutrients that stimulate enzyme production,

temperature, pH, oxygen content, enzyme cofactors, and enzyme inhibitors [34].

Extracellular phosphatases are responsible for mineralizing organic P compounds, all of which are ester phosphates (C–O–P bonds), to soluble, ionic phosphates (predominantly $H_2PO_4^-$ in acidic soils and $HPO_4^{2-}$ in alkaline soils) (Figure 2(a)). Similarly, extracellular sulfatases are responsible for the breakdown of ester sulfates (C–O–S bonds) to soluble $SO_4^{2-}$ (Figure 2(b)) and extracellular urease converts urea to $NH_4^+$ (Figure 2(c)). In the case of plant-derived extracellular phosphatases, the enzyme appears to fulfill two functions: (1) hydrolysis of ester phosphates to $H_2PO_4^-$ or $HPO_4^{2-}$ and (2) carrier protein that transports phosphate ions across the plasma membrane [35].

*2.2. Soil Biota Involved in the Mineralization Process.* The decomposition process liberates monomeric compounds that can diffuse through the soil pore water and be absorbed through the membrane of microbial cells for further hydrolysis *in vivo*, releasing energy and precursors for metabolic processes (Figure 3). As described in Whalen et al. [16],

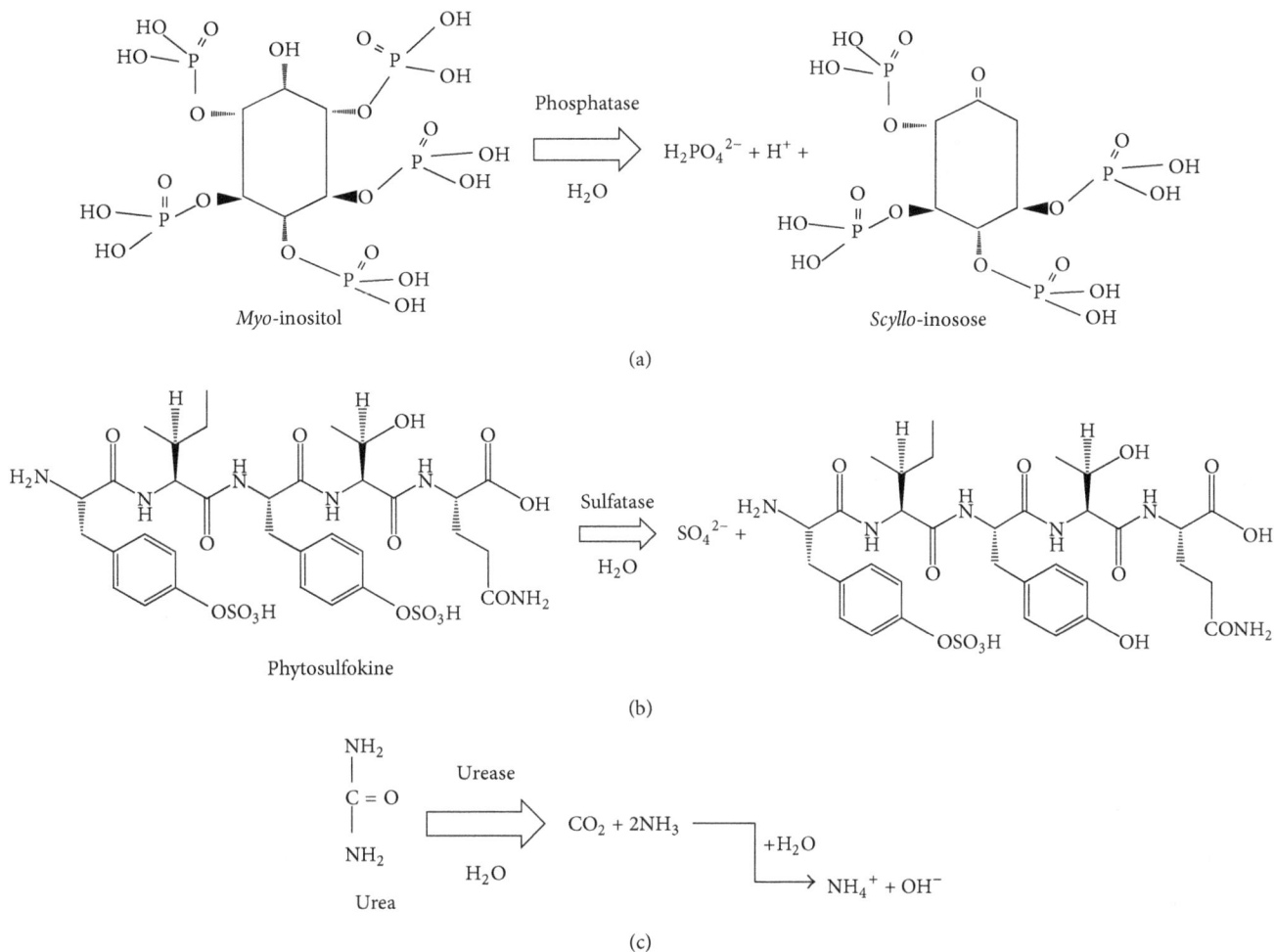

FIGURE 2: Extracellular enzymes are responsible for hydrolysing organic compounds in soil pore water, for example, (a) *myo*-inositol hydrolysis by phosphatase releases $H_2PO_4^{2-}$ and *scyllo*-inosose, (b) phytosulfokine hydrolysis by sulfatase releases $SO_4^{2-}$ and a derivative of phytosulfokine, and (c) urea hydrolysis by urease releases carbon dioxide and ammonia, which is converted to $NH_4^+$ in the presence of water.

energy-rich compounds like cellobiose are hydrolysed to glucose to fuel reactions in the tricarboxylic acid cycle when oxygen is present, or pyruvate molecules synthesized from glucose are fermented when anaerobic conditions prevail. Amino acids containing N and S C-N and C-S bonds are converted to $NH_4^+$ and $SO_4^{2-}$ through the action of intracellular amino acid hydrolases. As precursors for protein synthesis and other metabolic pathways, these soluble ions could be immobilized in the microbial cell or released into the soil pore water if microbial requirements were already met. Lysis of microbial cells due to soil disruption (e.g., freezing and thawing, wetting and drying) or grazing by predators (e.g., protists and nematodes) is another way that $NH_4^+$ and $SO_4^{2-}$ within microbial cells are released into soil pore water.

In aerated soils, the $NH_4^+$ produced in the N mineralization process is sequentially oxidized to $NO_3^-$ by two groups of organisms, the ammonia oxidizers and nitrifiers (Figure 4). Ammonia oxidizers may be bacteria or archaea that use ammonia monooxygenase to produce $NH_2OH$ and hydroxylamine oxidoreductase to produce $NO_2^-$, in

the process deriving energy for growth from the oxidation of these inorganic compounds and acquiring C from $CO_2$ (autotrophic ammonia oxidizers; [36]) or organic C sources (heterotrophic ammonia oxidizers; [37]). Nitrifiers include autotrophic and heterotrophic microorganisms that convert $NO_2^-$ to $NO_3^-$ with nitrite oxidoreductase [17]. The $NO_3^-$ released into soil pore water is readily available to crops due to the high mobility in soil pore water of the $NO_3^-$ molecule, relative to $NH_4^+$. However, it is more demanding energetically for plants to use $NO_3^-$ for protein synthesis because it has to be first reduced to $NH_4^+$ within leaf cells before glutamine production and protein synthesis occur [38].

## 3. Synchronizing the Decomposition and Nutrient Mineralization Processes with Crop Nutrient Requirements

Crops require N to synthesize proteins, chlorophyll, and other N-rich compounds, P for energy relations and cell

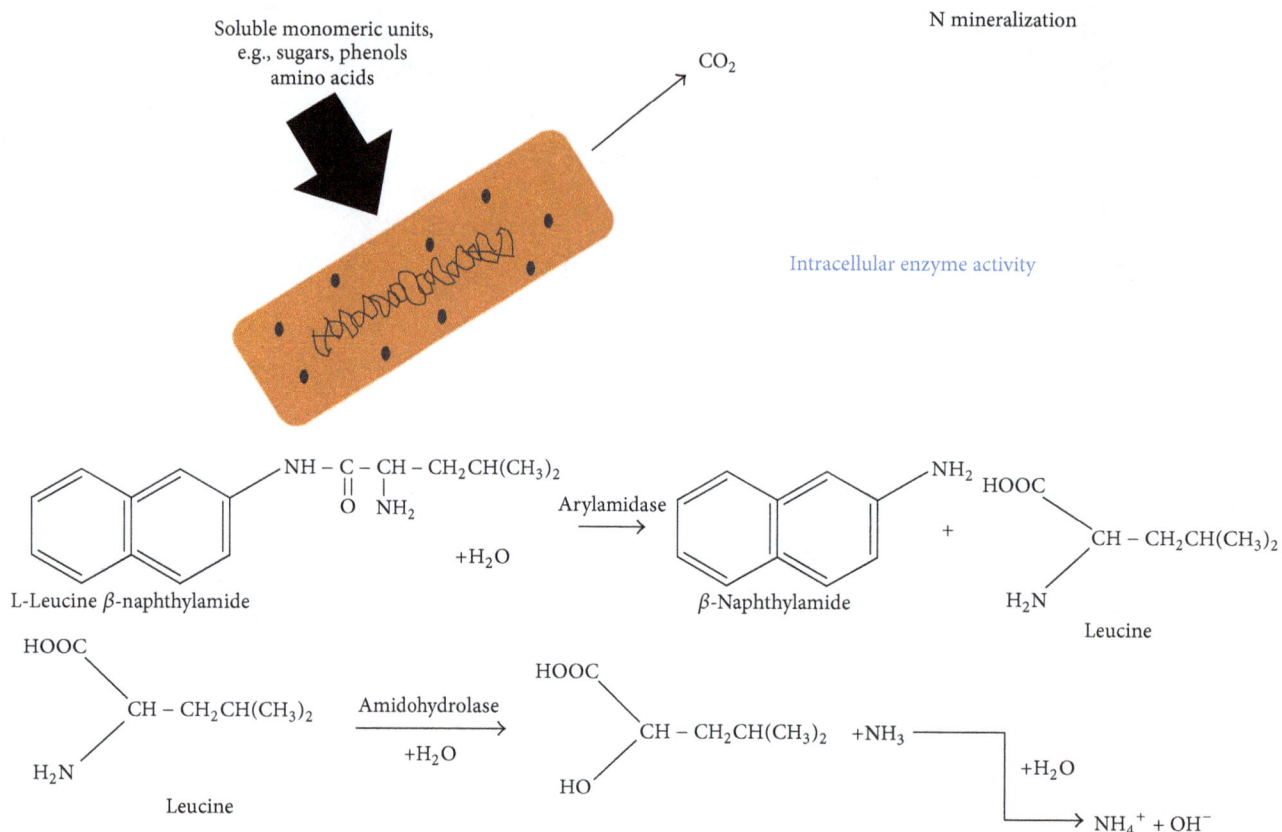

FIGURE 3: Following the absorption of organic N compounds such as amino acids and amino sugars through the cell membrane, intracellular enzymes are responsible for N mineralization, which yields $NH_4^+$ as the end product. The $NH_4^+$ may be retained in the microbial cell for protein synthesis and other metabolic processes or released from the cell into soil pore water where it can undergo other transformations (e.g., immobilization, plant $NH_4^+$ uptake, and $NH_4^+$ fixation).

division, and S for protein synthesis, glucosinolate synthesis as an anti-herbivore defense, and production of organosulfur compounds that gives food their distinctive odors and flavors. There is a constant but variable requirement for these nutrients during the growing season. Crop nutrient demands can be predicted roughly from the pattern of biomass accumulation, since nutrient uptake is proportional to the plant's photosynthetic activity, especially during the vegetative growth stage (Figure 5). Estimates of nutrient uptake can be further refined by considering root distribution, biomass, and maximum rooting depth [39].

In crops that derive most or all of their N, P, and S requirements from soil, any impediment to acquiring nutrients from soil pore water could lead to nutrient deficiency at the critical growth stage. The limited mobility of $H_2PO_4^-/HPO_4^{2-}$ and $NH_4^+$ arises from the fact that these ions bind readily to exchange sites on soil organo-mineral surfaces; phosphate ions precipitate with Al, Fe, and Ca minerals as well. Although $NO_3^-$ and $SO_4^{2-}$ are mobile, moving by mass flow to the roots, rainfall or irrigation events could cause these ions to leach below the root zone where they are inaccessible to the crop [17].

Could soil biota-mediated decomposition and mineralization supply ample N, P, and S at critical crop growth stages?

There are two additional questions that we should consider to narrow the scope of investigation. First, what are the N, P, and S requirements of the crop at the critical growth stage? The assumption is that adequate N, P, and S concentrations in the plant at this growth stage are a good indication of the crop's yield potential. Plant tissue analysis coupled with the Diagnosis and Recommendation Integrated System (DRIS; [40]) or compositional nutrient diagnosis (CND; [41]) is helpful in determining the nutritional requirements for high-yielding crops.

Second, is the quantity of N, P, and S in the organic residues sufficient to meet crop requirements? Taking a mass balance approach, the nutrient stock of the residue (e.g., N concentration in $kg\,Mg^{-1}$ × residue mass in $Mg^{-1}$) can be compared to the nutrient removal by the crop (e.g., N concentration in $kg\,Mg^{-1}$ × crop biomass in $Mg^{-1}$). However, what proportion of N, P, and S released from the decomposition and mineralization processes is assimilated by the crop at the critical growth stage? We realize that plant-available N, P, and S are subject to microbially mediated reactions (e.g., immobilization, denitrification), chemical adsorption and precipitation reactions, and physical transport in the soil profile, all of which diminish their solubility and accessibility for plant uptake. If the question is asked for a particular crop,

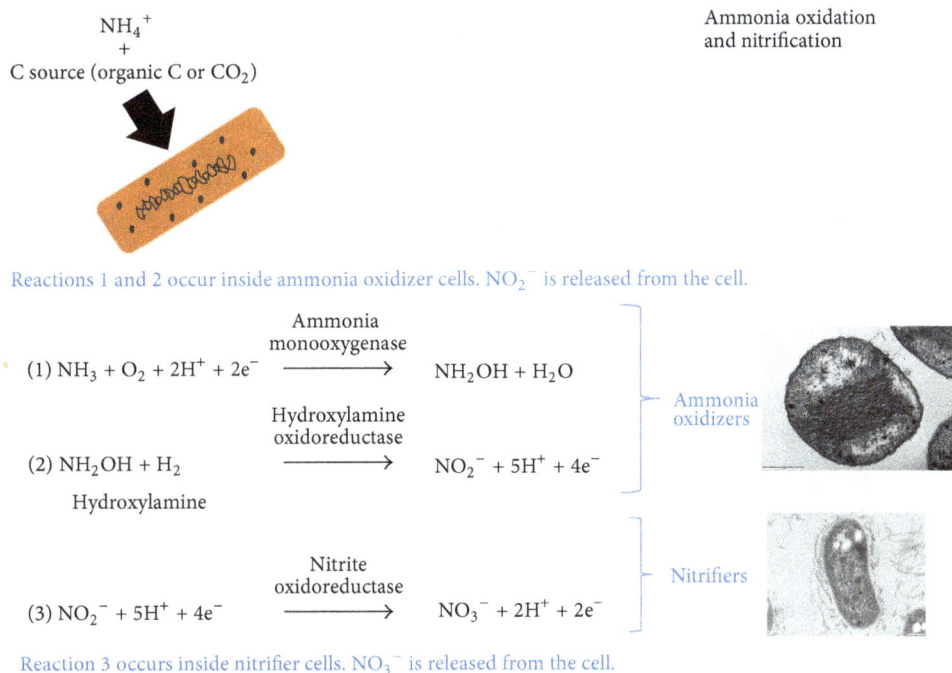

$NH_4^+$
+
C source (organic C or $CO_2$)

Ammonia oxidation
and nitrification

Reactions 1 and 2 occur inside ammonia oxidizer cells. $NO_2^-$ is released from the cell.

Ammonia
monooxygenase
(1) $NH_3 + O_2 + 2H^+ + 2e^- \longrightarrow NH_2OH + H_2O$

Hydroxylamine
oxidoreductase
(2) $NH_2OH + H_2 \longrightarrow NO_2^- + 5H^+ + 4e^-$
Hydroxylamine

Ammonia
oxidizers

Nitrite
oxidoreductase
(3) $NO_2^- + 5H^+ + 4e^- \longrightarrow NO_3^- + 2H^+ + 2e^-$

Nitrifiers

Reaction 3 occurs inside nitrifier cells. $NO_3^-$ is released from the cell.

FIGURE 4: Ammonia oxidizers absorb $NH_4^+$ from soil pore water and convert it to hydroxylamine and nitrite. Ammonia oxidizers release nitrate into soil pore water, where it is then absorbed by nitrifiers, who convert it to nitrate. Once released into soil pore water, nitrate can undergo other transformations (e.g., immobilization, plant $NO_3^-$ uptake, leaching, and denitrification). Top image: transmission electron micrograph of *Nitrosococcus oceani* ATCC 19707 (Klotz et al. [18], copyright American Society of Microbiology). Bottom image: transmission electron micrograph of *Nitrobacter winogradskyi* Nb-255 (Starkenburg et al. [19], image by W. J. Hickey, University of Wisconsin-Madison).

organic residue, and soil type, it could be studied with isotopic tracers ($^{15}N$ stable isotopes, $^{32}P$, $^{33}P$, and $^{35}S$ radioisotopes), but the major limitation is that the experiment needs to be repeated for each crop, organic residue, and soil type of interest. Also, the radioisotope work cannot be done in the field for safety reasons, since the relatively short half-life of these radioisotopes (half-life of $^{32}P$ = 14.29 d; $^{33}P$ = 25.3 d; $^{35}S$ = 87.2 d) means that large amounts are needed to monitor P and S transformation rates. Alternative approaches are needed to quantify how soil biota-mediated decomposition and mineralization could contribute to crop nutrition.

## 4. Experimental Approaches to Evaluate Soil Biota-Mediated Decomposition and Mineralization

Many physical, chemical, and biological tests exist that describe, to some extent, the process of soil biota-mediated decomposition and mineralization. Field-based litterbag studies provide information on the mass loss from organic residues and net release of N, P, and S into soil pore water [42]. Laboratory-based incubation studies under controlled conditions predict the mineralization rate and quantity of soluble N, P, and S generated in soil that was historically or recently amended with organic residues [43, 44]. Biological assays that consider both extracellular and intracellular enzyme activity provide insight into the capacity of the microbial community to hydrolyze organic polymers in soil

pore water and within their cells [34, 45]. While each test method is useful, none can fully describe the processes that transform organic residue into soluble, plant-available nutrients. Experimental approaches are needed that consider (1) the physical barriers to decomposition based on particle size, (2) the chemical barriers to decomposition based on residue chemistry, and (3) the biological barriers to mineralization, arising from biochemical capacity of extracellular and intracellular enzymes, which are strongly controlled by soil temperature and moisture conditions. A conceptual model of the physical, chemical, and biological barriers to soil biota-mediated decomposition and mineralization are illustrated in Figure 6, and methods proposed to evaluate them are listed in Table 2.

## 5. Agroecosystem Management to Promote Soil Biota-Mediated Decomposition and Mineralization for the Benefit of Crops

Understanding the physical, chemical, and biological barriers to decomposition and nutrient mineralization allows agricultural managers to implement practices that will facilitate soil biological activity. Three control points (Figure 7) are envisioned in this process: (1) control the substrate quantity and quality by selecting appropriate organic residue(s), (2) control the transformation from substrate to product, based on knowledge of the susceptibility of residue to physical breakdown and enzymatic hydrolysis, and (3) control the

TABLE 2: Methods proposed to assess physical, chemical, and biological barriers to soil biota-mediated decomposition and nutrient mineralization.

| Physical methods | Chemical methods (in whole soil or soil extracts, for example, hot water extractable organic matter) | Biological methods |
|---|---|---|
| Mass loss (litterbags) Light fraction of organic matter Particulate organic matter (whole soil) Particulate organic matter (associated with aggregate size fractions) | *Qualitative/semiquantitative* Near infrared spectroscopy Fourier-transform infrared spectroscopy Nuclear magnetic resonance spectroscopy (solid-state and soil pore water) *Quantitative (in whole soil or soil extracts)* Organic C, N, P, and S and elemental ratios Fiber analysis (acid detergent and neutral detergent fibers) Cellulose, lignin, and so forth by pyrolysis-gas chromatography/mass spectrometry (Py-GC/MS) Proteins, peptides, and so forth by electrospray ionization-liquid chromatography/mass spectrometry (ESI-LC/MS) Ester phosphates by LC-quadrupole time-of-flight (QTOF)/MS Ester sulfates by matrix-assisted laser desorption-ionization time-of-flight/mass spectrometry (MALDI-TOF/MS) | Extracellular enzyme assays (cellobiohydrolase, peroxidase, phenol oxidase, laccase, protease, phosphatase, sulfatase, etc.) Soil respiration ($CO_2$) Net N, P, and S mineralization Intracellular enzyme assays (dehydrogenase, $\beta$-glucosidase) Concentrations of soluble $NH_4^+$, $NO_3^-$, $H_2PO_4^-$, $HPO_4^{2-}$, and $SO_4^{2-}$ ions in soil Concentrations of N, P, and S in plants at critical growth stages |

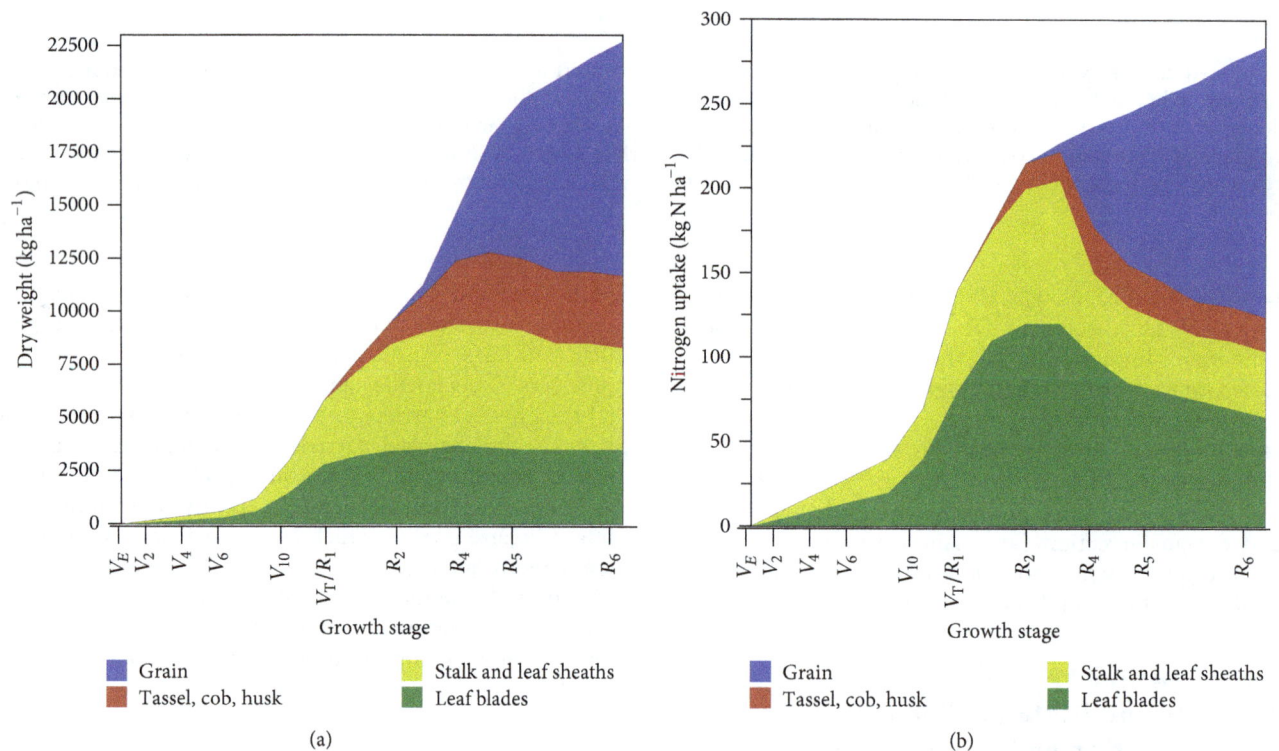

FIGURE 5: Pattern of (a) dry matter accumulation and (b) nitrogen uptake in maize hybrids (modified from [20]). The graphs illustrate the partitioning of dry matter and nitrogen in above-ground components—leaves, stalks, reproductive components, and grain—by growth stages. As described by Abdendroth et al. [21], $V_E$ refers to vegetative emergence and each subsequent vegetative growth stage is designated $V_n$ where $n$ is the number of leaves with visible collar. The $V_T$ stage indicates tasselling, the final vegetative growth stage when the whole tassel is visible. The subsequent reproductive ($R_n$) growth stages are numbered $R_1$ (silking, which coincides with pollen shed) to $R_6$ (physiological maturity).

transfer of product (soluble ions) to the crop. Considering these control points allows agricultural producers to select the right time to apply organic residues, based on expected decomposition and mineralization rates, as well as the crop demand for nutrients during the growing season.

*5.1. Selecting Appropriate Organic Residues.* The quantity of nutrients provided by organic residues is calculated as the total nutrient input of the material, which is appropriate for senescent crop residues and other large, relatively undecomposed materials. When applying partially decomposed

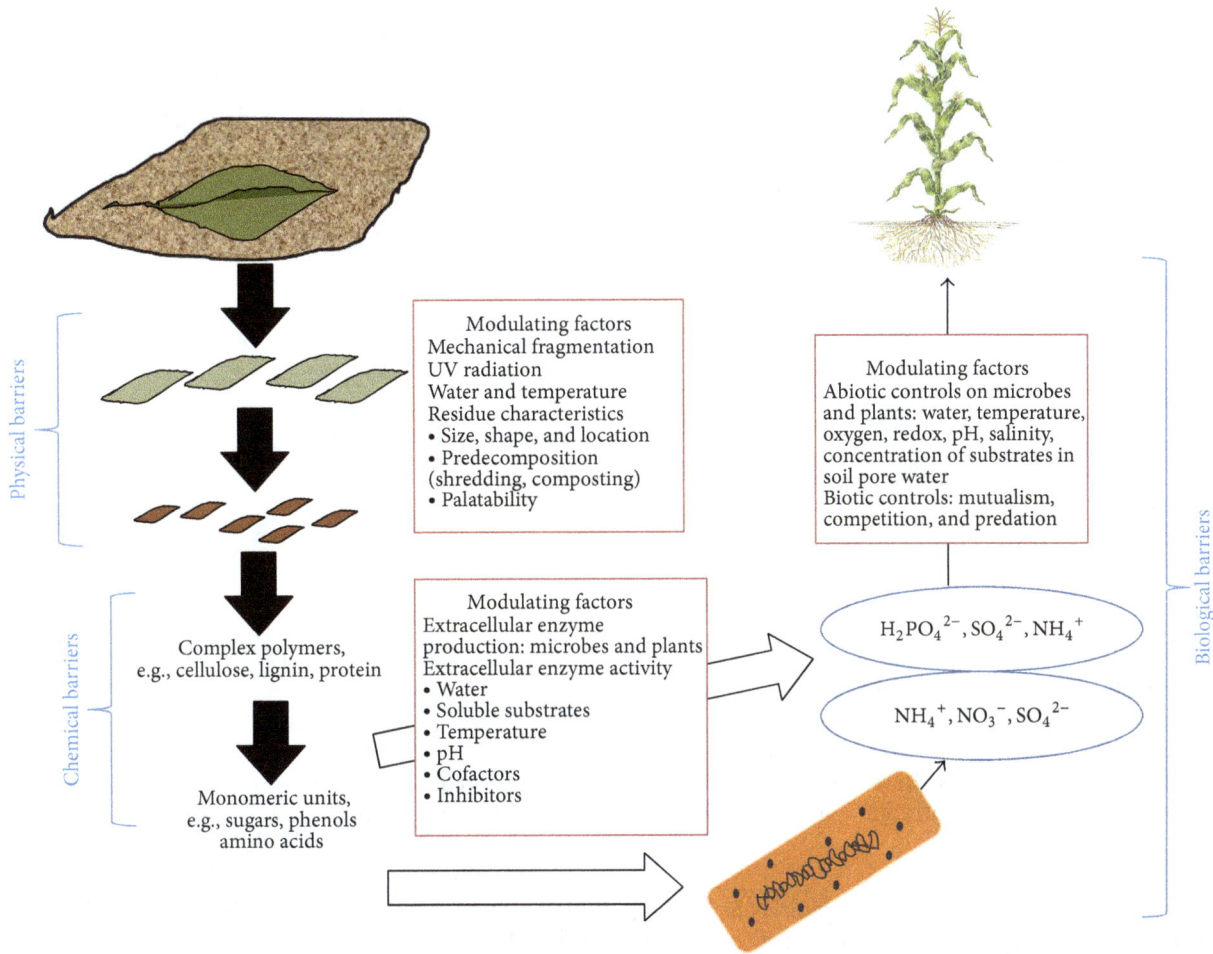

FIGURE 6: Conceptual model of the physical, chemical, and biological barriers to soil biota-mediated decomposition and mineralization. Modulating factors can change the rate at which organic residues pass each barrier and eventually release soluble ions that are taken up by plants.

residues such as compost and animal manure, we must consider the fraction of immediately plant-available nutrients as well. For instance, animal manure contains about 20 to 75% $NH_4^+$, with lower values for solid cattle manure mixed with woodchip bedding and the highest proportion of $NH_4^+$ in liquid pig manure with 3% solids [46]. The rest is organic N, some of which will be decomposed and mineralized during the application year, and the remainder becomes plant-available in a future growing season. In Quebec, Canada, it is estimated that 65 to 90% of the organic N in manure with C : N ratio < 10 is plant-available in the application year, depending on the soil type, crop, application time, and application method. More variability is expected for manure with a C : N ratio > 10, such that 15 to 75% of the organic N is mineralized within one year of application and the undecomposed fraction contributes to a pool of residual organic N that can be mineralized in subsequent growing seasons [46]. There are short windows of opportunity for agricultural producers to apply organic inputs (senescent crop residues, compost, or animal manure), which also affects the decomposition and mineralization process. For instance,

decomposition of surface-applied organic inputs begins after they are plowed down prior to spring planting of annual crops. Although liquid animal manure could be sidedressed between crop rows, either on the soil surface or injected, or broadcast onto forage fields after hay harvesting, there is only a short period in the growing season when this is possible. Producers are aware that heavy agricultural machinery can harm the crop and cause compaction when driven over wet soils and are sensitive to these potential impacts when deciding on when to apply organic residues and how much.

The quality of organic residues is generally described with simple indicators like the C : N ratio and lignin content, often determined by fiber analysis [30]. The chemistry of organic residues is a function of the crop species, the environment in which it is grown, and the agronomic practices followed when it is added to the agroecosystem. For example, if the organic residues are from a green manure crop, the residue quality will be impacted by the crop species in the green manure ($N_2$-fixing crops typically have a lower C : N ratio than nonleguminous crops), the nutrient content of the green manure residues, and the age of the green manure at

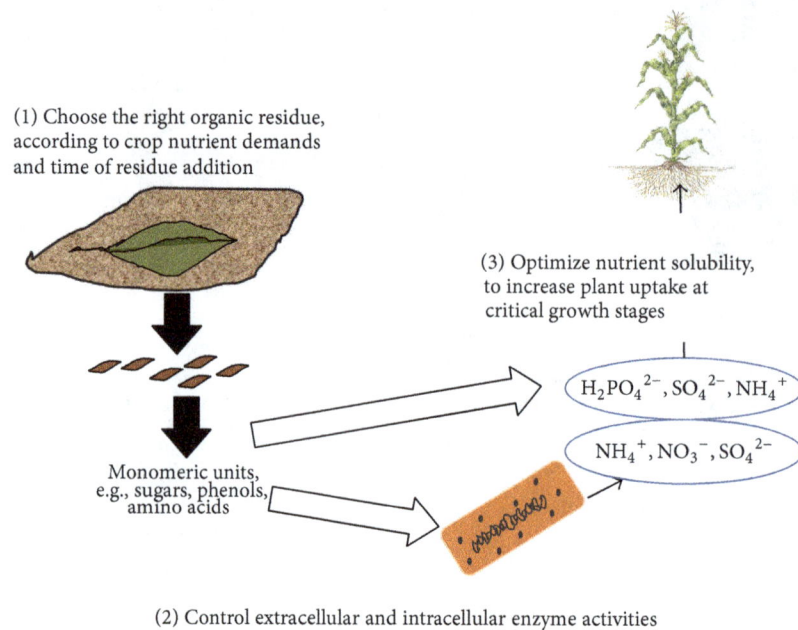

FIGURE 7: Control points where agricultural managers can implement practices to facilitate soil biological activity, leading to nutrient release from organic residues and improving nutrient uptake by crops.

termination, since this controls the residue size and fiber content, including lignin concentration and the $C:N$ ratio [47].

*5.2. Moderating the Organic Residue Decomposition and Mineralization Rates.* As illustrated in Figure 6, soil biota are involved at every step of the decomposition and mineralization process, although their activities can be enhanced or reduced by agronomic practices. For instance, litter fragmentation and comminution are mediated by earthworms and other soil meso- and macrofauna when soil moisture and temperature are appropriate for their activities. Earthworms process appreciable quantities of organic residues ($2$ to $17\,\text{Mg ha}^{-1}\,\text{y}^{-1}$ in temperate agroecosystems; [28]), mostly during the spring and autumn months when soils are sufficiently moist and the temperature is typically less than $20°C$, which is favorable for earthworms. Although tillage operations can effectively perform the same function as earthworms, increasing the tillage intensity is generally damaging to earthworm populations [48] and energetically costly, so there must be a trade-off between relying on soil biota and tillage to physically reduce organic residues.

Exposure to light and ultraviolet (UV) radiation contributes to organic residue decomposition. Photooxidation by UV-B light ($302\,\text{nm}$) increased the solubility of soil organic matter, leading to enhanced oxidation of lignin derived phenols and photochemical degradation of macromolecular soil organic matter species [49]. While biodegradation was the dominant process in litter decomposition, photooxidation and UV radiation altered litter chemistry and destabilized soil organic matter, rendering it more susceptible to decomposition. In semiarid Patagonian steppe ecosystem, Austin and Vivanco [50] concluded that photodegradation had

a dominant control of above-ground litter decomposition. If this phenomenon is important in agroecosystems, exposing senescent organic residues to sunlight for some time, before they are either incorporated by tillage or left on the soil surface in a no-tillage system, would facilitate their decomposition.

Finally, agronomic practices that support a large, diverse microbial community are expected to boost production of extracellular enzymes, which are essential for removing chemical barriers to decomposition. These practices include (1) regular application of organic residues with complex chemistry to support a metabolically diverse microbial community, (2) adding organic residues that are partially decomposed, such as compost and animal manure, to stimulate production of extracellular and intracellular enzymes by soil microorganisms, (3) maintaining vegetative cover to provide energy, in the form of root exudates, to free-living and symbiotic microorganisms that produce extracellular enzymes, and (4) maintaining vegetative cover, to stimulate enzyme release from plant roots [34].

Agricultural practices that modulate soil moisture content could also be important to optimize the activity of hydrolytic extracellular enzymes, as well as those that function intracellularly (e.g., within the cells of ammonia oxidizers and nitrifiers). A target of 60% water-filled pore space would be selected to maximize decomposition by aerobic soil microorganisms [51]. These practices include (1) irrigation, (2) installation of tile drainage lines to remove excess water from fields, particularly in clayey soils that tend to waterlogging, (3) retaining mulch and other residues at the soil surface, to improve water infiltration and slow transpiration, and (4) synchronizing organic residue incorporation with rainfall, based on the amount and intensity of the antecedent and future rainfall events. Global changes in weather (precipitation amount, frequency, temperature, and extreme weather events) make it ever more important to consider how to manage water for agriculture, since all biological activities leading to nutrient mineralization in relation to crop nutrient demands depend on the soil water status.

*5.3. Improving Uptake of Soluble Ions by the Crop.* The final control point involves getting soluble nutrients to the plant, and the management strategy will depend on nutrient mobility in the soil pore water. Mobile $NO_3^-$ and $SO_4^{2-}$ ions are transported to the rhizosphere by mass flow, meaning that an extensive root zone and evapotranspiration will facilitate their uptake. Immobile $NH_4^+$ and $H_2PO_4^-/HPO_4^{2-}$ ions are captured through root interception and diffusion, which requires close contact between the ion and the root surface [17]. A well-developed root system will improve acquisition of nutrients as they are released from organic residues, and this is achieved by avoiding compaction in the crop row. In some cropping systems and soil type, seedbed preparation implies tillage of the entire field, while zone tillage and ridge tillage may be adopted for widely-spaced row crops like corn and soybeans.

For maximum nutrient use efficiency, the root system should be sufficiently large and colonized by mycorrhizae to intercept soluble nutrients [52]. Practically, this implies that agricultural managers should slow the mineralization process during the preplanting period and early vegetative growth stages to avoid nutrient losses and promote mineralization during the exponential vegetative growth and reproductive growth stages, after mycorrhizal colonization has occurred and roots have greater capacity to absorb nutrients from soil pore water. Thus, crop residues that take a long time to decompose should be incorporated weeks to months before planting crops, to avoid early-season N and P deficiencies in the crop that are a result of N and P immobilization by microbial biomass. Early-season S deficiency is less often reported and would also be alleviated by applying S-containing amendments well in advance of periods of appreciable crop S demand. In contrast, well-decomposed residues or those with appreciable $NH_4^+$ content like compost and animal manure should be applied during the growing season by side-dressing or top-dressing. Producers opting for in-season application of organic residues should be aware of a possible increase in crop susceptibility to diseases caused by fungal pathogens. While phytosanitation issues should be minimal if organic residues have undergone thermophilic decomposition (e.g., composting), the other risk is that well-nourished plants with scant root colonization by arbuscular mycorrhizal fungi may be susceptible to pathogens, given that mycorrhiza confer protection against plant diseases through physical and biochemical mechanisms [5, 6].

## 6. Conclusions

The concepts presented in this review are aimed at the development of a robust model of the soil biota-mediated decomposition and mineralization of organic residues applied to agricultural soils. This model will be parameterized and validated with experimental data that describes the physical, chemical, and biological barriers to transforming organically bound nutrients into plant-available N, P, and S. It is necessary to describe how those fundamental reactions are modulated by abiotic conditions, such as soil temperature, moisture/oxygen supply, texture and pH, and biotic factors such as the crop species grown and its nutrient uptake pattern during the growing season. This permits the model to simulate site-specific controls on nutrient transfer from organic residues to plants, allowing agricultural producers to adjust their management practices accordingly. It takes time for soil biota to decompose organic residues and release plant-available nutrients, and there will be periods when the crop cannot obtain sufficient nutrients from soil biological activity. This is when supplemental mineral fertilizer will be most effective in boosting crop production, although the amount applied should match the crop needs during those periods and account for the fact that nutrients will be released from organic residues later during the growing season. Careful management of nutrients from all available sources, as described here, is consistent with the goals of sustainable agriculture.

## Conflict of Interests

The author declares that there is no conflict of interests regarding the publication of this paper.

## Acknowledgments

The author thanks the students in Soil Ecology Research Group for helpful discussions and two anonymous reviewers for their insightful critiques of an earlier version of this paper. This work was supported by Grant no. 2383823-10 from the Natural Sciences and Engineering Research Council of Canada (NSERC).

## References

[1] United Nations, *Population, Development and the Environment 2013*, United Nations, Department of Economic and Social Affairs, Population Divisions, 2013, http://www.unpopulation.org/.

[2] J. W. A. Langeveld, J. Dixon, H. van Keulen, and P. M. F. Quist-Wessel, "Analyzing the effect of biofuel expansion on land use in major producing countries: evidence of increased multiple cropping," *Biofuels, Bioproducts and Biorefining*, vol. 8, no. 1, pp. 49–58, 2014.

[3] J. Rockström, W. Steffen, K. Noone et al., "Planetary boundaries: exploring the safe operating space for humanity," *Ecology and Society*, vol. 14, no. 2, article 32, 2009.

[4] W. de Vries, J. Kros, C. Kroeze, and S. P. Seitzinger, "Assessing planetary and regional nitrogen boundaries related to food security and adverse environmental impacts," *Current Opinion in Environmental Sustainability*, vol. 5, no. 3-4, pp. 392–402, 2013.

[5] G. M. Abdel-Fattah, S. A. El-Haddad, E. E. Hafez, and Y. M. Rashad, "Induction of defense responses in common

bean plants by arbuscular mycorrhizal fungi," *Microbiological Research*, vol. 166, no. 4, pp. 268–281, 2011.

[6] S. C. Jung, A. Martinez-Medina, J. A. Lopez-Raez, and M. J. Pozo, "Mycorrhiza-induced resistance and priming of plant defenses," *Journal of Chemical Ecology*, vol. 38, no. 6, pp. 651–664, 2012.

[7] Y. Y. Song, M. Ye, C. Y. Li et al., "Priming of anti-herbivore defense in tomato by arbuscular mycorrhizal fungus and involvement of the jasmonate pathway," *Journal of Chemical Ecology*, vol. 39, no. 7, pp. 1036–1044, 2013.

[8] R. Puga-Freitas, S. Barot, L. Taconnat, J.-P. Renou, and M. Blouin, "Signal molecules mediate the impact of the earthworm *Aporrectodea caliginosa* on growth, development and defence of the plant *Arabidopsis thaliana*," *PLoS ONE*, vol. 7, no. 12, Article ID e49504, 2012.

[9] M. D. Whiteside, M. O. Garcia, and K. K. Treseder, "Amino acid uptake in arbuscular mycorrhizal plants," *PLoS ONE*, vol. 7, no. 10, Article ID e47643, 2012.

[10] R. D. Bardgett, T. C. Streeter, and R. Bol, "Soil microbes compete effectively with plants for organic-nitrogen inputs to temperate grasslands," *Ecology*, vol. 84, no. 5, pp. 1277–1287, 2003.

[11] K. F. Månsson, M. O. Olsson, U. Falkengren-Grerup, and G. Bengtsson, "Soil moisture variations affect short-term plant-microbial competition for ammonium, glycine, and glutamate," *Ecology and Evolution*, vol. 4, no. 7, pp. 1061–1072, 2014.

[12] J. R. Reeve, J. L. Smith, L. Carpenter-Boggs, and J. P. Reganold, "Soil-based cycling and differential uptake of amino acids by three species of strawberry (*Fragaria* spp.) plants," *Soil Biology and Biochemistry*, vol. 40, no. 10, pp. 2547–2552, 2008.

[13] V. J. Nikiforova, M. Bielecka, B. Gakière et al., "Effect of sulfur availability on the integrity of amino acid biosynthesis in plants," *Amino Acids*, vol. 30, no. 2, pp. 173–183, 2006.

[14] A. E. Richardson, P. A. Hadobas, and J. E. Hayes, "Acid phosphomonoesterase and phytase activities of wheat (*Triticum aestivum* L.) roots and utilization of organic phosphorus substrates by seedlings grown in sterile culture," *Plant, Cell & Environment*, vol. 23, no. 4, pp. 397–405, 2000.

[15] C. Paungfoo-Lonhienne, J. Visser, T. G. A. Lonhienne, and S. Schmidt, "Past, present and future of organic nutrients," *Plant and Soil*, vol. 359, no. 1-2, pp. 1–18, 2012.

[16] J. K. Whalen, M. L. Kernecker, B. W. Thomas, V. Sachdeva, and C. Ngosong, "Soil food web controls on nitrogen mineralization are influenced by agricultural practices in humid temperate climates," *CAB Reviews: Perspectives in Agriculture, Veterinary Science, Nutrition and Natural Resources*, vol. 8, no. 23, pp. 1–18, 2013.

[17] J. K. Whalen and L. Sampedro, *Soil Ecology and Management*, CABI Publishers, Wallingford, UK, 2010.

[18] M. G. Klotz, D. J. Arp, P. S. G. Chain et al., "Complete genome sequence of the marine, chemolithoautotrophic, ammonia-oxidizing bacterium *Nitrosococcus oceani* ATCC 19707," *Applied and Environmental Microbiology*, vol. 72, no. 9, pp. 6299–6315, 2006.

[19] S. R. Starkenburg, P. S. G. Chain, L. A. Sayavedra-Soto et al., "The genome sequence of the chemolithoautotrophic nitrite-oxidizing bacterium Nitrobacter winogradskyi Nb-255," *Applied and Environmental Microbiology*, vol. 72, no. 3, pp. 2050–2063, 2006.

[20] R. R. Bender, J. W. Haegele, M. L. Ruffo, and F. E. Below, "Nutrient uptake, partitioning, and remobilization in modern, transgenic insect-protected maize hybrids," *Agronomy Journal*, vol. 105, no. 1, pp. 161–170, 2010.

[21] L. J. Abdendroth, R. W. Elmore, M. J. Boyer, and S. K. Marlay, "Corn growth and development," PMR 1009, Iowa State University Extension, Ames, Iowa, USA, 2011.

[22] S. F. Yanni, J. K. Whalen, and B. L. Ma, "Crop residue chemistry, decomposition rates, and $CO_2$ evolution in Bt and non-Bt corn agroecosystems in North America: a review," *Nutrient Cycling in Agroecosystems*, vol. 87, no. 2, pp. 277–293, 2010.

[23] F. C. Hoyle and D. V. Murphy, "Influence of organic residues and soil incorporation on temporal measures of microbial biomass and plant available nitrogen," *Plant and Soil*, vol. 347, no. 1-2, pp. 53–64, 2011.

[24] A. E. Miller, J. P. Schimel, T. Meixner, J. O. Sickman, and J. M. Melack, "Episodic rewetting enhances carbon and nitrogen release from chaparral soils," *Soil Biology and Biochemistry*, vol. 37, no. 12, pp. 2195–2204, 2005.

[25] A. Lamparter, J. Bachmann, M.-O. Goebel, and S. K. Woche, "Carbon mineralization in soil: impact of wetting-drying, aggregation and water repellency," *Geoderma*, vol. 150, no. 3-4, pp. 324–333, 2009.

[26] V. Poirier, D. A. Angers, P. Rochette, and J. K. Whalen, "Initial soil organic carbon concentration influences the short-term retention of crop-residue carbon in the fine fraction of a heavy clay soil," *Biology and Fertility of Soils*, vol. 49, no. 5, pp. 527–535, 2013.

[27] O. P. Schulmann and A. V. Tiunov, "Leaf litter fragmentation by the earthworm *Lumbricus terrestris* L.," *Pedobiologia*, vol. 43, no. 5, pp. 453–458, 1999.

[28] J. K. Whalen and R. W. Parmelee, "Quantification of nitrogen assimilation efficiencies and their use to estimate organic matter consumption by the earthworms *Aporrectodea tuberculata* (Eisen) and *Lumbricus terrestris* L.," *Applied Soil Ecology*, vol. 13, no. 3, pp. 199–208, 1999.

[29] M. H. Beare, R. W. Parmelee, P. F. Hendrix, W. Cheng, D. C. Coleman, and D. A. Crossley Jr., "Microbial and faunal interactions and effects on litter nitrogen and decomposition in agroecosystems," *Ecological Monographs*, vol. 62, no. 4, pp. 569–591, 1992.

[30] S. Gul and J. Whalen, "Plant life history and residue chemistry influences emissions of $CO_2$ and $N_2O$ from soil—perspectives for genetically modified cell wall mutants," *Critical Reviews in Plant Sciences*, vol. 32, no. 5, pp. 344–368, 2013.

[31] S. Gul, S. F. Yanni, and J. K. Whalen, "Lignin controls on soil ecosystem services: implications for biotechnological advances in biofuel crops," in *Lignin: Structural Analysis, Applications in Biomaterials and Ecological Significance*, F. Lu, Ed., pp. 375–416, Nova Science, Hauppauge, NY, USA, 2014.

[32] S. Manzoni, S. M. Schaeffer, G. Katul, A. Porporato, and J. P. Schimel, "A theoretical analysis of microbial eco-physiological and diffusion limitations to carbon cycling in drying soils," *Soil Biology and Biochemistry*, vol. 73, no. 6, pp. 69–83, 2014.

[33] J. M. Steinweg, J. S. Dukes, and M. D. Wallenstein, "Modeling the effects of temperature and moisture on soil enzyme activity: linking laboratory assays to continuous field data," *Soil Biology & Biochemistry*, vol. 55, no. 12, pp. 85–92, 2012.

[34] R. G. Burns, J. L. DeForest, J. Marxsen et al., "Soil enzymes in a changing environment: current knowledge and future directions," *Soil Biology and Biochemistry*, vol. 58, pp. 216–234, 2013.

[35] R. B. Lee, "Phosphate influx and extracellular phosphatase-activity in barley roots and rose cells," *New Phytologist*, vol. 109, no. 2, pp. 141–148, 1988.

[36] D. J. Arp, "Nitrification," in *eLS*, John Wiley & Sons, Chichester, UK, 2009.

[37] S. Otte, J. Schalk, J. G. Kuenen, and M. S. M. Jetten, "Hydroxylamine oxidation and subsequent nitrous oxide production by the heterotrophic ammonia oxidizer *Alcaligenes faecalis*," *Applied Microbiology and Biotechnology*, vol. 51, no. 2, pp. 255–261, 1999.

[38] A. Ali, S. Sivakami, and N. Raghuram, "Effect of nitrate, nitrite, ammonium, glutamate, glutamine and 2-oxoglutarate on the RNA levels and enzyme activities of nitrate reductase and nitrite reductase in rice," *Physiology and Molecular Biology of Plants*, vol. 13, no. 1, pp. 17–25, 2007.

[39] R. E. McMurtrie, C. M. Iversen, R. C. Dewar et al., "Plant root distributions and nitrogen uptake predicted by a hypothesis of optimal root foraging," *Ecology and Evolution*, vol. 2, no. 6, pp. 1235–1250, 2012.

[40] S. T. Dara, P. E. Fixen, and R. H. Gelderman, "Sufficiency level and diagnosis and recommendation integrated system approaches for evaluating the nitrogen status of corn," *Agronomy Journal*, vol. 84, no. 6, pp. 1006–1010, 1992.

[41] L. E. Parent, A. N. Cambouris, and A. Muhawenimana, "Multivariate diagnosis of nutrient imbalance in potato crops," *Soil Science Society of America Journal*, vol. 58, no. 5, pp. 1432–1438, 1994.

[42] J. M. Blair, "Nitrogen, sulfur and phosphorus dynamics in decomposing deciduous leaf litter in the southern appalachians," *Soil Biology and Biochemistry*, vol. 20, no. 5, pp. 693–701, 1988.

[43] J. K. Whalen, C. Chang, and B. M. Olson, "Nitrogen and phosphorus mineralization potentials of soils receiving repeated annual cattle manure applications," *Biology and Fertility of Soils*, vol. 34, no. 5, pp. 334–341, 2001.

[44] M. St. Luce, J. K. Whalen, N. Ziadi, and B. J. Zebarth, "Nitrogen dynamics and indices to predict soil nitrogen supply in humid temperate soils," *Advances in Agronomy*, vol. 112, pp. 55–102, 2011.

[45] B. A. Caldwell, "Enzyme activities as a component of soil biodiversity: A review," *Pedobiologia*, vol. 49, no. 6, pp. 637–644, 2005.

[46] Centre de Référence en Agriculture et Agroalimentaire du Québec (CRAAQ), *Guide de référence en fertilisation*, Centre de Référence en Agriculture et Agroalimentaire du Québec (CRAAQ), Québec, Canada, 2nd edition, 2010.

[47] R. Dinesh and R. P. Dubey, "Nitrogen mineralization rates and kinetics in soils freshly amended with green manures," *Journal of Agronomy and Crop Science*, vol. 181, no. 1, pp. 49–53, 1998.

[48] K. Y. Chan, "An overview of some tillage impacts on earthworm population abundance and diversity—implications for functioning in soils," *Soil and Tillage Research*, vol. 57, no. 4, pp. 179–191, 2000.

[49] X. Feng, K. M. Hills, A. J. Simpson, J. K. Whalen, and M. J. Simpson, "The role of biodegradation and photo-oxidation in the transformation of terrigenous organic matter," *Organic Geochemistry*, vol. 42, no. 3, pp. 262–274, 2011.

[50] A. T. Austin and L. Vivanco, "Plant litter decomposition in a semi-arid ecosystem controlled by photodegradation," *Nature*, vol. 442, no. 7102, pp. 555–558, 2006.

[51] D. M. Linn and J. W. Doran, "Effect of water-filled pore space on carbon dioxide and nitrous oxide production in tilled and non-tilled soils," *Soil Science Society of America Journal*, vol. 48, no. 6, pp. 1267–1272, 1984.

[52] A. Elbon and J. K. Whalen, "Phosphorus supply to vegetable crops from arbuscular mycorrhizal fungi: a review," *Biological Agriculture & Horticulture*, 2014.

# 10

# Reducing Runoff Loss of Applied Nutrients in Oil Palm Cultivation Using Controlled-Release Fertilizers

A. Bah,[1] M. H. A. Husni,[1] C. B. S. Teh,[1] M. Y. Rafii,[2] S. R. Syed Omar,[3] and O. H. Ahmed[4]

[1]Department of Land Management, Faculty of Agriculture, Universiti Putra Malaysia, 43400 Serdang, Selangor, Malaysia
[2]Institute of Tropical Agriculture, Universiti Putra Malaysia, 43400 Serdang, Selangor, Malaysia
[3]Diversatech (M) Fertilizer Sdn. Bhd., Bandar Baru Bangi, 43650 Selangor, Malaysia
[4]Department of Crop Science, Faculty of Agriculture and Food Sciences, Universiti Putra Malaysia, Bintulu Campus Sarawak, 97008 Bintulu, Sarawak, Malaysia

Correspondence should be addressed to M. H. A. Husni; husni@upm.edu.my

Academic Editor: Heike Bucking

Controlled-release fertilizers are expected to minimize nutrient loss from crop fields due to their potential to supply plant-available nutrients in synchrony with crop requirements. The evaluation of the efficiency of these fertilizers in tropical oil palm agroecological conditions is not yet fully explored. In this study, a one-year field trial was conducted to determine the impact of fertilization with water soluble conventional mixture and controlled-release fertilizers on runoff loss of nutrients from an immature oil palm field. Soil and nutrient loss were monitored for one year in 2012/2013 under erosion plots of 16 m$^2$ on 10% slope gradient. Mean sediments concentration in runoff amounted to about 6.41 t ha$^{-1}$. Conventional mixture fertilizer posed the greatest risk of nutrient loss in runoff following fertilization due to elevated nitrogen (6.97%), potassium (13.37%), and magnesium (14.76%) as percentage of applied nutrients. In contrast, this risk decreased with the application of controlled-release fertilizers, representing 0.75–2.44% N, 3.55–5.09% K, and 4.35–5.43% Mg loss. Meanwhile, nutrient loss via eroded sediments was minimal compared with loss through runoff. This research demonstrates that the addition of controlled-release fertilizers reduced the runoff risks of nutrient loss possibly due to their slow-release properties.

## 1. Introduction

Oil palm is mainly cultivated on highly weathered soils which belong to the orders Ultisols and Oxisols. These soils are predominantly acidic and low in fertility [1]. Fertilizers are crucial in oil palm production, accounting for 50–70% of field operational costs and about 25% of the total cost of production [2, 3]. Mineral fertilizers, mainly conventional forms, account for more than 90% of fertilizers used by all types of farming systems in Malaysia [4]. The oil palm is a heavy feeder and requires quite large quantities of fertilizers to produce good yield [5]. Fertilizer management on undulating, hilly soils used for oil palm cultivation is very important because of the need to maintain fertility of the soil and to as well minimize soil erosion and nutrient loss. Frequent application of large amounts of chemical fertilizers

coupled with high rainfall intensity tends to increase the risk of nutrient loss. The loss of nutrients through leaching and runoff reduces both crop productivity and economic gains. Furthermore, excess nutrient loading to ground or underground water bodies can impair designated uses of water [6, 7]. There is a need to develop alternatives from the fertilizer industry to make mineral fertilization for high value crops such as oil palm more economically viable and ecologically compatible.

Approaches to improve crop nutrient use efficiency have been proposed [8, 9]. Among the management practices to both improve fertilizer efficiency and reduce environmental pollution, the use of slow-release fertilizers (SRFs) and controlled-release fertilizers (CRFs) seems to be promising for widespread use in agriculture [10–13]. As compared to conventional fertilizers, the gradual release of nutrients from

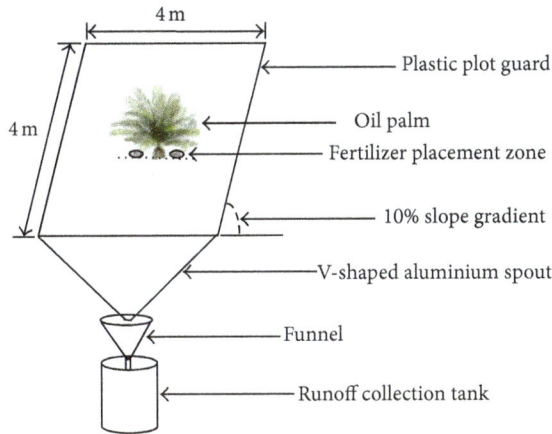

FIGURE 1: Experimental plot setup.

TABLE 1: Fertilizer application treatments.

| Fertilizer treatments | Nutrient rate (kg ha$^{-1}$ yr$^{-1}$) | | | |
|---|---|---|---|---|
| | N | P | K | Mg |
| Control | 0.00 | 0.00 | 0.00 | 0.00 |
| Mixture | 48.96 | 13.07 | 81.60 | 5.86 |
| CRFB-60% | 32.64 | 8.71 | 54.40 | 3.91 |
| CRFG-60% | 32.64 | 8.71 | 54.40 | 3.91 |
| CRFB-100% | 48.96 | 13.07 | 81.60 | 5.86 |
| CRFG-100% | 48.96 | 13.07 | 81.60 | 5.86 |

CRFs could be synchronized with plant needs and minimize nutrient loss through runoff and leaching to ultimately improve fertilizer use efficiency. According to the International Fertilizer Industry Association [14], controlled-release nitrogen fertilizers have agronomic advantages, especially in the tropics and in regions with light-textured soils and under heavy rainfall or irrigation, where N losses are particularly high.

Presently, Malaysia's oil palm industry is faced with a growing challenge of labor shortage partly due to frequent application of straight fertilizers in 4–6 splits per annum. Therefore, one of the possible approaches to address this challenge is through the adoption of improved fertilizer technologies such as the use of CRFs where fertilizer application rounds can be reduced to two splits per annum. The evaluation of efficiency of CRF is necessary to help oil palm plantations make informed decisions. The objective of this study was to evaluate whether application of controlled-release fertilizers (AJIB CRF) instead of water soluble conventional mixture fertilizer can reduce the risk of nutrient loss via runoff and erosion.

## 2. Materials and Methods

*2.1. Experimental Site Details.* The study was carried out in 2012/2013 on a newly established oil palm farm at the experimental station of Universiti Putra Malaysia Agriculture Park in Puchong, Selangor (02°N 59.035′, 101°E 38.913′). Initially, the field was a fallow grassland. The area has a humid tropical climate with a mean annual rainfall of 2700 mm and temperature of 25.3°C [15]. Experimental plots measuring 4 m × 4 m and delineated along uniform land slope of 10% were demarcated with transparent plastic sheets inserted 5 cm deep into the soil and 15 cm above soil surface to prevent lateral flow and control the risk of plot contamination (Figure 1). Twelve-month-old oil palm clones (AA Hybrida IS) obtained from Applied Agricultural Resources Company were planted using the 9 m × 9 m × 9 m triangular system. Daily precipitation records were taken throughout the experimental period using rain gauge. Runoff was routed via

a V-shaped stainless steel aluminum spout attached to a funnel-fitted tank. The experiment was arranged in three blocks (replicates) with six treatments established perpendicular to the slope.

*2.2. Experimental Soil.* Soil particle size distribution measurement was based on the pipette method [16], while the bulk density of the soil was determined using core rings as outlined by Teh and Talib [17]. Soil pH, EC, and CEC were determined following standard protocols [18–20]. Total soil carbon and nitrogen were analysed using TruMac CNS Analyzer (LECO, St. Joseph), which utilises the dry combustion principle. Soil exchangeable K, Ca, Mg, and CEC using 1 M NH$_4$OAc (pH 7.0) [21, 22] and P using Bray II method [23] were determined.

*2.3. Fertilizer Treatments.* The treatments evaluated in this study were as follows:

    (i) control (no fertilizer),

    (ii) mixture (water soluble mixed fertilizer),

    (iii) CRFB-60% (60% dosage of briquette CRF),

    (iv) CRFG-60% (60% dosage of granular CRF),

    (v) CRFB-100% (full dosage of briquette CRF),

    (vi) CRFG-100% (full dosage of granular CRF).

T2 consists of a mixture of water soluble straight fertilizers, namely, ammonium sulfate (21% N), Christmas Island Rock Phosphate (33% P$_2$O$_5$), Muriate of Potash (60% K$_2$O), and Kieserite (27% MgO). The grade of the coated CRFs (AJIB CRF) is 10-6-20-2. Briquette CRF is spherical in shape with a size of about 5 cm in length, while granular CRF is pelletized and is approximately 1 cm in length. The full dosage of fertilizer application was based on standard recommended rates for immature oil palm. The treatments were arranged in a randomized complete block design (RCBD) with three replications (Table 1). The fertilizers were surface-placed under the plant canopy in two splits per year at six months' intervals.

*2.4. Runoff Collection and Analysis.* Following each rainfall event, surface runoff water plus eroded sediments per plot was collected for one year (October 2012 to September 2013) using tanks. For each rainfall event, the total water volume per each tank was recorded and 1 L subsamples were collected

TABLE 2: Soil textural classification of study site (0–20 cm) and sediments distribution in runoff.

| Sample | Textural distribution (%) | | | Soil texture class (USDA) |
| --- | --- | --- | --- | --- |
| | Sand | Silt | Clay | |
| Soil | 63 | 6 | 31 | Sandy clay loam |
| Sediments | 32 | 40 | 28 | Clay loam |

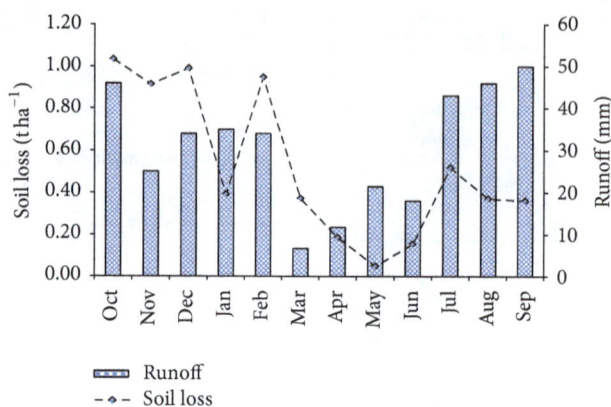

FIGURE 2: Monthly soil loss and runoff during study period (2012/2013).

in detergent-washed polypropylene bottles, after which they were transported to the laboratory, filtered, and analyzed for N, P, K, and Mg content. All tanks were emptied and thoroughly cleaned with distilled water after each sample collection event. The sediments retained after filtration of runoff samples with Whatman number 5 were oven-dried at 60°C for 24 h and weighed with an electronic balance. Sediment loss from each plot was calculated by multiplying the volume of runoff with the nutrients concentration in the analyte. Soil loss was converted into sediment yield in t ha$^{-1}$. Dried sediments and runoff filtrates were separately analyzed for their N, P, K, and Mg contents following standard procedures as stated above.

Analysis of variance was used to detect treatment effect, whereas treatment means were compared using Duncan's new multiple range test (DNMRT) at $P \leq 0.05$ with statistical analysis system (SAS) software (Version 9.3).

## 3. Results and Discussion

*3.1. Rainfall, Runoff, and Soil Loss.* The texture of the soil in the study area is classified as sandy clay loam (Typic Kandiudult) having a deep kandic horizon with clay content between 18 and 35% [24]. Based on texture and its friable nature, this soil is prone to erosion but may also be highly permeable to enhance water infiltration. Although soil consisted of 63% sand, clay 31%, and 6% silt, there was substantial loss of silt (40%) in the eroded sediments compared to sand (32%) and clay (28%) as shown in Table 2. The higher loss of silt particles in the eroded sediments relates to the lightness of this particle.

The annual recorded precipitation was 2635 mm. Highest observed rainfall occurred in the months of October (316 mm) and November (302 mm) in 2012. A short dry spell was observed in February, June, and August, 2013, as a result of seasonal changes in rainfall distribution patterns in the area. Figure 2 shows monthly generated runoff and soil loss due at the experimental site. A total of 371 mm of runoff volume was recorded during the study period. Also, runoff volume as percentage of rainfall ranged from 5 to 20% with a mean value of 14%. This result relates to the relatively high infiltration (5.40 mm hr$^{-1}$) of the soil. The annual sediment loss was 6.41 t ha$^{-1}$ and maximum loss (1.05 t ha$^{-1}$) occurred in the month of November 2012 (Figure 2).

Soil erosion studies in oil palm fields in Malaysia have shown a wide range of soil loss. According to reports by PORIM [25] and Hartemink [26] soil loss ranges from 13 to 78 Mg ha$^{-1}$ yr$^{-1}$ for Oxisols and 1 to 28 Mg ha$^{-1}$ yr$^{-1}$ for Ultisols. The results of the present study are comparable to

that of Corley and Tinker [27] who reported soil loss of less than 9 t ha$^{-1}$. Similar findings have also been previously reported [28, 29], where soil erosion losses from oil palm fields were in the range of 7.47–14.92 t ha$^{-1}$. Loss of nutrients through runoff and sediments relates to soil texture, age of palms, soil topography and infiltrability, land clearing, inadequate soil conservation measures, and the lag time between fertilizer application and rainfall [30, 31].

*3.2. Nutrient Loss in Runoff and Eroded Sediments.* As shown in Table 3, net annual losses of nutrients in runoff varied significantly between soluble mixture and CRF treatments, but the differences were not significant between briquette and granular CRFs. Comparison of the overall nutrients concentration of the different treatments indicated that runoff losses of nutrients were highest in mixture fertilizer compared with the controlled-release fertilizers (AJIB CRF) and the control. The loss of nutrients from applied mixture treatments was rapid at the initial stage following first fertilization in October, after which it declined rapidly and picked up again in June following second fertilization (Figure 3). Cumulatively, the annual loss of N through runoff in the plots treated with mixture fertilizers accounted was 4.78 kg ha$^{-1}$ whereas the N loss from applied CRF fertilizers ranged between 1.73 and 2.56 kg ha$^{-1}$. The net losses of K and Mg were greatest for mixture fertilizer treatment, accounting for 13% and 15%, respectively, of the added fertilizer. The average annual loss of P in the runoff due to fertilization was relatively low with full dosage of granular CRF accounting for the highest loss (4%). The losses of nutrients from the reduced dosages of CRFs (CRFB-60% and CRFG-60%) were consistently lower than their full dosage counterparts, although the latter showed better plant response in terms of N and Mg uptake. By comparing the nutrients lost through surface runoff with those lost in the eroded sediments (Table 3), it is clear that runoff losses of nutrients in all treatments were substantially greater than those lost in displaced sediments.

Nutrient loss data in this current study are consistent with other studies conducted in Malaysia. For example, studies conducted by Maene et al. [28] showed that 11% N, 3% P, 5%

TABLE 3: Summary of net nutrient loss in oil palm ecosystem through surface runoff and eroded sediments.

| Fertilizer treatments | Net loss in runoff $(\text{kg ha}^{-1}\,\text{yr}^{-1})$ | Net loss in eroded sediments $(\text{kg ha}^{-1}\,\text{yr}^{-1})$ | Net total loss $(\text{kg ha}^{-1}\,\text{yr}^{-1})$ | Net loss as % of nutrient applied |
|---|---|---|---|---|
| | | Nitrogen (N) | | |
| Control | $1.20^c \pm 0.04$ | $0.17^c \pm 0.01$ | $1.37^c$ | — |
| Mixture | $3.85^a \pm 0.22$ | $0.93^a \pm 0.27$ | $4.78^a$ | 6.97 |
| CRFB-60% | $1.33^{bc} \pm 0.02$ | $0.40^{ab} \pm 0.24$ | $1.73^{bc}$ | 0.75 |
| CRFG-60% | $1.39^{bc} \pm 0.24$ | $0.39^{ab} \pm 0.20$ | $1.78^{bc}$ | 0.84 |
| CRFB-100% | $1.81^b \pm 0.26$ | $0.41^{ab} \pm 0.12$ | $2.22^b$ | 1.74 |
| CRFG-100% | $1.98^b \pm 0.30$ | $0.58^{ab} \pm 0.16$ | $2.56^b$ | 2.44 |
| | | Phosphorus (P) | | |
| Control | $0.40^{cb} \pm 0.16$ | $0.12^b \pm 0.01$ | $0.52^b$ | — |
| Mixture | $0.70^{ab} \pm 0.09$ | $0.30^a \pm 0.07$ | $1.00^a$ | 3.74 |
| CRFB-60% | $0.45^b \pm 0.14$ | $0.24^{ab} \pm 0.05$ | $0.69^{ab}$ | 1.31 |
| CRFG-60% | $0.49^b \pm 0.01$ | $0.29^a \pm 0.03$ | $0.78^{ab}$ | 2.02 |
| CRFB-100% | $0.64^{ab} \pm 0.03$ | $0.31^a \pm 0.02$ | $0.95^a$ | 3.28 |
| CRFG-100% | $0.92^a \pm 0.19$ | $0.15^b \pm 0.06$ | $1.07^a$ | 4.22 |
| | | Potassium (K) | | |
| Control | $3.70^c \pm 0.17$ | $0.45^c \pm 0.11$ | $3.70^c$ | — |
| Mixture | $12.48^a \pm 0.87$ | $2.14^a \pm 0.37$ | $14.62^a$ | 13.37 |
| CRFB-60% | $5.56^b \pm 0.13$ | $1.08^b \pm 0.21$ | $6.64^b$ | 3.60 |
| CRFG-60% | $5.35^b \pm 0.17$ | $1.25^b \pm 0.37$ | $6.60^b$ | 3.55 |
| CRFB-100% | $5.96^b \pm 0.16$ | $1.32^b \pm 0.17$ | $7.28^b$ | 4.38 |
| CRFG-100% | $6.83^b \pm 0.35$ | $1.03^b \pm 0.23$ | $7.86^b$ | 5.09 |
| | | Magnesium (Mg) | | |
| Control | $0.51^b \pm 0.04$ | $0.11^c \pm 0.02$ | $0.62^c$ | — |
| Mixture | $1.20^a \pm 0.14$ | $0.28^a \pm 0.05$ | $1.48^a$ | 14.76 |
| CRFB-60% | $0.70^{ab} \pm 0.08$ | $0.19^{ab} \pm 0.04$ | $0.89^b$ | 4.72 |
| CRFG-60% | $0.58^b \pm 0.03$ | $0.29^a \pm 0.10$ | $0.87^b$ | 4.35 |
| CRFB-100% | $0.76^{ab} \pm 0.07$ | $0.17^{ab} \pm 0.02$ | $0.93^b$ | 5.43 |
| CRFG-100% | $0.68^{ab} \pm 0.02$ | $0.21^{ab} \pm 0.09$ | $0.89^b$ | 4.65 |

All analyses are mean ± standard error of mean (SEM). Means not sharing a common letter are significantly different by DNMRT ($P \leq 0.05$).

K, 6% Mg, and 5% Ca of applied fertilizers were lost through surface runoff alone during a low rainfall (1426 mm) on a 9% slope. The findings further indicated that the harvesting paths are the most susceptible areas to surface runoff due to compaction of the soil by machinery. Another study [32] revealed that 5–8% N, 10–15% K, 4–6% Mg, and <2% P were lost through runoff. This suggests that soluble nutrients such as N, K, and Mg are more susceptible to runoff losses. According to Wallace et al. [33], large losses of nutrients via surface runoff are still possible when a large rainfall event occurs soon after application of a fertilizer amendment.

The macronutrient release pattern varied substantially between mixture and CRF fertilizers (Figure 3). The nutrients released from mixture fertilizer (except for P) were higher and varied more after fertilization. Coated fertilizers (CRFs) gradually released nutrients compared to that of the mixture fertilizer. Mixture fertilizer showed lower P release in runoff due to low solubility of its P content from the phosphate rock source in contrast with CRF treatments, which are formulated with diammonium phosphate (DAP) as P source.

An examination of the nutrient release pattern shows that, immediately after fertilizer application in October 2012 and April 2013, the mean soluble N, K, and Mg concentrations in runoff were greatest for mixture fertilizer treatment and then followed by CRF treatments (Figure 3). Mean loss of nutrients in runoff from fertilized plots decreased significantly from the months of January to May, such that losses were similar to those observed for the unfertilized treatment. This trend is more obvious for nutrients such as N, K, and Mg than for P.

*3.3. Effect of Rainfall Intensity.* The effects of rainfall intensity on nutrient loss were explored for the months of November (2012) and April (2013) and the findings suggest that 76% of the rainfall occurred at intensities below 10 mm hr$^{-1}$ and 13% in excess of 20 mm hr$^{-1}$ (Figure 4). Research findings suggest that rainfall intensity is one of the most important factors that contribute to soil loss and runoff [34, 35]. Soil erosion due to rainfall involves series of complex processes commencing from detachment by raindrop impact, transport of entrained soil particles by rain splash or surface flow, and

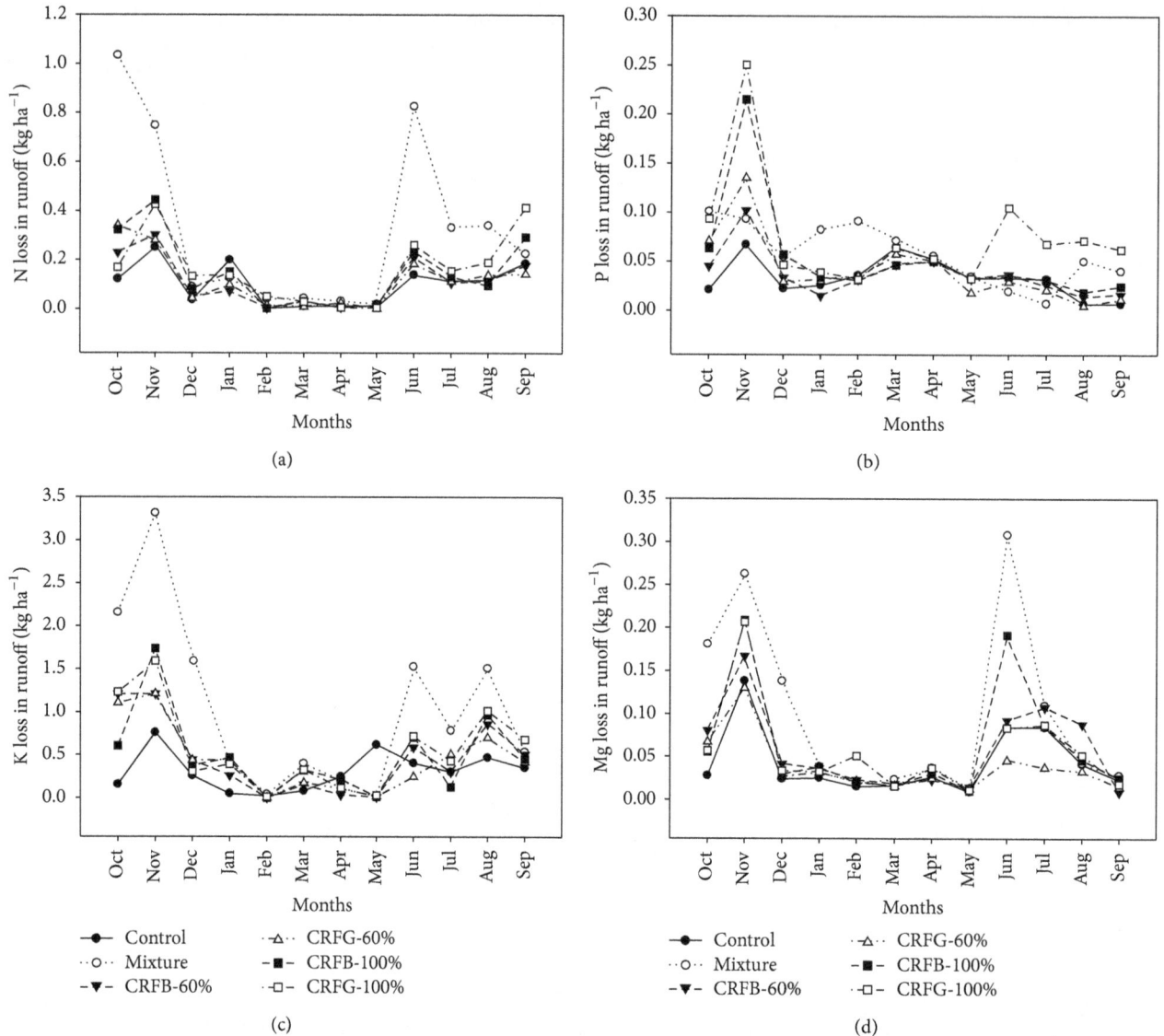

(a)

(b)

(c)

(d)

FIGURE 3: Monthly runoff loss patterns of nutrients as impacted by fertilizer application.

then deposition. There is strong relationship between rainfall, runoff, and soil erosion by water. The occurrence of runoff, whether small or huge, always results in decline of soil fertility due to loss of topsoil and nutrients, loss of organic matter, and the consequent loss of the soil's capacity to retain nutrients and water.

Nutrient runoff potential during storm events is influenced by numerous factors, including amount and intensity of rainfall, antecedent rainfall conditions, timing and rate of fertilization, ponding, and irrigation management practices [36]. Rainfall intensity is considered a key determinant factor that influences the loss of soil and nutrients. As shown in Figure 4, the loss of soil through runoff significantly increased with increase in rainfall intensity. A recent study on the interactions between rainfall intensities and soil surface interrill erosion [37] showed that the soil erosion rates spiked following sharp increase in rainfall intensity. Attempts were

conducted to evaluate P and N in surface runoff in relation to rainfall intensity and hydrology for two soils along a single hill slope and the results revealed that nutrient loss was significantly greater under the high intensity rainfall due to larger runoff volumes [38].

Linear relationships were observed between the rainfall intensity and nutrient (N) loss (Figure 5). The coefficients of determination ($R^2$) of the linear regressions were high ($P < 0.01$) for all treatments, ranging from 0.778 to 0.939, indicating the influence of rainfall intensity on nutrient loss. A study on nutrient loss of a limed soil in a reservoir area found a positive regressive relationship between nutrient loss and rainfall intensity [39]. Previously, it has also been found that the loss of P has a strongly positive relationship with rainfall intensity [40]. A study on the effect of rainfall intensity and soil surface cover on losses of sediment and soil organic carbon (SOC) via surface runoff found that higher

TABLE 4: Frond number of immature oil palms at different months after fertilization (MAF).

| Fertilizer treatments | Frond number plant$^{-1}$ | | | | |
|---|---|---|---|---|---|
| | 0 MAF | 3 MAF | 6 MAF | 9 MAF | 12 MAF |
| Control | 13$^a$± 0.12 | 17$^c$± 0.50 | 20$^b$± 0.49 | 23$^c$± 0.99 | 26$^c$± 1.53 |
| Mixture | 13$^a$± 0.29 | 19$^{ab}$± 1.15 | 24$^{ab}$± 1.00 | 28$^b$± 0.55 | 32$^b$± 0.75 |
| CRFB-60% | 12$^a$± 0.27 | 19$^{ab}$± 0.27 | 26$^a$± 0.82 | 29$^{ab}$± 0.56 | 33$^{ab}$± 0.67 |
| CRFG-60% | 12$^a$± 0.27 | 18$^{ab}$± 0.81 | 27$^a$± 0.46 | 31$^a$± 0.22 | 35$^a$± 1.20 |
| CRFB-100% | 12$^a$± 0.12 | 19$^{ab}$± 0.87 | 25$^{ab}$± 1.35 | 31$^a$± 1.05 | 34$^{ab}$± 1.33 |
| CRFG-100% | 13$^a$± 0.24 | 20$^a$± 0.43 | 26$^a$± 1.98 | 32$^a$± 0.66 | 36$^a$± 0.48 |

All analyses are mean ± standard error of mean (SEM). Means not sharing a common letter are significantly different by DNMRT ($P \leq 0.05$).

TABLE 5: Effects of fertilizer treatments on leaf and rachis nutrients concentration of oil palm at 12 MAF.

| Fertilizer treatments | Leaf nutrients concentration (%) | | | |
|---|---|---|---|---|
| | N | P | K | Mg |
| Control | 2.03$^d$± 0.03 | 0.12$^c$± 0.01 | 1.10$^d$± 0.01 | 0.26$^c$± 0.02 |
| Mixture | 2.28$^{bc}$± 0.03 | 0.20$^{ab}$± 0.02 | 1.55$^{bc}$± 0.01 | 0.42$^b$± 0.01 |
| CRFB-60% | 2.27$^{ab}$± 0.13 | 0.21$^{ab}$± 0.01 | 1.57$^{bc}$± 0.04 | 0.43$^b$± 0.02 |
| CRFG-60% | 2.30$^{bc}$± 0.09 | 0.20$^{ab}$± 0.01 | 1.53$^c$± 0.03 | 0.44$^b$± 0.01 |
| CRFB-100% | 2.38$^a$± 0.10 | 0.24$^a$± 0.01 | 1.66$^{ab}$± 0.01 | 0.56$^a$± 0.02 |
| CRFG-100% | 2.46$^a$± 0.03 | 0.25$^a$± 0.01 | 1.75$^a$± 0.04 | 0.55$^a$± 0.01 |

All analyses are mean ± standard error of mean (SEM). Means not sharing a common letter are significantly different by DNMRT ($P \leq 0.05$).

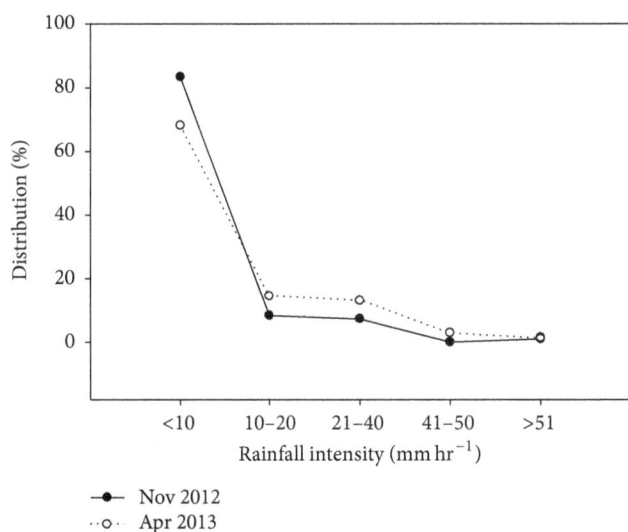

FIGURE 4: Rainfall intensity distribution for the months of November (2012) and April (2013).

rainfall intensity and lower vegetative cover produced higher sediment and consequently higher nutrient loss [41].

*3.4. Oil Palm Response to Fertilizer Treatments.* Oil palm response to fertilizer treatments during the first six months after fertilization (MAF) appears to show no significant variation ($P \leq 0.05$) in terms of frond production. However, at 12 MAF, plants that received treatments of granular CRFs showed significant improvement in frond number by 64–66% compared to mixture (59%) and control (52%) treatments (Table 4).

The effects of fertilizer treatments on foliar nutrient content of frond [#]3 are given. In general, leaf nutrient concentrations for N, P, K, and Mg for fertilized plants were higher at 12 MAF compared to 0, 3, 6, and 9 MAF. This is probably due to the cumulative effect of the fertilizers which were applied in two rounds per annum at six months' interval. Plant response in terms of nutrients uptake appears to be higher with application of full dose CRF treatments (Table 5). For full dose granular CRF treated plants, foliar N increased from 2.10% to 2.46%, while foliar K and Mg increased from 1.35% to 1.75% and 0.25% to 0.55%, respectively. Leaf analyses for the purpose of diagnosing nutritional status of oil palm have proven successful in most industrial plantations worldwide for several decades, due to the fact that the obtained information serves as a reliable indicator for annual guidance of mineral fertilization [42]. It has been reported that foliar concentrations of N, Mg, B, Cu, Fe, and Mo were significantly affected by controlled-release fertilizer treatments on container-grown ponderosa pine seedlings [43].

## 4. Conclusion

Controlled-release fertilizers (AJIB CRF) application in an immature oil palm field potentially decreased surface runoff loss of nutrients. This could be attributed to the fact that nutrient elements in CRFs are readily and slowly available for plant uptake over a given period. It is imperative to note that fertilizers should be applied during periods of less rainfall events in order to limit nutrient and soil loss due to the effect of high rainfall intensity. Considering the long term economic

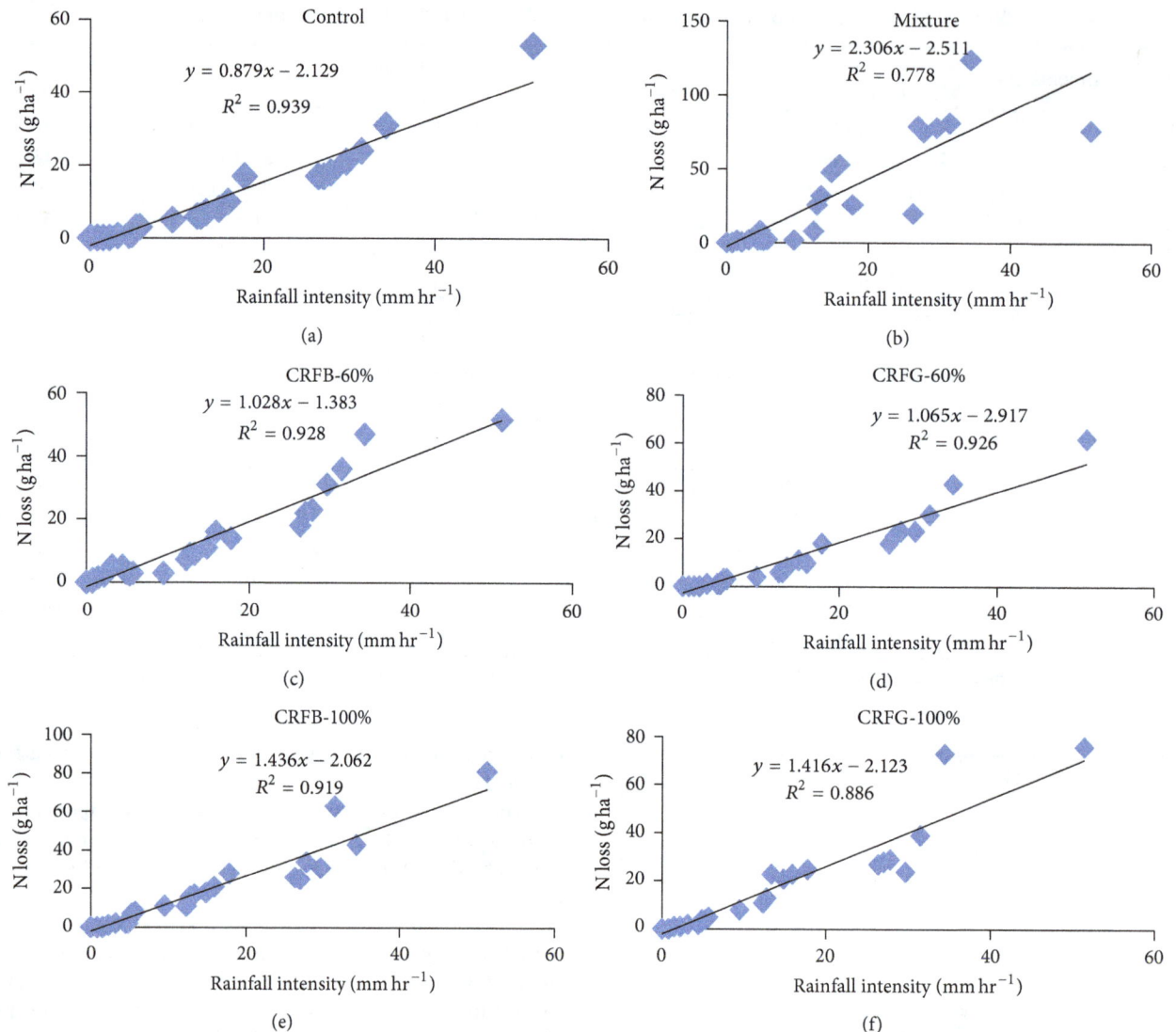

FIGURE 5: Relationship between rainfall intensity and nutrient (N) loss for different fertilizer treatments following each rainfall event.

viability and environmental factors associated with nutrient loss, oil palm plantations may consider application of CRFs as an improved fertilizer management option. More research is needed to better elucidate mechanism of nutrient loss from oil palm ecosystems during specific storm events.

## Conflict of Interests

The authors declare that there is no conflict of interests regarding the publication of this paper.

## Acknowledgments

The authors wish to acknowledge Diversatech (M) Fertilizer Sdn. Bhd., Universiti Putra Malaysia, and Commonwealth Scholarship and Fellowship Plan (CSFP) through Ministry of Education Malaysia (MOE) for financial and technical support. Thanks are due to Madam Nor Asma Mohd Zaki for her help in the laboratory work.

## References

[1] J. Shamshuddin and M. Anda, "Enhancing the productivity of ultisols and oxisols in Malaysia using basalt and/or compost," *Pedologist*, vol. 55, no. 3, pp. 382–391, 2012.

[2] J. P. Caliman, R. Carcasses, N. Perel et al., "Agri-environmental indicators for sustainable palm oil production," *Palmas*, vol. 28, pp. 434–445, 2007.

[3] K. J. Goh and R. Härdter, "General oil palm nutrition," in *Oil Palm: Management for Large and Sustainable Yields*, T. Fairhurst and R. Härdter, Eds., pp. 191–230, PPI/PPIC and IPI, Singapore, 2003.

[4] Food and Agriculture Organization of the United Nations (FAO), *Fertilizer Use by Crop in Malaysia*, FAO, Rome, Italy, 2004.

[5] I. Comte, F. Colin, J. K. Whalen, O. Grünberger, and J.-P. Caliman, "Agricultural practices in oil palm plantations and their impact on hydrological changes, nutrient fluxes and water quality in indonesia. A review," *Advances in Agronomy*, vol. 116, pp. 71–124, 2012.

[6] S. Bricker, B. W. Longstaff, A. Dennison, K. Jones, C. Boicourt, and J. Woerner, "Effects of nutrient enrichment in the nations estuaries: a decade of change," in *NOAA Coastal Ocean Program Decision Analysis*, Series No. 26, National Centers for Coastal Ocean Science, Silver Spring, Md, USA, 2007.

[7] A. M. Freeman, E. C. Lamon III, and C. A. Stow, "Nutrient criteria for lakes, ponds, and reservoirs: a Bayesian TREED model approach," *Ecological Modelling*, vol. 220, no. 5, pp. 630–639, 2009.

[8] E. E. Aziz and S. M. El-Asry, "Efficiency of slow release Urea fertilizer on herb yield and essential oil production of lemon balm (*Melissa officinalis* L.) plant," *American-Eurasian Journal of Agricultural & Environmental Sciences*, vol. 5, no. 2, pp. 141–147, 2009.

[9] R. Prasad, "Efficient fertilizer use: the key to food security and better environment," *Journal of Tropical Agriculture*, vol. 47, no. 1-2, pp. 1–17, 2009.

[10] P. P. Motavalli, K. W. Goyne, and R. P. Udawatta, "Environmental impacts of enhanced-efficiency nitrogen fertilizers," *Crop Management*, vol. 7, no. 1, 2008.

[11] K. A. Nelson, P. C. Scharf, L. G. Bundy, and P. Tracy, "Agricultural management of enhanced-efficiency fertilizers in the north-central United States," *Crop Management*, vol. 7, no. 1, 2008.

[12] A. Jarosiewicz and M. Tomaszewska, "Controlled-release NPK fertilizer encapsulated by polymeric membranes," *Journal of Agricultural and Food Chemistry*, vol. 51, no. 2, pp. 413–417, 2003.

[13] A. D. Blaylock, J. Kaufmann, and R. D. Dowbenko, "Nitrogen fertilizer technologies," in *Proceedings of the Western Nutrient Management Conference*, vol. 6, Salt Lake City, Utah, USA, 2005.

[14] International Fertilizer Industry Association (IFA), "Mineral Fertilizer Use and the Environment," 2000.

[15] T. Kenzo, R. Yoneda, Y. Matsumoto, M. A. Azani, and N. M. Majid, "Leaf photosynthetic and growth responses on four tropical tree species to different light conditions in degraded tropical secondary forest, Peninsular Malaysia," *Japan Agricultural Research Quarterly*, vol. 42, no. 4, pp. 299–306, 2008.

[16] G. W. Gee and J. W. Bauder, "Particle size analysis," in *Methods of Soil Analysis. Part 1. Physican and Mineralogical Methods*, A. Klute, Ed., pp. 383–411, ASA-SSSA, Madison, Wis, USA, 1986.

[17] C. B. S. Teh and J. Talib, *Soil Physics Analysis*, vol. 1, Universiti Putra Malaysia Press, Serdang, Malaysia, 2006.

[18] A. A. Ibitoye, *Laboratory Manual on Basic Soil Analysis*, Foladave, Akure, Nigeria, 2nd edition, 2006.

[19] J. B. Jones Jr., *Laboratory Guide for Conducting Soil Tests and Plant Analysis*, pp. 81-82, CRC Press, Boca Ranton, Fla, USA, 2001.

[20] A. Cottenie, *Soil and Plant Testing as a Basis of Fertilizer Recommendations*, FAO Soil Bulletin 38/2, FAO, Rome, Italy, 1980.

[21] K. H. Tan, *Soil Sampling, Preparation and Analysis*, Taylor and Francis/CRC Press, Boca Raton, Fla, USA, 2nd edition, 2005.

[22] G. W. Thomas, "Exchangeable cations," in *Methods of Soil Analyes*, A. L. Page, R. H. Miller, and D. R. Keeny, Eds., pp. 159–165, American Society of Agronomy, Madison, Wis, USA, 1982.

[23] R. H. Bray and L. T. Kurtz, "Determination of total, organic and available forms of phosphorus in soils," *Soil Science*, vol. 59, pp. 39–45, 1945.

[24] S. Paramananthan, *Soils of Malaysia: Their Characteristics and Identification*, Academy of Sciences Malaysia, 2000.

[25] PORIM, *Environmental Impacts of Oil Palm Plantations in Malaysia*, Palm Oil Research Institute of Malaysia, 1994.

[26] A. E. Hartemink, "Soil erosion: perennial crop plantations," in *Encyclopedia of Soil Science*, Marcel Dekker, New York, NY, USA, 2006.

[27] R. H. V. Corley and P. B. Tinker, *The Oil Palm*, John Wiley & Sons/CRC Press, Hoboken, NJ, USA, 4th edition, 2003.

[28] L. M. Maene, K. C. Tong, T. S. Ong, and A. M. Mokhtaruddin, "Surface wash under mature oil palm," in *Proceedings of the Symposium on Water in Malaysian Agriculture*, pp. 203–216, MSSS, Kuala Lumpur, Malaysia, 1979.

[29] K. J. Goh, C. B. Teo, P. S. Chew, and S. B. Chiu, "Fertilizer management in oil palm: agronomic principles and field practices," in *Fertilizer Management for Oil Palm Plantations*, vol. 44, pp. 20–21, ISP North-east Branch, Sandakan, Malaysia, 1999.

[30] M. Banabas, M. A. Turner, D. R. Scotter, and P. N. Nelson, "Losses of nitrogen fertiliser under oil palm in Papua New Guinea: 1. Water balance, and nitrogen in soil solution and runoff," *Australian Journal of Soil Research*, vol. 46, no. 4, pp. 332–339, 2008.

[31] K. J. Goh, R. Härdter, and T. Fairhurst, "Fertilizing for Maximum return," in *Oil Palm: Management for Large and Sustainable Yields*, T. Fairhurst and R. Härdter, Eds., pp. 279–306, PPI/PPIC and IPI, Singapore, 2003.

[32] K. K. Kee and P. S. Chew, "A13: Nutrient Losses through Surface Runoff and Erosion- Implications for Improved Fertilizer Efficiency in mature Oil Palm," Applied Agricultural Research Sdn. Bhd., Locked Bag no. 212, 1996.

[33] C. B. Wallace, M. G. Burton, S. G. Hefner, and T. A. DeWitt, "Effect of preceding rainfall on sediment, nutrients, and bacteria in runoff from biosolids and mineral fertilizer applied to a hayfield in a mountainous region," *Agricultural Water Management*, vol. 130, pp. 113–118, 2013.

[34] F. M. Ziadat and A. Y. Taimeh, "Effect of rainfall intensity, slope, land use and antecedent soil moisture on soil erosion in an arid environment," *Land Degradation & Development*, vol. 24, no. 6, pp. 582–590, 2013.

[35] J. F. Martínez-Murillo, E. Nadal-Romero, D. Regüés, A. Cerdà, and J. Poesen, "Soil erosion and hydrology of the western Mediterranean badlands throughout rainfall simulation experiments: a review," *Catena*, vol. 106, pp. 101–112, 2013.

[36] J. S. Kim, S. Y. Oh, and K. Y. Oh, "Nutrient runoff from a Korean rice paddy watershed during multiple storm events in the growing season," *Journal of Hydrology*, vol. 327, no. 1-2, pp. 128–139, 2006.

[37] S. I. Ahmed, R. P. Rudra, B. Gharabaghi, K. Mackenzie, and W. T. Dickinson, "Within-storm rainfall distribution effect on soil erosion rate," *ISRN Soil Science*, vol. 2012, Article ID 310927, 7 pages, 2012.

[38] P. J. A. Kleinman, M. S. Srinivasan, C. J. Dell, J. P. Schmidt, A. N. Sharpley, and R. B. Bryant, "Role of rainfall intensity and hydrology in nutrient transport via surface runoff," *Journal of Environmental Quality*, vol. 35, no. 4, pp. 1248–1259, 2006.

[39] T. Fu, J. P. Ni, C. F. Wei, and D. T. Xie, "Research on nutrient loss from terra gialla soil in Three Gorges Region under different rainfall intensity," *Journal of Soil and Water Conservation*, vol. 16, no. 2, pp. 33–35, 83, 2002.

[40] X. Chen, S. Q. Jiang, K. Z. Zhang, and Z. P. Bian, "Law of phosphorus loss and its affecting factors in red soil slope land," *Journal of Soil Erosion and Soil and Water Conservation*, vol. 5, no. 3, pp. 38–41, 1999.

[41] K. Jin, W. M. Cornelis, D. Gabriels et al., "Residue cover and rainfall intensity effects on runoff soil organic carbon losses," *Catena*, vol. 78, no. 1, pp. 81–86, 2009.

[42] J. P. Caliman, B. Dubos, B. Tailliez, P. Robin, X. Bonneau, and I. de Barros, "Oil palm mineral nutrition management: current situation and prospects," in *Proceedings of the 14th International Oil Palm Conference*, p. 33, Cartagena de Indias, Columbia, September 2003.

[43] Z. Fan, J. A. Moore, and D. L. Wenny, "Growth and nutrition of container-grown ponderosa pine seedlings with controlled-release fertilizer incorporated in the root plug," *Annals of Forest Science*, vol. 61, no. 2, pp. 117–124, 2004.

# 11

# Screening of Pearl Millet $F_1$ Hybrids for Heat Tolerance at Early Seedling Stage

**Ashok Kumar Yadav,[1,2] Rajesh Kumar Arya,[2] and M. S. Narwal[2]**

[1] CSIR-Institute of Himalayan Bioresource Technology, Palampur, Himachal Pradesh 176061, India
[2] Department of Genetics and Plant Breeding, CCS HAU, Hisar 125004, India

Correspondence should be addressed to Ashok Kumar Yadav; ashok@ihbt.res.in

Academic Editor: Clifford Gold

Ten pearl millet genotypes selected on the basis of response to supra-optimal temperature tolerance were crossed in a half-diallel mating system. The 45 $F_1$ hybrids produced were tested along with parents for heat tolerance and related traits at seedling stage. Field screening and laboratory screening techniques were simultaneously used for the evaluation of $F_1$ hybrids and their parents. Heat tolerance was measured as seedling thermotolerance index (STI) and seed to seedling thermotolerance index (SSTI) under field conditions, but membrane thermostability (MTS) in the laboratory. The hybrid H77/29-2 × CVJ-2-5-3-1-3 showed highest STI value followed by H77/833-2 × 96AC-93. The genotype H77/833-2 × 96AC-93 had the highest worth for SSTI. These three indices were highly correlated among themselves. STI values were invariably high, whereas SSTI has lower values, as it also covers the effect of under soil mortality (USM). It was seen that the heat tolerance indices STI and SSTI were not showing any perceptible pooled correlation with developmental traits except germination and emergence rate. Based on our results, it could be suggested that membrane thermostability (MTS) may be used for screening large number of genotypes. Field based indices STI and SSTI may be used for evaluation of hybrids and varieties before they are released.

## 1. Introduction

Pearl millet [*Pennisetum glaucum* (L.) R. Br.] is a multipurpose cereal grown for grain, stover, and green fodder. It is the most important staple crop in the semiarid and arid regions of Asia and sub-Saharan Africa. It shall continue to play a prominent role in the integrated agricultural and livestock economy of the country particularly in rainfed areas due to its drought hardiness. Pearl millet provides an excellent nutritious food because of high biological value, more protein (11.6%), fat (5%), and mineral content (2.3%). It also contains adequate amount of essential amino acids such as tryptophan, threonine, arginine, and lysine. Pearl millet is also high in lipids and calcium and low in crude fibre.

Adequate crop stand establishment problems are severe constraints to get good production of pearl millet. Failure to obtain adequate plant populations is often associated with high temperatures occurring during the germination and seedling establishment period of pearl millet. The temperature is one of the key climatic factors and has profound effect on the growth and development of the pearl millet. The soil temperatures in farmers' field in India and Africa commonly exceed 45°C and the temperatures as high as 60°C have occasionally been measured [1–3]. The supraoptimal temperatures are known to be the most important cause of poor crop stands in farmer's field. This can only be managed through developing hybrid varieties which can tolerate high temperature during germination and early seedling stages. Genetic variability for adaptation to high temperature exists in crop plants. Screening of $F_1$ crosses and identification of the superior cross combinations for seedling heat tolerance is essential for effective manipulation through hybrid breeding. The objective of this study was to understand the response of different cross combinations for heat tolerance during

TABLE 1: Description of pearl millet genotypes used as parents for hybridization.

| Sr. No. | Genotype | Response | Status | Origin and characteristics |
|---|---|---|---|---|
| 1 | H77/833-2 | Tolerant | Inbred | Bred at CCSHAU, Hisar, by selfing and selection within a Rajasthani landrace population |
| 2 | H77/29-2 | Susceptible | Inbred | Bred at CCSHAU, Hisar. It is medium dwarf, nodal pigmentation—violet brown, resistant to downy mildew, shoot fly, grey weevil, chaffer beetle, and leaf roller |
| 3 | G73-107 | Tolerant | Inbred | Bred at CCSHAU, Hisar, and has profuse tillering, narrow leaves, and typical orange brown anthers |
| 4 | 77/245 | Susceptible | Inbred | Bred at CCSHAU, Hisar, by selfing and selection. It is late flowering and high tillering |
| 5 | CVJ-2-5-3-1-3 | Tolerant | Inbred | Dwarf, developed at ICRISAT, Patancheru |
| 6 | 1305 | Susceptible | Inbred | Bred at CCSHAU, Hisar, late flowering, medium dwarf, semicompact earhead |
| 7 | (77/371 × BSECT CP-1) | Tolerant | Inbred | Bred at CCSHAU, Hisar, using BSECT CP-1 from ICRISAT, Patancheru as male parent |
| 8 | 96AC-93 | Tolerant | Inbred | Resistant to downy mildew; bred at CCSHAU, Hisar, by selfing and selection |
| 9 | Togo-II | Tolerant | Inbred | Selection from ICRISAT material-Togo; bred at CCSHAU, Hisar |
| 10 | 99HS-18 | Tolerant | Inbred | Bred at CCS Haryana Agricultural University, Hisar |

TABLE 2: Details of different environments created through sowing dates.

| Environment | Nature of environment | Date of sowing | Site | Location |
|---|---|---|---|---|
| Environment-1 ($E_1$) | Heat stress | May 21 | Research area, Department of Genetics and Plant Breeding | CCS HAU, Hisar, Haryana |
| Environment-2 ($E_2$) | Heat stress | June 13 | Research area, Department of Genetics and Plant Breeding | CCS HAU, Hisar, Haryana |
| Environment-3 ($E_3$) (during monsoon season) | Nonstress | July 17 | Research area, Department of Genetics and Plant Breeding | CCS HAU, Hisar, Haryana |

germination and early seedling stages in pearl millet and to identify some promising $F_1$ hybrids improvement for this trait particularly for commercial cultivation.

## 2. Materials and Methods

*2.1. Plant Material.* A field screening technique, that is, thermotolerance index [4], was applied to screen the pearl millet $F_1$ hybrids for heat tolerance at seedling stage. This technique could be used in the hottest months (May-June) only with no rainfall. Since the technique was applicable to the very young seedlings of pearl millet, it was also of concern to test the association of seedling thermotolerance to other phenotypic characteristics of seedlings. The membrane thermostability (based on quantification of electrolyte leakage) was also conducted for screening the pearl millet $F_1$ hybrids at seedling stage. Ten pearl millet genotypes, namely, H77/833-2, H77/29-2, G73-107, 77/245, CVJ-2-5-3-1-3, 1305, 77/371 × BSECT CP-1, 96AC-93, Togo-II, and 99HS-18, were selected from the germplasm on the basis of different response to supraoptimal temperature tolerance (Table 1) in separate experiment [5]. These were crossed in a half-diallel mating system (excluding reciprocals). The 45 $F_1$ hybrids

produced were tested along with parents for heat tolerance and related traits at seedling stage.

*2.2. Experimental Site and Conditions of Growth.* The experiment was conducted at the research area of Department of Genetics and Plant Breeding, Chaudhary Charan Singh Haryana Agricultural University, Hisar (Lat.: 29° 10′N, Long.: 75° 46′E, and 215.2 m above mean sea level), located in subtropical region of Haryana, India. The 45 $F_1$ hybrids produced were tested along with parents for heat tolerance and related traits in three replicated plots under supraoptimal temperature exposure at seedling stage in two stress environments (environment-1 and environment-2) created through different dates of sowing on May 21 and June 13 (the hottest period in North India), respectively, so as to conduct the experiment under different range of temperatures with no rainfall. A presowing medium light irrigation was done through flooding and the field was precisely levelled before sowing. After sowing no further irrigation was given to the stressed environments (environment-1 and environment-2). One nonstress environment (Table 2) was created through sowing on July 17 (normal sowing period during monsoon). Each genotype was grown in 3 rows each of 3 m length spaced

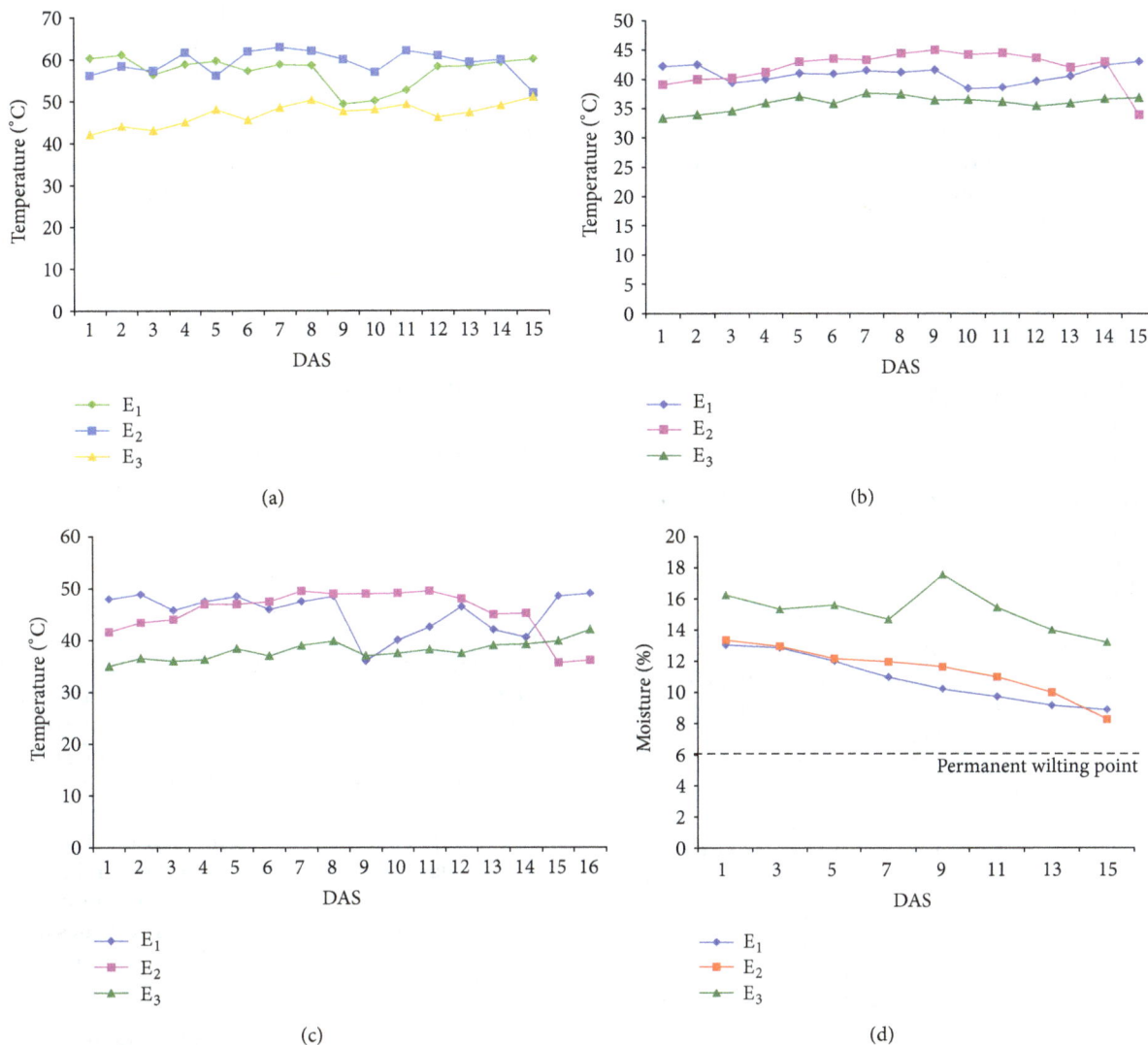

(a)

(b)

(c)

(d)

FIGURE 1: (a) Soil surface temperature, (b) air temperature, (c) temperature 5 cm above the soil surface, and (d) soil moisture status in different environments.

30 cm apart with 10 cm plant to plant spacing. The seeds were sown manually.

The soil of experimental field was sandy loam in texture. The water holding capacity, field capacity, and permanent wilting point of the experimental field were 40, 17, and 6 percent, respectively [6]. Irrigation was withheld during the course of experiment; however, drought stress-free conditions were ensured through monitoring at regular interval (on alternate days) by measuring the moisture status of the soil by gravimetric method. The soil samples of about 50 g were taken up to 20 cm depth from the experimental plot. The moisture content on dry weight basis may be calculated [7] using the following formula:

Soil moisture (%)

$$= \frac{(\text{Weight of wet soil} - \text{Weight of dry soil})}{\text{Weight of dry soil}} \times 100. \quad (1)$$

The heat stress period was terminated by irrigating the experimental field before reaching the soil moisture level at permanent wilting point. A condition of no drought was maintained in order to determine the exclusive effect of temperature at seedling stage. This was maintained by measuring the moisture status of the soil by gravimetric method on alternate days. The soil surface temperature was measured by a soil thermometer at soil surface; temperature at 5 cm above the soil surface and air temperature at 1 m above the soil surface were recorded between 2:00 p.m. and 2:30 p.m. daily (Figure 1).

Emphasis of observation is centred on germination, survival, and mortality of the seedlings.

### 2.3. Plant Characteristics/Screening Techniques.

The emphasis was given to the observations of germination, survival, and mortality of the seedlings. The seedlings were inspected every day in the morning. The seedlings which had died due to heat stress were noted in each genotype on daily basis. Observations continued till the surviving seedlings were established (15 DAE) and there was no further mortality. Moreover, the soil moisture was depleting near to the permanent wilting point and drought stress could also come into play along with heat stress to further continue the experiment. Since experiment proceeded without any stress after the establishment of seedling [4], the pearl millet plants show great resilience and recovery. The main impact of supraoptimal temperature at germination and seedling stage was observed on the seedling survival and resulted in poor plant stand/population of pearl millet. Poor crop stand reduces the productivity (fodder and grain) drastically, whereas after seedling establishment no impact of heat stress at seedling stage was observed on later growth stages.

*2.3.1. Thermotolerance Characteristics.* For heat tolerance characteristics 3 indices, namely, seedling thermotolerance index (STI) [4], seed to seedling thermotolerance index (SSTI) [8], and membrane thermostability (MTS), were applied to evaluate the hybrids and parents. STI was calculated as the ratio of number of seedlings survived to the number of seedling emergences, expressed in percentage. Seed to seedling thermotolerance index (SSTI) was calculated as the ratio of seedling survival to the number of seedlings expected to emerge, expressed in percentage [8]. The SSTI is an extension of STI by taking expected germination into account. It was necessary to correct the effect of under soil mortality (USM). The loss of germination due to heat (i.e., under soil mortality) can be measured only after a standard germination test. Therefore, the germination of seed from the same lot under monsoon environment (main pearl millet growing season, where no under soil mortality occurred due to heat stress) was taken as expected germination for calculation of SSTI.

*2.3.2. Membrane Thermostability.* These genotypes were also screened through laboratory technique (membrane thermostability) [9, 10]. The young leaves of 21-day-old seedlings of each genotype in three replication plots were used for membrane thermostability test. A middle portion of 10 cm from each leaf was taken and the midrib was removed and two equal halves of the 10 cm each were made. Each half was taken to provide replications. Each leaf portion was cut into pieces and thoroughly washed and rinsed with distilled water two times to remove the electrolytes adhering to the plant tissue, as well as electrolyte released from the cutting of plant tissue, and put them into test tubes. After final rinsing, tubes were drained, maintaining sufficient water to prevent desiccation of plant material during heat treatment. These test tubes were covered with aluminium foil and incubated in water bath at 55°C for 15 minutes for giving heat treatment. After the heat treatment period, 10 mL of distilled water was added to the test tubes. Then all the test tubes were held at 10°C overnight to allow diffusion of electrolytes from plant material. Test tubes were brought out of refrigerator and shaken well to mix the contents. An initial electrical conductivity (EC) of both replications of each treatment ($T_1$)

was determined with electrical conductivity meter. Then these test tubes were autoclaved at 15 psi for 2 minutes for complete electrolyte leakage from the leaf samples. These test tubes were then cooled at room temperature and contents mixed and final EC was taken for all the treatments ($T_2$). Membrane thermostability (MTS) was calculated according to Ibrahim and Quick [9] and Yadav et al. [10] as follows:

$$\text{Membrane Thermostability (MTS)} = \left(1 - \frac{T_1}{T_2}\right) \times 100, \tag{2}$$

where $T_1$ is conductivity reading after heat treatment and $T_2$ is conductivity reading after autoclaving.

Completely randomized design (CRD) for membrane thermostability was conducted to detect variations among the genotypes.

*2.3.3. Morphological Characteristics.* The seedling growth characteristics, namely, germination (%), emergence rate (ER), number of leaves/seedling (two-week stage), seedling height (two-week stage), seedling fresh weight (g) (four-week stage), and seedling dry weight (g) (four-week stage), were recorded. The data for seedling height and number of leaves per seedling was recorded on 10 random seedlings per plot. Emergence rate (an index for speed of germination) can be used as a tool in pearl millet breeding programs for evaluation of seedling vigour. Speed of germination is one of the oldest concepts of seedling vigour. The emergence rate (ER) was calculated by dividing the number of normal seedlings obtained at each counting by the number of days seeds had been in the seed bed (DAS). The values obtained at each count were then summed at the end of the germination test to obtain the emergence rate [11]:

$$\text{Emergence Rate (ER)} = \frac{\text{Seedling count}}{\text{Days to first count}} + \cdots + \frac{\text{Seedling count}}{\text{Days to final count}}. \tag{3}$$

After termination of the heat stress on the seedlings by irrigating the field, the experiment was continued to study the effect of seedling heat stress on maturity traits on the survived plants. The maturity traits observed were panicle emergence, effective tillers/plant, plant height, ear length, ear weight/plant, dry fodder yield/plant, grain yield/plant, and total biological yield/plant.

*2.4. Statistical Analysis.* Analysis of variance for randomized block design (RBD) for field traits, that is, germination, emergence rate, number of leaves/seedling, seedling height, seedling fresh weight, and seedling dry weight, and completely randomized design (CRD) for membrane thermostability test was conducted under each environment on mean basis [12]. Analyses were done using the software OPSTAT developed at CCS HAU, Hisar. Correlations among the various characters studied were calculated to establish the nature and magnitude of associations among the traits [8, 13] with the software SPSS (SPSS Inc.). Multivariate clustering was done based on paired group cluster method using similarity coefficient under Past 1.40 software [14]. Multivariate analysis based on the Euclidean distance matrix was estimated among genotypes to distinguish genotypes from each other. The grouping of genotypes was generated using paired group cluster method.

# 3. Results and Discussion

The excessive heat (soil temperature up to 63.0°C in $E_2$) about 18°C higher than the normal atmospheric temperature (Figures 1(a) and 1(b)) was responsible for mortality of seedlings. Many seedlings were found desiccated at the point of contact besides the desiccation at the tips. The temperature around the tip level (5 cm above the soil surface) also reached up to 50°C (Figure 1(c)). It could therefore be concluded that the seedlings had to survive in a really high temperature. The moisture level remained above the permanent wilting point (Figure 1(d)); therefore, drought effect was not in the play. The heat tolerance was measured as STI (Table 3) through the field screening technique of Peacock et al. [4]. Another index known as SSTI was also calculated as an improvement and extension. The analysis of variance (RBD) for 8 characters in 3 environments showed significant genotypic differences. No variability was observed for STI and SSTI in $E_3$ only. In this environment no seedling mortality was observed and the germination was taken as standard. The values for STI and SSTI were therefore equal and the highest for all the genotypes in this environment and hence no variation was observed.

The hybrids were evaluated on the basis of mean performance for various characters in each environment (Table 3). On the mean basis, STI value ranged from 53.32 to 82.24 with an average of 70.72. The hybrid H77/29-2 × CVJ-2-5-3-1-3 was the best scorer for heat tolerance followed by H77/833-2 × 96AC-93 and (77/371 × BSECT CP-1) × Togo-II. The hybrid H77/833-2 × 1305 was found susceptible to heat stress. Seed to seedling thermotolerance index (SSTI) which is expected to be more rigorous index deviated the values from STI according to variation in germination under normal environment ($E_3$) as the SSTI also covers/includes the under surface mortality (USM) during germination. The reductions in values were not strictly proportional to the STI. The reduction is dependent on thermosensitivity of germination. We also came across such hybrids which showed poor germination and high STI. SSTI values ranged from 46.05 to 72.87 with an average of 60.90. The genotype H77/833-2 × 96AC-93 had the highest worth for SSTI on mean basis followed by G73-107 × (77/371 × BSECT CP-1) and H77/833-2 × (77/371 × BSECT CP-1) (Table 3). The analysis of variance for ten different characters in three different environments (two stressed and one normal) was carried out and the results showed significant genotypic differences for all the characters (Table 4).

A caution is needed for evaluating such genotypes under STI. The SSTI gave a highly reduced value putting these crosses in the susceptible list. It may be concluded therefore

that SSTI is more dependable and desirable tool for screening and evaluating the genotypes for heat tolerance. The germination was lower in $E_2$ than in $E_1$. This period has the highest range of temperature with a high average. Genotype-wise, the highest germination (62.09) was recorded for the genotype H77/29-2 × (77/371 × BSECT CP-1) on mean basis, which gives a preliminary indication of resistance to heat, while the genotype 1 × 6 had the lowest germination (35.33). The hybrid H77/833-2 × G73-107 was adjudged as the fastest emergence of seedlings. The genotype (77/371 × BSECT CP-1) × 96AC-93 had the highest fresh weight/seedling. The genotype H77/29-2 × Togo-II had the highest seedling dry weight. $E_1$ had lower dry weight/seedling than $E_2$ indicating the more heat stress during the $E_1$.

*3.1. Membrane Thermostability.* Membrane thermostability (MTS) has long been recognised as a parameter of heat tolerance. All the hybrids were put to membrane thermostability test in the laboratory under standard protocol. A completely randomised design (CRD) analysis was carried out for membrane thermostability (MTS). The mean sum of squares due to genotypes was highly significant indicating that enough genetic variability was present for membrane thermostability (MTS) in each environment, which justified further analysis of the data. The range of values lay between 50.73 and 77.42. Among the parents H77/833-2, G73-107, CVJ-2-5-3-1-3, 77/371 × BSECT CP-1, 96AC-93, Togo-II, and 99HS-18 were verified as resistant genotypes and H77/29-2, 77/245, and 1305 genotypes were found susceptible as mentioned in Table 1 (on the basis of previous experiment). Among all the hybrids, H77/29-2 × CVJ-2-5-3-1-3, CVJ-2-5-3-1-3 × 96AC-93, and (77/371 × BSECT CP-1) × Togo-II were three highly tolerant genotypes on the basis of membrane thermostability (Table 5).

*3.2. Correlations.* The pooled correlations among the characters were calculated (Table 6) around heat tolerance indices (STI, SSTI, and membrane thermostability) with a view to find whether there was any association of these characters with seedling as well as maturity traits. It was observed that STI and SSTI recorded the highest significant correlation among themselves as they are related indices for heat tolerance. Membrane thermostability also showed high significant correlation with STI as well as SSTI. STI had a significant positive association with germination of seed only among seedling and maturity traits studied. SSTI, in addition, was also positively correlated with germination and emergence rate and negatively associated with number of effective tillers/plant. Membrane thermostability also showed significant positive correlation with seed germination among seedling and maturity traits studied. While evaluating the diallel progenies it was also seen that the heat tolerance indices, STI, SSTI, and membrane thermostability, were not showing any perceptible correlation with the rest of the developmental traits (seedling and maturity traits). The stress period was terminated by irrigating the field experiment after 15 days from date of sowing when the seedlings get established and further mortality stopped due to heat. Moreover, the soil

TABLE 3: Performance of the parents and the crosses for different characters.

(a)

| Sr. No. | Genotypes | Germination (%) | | | | Rate of emergence* | | | | STI* (%) | | | | SSTI* (%) | | | | Number of leaves/seedling | | | |
|---|---|---|---|---|---|---|---|---|---|---|---|---|---|---|---|---|---|---|---|---|---|
| | | $E_1$ | $E_2$ | $E_3$ | Mean | $E_1$ | $E_2$ | $E_3$ | Mean | $E_1$ | $E_2$ | $E_3$ | Mean | $E_1$ | $E_2$ | $E_3$ | Mean | $E_1$ | $E_2$ | $E_3$ | Mean |
| 1 | H77/833-2 | 39.41 | 37.37 | 61.64 | 46.14 | 40.37 | 41.09 | 42.79 | 41.42 | 78.30 | 73.87 | 90.00 | 80.72 | 44.98 | 41.33 | 90.00 | 58.77 | 6.61 | 5.02 | 8.00 | 6.54 |
| 2 | H77/29-2 | 55.13 | 47.26 | 60.92 | 54.43 | 28.15 | 49.80 | 32.58 | 36.84 | 48.35 | 44.95 | 90.00 | 61.10 | 44.41 | 36.02 | 90.00 | 56.81 | 3.63 | 2.94 | 4.27 | 3.61 |
| 3 | G73-107 | 50.37 | 48.26 | 59.18 | 52.60 | 45.34 | 41.32 | 44.81 | 43.82 | 61.87 | 54.13 | 90.00 | 68.66 | 52.13 | 44.60 | 90.00 | 62.24 | 8.78 | 7.10 | 9.35 | 8.41 |
| 4 | 77/245 | 47.47 | 46.24 | 55.65 | 49.78 | 36.25 | 46.04 | 40.47 | 40.92 | 43.85 | 35.80 | 90.00 | 56.55 | 38.03 | 30.40 | 90.00 | 52.81 | 3.77 | 2.92 | 4.37 | 3.69 |
| 5 | CVJ-2-5-3-1-3 | 48.24 | 48.32 | 68.97 | 55.18 | 50.44 | 49.07 | 49.60 | 49.70 | 71.93 | 68.13 | 90.00 | 76.69 | 49.59 | 47.85 | 90.00 | 62.48 | 8.02 | 5.80 | 9.23 | 7.68 |
| 6 | 1305 | 43.26 | 38.91 | 54.04 | 45.40 | 33.76 | 50.56 | 35.73 | 40.02 | 39.03 | 32.81 | 90.00 | 53.95 | 32.11 | 24.52 | 90.00 | 48.88 | 3.50 | 2.84 | 4.14 | 3.49 |
| 7 | 77/371 × BSECT CP-1 | 47.47 | 45.02 | 63.20 | 51.90 | 49.34 | 50.75 | 50.65 | 50.25 | 69.52 | 67.06 | 90.00 | 75.52 | 50.56 | 46.71 | 90.00 | 62.42 | 6.21 | 5.42 | 7.00 | 6.21 |
| 8 | 96AC-93 | 46.51 | 38.18 | 57.22 | 47.30 | 43.12 | 49.15 | 47.53 | 46.60 | 66.60 | 60.01 | 90.00 | 72.20 | 52.12 | 39.20 | 90.00 | 60.44 | 7.01 | 5.90 | 8.99 | 7.30 |
| 9 | Togo-II | 46.13 | 39.37 | 52.76 | 46.09 | 42.45 | 47.37 | 47.97 | 45.93 | 71.98 | 68.61 | 90.00 | 76.86 | 59.80 | 47.67 | 90.00 | 65.82 | 6.81 | 6.29 | 8.53 | 7.21 |
| 10 | 99HS-18 | 45.75 | 38.23 | 62.47 | 48.82 | 43.17 | 46.59 | 47.46 | 45.74 | 75.94 | 73.19 | 90.00 | 79.71 | 51.54 | 41.72 | 90.00 | 61.09 | 7.45 | 5.54 | 9.15 | 7.38 |
| 1 | H77/833-2 × H77/29-2 | 57.87 | 52.95 | 67.33 | 59.38 | 49.64 | 50.15 | 49.89 | 49.89 | 68.03 | 65.04 | 90.00 | 74.36 | 58.08 | 51.34 | 90.00 | 66.47 | 7.11 | 5.85 | 7.80 | 6.92 |
| 2 | H77/833-2 × G73-107 | 54.72 | 44.17 | 64.05 | 54.31 | 54.48 | 54.22 | 52.39 | 53.70 | 72.31 | 67.69 | 90.00 | 76.67 | 59.80 | 45.56 | 90.00 | 65.12 | 11.01 | 7.05 | 13.40 | 10.49 |
| 3 | H77/833-2 × 77/245 | 40.76 | 37.07 | 55.04 | 44.29 | 52.17 | 54.46 | 52.38 | 53.00 | 66.27 | 62.24 | 90.00 | 72.84 | 46.90 | 40.37 | 90.00 | 59.09 | 7.40 | 5.25 | 9.65 | 7.44 |
| 4 | H77/833-2 × CVJ-2-5-3-1-3 | 48.43 | 43.00 | 58.56 | 50.00 | 48.62 | 49.56 | 49.13 | 49.11 | 70.61 | 65.19 | 90.00 | 75.27 | 55.75 | 46.32 | 90.00 | 64.03 | 9.98 | 8.85 | 12.58 | 10.47 |
| 5 | H77/833-2 × 1305 | 32.96 | 27.80 | 45.21 | 35.33 | 50.76 | 49.07 | 52.38 | 50.74 | 37.32 | 32.65 | 90.00 | 53.32 | 27.65 | 20.50 | 90.00 | 46.05 | 11.92 | 9.34 | 16.00 | 12.42 |
| 6 | H77/833-2 × (77/371 × BSECT CP-1) | 56.36 | 55.39 | 62.58 | 58.11 | 44.89 | 52.19 | 50.15 | 49.08 | 69.81 | 64.82 | 90.00 | 74.88 | 61.46 | 56.80 | 90.00 | 69.42 | 8.98 | 7.26 | 15.25 | 10.50 |
| 7 | H77/833-2 × 96AC-93 | 57.20 | 55.15 | 63.35 | 58.57 | 48.00 | 51.86 | 50.75 | 50.21 | 78.33 | 77.86 | 90.00 | 82.06 | 66.03 | 62.56 | 90.00 | 72.87 | 11.07 | 8.73 | 11.93 | 10.58 |
| 8 | H77/833-2 × Togo-II | 48.62 | 34.53 | 57.09 | 46.75 | 50.36 | 48.29 | 49.11 | 49.25 | 61.90 | 55.00 | 90.00 | 68.97 | 52.14 | 33.39 | 90.00 | 58.51 | 9.79 | 6.70 | 11.07 | 9.19 |
| 9 | H77/833-2 × 99HS-18 | 45.75 | 44.18 | 54.47 | 48.13 | 50.41 | 50.74 | 50.06 | 50.40 | 58.00 | 50.73 | 90.00 | 66.25 | 48.46 | 41.33 | 90.00 | 59.93 | 9.78 | 7.33 | 11.35 | 9.49 |
| 10 | H77/29-2 × G73-107 | 56.15 | 49.62 | 62.73 | 56.17 | 50.23 | 55.35 | 51.65 | 52.41 | 61.43 | 55.01 | 90.00 | 68.81 | 55.40 | 44.41 | 90.00 | 63.27 | 10.00 | 7.20 | 13.63 | 10.28 |
| 11 | H77/29-2 × 77/245 | 54.72 | 51.64 | 62.81 | 56.39 | 47.45 | 50.31 | 50.28 | 49.35 | 55.72 | 51.75 | 90.00 | 65.82 | 49.01 | 43.45 | 90.00 | 60.82 | 8.80 | 6.20 | 8.97 | 7.99 |
| 12 | H77/29-2 × CVJ-2-5-3-1-3 | 54.94 | 51.24 | 65.36 | 57.18 | 48.04 | 53.16 | 49.57 | 50.26 | 75.96 | 80.76 | 90.00 | 82.24 | 60.93 | 56.80 | 90.00 | 69.24 | 9.81 | 9.32 | 14.51 | 11.21 |
| 13 | H77/29-2 × 1305 | 38.04 | 38.26 | 55.11 | 43.80 | 35.96 | 47.24 | 43.87 | 42.36 | 63.62 | 70.75 | 90.00 | 74.79 | 42.30 | 45.17 | 90.00 | 59.16 | 9.26 | 7.25 | 14.97 | 10.49 |
| 14 | H77/29-2 × (77/371 × BSECT CP-1) | 63.69 | 48.47 | 74.12 | 62.09 | 48.47 | 53.51 | 52.46 | 51.48 | 63.77 | 60.31 | 90.00 | 71.36 | 56.83 | 42.30 | 90.00 | 63.04 | 9.77 | 6.71 | 13.80 | 10.09 |
| 15 | H77/29-2 × 96AC-93 | 53.14 | 46.57 | 67.02 | 55.58 | 45.14 | 52.95 | 51.02 | 49.71 | 56.12 | 48.56 | 90.00 | 64.89 | 46.13 | 36.03 | 90.00 | 57.39 | 8.46 | 6.79 | 13.97 | 9.74 |
| 16 | H77/29-2 × Togo-II | 58.27 | 55.94 | 66.51 | 60.24 | 50.72 | 52.79 | 52.71 | 52.07 | 65.63 | 53.71 | 90.00 | 69.78 | 57.63 | 46.52 | 90.00 | 64.72 | 8.73 | 7.19 | 10.40 | 8.77 |
| 17 | H77/29-2 × 99HS-18 | 57.41 | 46.38 | 79.18 | 60.99 | 48.47 | 50.24 | 50.21 | 49.64 | 61.79 | 56.03 | 90.00 | 69.27 | 48.82 | 37.25 | 90.00 | 58.69 | 10.12 | 7.30 | 16.00 | 11.14 |
| 18 | G73-107 × 77/245 | 51.15 | 46.46 | 61.03 | 52.88 | 49.77 | 52.42 | 54.50 | 52.23 | 70.06 | 60.63 | 90.00 | 73.56 | 56.85 | 45.94 | 90.00 | 64.26 | 10.21 | 7.90 | 12.92 | 10.34 |
| 19 | G73-107 × CVJ-2-5-3-1-3 | 51.34 | 48.94 | 64.27 | 54.85 | 49.01 | 53.17 | 51.81 | 51.33 | 66.37 | 61.24 | 90.00 | 72.54 | 52.53 | 47.10 | 90.00 | 63.21 | 10.39 | 8.63 | 12.54 | 10.52 |
| 20 | G73-107 × 1305 | 48.81 | 45.61 | 75.97 | 56.80 | 47.17 | 47.91 | 47.75 | 47.61 | 67.04 | 63.80 | 90.00 | 73.61 | 45.56 | 41.15 | 90.00 | 58.90 | 11.40 | 8.82 | 16.57 | 12.26 |
| 21 | G73-107 × (77/371 × BSECT CP-1) | 56.36 | 55.19 | 62.83 | 58.12 | 49.86 | 52.19 | 52.03 | 51.36 | 79.90 | 62.03 | 90.00 | 77.31 | 66.20 | 54.36 | 90.00 | 70.19 | 9.77 | 7.03 | 13.95 | 10.25 |
| 22 | G73-107 × 96AC-93 | 57.40 | 45.10 | 64.15 | 55.55 | 51.29 | 49.82 | 51.95 | 51.02 | 69.78 | 55.56 | 90.00 | 71.78 | 61.56 | 40.19 | 90.00 | 63.92 | 10.77 | 8.60 | 12.64 | 10.67 |
| 23 | G73-107 × Togo-II | 44.60 | 43.57 | 60.44 | 49.53 | 43.96 | 51.38 | 49.51 | 48.29 | 52.79 | 48.35 | 90.00 | 63.71 | 40.18 | 36.03 | 90.00 | 55.40 | 9.06 | 8.81 | 11.89 | 9.92 |
| 24 | G73-107 × 99HS-18 | 54.73 | 43.56 | 63.74 | 54.01 | 51.02 | 49.55 | 50.55 | 50.37 | 64.60 | 51.97 | 90.00 | 68.86 | 55.34 | 37.04 | 90.00 | 60.79 | 11.71 | 7.66 | 16.35 | 11.91 |
| 25 | 77/245 × CVJ-2-5-3-1-3 | 47.66 | 46.66 | 59.21 | 51.17 | 51.70 | 49.18 | 49.00 | 49.96 | 59.28 | 58.59 | 90.00 | 69.29 | 47.85 | 46.13 | 90.00 | 61.33 | 9.84 | 9.20 | 11.94 | 10.33 |
| 26 | 77/245 × 1305 | 45.75 | 40.60 | 60.52 | 48.95 | 51.28 | 52.97 | 52.03 | 52.10 | 51.02 | 49.90 | 90.00 | 63.64 | 39.79 | 34.64 | 90.00 | 54.81 | 12.79 | 8.23 | 15.50 | 12.17 |
| 27 | 77/245 × (77/371 × BSECT CP-1) | 50.17 | 49.29 | 60.22 | 53.22 | 51.08 | 54.73 | 52.74 | 52.85 | 53.08 | 50.44 | 90.00 | 64.51 | 44.98 | 41.91 | 90.00 | 58.96 | 8.33 | 5.88 | 9.39 | 7.87 |

(a) Continued.

| Sr. No. | Genotypes | Germination (%) | | | | Rate of emergence* | | | | STI* (%) | | | | SSTI* (%) | | | | Number of leaves/seedling | | | |
|---|---|---|---|---|---|---|---|---|---|---|---|---|---|---|---|---|---|---|---|---|---|
| | | $E_1$ | $E_2$ | $E_3$ | Mean | $E_1$ | $E_2$ | $E_3$ | Mean | $E_1$ | $E_2$ | $E_3$ | Mean | $E_1$ | $E_2$ | $E_3$ | Mean | $E_1$ | $E_2$ | $E_3$ | Mean |
| 28 | 77/245 × 96AC-93 | 54.72 | 47.05 | 62.21 | 54.66 | 51.35 | 51.97 | 50.06 | 51.13 | 52.47 | 48.43 | 90.00 | 63.63 | 47.08 | 37.84 | 90.00 | 58.31 | 7.95 | 6.62 | 10.58 | 8.38 |
| 29 | 77/245 × Togo-II | 52.52 | 44.92 | 58.89 | 52.11 | 51.88 | 48.61 | 50.74 | 50.41 | 48.85 | 42.63 | 90.00 | 60.49 | 44.21 | 33.79 | 90.00 | 56.00 | 9.77 | 7.51 | 12.43 | 9.90 |
| 30 | 77/245 × 99HS-18 | 57.40 | 40.99 | 68.72 | 55.70 | 52.49 | 48.98 | 52.58 | 51.35 | 56.28 | 46.16 | 90.00 | 64.15 | 48.63 | 30.14 | 90.00 | 56.26 | 10.45 | 8.63 | 11.84 | 10.31 |
| 31 | CVJ-2-5-3-1-3 × 1305 | 43.64 | 39.44 | 57.00 | 46.69 | 49.74 | 45.17 | 46.50 | 47.14 | 60.97 | 55.79 | 90.00 | 68.92 | 45.94 | 38.42 | 90.00 | 58.12 | 9.36 | 8.48 | 14.03 | 10.63 |
| 32 | CVJ-2-5-3-1-3 × (77/371 × BSECT CP-1) | 43.45 | 38.08 | 67.74 | 49.76 | 52.72 | 49.96 | 50.35 | 51.01 | 66.50 | 57.66 | 90.00 | 71.39 | 42.88 | 33.81 | 90.00 | 55.56 | 8.00 | 7.12 | 10.13 | 8.41 |
| 33 | CVJ-2-5-3-1-3 × 96AC-93 | 48.62 | 41.97 | 57.43 | 49.34 | 48.84 | 49.80 | 50.58 | 49.74 | 74.77 | 67.60 | 90.00 | 77.45 | 59.57 | 47.09 | 90.00 | 65.55 | 9.33 | 7.38 | 14.78 | 10.50 |
| 34 | CVJ-2-5-3-1-3 × Togo-II | 46.70 | 43.43 | 62.33 | 50.82 | 43.99 | 48.13 | 46.31 | 46.14 | 58.06 | 66.14 | 90.00 | 71.40 | 44.22 | 44.79 | 90.00 | 59.67 | 9.63 | 8.55 | 11.80 | 9.99 |
| 35 | CVJ-2-5-3-1-3 × 99HS-18 | 54.93 | 50.80 | 72.73 | 59.49 | 52.76 | 49.84 | 51.62 | 51.41 | 73.22 | 65.90 | 90.00 | 76.37 | 55.34 | 47.66 | 90.00 | 64.33 | 9.98 | 8.07 | 13.38 | 10.47 |
| 36 | 1305 × (77/371 × BSECT CP-1) | 46.13 | 45.39 | 57.17 | 49.56 | 49.86 | 49.55 | 50.80 | 50.07 | 51.37 | 48.70 | 90.00 | 63.36 | 42.10 | 39.36 | 90.00 | 57.15 | 8.63 | 6.30 | 9.13 | 8.02 |
| 37 | 1305 × 96AC-93 | 54.73 | 51.77 | 68.30 | 58.27 | 52.24 | 47.79 | 49.19 | 49.74 | 71.17 | 69.65 | 90.00 | 76.94 | 56.40 | 52.20 | 90.00 | 66.20 | 8.75 | 6.94 | 14.76 | 10.15 |
| 38 | 1305 × Togo-II | 42.30 | 34.20 | 52.62 | 43.04 | 43.09 | 45.61 | 44.45 | 44.38 | 57.81 | 43.73 | 90.00 | 63.85 | 45.75 | 28.87 | 90.00 | 54.87 | 6.37 | 5.66 | 7.94 | 6.66 |
| 39 | 1305 × 99HS-18 | 54.33 | 45.07 | 69.47 | 56.29 | 52.25 | 47.53 | 49.91 | 49.89 | 72.87 | 69.68 | 90.00 | 77.52 | 55.76 | 44.79 | 90.00 | 63.52 | 11.15 | 8.18 | 14.10 | 11.14 |
| 40 | (77/371 × BSECT CP-1) × 96AC-93 | 55.95 | 44.29 | 74.61 | 58.28 | 52.16 | 49.55 | 51.21 | 50.97 | 70.87 | 63.68 | 90.00 | 74.85 | 54.32 | 40.38 | 90.00 | 61.57 | 9.55 | 7.53 | 14.62 | 10.57 |
| 41 | (77/371 × BSECT CP-1) × Togo-II | 52.52 | 47.28 | 61.06 | 53.62 | 54.92 | 48.08 | 51.32 | 51.44 | 80.25 | 75.05 | 90.00 | 81.77 | 63.10 | 54.13 | 90.00 | 69.08 | 8.59 | 6.93 | 9.90 | 8.48 |
| 42 | (77/371 × BSECT CP-1) × 99HS-18 | 55.95 | 42.98 | 62.95 | 53.96 | 47.29 | 48.87 | 50.31 | 48.83 | 73.66 | 69.45 | 90.00 | 77.70 | 63.12 | 45.56 | 90.00 | 66.23 | 9.03 | 6.22 | 11.25 | 8.83 |
| 43 | 96AC-93 × Togo-II | 50.17 | 39.77 | 64.86 | 51.60 | 50.19 | 51.95 | 52.37 | 51.50 | 59.22 | 54.86 | 90.00 | 68.03 | 46.89 | 34.84 | 90.00 | 57.24 | 10.07 | 7.28 | 11.67 | 9.67 |
| 44 | 96AC-93 × 99HS-18 | 55.95 | 48.12 | 69.24 | 57.77 | 50.36 | 53.85 | 52.05 | 52.09 | 74.11 | 53.78 | 90.00 | 72.63 | 58.73 | 39.79 | 90.00 | 62.84 | 10.13 | 6.52 | 13.34 | 10.00 |
| 45 | Togo-II × 99HS-18 | 46.70 | 44.97 | 72.93 | 54.87 | 47.18 | 43.19 | 47.96 | 46.11 | 56.88 | 54.31 | 90.00 | 67.06 | 39.80 | 36.65 | 90.00 | 55.48 | 9.39 | 6.53 | 14.62 | 10.18 |
| | Mean | 50.43 | 44.82 | 62.71 | 52.65 | 47.92 | 49.81 | 49.23 | 48.99 | 63.77 | 58.41 | 90.00 | 70.72 | 50.79 | 41.90 | 90.00 | 60.90 | 8.98 | 7.01 | 11.68 | 9.22 |
| | SE | 0.97 | 1.06 | 1.61 | — | 2.13 | 1.91 | 1.90 | — | 1.98 | 1.87 | — | — | 1.49 | 1.46 | — | — | 0.36 | 0.47 | 0.68 | — |
| | CD | 2.72 | 2.96 | 4.53 | — | 5.97 | 5.37 | 5.34 | — | 5.55 | 5.25 | — | — | 4.21 | 4.09 | — | — | 1.03 | 1.33 | 1.92 | — |

*Transformed data.

(b)

| Sr. No. | Genotypes | Seedling height (cm) | | | | Fresh weight (g)/seedling | | | | Dry weight (g)/seedling | | | | Panicle emergence (days) | | | | Effective tillers/plant | | | | Plant height (cm) | | | |
|---|---|---|---|---|---|---|---|---|---|---|---|---|---|---|---|---|---|---|---|---|---|---|---|---|---|
| | | $E_1$ | $E_2$ | $E_3$ | Mean | $E_1$ | $E_2$ | $E_3$ | Mean | $E_1$ | $E_2$ | $E_3$ | Mean | $E_1$ | $E_2$ | $E_3$ | Mean | $E_1$ | $E_2$ | $E_3$ | Mean | $E_1$ | $E_2$ | $E_3$ | Mean |
| 1 | H77/833-2 | 9.08 | 7.85 | 22.45 | 13.13 | 8.59 | 15.37 | 19.24 | 14.40 | 1.54 | 1.43 | 1.60 | 1.52 | 62.67 | 48.00 | 38.33 | 49.67 | 2.12 | 2.27 | 2.49 | 2.29 | 156.44 | 154.58 | 132.67 | 147.90 |
| 2 | H77/29-2 | 11.25 | 11.75 | 17.89 | 13.63 | 3.17 | 24.00 | 28.11 | 18.43 | 0.93 | 1.91 | 2.52 | 1.79 | 67.00 | 53.33 | 51.00 | 57.11 | 1.85 | 2.10 | 2.34 | 2.10 | 188.50 | 185.65 | 153.89 | 176.01 |
| 3 | G73-107 | 11.66 | 14.75 | 19.66 | 15.36 | 9.43 | 17.44 | 49.45 | 25.44 | 1.43 | 1.29 | 3.66 | 2.13 | 67.00 | 54.33 | 53.00 | 58.11 | 2.18 | 2.19 | 2.38 | 2.25 | 150.09 | 149.65 | 135.03 | 144.92 |
| 4 | 77/245 | 14.38 | 13.73 | 20.85 | 16.32 | 6.07 | 18.63 | 59.57 | 28.09 | 1.17 | 1.80 | 5.09 | 2.69 | 61.00 | 48.00 | 51.00 | 53.33 | 2.32 | 2.21 | 2.53 | 2.35 | 140.36 | 138.02 | 117.42 | 131.93 |
| 5 | CVJ-2-5-3-1-3 | 12.14 | 14.37 | 18.07 | 14.86 | 11.63 | 16.10 | 41.66 | 23.13 | 1.71 | 1.41 | 4.07 | 2.40 | 63.00 | 49.33 | 46.00 | 52.78 | 2.39 | 2.27 | 2.70 | 2.45 | 199.43 | 172.61 | 149.70 | 173.91 |
| 6 | 1305 | 8.58 | 11.37 | 17.37 | 12.44 | 5.40 | 9.67 | 48.61 | 21.23 | 1.25 | 0.88 | 4.17 | 2.10 | 65.33 | 50.33 | 48.00 | 54.56 | 2.44 | 2.13 | 2.57 | 2.38 | 155.27 | 154.05 | 122.74 | 144.02 |

(b) Continued.

| Sr. No. | Genotypes | Seedling height (cm) | | | | Fresh weight (g)/seedling | | | | Dry weight (g)/seedling | | | | Panicle emergence (days) | | | | Effective tillers/plant | | | | Plant height (cm) | | | |
|---|---|---|---|---|---|---|---|---|---|---|---|---|---|---|---|---|---|---|---|---|---|---|---|---|---|
| | | $E_1$ | $E_2$ | $E_3$ | Mean | $E_1$ | $E_2$ | $E_3$ | Mean | $E_1$ | $E_2$ | $E_3$ | Mean | $E_1$ | $E_2$ | $E_3$ | Mean | $E_1$ | $E_2$ | $E_3$ | Mean | $E_1$ | $E_2$ | $E_3$ | Mean |
| 7 | 77/371 × BSECT CP-1 | 13.77 | 14.87 | 18.92 | **15.85** | 9.28 | 26.60 | 33.30 | **23.06** | 1.50 | 2.58 | 3.01 | **2.36** | 57.33 | 45.33 | 47.00 | **49.89** | 2.42 | 2.20 | 2.93 | **2.52** | 151.38 | 145.50 | 131.50 | **142.79** |
| 8 | 96AC-93 | 13.03 | 12.59 | 18.52 | **14.71** | 10.56 | 14.45 | 59.72 | **28.25** | 1.85 | 1.47 | 5.86 | **3.06** | 55.00 | 48.33 | 38.33 | **47.22** | 2.56 | 3.04 | 3.18 | **2.93** | 155.77 | 140.32 | 100.00 | **132.03** |
| 9 | Togo-II | 10.56 | 12.20 | 22.72 | **15.16** | 9.24 | 14.66 | 91.13 | **38.34** | 1.46 | 1.29 | 8.83 | **3.86** | 61.67 | 45.33 | 46.00 | **51.00** | 1.85 | 1.49 | 1.24 | **1.53** | 145.78 | 120.73 | 100.33 | **122.28** |
| 10 | 99HS-18 | 12.77 | 10.75 | 18.08 | **13.87** | 12.41 | 28.33 | 70.70 | **37.15** | 2.28 | 2.26 | 6.15 | **3.56** | 61.00 | 50.00 | 41.67 | **50.89** | 2.43 | 1.73 | 3.24 | **2.47** | 126.27 | 103.55 | 101.73 | **110.52** |
| 1 | H77/833-2 × H77/29-2 | 14.32 | 14.98 | 22.33 | **17.21** | 13.50 | 44.05 | 83.73 | **47.09** | 1.59 | 4.99 | 7.82 | **4.80** | 57.67 | 44.33 | 38.00 | **46.67** | 3.09 | 2.90 | 2.48 | **2.82** | 210.10 | 180.71 | 158.47 | **183.09** |
| 2 | H77/833-2 × G73-107 | 16.75 | 16.35 | 22.05 | **18.38** | 12.36 | 32.11 | 57.89 | **34.12** | 1.34 | 3.13 | 5.47 | **3.31** | 58.33 | 47.67 | 38.33 | **48.11** | 2.62 | 2.95 | 2.36 | **2.65** | 212.53 | 204.49 | 158.08 | **191.70** |
| 3 | H77/833-2 × 77/245 | 15.77 | 15.22 | 22.63 | **17.87** | 10.58 | 69.33 | 87.19 | **55.70** | 2.55 | 5.53 | 7.82 | **5.30** | 52.67 | 46.33 | 46.00 | **48.33** | 2.73 | 2.01 | 2.65 | **2.46** | 207.41 | 195.11 | 152.08 | **184.87** |
| 4 | H77/833-2 × CVJ-2-5-3-1-3 | 17.91 | 17.57 | 21.65 | **19.04** | 20.21 | 80.30 | 52.76 | **51.09** | 3.54 | 7.79 | 4.77 | **5.37** | 55.33 | 42.33 | 37.67 | **45.11** | 1.65 | 3.30 | 3.51 | **2.82** | 218.27 | 200.72 | 171.10 | **196.70** |
| 5 | H77/833-2 × 1305 | 17.00 | 18.42 | 28.41 | **21.28** | 14.27 | 62.28 | 91.34 | **55.96** | 5.70 | 5.79 | 8.13 | **6.54** | 58.00 | 42.67 | 33.33 | **44.67** | 3.40 | 4.26 | 4.51 | **4.06** | 198.16 | 194.44 | 160.37 | **184.32** |
| 6 | H77/833-2 × (77/371 × BSECT CP-1) | 14.53 | 15.15 | 31.20 | **20.29** | 7.89 | 46.55 | 104.28 | **52.91** | 2.11 | 4.08 | 8.58 | **4.92** | 52.33 | 41.67 | 32.67 | **42.22** | 2.03 | 2.83 | 2.67 | **2.51** | 206.77 | 191.38 | 164.70 | **187.62** |
| 7 | H77/833-2 × 96AC-93 | 18.35 | 17.83 | 24.27 | **20.15** | 25.32 | 66.49 | 161.06 | **84.29** | 3.50 | 6.50 | 14.73 | **8.24** | 54.67 | 42.33 | 37.00 | **44.67** | 2.43 | 2.39 | 2.56 | **2.46** | 222.78 | 204.34 | 157.35 | **194.82** |
| 8 | H77/833-2 × Togo-II | 15.64 | 15.63 | 25.00 | **18.76** | 18.82 | 47.72 | 116.60 | **61.05** | 2.96 | 4.69 | 9.72 | **5.79** | 56.67 | 45.33 | 44.00 | **48.67** | 1.69 | 2.60 | 2.11 | **2.13** | 241.28 | 237.56 | 193.89 | **224.24** |
| 9 | H77/833-2 × 99HS-18 | 15.93 | 18.00 | 20.49 | **18.14** | 15.71 | 48.02 | 140.24 | **67.99** | 2.03 | 5.44 | 15.47 | **7.65** | 60.33 | 46.00 | 40.00 | **48.78** | 2.61 | 2.25 | 2.56 | **2.47** | 234.66 | 210.81 | 179.33 | **208.27** |
| 10 | H77/29-2 × G73-107 | 14.95 | 15.68 | 20.75 | **17.13** | 9.59 | 51.57 | 85.17 | **48.77** | 2.18 | 3.82 | 6.66 | **4.22** | 61.33 | 50.67 | 48.33 | **53.44** | 3.00 | 2.73 | 2.39 | **2.71** | 236.51 | 219.17 | 170.83 | **208.84** |
| 11 | H77/29-2 × 77/245 | 15.57 | 15.37 | 26.13 | **19.02** | 18.51 | 57.56 | 57.29 | **44.45** | 3.66 | 5.62 | 5.24 | **4.84** | 56.33 | 48.00 | 46.33 | **50.22** | 1.33 | 2.33 | 2.97 | **2.21** | 197.92 | 199.14 | 152.50 | **183.19** |
| 12 | H77/29-2 × CVJ-2-5-3-1-3 | 18.44 | 20.33 | 28.93 | **22.57** | 16.51 | 80.61 | 148.85 | **81.99** | 2.64 | 5.97 | 10.54 | **6.38** | 63.33 | 46.00 | 42.33 | **50.56** | 1.70 | 2.11 | 2.17 | **1.99** | 246.29 | 233.39 | 197.00 | **225.56** |
| 13 | H77/29-2 × 1305 | 13.25 | 16.77 | 30.99 | **20.34** | 17.56 | 72.43 | 140.48 | **76.82** | 3.18 | 6.35 | 12.32 | **7.28** | 63.00 | 46.67 | 36.33 | **48.67** | 2.98 | 2.61 | 3.61 | **3.07** | 216.66 | 193.71 | 174.22 | **194.86** |
| 14 | H77/29-2 × (77/371 × BSECT CP-1) | 19.58 | 18.87 | 31.05 | **23.17** | 11.18 | 46.75 | 144.11 | **67.34** | 2.42 | 4.57 | 13.18 | **6.72** | 59.00 | 45.00 | 38.00 | **47.33** | 1.64 | 2.09 | 2.40 | **2.04** | 236.61 | 199.54 | 192.50 | **209.55** |
| 15 | H77/29-2 × 96AC-93 | 15.12 | 17.17 | 27.79 | **20.02** | 26.68 | 44.08 | 113.93 | **61.56** | 3.56 | 5.00 | 12.11 | **6.89** | 56.33 | 45.33 | 42.00 | **47.89** | 2.21 | 1.95 | 2.82 | **2.32** | 219.49 | 216.35 | 172.13 | **202.66** |

(b) Continued.

| Sr. No. | Genotypes | Seedling height (cm) | | | | Fresh weight (g)/seedling | | | | Dry weight (g)/seedling | | | | Panicle emergence (days) | | | | Effective tillers/plant | | | | Plant height (cm) | | | |
|---|---|---|---|---|---|---|---|---|---|---|---|---|---|---|---|---|---|---|---|---|---|---|---|---|---|
| | | $E_1$ | $E_2$ | $E_3$ | Mean | $E_1$ | $E_2$ | $E_3$ | Mean | $E_1$ | $E_2$ | $E_3$ | Mean | $E_1$ | $E_2$ | $E_3$ | Mean | $E_1$ | $E_2$ | $E_3$ | Mean | $E_1$ | $E_2$ | $E_3$ | Mean |
| 16 | H77/29-2 × Togo-II | 15.19 | 17.38 | 21.59 | 18.05 | 24.37 | 98.67 | 161.62 | 94.89 | 3.61 | 9.61 | 15.27 | 9.50 | 61.67 | 47.00 | 42.00 | 50.22 | 2.34 | 2.72 | 2.30 | 2.46 | 263.38 | 241.45 | 205.53 | 236.78 |
| 17 | H77/29-2 × 99HS-18 | 19.38 | 17.87 | 31.51 | 22.92 | 15.74 | 62.18 | 113.18 | 63.70 | 2.90 | 6.06 | 9.56 | 6.17 | 60.67 | 47.67 | 40.67 | 49.67 | 1.72 | 2.10 | 1.99 | 1.94 | 246.89 | 218.54 | 177.89 | 214.44 |
| 18 | G73-107 × 77/245 | 18.56 | 18.13 | 22.10 | 19.60 | 24.35 | 85.50 | 78.28 | 62.71 | 3.38 | 6.33 | 6.12 | 5.28 | 55.00 | 46.33 | 44.00 | 48.44 | 1.99 | 2.55 | 3.07 | 2.54 | 199.36 | 192.77 | 160.95 | 184.36 |
| 19 | G73-107 × CVJ-2-5-1-3 | 15.17 | 17.69 | 24.07 | 18.97 | 15.56 | 77.28 | 91.36 | 61.40 | 2.39 | 8.76 | 9.17 | 6.77 | 59.00 | 43.00 | 34.00 | 45.33 | 2.33 | 2.66 | 3.15 | 2.72 | 234.18 | 223.33 | 193.17 | 216.89 |
| 20 | G73-107 × 1305 | 12.17 | 15.07 | 28.32 | 18.52 | 17.77 | 51.89 | 111.49 | 60.38 | 2.89 | 5.88 | 11.97 | 6.91 | 67.33 | 50.00 | 42.33 | 53.22 | 2.86 | 2.10 | 3.12 | 2.69 | 207.88 | 199.28 | 170.67 | 192.61 |
| 21 | G73-107 × (77/371 × BSECT CP-1) | 18.55 | 19.46 | 27.30 | 21.77 | 18.01 | 34.82 | 104.15 | 52.33 | 2.39 | 2.77 | 7.53 | 4.23 | 55.00 | 44.67 | 34.33 | 44.67 | 2.43 | 1.97 | 2.73 | 2.38 | 238.17 | 202.63 | 174.25 | 205.02 |
| 22 | G73-107 × 96AC-93 | 17.18 | 18.40 | 28.61 | 21.40 | 13.45 | 93.33 | 146.69 | 84.49 | 2.97 | 8.45 | 12.78 | 8.07 | 56.00 | 50.33 | 46.33 | 50.89 | 2.70 | 2.43 | 3.00 | 2.71 | 285.37 | 214.61 | 168.75 | 222.91 |
| 23 | G73-107 × Togo-II | 16.80 | 18.08 | 30.87 | 21.92 | 11.12 | 46.78 | 93.19 | 50.36 | 1.86 | 4.73 | 9.85 | 5.48 | 60.33 | 47.67 | 42.00 | 50.00 | 2.10 | 2.09 | 2.31 | 2.17 | 248.64 | 246.09 | 216.25 | 236.99 |
| 24 | G73-107 × 99HS-18 | 13.98 | 16.75 | 33.70 | 21.48 | 27.53 | 86.72 | 119.18 | 77.81 | 3.62 | 6.91 | 9.81 | 6.78 | 56.67 | 46.00 | 35.00 | 45.89 | 1.70 | 2.47 | 2.70 | 2.29 | 246.77 | 205.67 | 178.55 | 210.33 |
| 25 | 77/245 × CVJ-2-5-3-1-3 | 15.00 | 17.28 | 22.03 | 18.10 | 22.82 | 46.79 | 114.92 | 61.51 | 3.28 | 4.24 | 11.16 | 6.23 | 60.00 | 43.33 | 42.00 | 48.44 | 3.00 | 2.31 | 3.70 | 3.00 | 212.57 | 204.97 | 177.25 | 198.26 |
| 26 | 77/245 × 1305 | 14.38 | 15.77 | 20.80 | 16.98 | 9.49 | 104.39 | 120.53 | 78.14 | 1.21 | 9.15 | 9.92 | 6.76 | 57.67 | 48.00 | 35.00 | 46.89 | 1.68 | 2.46 | 3.16 | 2.44 | 197.89 | 177.90 | 155.17 | 176.99 |
| 27 | 77/245 × (77/371 × BSECT CP-1) | 15.05 | 16.97 | 28.46 | 20.16 | 32.22 | 85.18 | 126.21 | 81.20 | 4.69 | 8.27 | 11.41 | 8.12 | 54.00 | 43.67 | 38.33 | 45.33 | 1.96 | 2.25 | 2.42 | 2.21 | 209.28 | 194.85 | 171.92 | 192.02 |
| 28 | 77/245 × 96AC-93 | 16.53 | 17.97 | 21.17 | 18.56 | 19.34 | 60.56 | 115.43 | 65.11 | 3.52 | 5.63 | 9.11 | 6.08 | 53.33 | 42.67 | 38.33 | 44.78 | 1.86 | 1.69 | 1.98 | 1.84 | 220.03 | 190.54 | 167.67 | 192.75 |
| 29 | 77/245 × Togo-II | 17.23 | 17.18 | 31.63 | 22.01 | 17.49 | 82.50 | 154.43 | 84.81 | 2.29 | 8.06 | 14.12 | 8.16 | 59.67 | 47.00 | 40.67 | 49.11 | 1.74 | 2.45 | 2.98 | 2.39 | 220.75 | 235.03 | 198.75 | 218.18 |
| 30 | 77/245 × 99HS-18 | 20.16 | 19.70 | 28.54 | 22.80 | 32.17 | 130.27 | 35.31 | 65.91 | 3.95 | 9.65 | 2.85 | 5.49 | 57.33 | 44.67 | 38.33 | 46.78 | 1.48 | 2.28 | 2.09 | 1.95 | 221.27 | 214.62 | 169.00 | 201.63 |
| 31 | CVJ-2-5-3-1-3 × 1305 | 13.58 | 14.22 | 21.79 | 16.53 | 12.54 | 28.03 | 101.94 | 47.50 | 1.78 | 2.54 | 8.01 | 4.11 | 60.33 | 44.67 | 33.67 | 46.22 | 3.12 | 1.93 | 3.30 | 2.78 | 222.69 | 195.47 | 172.50 | 196.89 |
| 32 | CVJ-2-5-3-1-3 × (77/371 × BSECT CP-1) | 13.35 | 17.42 | 24.00 | 18.26 | 9.22 | 52.26 | 128.06 | 63.18 | 2.36 | 4.58 | 10.54 | 5.83 | 56.33 | 44.33 | 36.00 | 45.56 | 3.21 | 3.15 | 2.59 | 2.98 | 221.04 | 195.24 | 170.87 | 195.72 |
| 33 | CVJ-2-5-3-1-3 × 96AC-93 | 15.52 | 16.80 | 32.60 | 21.64 | 16.67 | 100.94 | 119.41 | 79.01 | 2.50 | 9.74 | 11.15 | 7.79 | 55.33 | 45.33 | 44.67 | 48.44 | 1.70 | 2.02 | 3.17 | 2.30 | 239.21 | 233.17 | 193.33 | 221.90 |
| 34 | CVJ-2-5-3-1-3 × Togo-II | 14.22 | 16.07 | 26.38 | 18.89 | 10.12 | 41.63 | 110.61 | 54.12 | 2.61 | 3.77 | 8.91 | 5.10 | 61.67 | 49.67 | 45.33 | 52.22 | 2.07 | 2.25 | 3.25 | 2.52 | 235.90 | 228.60 | 202.78 | 222.43 |
| 35 | CVJ-2-5-3-1-3 × 99HS-18 | 14.35 | 16.00 | 27.00 | 19.12 | 30.56 | 32.37 | 70.89 | 44.61 | 4.49 | 3.01 | 6.58 | 4.69 | 57.00 | 44.67 | 38.00 | 46.56 | 1.48 | 1.78 | 2.25 | 1.83 | 224.44 | 202.40 | 169.67 | 198.84 |
| 36 | 1305 × (77/371 × BSECT CP-1) | 14.99 | 17.67 | 20.58 | 17.74 | 18.82 | 46.65 | 79.20 | 48.22 | 3.56 | 4.26 | 7.24 | 5.02 | 56.67 | 42.67 | 38.00 | 45.78 | 2.71 | 2.42 | 3.05 | 2.73 | 220.11 | 192.95 | 164.17 | 192.41 |
| 37 | 1305 × 96AC-93 | 16.78 | 16.05 | 27.57 | 20.14 | 15.16 | 100.37 | 119.60 | 78.38 | 3.11 | 11.38 | 12.84 | 9.11 | 56.00 | 40.67 | 34.00 | 43.56 | 1.82 | 2.76 | 2.51 | 2.36 | 239.65 | 196.25 | 158.00 | 197.97 |
| 38 | 1305 × Togo-II | 11.77 | 16.80 | 22.49 | 17.02 | 9.02 | 46.76 | 93.88 | 49.89 | 2.41 | 4.56 | 7.95 | 4.97 | 65.33 | 55.33 | 47.33 | 56.00 | 2.14 | 2.66 | 2.96 | 2.59 | 218.94 | 202.17 | 192.22 | 204.44 |
| 39 | 1305 × 99HS-18 | 17.74 | 17.67 | 27.66 | 21.02 | 21.45 | 46.28 | 106.18 | 57.97 | 3.01 | 4.47 | 9.91 | 5.80 | 59.33 | 45.00 | 37.33 | 47.22 | 1.28 | 1.42 | 1.80 | 1.50 | 227.57 | 194.13 | 169.67 | 197.12 |
| 40 | (77/371 × BSECT CP-1) × 96AC-93 | 17.48 | 17.95 | 31.71 | 22.38 | 15.37 | 125.86 | 143.79 | 95.01 | 2.11 | 11.04 | 11.83 | 8.33 | 54.67 | 44.00 | 34.67 | 44.44 | 1.27 | 2.01 | 2.26 | 1.85 | 232.94 | 210.89 | 178.58 | 207.47 |

(b) Continued.

| Sr. No. | Genotypes | Seedling height (cm) $E_1$ | $E_2$ | $E_3$ | Mean | Fresh weight (g)/seedling $E_1$ | $E_2$ | $E_3$ | Mean | Dry weight (g)/seedling $E_1$ | $E_2$ | $E_3$ | Mean | Panicle emergence (days) $E_1$ | $E_2$ | $E_3$ | Mean | Effective tillers/plant $E_1$ | $E_2$ | $E_3$ | Mean | Plant height (cm) $E_1$ | $E_2$ | $E_3$ | Mean |
|---|---|---|---|---|---|---|---|---|---|---|---|---|---|---|---|---|---|---|---|---|---|---|---|---|---|
| 41 | (77/371 × BSECT CP-1) × Togo-II | 19.65 | 20.07 | 26.53 | **22.08** | 21.36 | 56.01 | 95.03 | **57.47** | 3.15 | 6.35 | 10.01 | **6.50** | 55.33 | 48.33 | 41.00 | **48.22** | 2.13 | 2.41 | 2.86 | **2.47** | 226.62 | 220.94 | 176.67 | **208.08** |
| 42 | (77/371 × BSECT CP-1) × 99HS-18 | 15.01 | 16.87 | 27.05 | **19.64** | 19.35 | 72.53 | 118.06 | **69.98** | 3.14 | 7.04 | 10.67 | **6.95** | 54.67 | 44.33 | 51.33 | **50.11** | 1.77 | 1.67 | 1.75 | **1.73** | 223.05 | 192.77 | 163.92 | **193.25** |
| 43 | 96AC-93 × Togo-II | 17.59 | 16.18 | 21.23 | **18.34** | 15.39 | 103.47 | 122.12 | **80.33** | 3.25 | 9.07 | 10.90 | **7.74** | 59.00 | 45.00 | 39.67 | **47.89** | 2.14 | 2.69 | 3.39 | **2.74** | 232.52 | 215.54 | 168.75 | **205.60** |
| 44 | 96AC-93 × 99HS-18 | 13.53 | 16.18 | 29.25 | **19.66** | 35.32 | 72.23 | 119.50 | **75.69** | 4.40 | 6.54 | 10.24 | **7.06** | 56.00 | 49.00 | 45.33 | **50.11** | 1.49 | 1.75 | 2.06 | **1.77** | 244.94 | 206.98 | 181.11 | **211.01** |
| 45 | Togo-II × 99HS-18 | 11.17 | 12.78 | 25.82 | **16.59** | 34.48 | 67.64 | 119.10 | **73.74** | 4.77 | 7.67 | 12.07 | **8.17** | 60.00 | 51.33 | 43.33 | **51.56** | 1.83 | 2.03 | 2.16 | **2.01** | 176.83 | 155.41 | 143.33 | **158.53** |
| | Mean | 15.13 | 16.18 | 24.92 | **18.74** | 16.56 | 57.73 | 98.54 | **57.61** | 2.72 | 5.38 | 8.85 | **5.65** | 58.68 | 46.53 | 41.16 | **48.79** | 2.18 | 2.33 | 2.68 | **2.40** | 212.97 | 195.47 | 165.29 | **191.24** |
| | SE | 1.01 | 0.73 | 0.98 | — | 0.79 | 1.52 | 3.17 | — | 0.14 | 0.14 | 0.28 | — | 3.51 | 3.65 | 4.28 | — | 0.36 | 0.39 | 0.46 | — | 9.85 | 11.73 | 12.11 | — |
| | CD | 2.84 | 2.05 | 2.76 | — | 2.21 | 4.27 | 8.90 | — | 0.38 | 0.40 | 0.79 | — | 1.25 | 1.30 | 1.52 | — | 0.13 | 0.14 | 0.16 | — | 3.51 | 4.18 | 4.31 | — |

(c)

| Sr. No. | Genotypes | Earhead length (cm) $E_1$ | $E_2$ | $E_3$ | Mean | Earhead weight (g)/plant $E_1$ | $E_2$ | $E_3$ | Mean | Dry fodder yield (g)/plant $E_1$ | $E_2$ | $E_3$ | Mean | Grain yield (g)/plant $E_1$ | $E_2$ | $E_3$ | Mean | Total biological yield (g)/plant $E_1$ | $E_2$ | $E_3$ | Mean |
|---|---|---|---|---|---|---|---|---|---|---|---|---|---|---|---|---|---|---|---|---|---|
| 1 | H77/833-2 | 16.42 | 16.67 | 17.44 | 16.84 | 15.04 | 13.49 | 15.26 | **14.60** | 27.83 | 24.05 | 26.34 | **26.07** | 8.36 | 7.45 | 9.08 | **8.30** | 42.87 | 37.54 | 41.60 | **40.67** |
| 2 | H77/29-2 | 16.93 | 18.17 | 19.50 | 18.20 | 10.12 | 8.59 | 10.16 | **9.62** | 34.33 | 28.09 | 25.91 | **29.44** | 6.49 | 5.97 | 6.51 | **6.32** | 44.45 | 36.68 | 36.07 | **39.06** |
| 3 | G73-107 | 20.00 | 20.28 | 20.77 | 20.35 | 22.80 | 19.66 | 23.12 | **21.86** | 46.95 | 40.23 | 42.29 | **43.16** | 12.81 | 11.70 | 14.63 | **13.05** | 69.75 | 59.89 | 65.41 | **65.02** |
| 4 | 77/245 | 15.23 | 16.30 | 17.28 | 16.27 | 17.96 | 24.89 | 27.02 | **23.29** | 54.23 | 48.51 | 44.98 | **49.24** | 7.48 | 14.64 | 17.43 | **13.19** | 72.19 | 73.41 | 72.00 | **72.53** |
| 5 | CVJ-2-5-3-1-3 | 22.82 | 26.92 | 27.87 | 25.87 | 24.11 | 26.43 | 29.01 | **26.52** | 58.57 | 50.24 | 48.36 | **52.39** | 12.06 | 13.35 | 17.69 | **14.37** | 82.68 | 74.79 | 79.25 | **78.91** |
| 6 | 1305 | 18.51 | 19.51 | 20.54 | 19.52 | 20.26 | 17.63 | 19.08 | **18.99** | 46.57 | 44.63 | 44.14 | **45.11** | 14.68 | 13.36 | 15.39 | **14.48** | 66.83 | 61.77 | 63.71 | **64.10** |
| 7 | 77/371 × BSECT CP-1 | 16.29 | 16.45 | 17.33 | 16.69 | 12.44 | 16.92 | 18.84 | **16.07** | 49.25 | 41.83 | 44.35 | **45.15** | 6.25 | 10.92 | 14.83 | **10.67** | 61.69 | 58.75 | 63.19 | **61.21** |
| 8 | 96AC-93 | 13.79 | 15.19 | 16.00 | 14.99 | 20.66 | 29.42 | 30.88 | **26.98** | 45.01 | 40.95 | 42.29 | **42.75** | 9.84 | 15.65 | 16.78 | **14.09** | 65.67 | 70.37 | 73.16 | **69.73** |
| 9 | Togo-II | 19.04 | 20.57 | 21.03 | 20.21 | 21.67 | 18.44 | 23.58 | **21.23** | 45.20 | 34.46 | 39.04 | **39.57** | 12.98 | 11.38 | 15.02 | **13.13** | 66.87 | 52.90 | 62.62 | **60.80** |
| 10 | 99HS-18 | 14.75 | 14.85 | 16.33 | 15.31 | 31.80 | 29.04 | 33.82 | **31.55** | 56.33 | 41.39 | 52.05 | **49.92** | 16.56 | 15.36 | 18.79 | **16.90** | 88.13 | 70.43 | 85.87 | **81.47** |
| 1 | H77/833-2 × H77/29-2 | 20.10 | 20.46 | 21.33 | 20.63 | 36.45 | 30.75 | 43.77 | **36.99** | 76.75 | 65.40 | 71.99 | **71.38** | 20.14 | 17.08 | 28.80 | **22.01** | 113.20 | 96.15 | 115.76 | **108.37** |
| 2 | H77/833-2 × G73-107 | 20.66 | 21.01 | 21.22 | 20.96 | 50.02 | 40.53 | 46.01 | **45.52** | 78.82 | 74.30 | 69.63 | **74.25** | 31.46 | 25.65 | 28.08 | **28.40** | 128.84 | 114.82 | 115.64 | **119.77** |
| 3 | H77/833-2 × 77/245 | 17.64 | 20.48 | 20.83 | 19.65 | 37.69 | 44.97 | 45.59 | **42.75** | 86.81 | 85.15 | 84.75 | **85.57** | 22.84 | 33.56 | 33.77 | **30.06** | 124.50 | 130.12 | 130.34 | **128.32** |
| 4 | H77/833-2 × CVJ-2-5-3-1-3 | 22.05 | 23.83 | 27.33 | 24.41 | 36.93 | 56.55 | 52.20 | **48.56** | 116.56 | 102.87 | 101.05 | **106.83** | 21.98 | 44.88 | 41.76 | **36.21** | 153.48 | 159.42 | 153.25 | **155.38** |
| 5 | H77/833-2 × 1305 | 22.09 | 24.23 | 24.91 | 23.74 | 60.08 | 58.42 | 69.33 | **62.61** | 152.38 | 143.11 | 138.89 | **144.79** | 37.58 | 40.57 | 55.91 | **44.69** | 212.46 | 201.53 | 208.22 | **207.40** |
| 6 | H77/833-2 × (77/371 × BSECT CP-1) | 17.89 | 18.23 | 19.92 | 18.68 | 35.93 | 42.88 | 49.03 | **42.62** | 89.22 | 65.22 | 74.29 | **76.25** | 21.78 | 32.24 | 30.70 | **28.24** | 125.15 | 108.11 | 123.33 | **118.86** |
| 7 | H77/833-2 × 96AC-93 | 17.39 | 19.60 | 20.11 | 19.03 | 45.57 | 43.50 | 48.18 | **45.75** | 79.16 | 60.68 | 72.90 | **70.92** | 24.11 | 25.29 | 31.91 | **27.10** | 124.74 | 104.19 | 121.08 | **116.67** |
| 8 | H77/833-2 × Togo-II | 19.27 | 23.89 | 24.06 | 22.41 | 41.85 | 58.28 | 62.15 | **54.09** | 123.19 | 102.79 | 110.23 | **112.07** | 24.33 | 42.54 | 46.31 | **37.73** | 165.04 | 161.07 | 172.38 | **166.16** |
| 9 | H77/833-2 × 99HS-18 | 19.41 | 22.45 | 23.30 | 21.72 | 42.27 | 41.39 | 45.29 | **42.98** | 92.11 | 81.20 | 79.69 | **84.33** | 31.54 | 31.84 | 32.53 | **31.97** | 134.38 | 122.59 | 124.98 | **127.32** |
| 10 | H77/29-2 × G73-107 | 19.68 | 21.09 | 21.58 | 20.79 | 37.25 | 30.42 | 34.88 | **34.18** | 73.13 | 59.56 | 61.74 | **64.81** | 28.22 | 20.01 | 23.57 | **23.93** | 110.38 | 89.98 | 96.62 | **98.99** |

(c) Continued.

| Sr. No. | Genotypes | Earhead length (cm) | | | | Earhead weight (g)/plant | | | | Dry fodder yield (g)/plant | | | | Grain yield (g)/plant | | | | Total biological yield (g)/plant | | | |
|---|---|---|---|---|---|---|---|---|---|---|---|---|---|---|---|---|---|---|---|---|---|
| | | $E_1$ | $E_2$ | $E_3$ | Mean | $E_1$ | $E_2$ | $E_3$ | Mean | $E_1$ | $E_2$ | $E_3$ | Mean | $E_1$ | $E_2$ | $E_3$ | Mean | $E_1$ | $E_2$ | $E_3$ | Mean |
| 11 | H77/29-2 × 77/245 | 18.47 | 20.48 | 22.00 | 20.32 | 24.67 | 37.59 | 54.66 | 38.98 | 76.72 | 74.49 | 89.21 | 80.14 | 13.56 | 23.50 | 37.96 | 25.00 | 101.39 | 112.08 | 143.87 | 119.11 |
| 12 | H77/29-2 × CVJ-2-5-3-1-3 | 22.18 | 25.85 | 27.67 | 25.23 | 36.74 | 43.72 | 44.83 | 41.76 | 100.53 | 94.67 | 86.64 | 93.95 | 24.17 | 33.12 | 33.45 | 30.25 | 137.28 | 138.39 | 131.47 | 135.71 |
| 13 | H77/29-2 × 1305 | 21.69 | 23.38 | 25.47 | 23.52 | 67.28 | 61.66 | 67.92 | 65.62 | 133.33 | 116.62 | 133.01 | 127.65 | 41.28 | 45.01 | 56.60 | 47.63 | 200.61 | 178.28 | 200.93 | 193.27 |
| 14 | H77/29-2 × (77/371 × BSECT CP-1) | 19.25 | 20.95 | 21.27 | 20.49 | 35.30 | 41.65 | 46.49 | 41.15 | 96.38 | 69.17 | 76.98 | 80.84 | 16.81 | 28.33 | 31.50 | 25.55 | 131.68 | 110.82 | 123.47 | 121.99 |
| 15 | H77/29-2 × 96AC-93 | 20.12 | 20.70 | 21.80 | 20.87 | 49.55 | 44.47 | 45.96 | 46.66 | 87.38 | 68.82 | 77.62 | 77.94 | 37.54 | 35.58 | 31.81 | 34.98 | 136.93 | 113.29 | 123.59 | 124.60 |
| 16 | H77/29-2 × Togo-II | 21.38 | 24.28 | 25.63 | 23.77 | 49.49 | 46.60 | 51.18 | 49.09 | 85.49 | 72.43 | 71.91 | 76.61 | 35.86 | 33.53 | 30.28 | 33.22 | 134.98 | 119.03 | 123.10 | 125.70 |
| 17 | H77/29-2 × 99HS-18 | 18.99 | 19.33 | 20.73 | 19.69 | 40.46 | 45.27 | 49.88 | 45.20 | 84.92 | 78.95 | 81.44 | 81.77 | 25.61 | 33.78 | 35.53 | 31.64 | 125.39 | 124.22 | 131.32 | 126.98 |
| 18 | G73-107 × 77/245 | 18.40 | 22.53 | 24.37 | 21.77 | 25.80 | 47.76 | 59.15 | 44.24 | 116.73 | 106.08 | 104.59 | 109.13 | 13.58 | 31.63 | 42.86 | 29.36 | 142.53 | 153.85 | 163.74 | 153.37 |
| 19 | G73-107 × CVJ-2-5-3-1-3 | 23.24 | 25.37 | 26.91 | 25.17 | 49.58 | 50.30 | 55.99 | 51.96 | 95.24 | 93.95 | 111.46 | 100.22 | 32.41 | 36.45 | 44.44 | 37.76 | 144.83 | 144.25 | 167.45 | 152.17 |
| 20 | G73-107 × 1305 | 22.28 | 24.33 | 25.83 | 24.15 | 63.40 | 57.94 | 63.73 | 61.69 | 124.90 | 101.58 | 119.61 | 115.36 | 41.99 | 41.99 | 50.18 | 44.72 | 188.30 | 159.52 | 183.34 | 177.06 |
| 21 | G73-107 × (77/371 × BSECT CP-1) | 19.43 | 21.28 | 22.20 | 20.97 | 39.36 | 39.80 | 57.70 | 45.62 | 116.97 | 98.58 | 103.03 | 106.19 | 18.74 | 23.69 | 41.81 | 28.08 | 156.33 | 138.39 | 160.73 | 151.82 |
| 22 | G73-107 × 96AC-93 | 21.18 | 22.18 | 23.17 | 22.17 | 42.70 | 58.43 | 62.01 | 54.38 | 103.19 | 98.72 | 101.59 | 101.17 | 27.03 | 44.60 | 43.79 | 38.47 | 145.89 | 157.15 | 163.59 | 155.55 |
| 23 | G73-107 × Togo-II | 22.37 | 23.65 | 27.33 | 24.45 | 57.36 | 51.75 | 68.22 | 59.11 | 120.98 | 108.08 | 129.96 | 119.67 | 39.84 | 36.70 | 53.72 | 43.42 | 178.35 | 159.82 | 198.18 | 178.78 |
| 24 | G73-107 × 99HS-18 | 20.25 | 20.81 | 23.22 | 21.43 | 35.58 | 42.29 | 66.12 | 48.00 | 123.50 | 96.85 | 115.16 | 111.84 | 21.96 | 27.82 | 50.09 | 33.29 | 159.08 | 139.14 | 181.29 | 159.83 |
| 25 | 77/245 × CVJ-2-5-3-1-3 | 21.09 | 23.60 | 26.75 | 23.81 | 67.60 | 65.58 | 72.99 | 68.72 | 152.25 | 118.61 | 141.62 | 137.49 | 42.25 | 47.52 | 58.39 | 49.39 | 219.85 | 184.18 | 214.61 | 206.21 |
| 26 | 77/245 × 1305 | 19.19 | 21.00 | 21.97 | 20.72 | 57.00 | 51.42 | 53.47 | 53.97 | 116.87 | 105.65 | 107.37 | 109.96 | 43.18 | 39.56 | 42.44 | 41.73 | 173.86 | 157.08 | 160.85 | 163.93 |
| 27 | 77/245 × (77/371 × BSECT CP-1) | 18.51 | 19.35 | 20.20 | 19.36 | 25.77 | 24.60 | 28.38 | 26.25 | 69.89 | 47.78 | 46.36 | 54.68 | 16.31 | 16.62 | 19.99 | 17.64 | 95.66 | 72.38 | 74.74 | 80.93 |
| 28 | 77/245 × 96AC-93 | 17.29 | 17.80 | 21.33 | 18.81 | 55.41 | 50.58 | 57.48 | 54.49 | 119.52 | 81.74 | 95.26 | 98.84 | 36.46 | 35.87 | 40.71 | 37.68 | 174.93 | 132.32 | 152.74 | 153.33 |
| 29 | 77/245 × Togo-II | 23.00 | 25.08 | 26.58 | 24.89 | 30.67 | 54.31 | 59.63 | 48.20 | 117.37 | 110.21 | 121.64 | 116.41 | 15.98 | 40.53 | 47.70 | 34.74 | 148.05 | 164.52 | 181.26 | 164.61 |
| 30 | 77/245 × 99HS-18 | 19.82 | 20.19 | 22.23 | 20.75 | 38.46 | 31.19 | 42.25 | 37.30 | 74.93 | 67.20 | 64.76 | 68.96 | 32.05 | 25.78 | 26.43 | 28.09 | 113.39 | 98.39 | 107.01 | 106.26 |

(c) Continued.

| Sr. No. | Genotypes | Earhead length (cm) | | | | Earhead weight (g)/plant | | | | Dry fodder yield (g)/plant | | | | Grain yield (g)/plant | | | | Total biological yield (g)/plant | | | |
|---|---|---|---|---|---|---|---|---|---|---|---|---|---|---|---|---|---|---|---|---|---|
| | | $E_1$ | $E_2$ | $E_3$ | Mean | $E_1$ | $E_2$ | $E_3$ | Mean | $E_1$ | $E_2$ | $E_3$ | Mean | $E_1$ | $E_2$ | $E_3$ | Mean | $E_1$ | $E_2$ | $E_3$ | Mean |
| 31 | CVJ-2-5-3-1-3 × 1305 | 24.06 | 25.53 | 30.67 | 26.75 | 41.78 | 47.96 | 62.81 | 50.85 | 138.22 | 114.25 | 122.52 | 125.00 | 24.72 | 32.41 | 49.07 | 35.40 | 180.00 | 162.22 | 185.32 | 175.85 |
| 32 | CVJ-2-5-3-1-3 × (77/371 × BSECT CP-1) | 21.42 | 21.73 | 22.38 | 21.84 | 45.22 | 52.35 | 56.40 | 51.32 | 116.73 | 96.58 | 102.48 | 105.26 | 26.14 | 37.66 | 41.78 | 35.19 | 161.95 | 148.93 | 158.88 | 156.59 |
| 33 | CVJ-2-5-3-1-3 × 96AC-93 | 23.50 | 24.77 | 26.00 | 24.76 | 46.76 | 42.80 | 60.90 | 50.15 | 119.75 | 98.32 | 108.68 | 108.91 | 30.97 | 29.31 | 44.45 | 34.91 | 166.51 | 141.11 | 169.58 | 159.07 |
| 34 | CVJ-2-5-3-1-3 × Togo-II | 20.93 | 26.66 | 29.89 | 25.83 | 40.58 | 37.27 | 51.90 | 43.25 | 125.88 | 107.17 | 112.17 | 115.07 | 27.42 | 25.70 | 41.85 | 31.66 | 166.45 | 144.43 | 164.07 | 158.32 |
| 35 | CVJ-2-5-3-1-3 × 99HS-18 | 20.28 | 23.11 | 23.86 | 22.42 | 32.87 | 25.13 | 53.56 | 37.19 | 100.55 | 97.33 | 103.64 | 100.51 | 20.04 | 15.81 | 42.51 | 26.12 | 133.42 | 122.46 | 157.20 | 137.69 |
| 36 | 1305 × (77/371 × BSECT CP-1) | 20.05 | 21.73 | 22.10 | 21.29 | 35.69 | 46.92 | 51.40 | 44.67 | 116.52 | 98.26 | 100.41 | 105.07 | 20.05 | 36.65 | 41.79 | 32.83 | 152.21 | 145.18 | 151.82 | 149.74 |
| 37 | 1305 × 96AC-93 | 18.22 | 20.03 | 20.47 | 19.57 | 51.15 | 40.23 | 48.13 | 46.50 | 99.65 | 76.75 | 75.71 | 84.04 | 34.56 | 26.47 | 31.55 | 30.86 | 150.80 | 116.98 | 123.84 | 130.54 |
| 38 | 1305 × Togo-II | 25.45 | 25.83 | 26.44 | 25.91 | 68.85 | 72.83 | 67.95 | 69.87 | 122.65 | 114.10 | 125.83 | 120.86 | 45.00 | 61.20 | 53.50 | 53.23 | 191.50 | 186.93 | 193.78 | 190.74 |
| 39 | 1305 × 99HS-18 | 20.70 | 24.08 | 25.93 | 23.57 | 52.19 | 43.32 | 48.00 | 47.84 | 109.29 | 76.77 | 89.68 | 91.92 | 39.24 | 33.32 | 38.40 | 36.99 | 161.49 | 120.09 | 137.69 | 139.76 |
| 40 | (77/371 × BSECT CP-1) × 96AC-93 | 20.56 | 21.15 | 21.57 | 21.09 | 31.58 | 44.63 | 47.17 | 41.13 | 91.66 | 75.40 | 81.00 | 82.68 | 18.80 | 35.42 | 33.47 | 29.23 | 123.24 | 120.03 | 128.17 | 123.81 |
| 41 | (77/371 × BSECT CP-1) × Togo-II | 18.68 | 22.53 | 24.50 | 21.90 | 44.22 | 64.58 | 70.38 | 59.73 | 141.53 | 123.69 | 137.97 | 134.39 | 22.91 | 46.46 | 55.86 | 41.74 | 185.75 | 188.27 | 208.35 | 194.12 |
| 42 | (77/371 × BSECT CP-1) × 99HS-18 | 19.95 | 20.26 | 22.09 | 20.77 | 27.20 | 29.09 | 35.43 | 30.57 | 73.92 | 46.92 | 50.23 | 57.02 | 17.66 | 21.87 | 21.47 | 20.33 | 101.12 | 76.01 | 85.66 | 87.60 |
| 43 | 96AC-93 × Togo-II | 19.53 | 20.27 | 25.11 | 21.64 | 59.72 | 43.91 | 71.26 | 58.30 | 155.37 | 139.27 | 144.32 | 146.32 | 37.80 | 25.38 | 56.55 | 39.91 | 215.10 | 183.18 | 215.58 | 204.62 |
| 44 | 96AC-93 × 99HS-18 | 19.52 | 20.21 | 22.42 | 20.72 | 34.65 | 27.58 | 44.96 | 35.73 | 59.59 | 51.86 | 60.94 | 57.46 | 27.07 | 21.54 | 24.09 | 24.23 | 94.24 | 79.44 | 105.90 | 93.19 |
| 45 | Togo-II × 99HS-18 | 19.32 | 20.08 | 26.58 | 21.99 | 53.81 | 49.95 | 54.36 | 52.70 | 102.99 | 88.76 | 92.44 | 94.73 | 39.86 | 38.42 | 36.68 | 38.32 | 156.80 | 138.71 | 146.80 | 147.43 |
| | Mean | 19.82 | 21.46 | 23.01 | 21.43 | 39.63 | 41.23 | 48.36 | 43.07 | 94.98 | 80.87 | 86.63 | 87.49 | 25.10 | 29.14 | 35.13 | 29.79 | 134.60 | 122.10 | 134.99 | 130.56 |
| | SE | 2.43 | 1.98 | 2.45 | — | 6.71 | 5.01 | 6.26 | — | 9.59 | 7.50 | 8.22 | — | 4.23 | 3.49 | 3.44 | — | 11.55 | 9.27 | 11.67 | — |
| | CD | 0.87 | 0.71 | 0.87 | — | 2.39 | 1.78 | 2.23 | — | 3.42 | 2.67 | 2.93 | — | 1.51 | 1.24 | 1.22 | — | 4.11 | 3.30 | 4.16 | — |

TABLE 4: Analysis of variance for 10 × 10 half-diallel for different characters in three environments.

(a)

| Environment | Source of variation | D.F | Germination (%) | Rate of emergence | STI (%) | SSTI (%) | Number of leaves/seedling | Seedling height (cm) | Fresh weight (g)/seedling | Dry weight (g)/seedling |
|---|---|---|---|---|---|---|---|---|---|---|
| | | | | | | Mean sum of squares | | | | |
| $E_1$ | Replicates | 2 | 6.52731 | 5.96768 | 7.83393 | 20.70093 | 0.18674 | 3.37641 | 0.33999 | 0.02590 |
| | Treatments | 54 | 109.70255** | 81.81153** | 308.63092** | 206.80900** | 11.00708** | 21.06317** | 9308.29527** | 3.19767** |
| | Parents | 9 | 51.18163** | 142.89925** | 591.14816** | 188.80254** | 10.84227** | 10.87906** | 225.61719** | 0.42162** |
| | Hybrids | 44 | 114.23573** | 33.98933** | 256.97199** | 206.33716** | 4.72874** | 13.91526** | 6741.23518** | 2.61839** |
| | Parents versus hybrids | 1 | 436.93104** | 1636.19916** | 38.96838 | 389.62793** | 288.73741** | 427.22798** | 2341.44289** | 53.67022** |
| | Error | 108 | 2.81184 | 13.56832 | 11.73684 | 6.74337 | 0.40104 | 3.06462 | 201.51619 | 0.05535 |
| $E_2$ | Replicates | 2 | 5.83819 | 37.07389* | 15.96635 | 1.98738 | 0.34925 | 4.48213 | 8.99529 | 0.07177 |
| | Treatments | 54 | 99.97443** | 28.25177** | 353.78177** | 190.74306** | 6.65284** | 17.64081** | 2576.94570** | 21.77519** |
| | Parents | 9 | 65.07784** | 37.18879** | 698.66289** | 179.85404** | 7.05379** | 14.08518** | 106.12519** | 0.77429** |
| | Hybrids | 44 | 105.69731** | 21.27302** | 291.02537** | 194.29240** | 3.29256** | 7.00003** | 1860.26552** | 14.83428** |
| | Parents versus hybrids | 1 | 162.23710** | 254.88361** | 1113326 | 132.57362** | 150.89673** | 517.83569** | 56348.25791** | 516.18348** |
| | Error | 108 | 3.34760 | 10.97039 | 10.49505 | 6.35570 | 0.67641 | 1.60576 | 6.94613 | 0.06206 |
| $E_3$ | Replicates | 2 | 8.94398 | 48.08815* | — | — | 0.51076 | 0.95989 | 37.68421 | 0.30220 |
| | Treatments | 54 | 126.15600** | 48.42216** | — | — | 28.11077** | 59.47316** | 3870.84450** | 34.45076** |
| | Parents | 9 | 71.35735** | 109.76388** | — | — | 14.68699** | 11.12764** | 1360.81650** | 13.07836** |
| | Hybrids | 44 | 132.17740** | 13.79360 | — | — | 15.54071** | 45.81632** | 2520.84767** | 23.78032** |
| | Parents versus hybrids | 1 | 354.40180** | 1020.00335** | — | — | 702.00767** | 1095.48362** | 85860.95705** | 696.30188** |
| | Error | 108 | 7.82907 | 10.85444 | — | — | 1.39771 | 2.91325 | 30.19169 | 0.24004 |

(b)

| Environment | Source of variation | D.F | Panicle emergence | Effective tillers/plant | Plant height | Ear length | Ear weight/plant | Dry fodder yield/plant | Grain yield/plant | Total biological yield/plant |
|---|---|---|---|---|---|---|---|---|---|---|
| | | | | | | Mean sum of squares | | | | |
| $E_1$ | Replicates | 2 | 31.86061** | 4.16348** | 167.29472* | 1.70641 | 10.77192 | 96.34539 | 3.92777 | 115.54459 |
| | Treatments | 54 | 41.52368** | 0.85657** | 3267.06089** | 16.02572** | 617.70406** | 3041.13204** | 342.85030** | 5756.17582** |
| | Parents | 9 | 45.11481** | 0.18538* | 1408.07476** | 22.40093** | 116.53959** | 270.14155** | 38.58650** | 613.21919** |
| | Hybrids | 44 | 32.01347** | 1.00797** | 1103.95657** | 10.10497** | 402.96671** | 1712.71411** | 241.36880** | 3029.99650** |
| | Parents versus hybrids | 1 | 427.65286** | 0.23609* | 115174.52616** | 219.16167** | 14576.62749** | 86430.43524** | 7546.41053** | 171994.67546** |
| | Error | 108 | 4.68159 | 0.04907 | 36.92099 | 2.24610 | 17.11430 | 34.98126 | 6.82350 | 50.75268 |
| $E_2$ | Replicates | 2 | 6.56364 | 0.02556 | 615.75565** | 8.28842** | 5.89094 | 22.68506 | 2.60877 | 15.07116 |
| | Treatments | 54 | 29.99618** | 0.69822** | 2689.98723** | 24.36886** | 605.92784** | 2472.59877** | 419.20073** | 5134.83489** |
| | Parents | 9 | 26.30000** | 0.48079** | 1655.55405** | 38.71874** | 141.22393** | 195.43682** | 31.32426** | 572.30706** |
| | Hybrids | 44 | 25.33165** | 0.73449** | 961.62857** | 14.63359** | 354.95366** | 1547.59288** | 262.62465** | 2934.92304** |
| | Parents versus hybrids | 1 | 268.50135** | 1.05929** | 88047.66697** | 323.57161** | 15831.12746** | 63667.31560** | 10799.43646** | 142993.70699** |
| | Error | 108 | 5.07598 | 0.05829 | 52.34122 | 1.49406 | 9.55133 | 21.42411 | 4.62311 | 32.73886 |
| $E_3$ | Replicates | 2 | 0.29697 | 0.21161 | 94.59847 | 1.43746 | 9.85141 | 46.80064 | 9.26334 | 33.45160 |
| | Treatments | 54 | 80.45499** | 0.94511** | 1924.20108** | 32.79225** | 720.49211** | 2989.53021** | 550.61308** | 6517.32994** |
| | Parents | 9 | 80.10741** | 0.94166** | 1160.37343** | 37.02185** | 163.05190** | 232.36440** | 45.53893** | 714.84982** |
| | Hybrids | 44 | 62.54377** | 0.95514** | 737.81513** | 21.88348** | 318.13193** | 1902.63916** | 315.72179** | 3686.89934** |
| | Parents versus hybrids | 1 | 871.67710** | 0.53496* | 60999.63199** | 474.71212** | 23441.30203** | 75627.22861** | 15431.49707** | 183278.59719** |
| | Error | 108 | 6.96981 | 0.07883 | 55.78505 | 2.28608 | 14.91014 | 25.74223 | 4.49588 | 51.87705 |

* Significant at P = 0.05.
** Significant at P = 0.01.

TABLE 5: Mean performance of the genotypes for membrane thermostability in two environments.

| Sr. No. | Genotypes | Membrane thermostability* | | |
| --- | --- | --- | --- | --- |
| | | $E_1$ | $E_2$ | Mean |
| 1 | H77/833-2 | 74.74 | 73.73 | **74.23** |
| 2 | H77/29-2 | 53.37 | 55.42 | **54.39** |
| 3 | G73-107 | 79.44 | 72.37 | **75.91** |
| 4 | 77/245 | 54.15 | 54.42 | **54.28** |
| 5 | CVJ-2-5-3-1-3 | 75.74 | 76.81 | **76.27** |
| 6 | 1305 | 52.28 | 51.69 | **51.99** |
| 7 | 77/371 × BSECT CP-1 | 74.21 | 73.58 | **73.90** |
| 8 | 96AC-93 | 72.22 | 75.63 | **73.92** |
| 9 | Togo-II | 74.35 | 75.98 | **75.16** |
| 10 | 99HS-18 | 76.39 | 74.65 | **75.52** |
| 1 | H77/833-2 × H77/29-2 | 72.11 | 68.66 | **70.39** |
| 2 | H77/833-2 × G73-107 | 75.29 | 70.72 | **73.00** |
| 3 | H77/833-2 × 77/245 | 66.70 | 67.42 | **67.06** |
| 4 | H77/833-2 × CVJ-2-5-3-1-3 | 74.03 | 71.78 | **72.90** |
| 5 | H77/833-2 × 1305 | 51.21 | 51.71 | **51.46** |
| 6 | H77/833-2 × (77/371 × BSECT CP-1) | 73.58 | 72.34 | **72.96** |
| 7 | H77/833-2 × 96AC-93 | 75.87 | 75.85 | **75.86** |
| 8 | H77/833-2 × Togo-II | 67.57 | 67.08 | **67.33** |
| 9 | H77/833-2 × 99HS-18 | 65.96 | 66.02 | **65.99** |
| 10 | H77/29-2 × G73-107 | 68.67 | 66.01 | **67.34** |
| 11 | H77/29-2 × 77/245 | 61.53 | 64.02 | **62.77** |
| 12 | H77/29-2 × CVJ-2-5-3-1-3 | 75.98 | 78.86 | **77.42** |
| 13 | H77/29-2 × 1305 | 68.73 | 68.92 | **68.82** |
| 14 | H77/29-2 × (77/371 × BSECT CP-1) | 66.20 | 65.18 | **65.69** |
| 15 | H77/29-2 × 96AC-93 | 59.90 | 58.30 | **59.10** |
| 16 | H77/29-2 × Togo-II | 70.06 | 68.21 | **69.13** |
| 17 | H77/29-2 × 99HS-18 | 67.38 | 65.07 | **66.22** |
| 18 | G73-107 × 77/245 | 73.36 | 70.89 | **72.12** |
| 19 | G73-107 × CVJ-2-5-3-1-3 | 68.37 | 69.60 | **68.99** |
| 20 | G73-107 × 1305 | 70.09 | 74.30 | **72.19** |
| 21 | G73-107 × (77/371 × BSECT CP-1) | 71.30 | 68.94 | **70.12** |
| 22 | G73-107 × 96AC-93 | 67.22 | 70.09 | **68.65** |
| 23 | G73-107 × Togo-II | 57.49 | 55.41 | **56.45** |
| 24 | G73-107 × 99HS-18 | 67.95 | 70.11 | **69.03** |
| 25 | 77/245 × CVJ-2-5-3-1-3 | 60.21 | 59.45 | **59.83** |
| 26 | 77/245 × 1305 | 54.23 | 53.99 | **54.11** |
| 27 | 77/245 × (77/371 × BSECT CP-1) | 52.24 | 52.17 | **52.20** |
| 28 | 77/245 × 96AC-93 | 56.39 | 54.24 | **55.31** |
| 29 | 77/245 × Togo-II | 49.96 | 51.50 | **50.73** |
| 30 | 77/245 × 99HS-18 | 59.70 | 62.19 | **60.95** |
| 31 | CVJ-2-5-3-1-3 × 1305 | 61.91 | 64.31 | **63.11** |
| 32 | CVJ-2-5-3-1-3 × (77/371 × BSECT CP-1) | 68.08 | 64.11 | **66.09** |
| 33 | CVJ-2-5-3-1-3 × 96AC-93 | 78.13 | 76.33 | **77.23** |
| 34 | CVJ-2-5-3-1-3 × Togo-II | 67.38 | 68.01 | **67.69** |
| 35 | CVJ-2-5-3-1-3 × 99HS-18 | 74.53 | 73.63 | **74.08** |
| 36 | 1305 × (77/371 × BSECT CP-1) | 64.88 | 64.47 | **64.68** |
| 37 | 1305 × 96AC-93 | 76.12 | 74.90 | **75.51** |
| 38 | 1305 × Togo-II | 69.11 | 71.12 | **70.12** |
| 39 | 1305 × 99HS-18 | 76.80 | 75.07 | **75.94** |

TABLE 5: Continued.

| Sr. No. | Genotypes | Membrane thermostability* | | |
|---|---|---|---|---|
| | | E₁ | E₂ | Mean |
| 40 | (77/371 × BSECT CP-1) × 96AC-93 | 75.79 | 74.62 | **75.21** |
| 41 | (77/371 × BSECT CP-1) × Togo-II | 76.06 | 78.73 | **77.40** |
| 42 | (77/371 × BSECT CP-1) × 99HS-18 | 75.06 | 74.70 | **74.88** |
| 43 | 96AC-93 × Togo-II | 68.81 | 66.15 | **67.48** |
| 44 | 96AC-93 × 99HS-18 | 73.68 | 69.27 | **71.47** |
| 45 | Togo-II × 99HS-18 | 65.60 | 63.08 | **64.34** |
| | Mean | **67.78** | **67.30** | **67.54** |
| | CD | **3.20** | **3.60** | — |
| | SE | **1.13** | **1.27** | — |
| | CV | **2.35** | **2.66** | — |

*Values are transformed values.

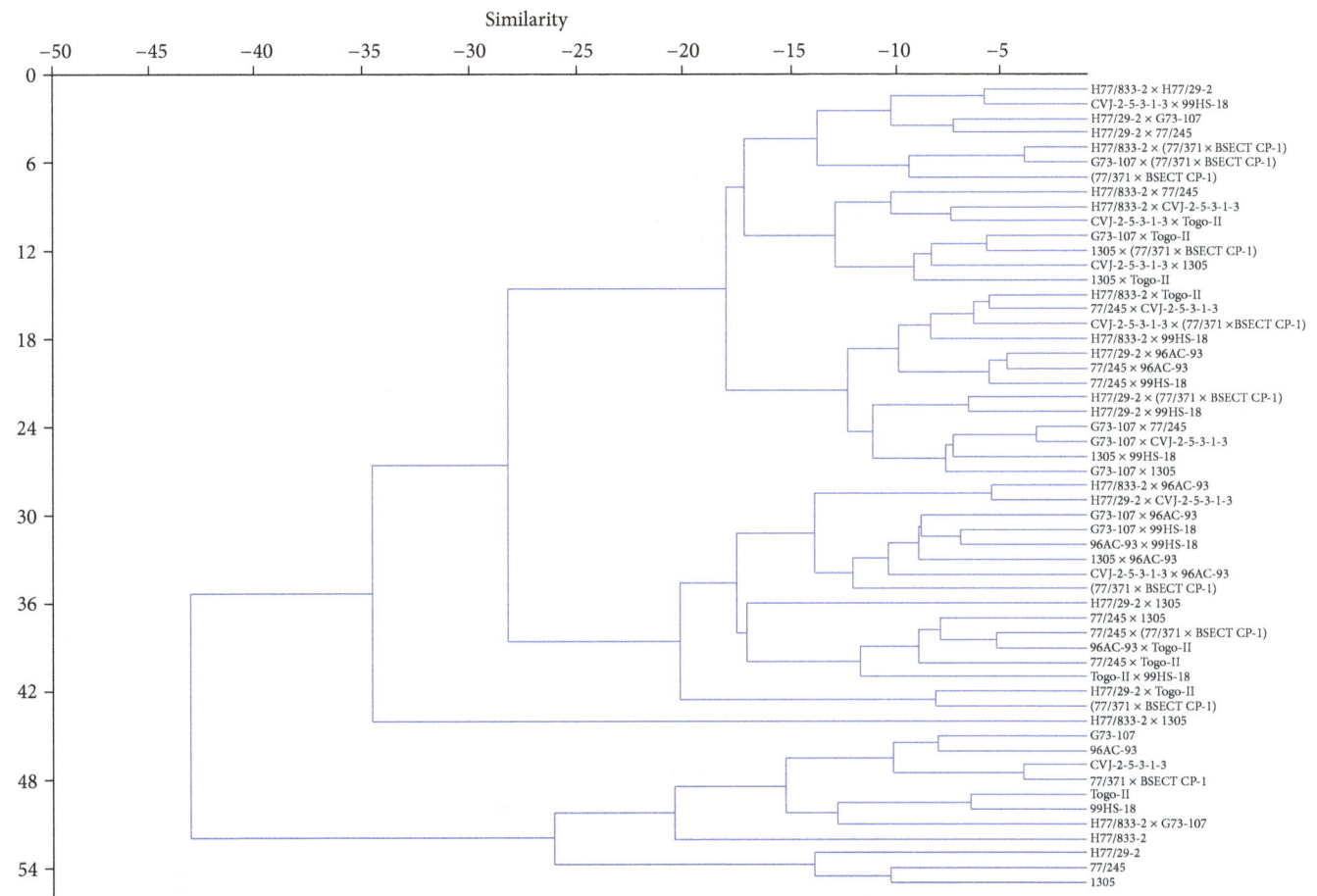

FIGURE 2: Dendrogram representing similarity among pearl millet genotypes (parents and hybrids) based on PAST cluster analysis of Euclidean distance based on phenotypic traits.

moisture was depleting near to the permanent wilting point and drought stress could also come into play along with heat stress to further continue the experiment. Since experiment proceeded without any stress after the establishment of seedling [4], the pearl millet plants show great resilience and recovery. The main impact of supraoptimal temperature at germination and seedling stage was observed on the seedling survival and resulted in poor plant stand/population of pearl millet. Poor crop stand reduces the productivity (fodder and grain) drastically, whereas after seedling establishment no impact of supraoptimal temperature at seedling stage was observed on later growth stages. Hence, the supraoptimal temperature at early seedling stage in pearl millet could have no impact on its later growth and development stages. The

TABLE 6: Pooled correlation coefficient among various characters in three environments.

| Characters | Germination | Rate of emergence | STI | SSTI | Leaves/ seedling | Seedling height | Fresh weight | Dry weight | MTS | Panicle emergence | Effective tillers/plant | Plant height | Ear length | Ear weight/plant | Dry fodder yield/plant | Grain yield/plant | Total biological yield/plant |
|---|---|---|---|---|---|---|---|---|---|---|---|---|---|---|---|---|---|
| Germination | 1.000 | 0.357** | 0.353** | 0.590** | 0.230 | 0.373** | 0.293* | 0.260 | 0.276* | -0.135 | -0.478** | 0.326* | -0.069 | -0.104 | -0.190 | -0.145 | -0.167 |
| Rate of emergence | | 1.000 | 0.233 | 0.351** | 0.605** | 0.585** | 0.551** | 0.524** | 0.139 | -0.593** | 0.032 | 0.517** | 0.217 | 0.388** | 0.361** | 0.348** | 0.377** |
| STI | | | 1.000 | 0.837** | 0.241 | 0.129 | 0.081 | 0.048 | 0.890** | -0.215 | -0.211 | 0.030 | -0.007 | -0.043 | -0.101 | -0.111 | -0.084 |
| SSTI | | | | 1.000 | 0.233 | 0.264 | 0.183 | 0.140 | 0.745** | -0.247 | -0.268* | 0.184 | -0.017 | -0.066 | -0.129 | -0.144 | -0.112 |
| Number of leaves | | | | | 1.000 | 0.683** | 0.633** | 0.604** | 0.176 | -0.474** | 0.109 | 0.576** | 0.496** | 0.641** | 0.607** | 0.630** | 0.630** |
| Seedling height | | | | | | 1.000 | 0.703** | 0.663** | 0.031 | -0.493** | -0.115 | 0.735** | 0.360** | 0.526** | 0.520** | 0.515** | 0.532** |
| Fresh weight | | | | | | | 1.000 | 0.957** | -0.016 | 0.425** | -0.082 | 0.661** | 0.351** | 0.581** | 0.496** | 0.576** | 0.533** |
| Dry weight | | | | | | | | 1.000 | -0.033 | -0.428** | -0.016 | 0.634** | 0.354** | 0.620** | 0.509** | 0.614** | 0.554** |
| MTS | | | | | | | | | 1.000 | -0.40 | -0.177 | -0.037 | 0.015 | -0.074 | -0.138 | -0.131 | -0.120 |
| Panicle emergence | | | | | | | | | | 1.000 | -0.150 | -0.269* | 0.022 | -0.329* | -0.337* | -0.280* | -0.341* |
| Effective tillers/plant | | | | | | | | | | | 1.000 | -0.012 | 0.170 | 0.366** | 0.405** | 0.323* | 0.401** |
| Plant height | | | | | | | | | | | | 1.000 | 0.650** | 0.603** | 0.596** | 0.619** | 0.610** |
| Ear length | | | | | | | | | | | | | 1.000 | 0.608** | 0.650** | 0.648** | 0.664** |
| Ear weight/plant | | | | | | | | | | | | | | 1.000 | 0.910** | 0.984** | 0.957** |
| Dry fodder yield/plant | | | | | | | | | | | | | | | 1.000 | 0.907** | 0.991** |
| Grain yield/plant | | | | | | | | | | | | | | | | 1.000 | 0.950** |
| Total biological yield/plant | | | | | | | | | | | | | | | | | 1.000 |

* Significant at $P = 0.05$. ** Significant at $P = 0.01$.

faster germination and early vigour in seedling could act as physical indicators for heat tolerance.

Most of the maturity traits showed significant correlation among themselves. Panicle emergence had a significant but negative correlation with plant height, ear weight/plant, dry fodder yield/plant, grain yield/plant, and total biological yield/plant, whereas grain yield/plant showed significant correlation with rate of emergence of seedlings, number of leaves/seedling, seedling height, fresh weight of seedling, dry weight of seedling among the seedling traits and number of effective tillers/plant, plant height, ear length, ear weight/plant, dry fodder yield/plant, and total biological yield/plant among the maturity traits.

Cluster analysis of the data (all the characters) based on similarity among genotypes gave the relative position of parents and hybrids in each group (Figure 2). The pearl millet genotypes were broadly grouped into four clusters. The parents are grouped into separate cluster. The resulting populations from selective crosses will be useful for deriving quantitative data which can be applied to quantitative genetics in order to understand the inheritance and basis of thermotolerance expression in pearl millet. On similar lines quantification of thermotolerance variations in pearl millet will contribute to the understanding of inheritance of thermotolerance expression at seedling stage through use of data in quantitative genetics.

## Conflict of Interests

The authors declare that there is no conflict of interests regarding the publication of this paper.

## References

[1] P. Soman, F. R. Bidinger, J. M. Peacock, and T. S. Walker, "Seedling establishment—a preliminary survey taken up in Aurepally during *kharif* 1981," ICRISAT Internal Report, ICRISAT, Patancheru, India, 1981.

[2] S. Baljeet, *Supra-optimal temperature tolerance in pearl millet inheritance pattern of some adaptive traits at seedling stage [M.Sc. Thesis]*, CCS HAU, Hisar, India, 1993.

[3] A. K. Yadav, M. S. Narwal, and B. Singh, "Field screening technique for heat effect on seedlings of pearl millet," in *Proceedings of the National Seminar on Transgenic Crops in Indian Agriculture: Status, Risk and Acceptance*, pp. 109–112, CCS HAU, Hisar, India, January 2006.

[4] J. M. Peacock, P. Soman, R. Jayachandran, A. V. Rani, C. J. Howarth, and A. Thomas, "Effect of high soil surface temperature on seedling survival in pearl millet," *Experimental Agriculture*, vol. 29, no. 2, pp. 215–225, 1993.

[5] A. K. Yadav, M. S. Narwal, and R. K. Arya, "Evaluation of pearl millet (*Pennisetum glaucum*) genotypes and validation of screening methods for supra-optimal temperature tolerance at seedling stage," *Indian Journal of Agricultural Sciences*, vol. 83, no. 3, pp. 260–271, 2013.

[6] S. Rajbir, "Performance of diclofop-methyl to control weeds in wheat (*Triticum aestivum*) under different irrigation levels," *Indian Journal of Agricultural Research*, vol. 41, no. 1, pp. 75–78, 2007.

[7] C. A. Black, *Methods of Soil Analysis: Part I Physical and Mineralogical Properties*, American Society of Agronomy, Madison, Wis, USA, 1965.

[8] A. K. Yadav, M. S. Narwal, and R. K. Arya, "Genetic dissection of temperature tolerance in pearl millet (*Pennisetum glaucum*)," *Indian Journal of Agricultural Sciences*, vol. 81, no. 3, pp. 203–213, 2011.

[9] A. M. H. Ibrahim and J. S. Quick, "Genetic control of high temperature tolerance in wheat as measured by membrane thermal stability," *Crop Science*, vol. 41, no. 5, pp. 1405–1407, 2001.

[10] A. K. Yadav, R. K. Arya, and S. M. Narwal, "Screening for supra-optimal temperature tolerance through membrane thermo-stability in pearl millet (*Pennisetum glaucum*)," *Forage Research*, vol. 35, no. 2, pp. 85–90, 2009.

[11] J. D. Maguire, "Speed of germination—aid in selection and evaluation for seedling emergence and vigour," *Crop Science*, vol. 2, no. 2, pp. 176–177, 1962.

[12] V. G. Panse and P. V. Sukhatme, *Statistical Method for Agricultural Workers*, ICAR, New Delhi, India, 4th edition, 1967.

[13] R. A. Fisher and F. Yates, *Statistical Tables for Biological Agricultural and Medical Research*, Oliver and Boyd, Edinburgh, UK, 6th edition, 1963.

[14] Ø. Hammer, D. A. T. Harper, and P. D. Ryan, "Past: paleontological statistics software package for education and data analysis," *Palaeontologia Electronica*, vol. 4, no. 1, 2001.

# 12

# Morphoagronomic Characterization of Tomato Plants and Fruit: A Multivariate Approach

**Aniela Pilar Campos de Melo,**[1] **Paulo Marçal Fernandes,**[2] **Fábio Venturoli,**[3]
**Carlos de Melo Silva-Neto,**[3] **and Aurélio Rubio Neto**[4]

[1] *Federal University of Goiás, Sector of Horticulture, Brazil*
[2] *Federal University of Goiás, Sector of Phytosanitary, Brazil*
[3] *Federal University of Goiás, Sector of Forestry, Brazil*
[4] *Federal Institute Goiano, Brazil*

Correspondence should be addressed to Aniela Pilar Campos de Melo; aniela.pcdmelo@gmail.com

Academic Editor: Gábor Kocsy

Consumers in the fresh fruit market choose fruits mainly following criteria related to the external appearance. However, the introduction of new material for planting depends on the productive capacity of the plant as well as on the formation of fruit that meets consumer desires. Given the above, the objective of this study was to morphoagronomically characterize tomato genotypes using multivariate statistics. The genotype seedlings (Ellus, Black Mauri, Green Zebra, Green Tomato, Pomodoro Marmande, Pomodoro Fiorentino, Pitanga, and Black Krim) were transplanted 30 days after sowing. The morphoagronomic characterization of the genotypes was carried out by evaluating plants and fruits. The data were analyzed using descriptive analysis, namely, position and variability measurements. In addition, a multivariate cluster analysis and a principal component analysis were carried out for plant and fruit attributes. The cluster and principal component analyses were efficient in characterizing plants and/or fruits of different tomato genotypes. Such efficiency enhances result interpretation and proposed inferences, with applied relevance for the producers. The genotype Ellus has a combination of morphoagronomic plant and fruit traits superior to other genotypes. Such superior traits enable a high productivity.

## 1. Introduction

The tomato (*Solanum lycopersicon* L.) was originated in South America and is the second most important vegetable in economic importance and consumption in the world, second only to potatoes [1]. Bioactive substances such as carotenoids (lycopene, $\beta$-carotene, and lutein), phenolics (flavonoids, phenolic acid, and tannins), and vitamins (C, E, and B) [2, 3] are obtained by the in *natura* consumption of fruits. These compounds positively affect health for having anti-inflammatory and anticancer effects and preventing chronic diseases (obesity, diabetes, coronary heart disease, and hypertension) [4–7].

Consumers in the fresh fruit market choose fruits mainly following criteria related to the external appearance [8].

Color, size, shape, and texture are the first attributes evaluated. Therefore, the introduction of new material for planting depends on the productive capacity of the plant as well as on the formation of fruit that meets consumer desires [9].

In general, the phytotechnical characterization of plants and fruits of different tomato genotypes is carried out by tests that distinguish the phytotechnical characteristics using only analysis of variance and mean comparison tests. Studies on the existing relationship among phytotechnical characteristics, among genotypes, and within genotypes are deficient in the scientific literature. Therefore, the proposed inferences are somewhat superficial.

An approach appropriate for a thorough study on such relationships is the use of multivariate techniques [10]. All variables are simultaneously assessed by exploratory or

TABLE 1: Height 35 DAT ($H$-35), number of flowers 35 DAT (FLO-35), number of fruits 35 DAT (FRU-35), number of fruits 80 DAT (FRU-80), estimated production (EST-PRO), estimated productivity (EST-PROD), and growth habit (GHAB) of tomato genotypes.

| Genotypes | $H$-35 | FLO-35 | FRU-35 | FRU-80 | EST-PRO | EST-PROD | GHAB |
|---|---|---|---|---|---|---|---|
| | cm | Plant | Plant | Plant | kg plant$^{-1}$ | ton ha$^{-1}$ | — |
| Ellus | 180.75 | 11.00 | 11.00 | 39.50 | 3.77 | 62.95 | Undetermined |
| Black Mauri | 175.87 | 11.87 | 12.62 | 79.00 | 1.38 | 23.04 | Undetermined |
| Green Zebra | 118.00 | 5.75 | 2.12 | 27.12 | 1.42 | 23.73 | Undetermined |
| Green Tomato | 168.12 | 11.25 | 9.50 | 24.62 | 1.47 | 24.53 | Undetermined |
| Pitanga | 142.25 | 13.50 | 4.12 | 20.87 | 1.63 | 27.25 | Undetermined |
| Black Krim | 79.50 | 31.00 | 13.25 | 19.87 | 1.37 | 22.84 | Determined |
| Pomodoro Fiorentino | 117.75 | 12.87 | 4.00 | 23.50 | 1.69 | 28.25 | Undetermined |
| Pomodoro Marmande | 103.50 | 10.37 | 5.87 | 13.87 | 1.27 | 21.29 | Undetermined |
| CV (%) | 5.94 | 24.26 | 64.90 | 19.22 | — | — | — |

CV: coefficient of variation.

inference techniques that enable checking possible links, similarities, or differences among variables [11]. Therefore, the interpretation of data structure is optimized, mainly providing a reduction in information and time loss.

Given the above, the objective of this study was to morphoagronomically characterize tomato genotypes using multivariate statistics (exploratory techniques such as cluster analysis and principal component analysis).

## 2. Material and Methods

The study was carried out in a greenhouse present in organic system (16°57'51.79'' S; 49°11'02.09'' O; 865 m altitude). The soil is classified as Red Oxisol and has the following physical and chemical attributes: texture (clay 26%, silt 7%, and sand 67%), pH (6.3), organic matter (27%), cation exchange capacity (CEC) (7.5), base saturation (81.36%), K (214 ppm), and Ca (4.0 mE/100 mL).

The genotype seedlings (Ellus, Black Mauri, Green Zebra, Green Tomato, Pomodoro Marmande, Pomodoro Fiorentino, Pitanga, and Black Krim) were transplanted 30 days after sowing. The plantation was fertilized applying a kilogram of poultry manure (26.4 g Kg$^{-1}$ of nitrogen, 84.0 g Kg$^{-1}$ of P$_2$O$_5$, 23.0 g Kg$^{-1}$ of K$_2$O, 112 g Kg$^{-1}$ of Ca, 6.4 g Kg$^{-1}$ of Mg, 2.5 g Kg$^{-1}$ of S, and a 7.25 pH) per meter. The lines were separated by a distance of 1 meter and the plants by 0.6 meters. The tomato support system consisted of a double rod with a narrow ribbon. The removal of excess shoots was held weekly as of 30 days after transplanting (DAT).

The morphoagronomic characterization of the genotypes was carried out by evaluating plants and fruits. The growth habit, shoot height, number of flowers, and number of fruits were determined in eight plants of each genotype, randomly picked, at 35 DAT.

At 80 DAT, 24 fruits were harvested from eight plants of each genotype and the following measures were obtained: fresh mass weight (gravimetry), longitudinal diameter (direct measurement with calipers), transversal diameter (direct measurement with calipers), epicarp, mesocarp, and endocarp coloring (visual method), number of locules, mesocarp

thickness (direct measurement with calipers), epicarp firmness (texturometry), and seed quantity per fruit. The number of fruits per plant was again counted 80 DAT to estimate the yield per plant (number of fruits * fruit mass) and production in tons per hectare (yield per plant * 16666/1000).

The data were analyzed using descriptive analysis, namely, position (mean) and variability (coefficient of variation) measurements. In addition, a multivariate cluster analysis and a principal component analysis (PCA using correlation matrices) were carried out for plant and fruit attributes. The cluster dendrogram was obtained through the UPGMA (Unweighted Pair Group Method using Arithmetic Averages), using the Euclidean distance as a similarity coefficient.

## 3. Results and Discussion

A large variation in the number of fruits per plant (35 DAT), number of locules, and firmness was observed among the evaluated genotypes (coefficient of variation of 64.90%, 37.40%, and 28.73%, resp.; Tables 1 and 2). Tables 1 and 2 exhibit parameters associated with height and flowers and fruits production of the tomato genotypes evaluated. Height, flower production, and fruit production had a high variability and thus may be used as indicating variables for material selection in genetic improvement programs [1].

The genotype Ellus stood out among the others for having a higher estimated fruit production per plant and area (Table 1) and for the fruit traits related to mesocarp mass and thickness (Table 2). The genotype Black Krim had the highest number of flowers, but it did not result in the greatest fixing and fruit formation. Some genotypes or/and cultivars are more sensible to abiotic factors in the reproductive phase, such as temperature and radiation [12, 13]. High temperatures cause flowers to fall, affect pollen viability and/or fertilization, and lead to a low fruit set.

The genotypes Ellus, Black Mauri, Pitanga, Black Krim, Pomodoro Fiorentino, and Pomodoro Marmande had fruit with a reddish epicarp, with tonal variations (Table 3). Specifically, materials for industrial processing (e.g., Black Krim) have peel redder than the pulp due to greater accumulation of lycopene in the epicarp [8].

TABLE 2: Biometrics of tomato fruits, of different genotypes: fresh weight (FW), longitudinal diameter (LD), transverse diameter (TD), longitudinal and transverse diameter ratio (LD/TD), firmness (FIR), mesocarp thickness (THICK), number of locules (LOC), and seed quantity (SEM).

| Genotypes | FW | LD | TD | LD/TD | FIR | THICK | LOC | SEM |
| --- | --- | --- | --- | --- | --- | --- | --- | --- |
| | g | mm | mm | — | kgf | mm | Fruit | Fruit |
| Ellus | 95.62 | 54.52 | 62.23 | 1.01 | 2.79 | 6.19 | 2.87 | 83.12 |
| Black Mauri | 17.50 | 28.54 | 38.03 | 0.75 | 2.06 | 2.85 | 2.17 | 50.69 |
| Green Zebra | 52.50 | 47.16 | 43.53 | 1.07 | 1.71 | 3.25 | 2.96 | 40.33 |
| Green Tomato | 59.79 | 50.62 | 40.18 | 1.25 | 1.89 | 3.79 | 4.92 | 95.56 |
| Pitanga | 78.33 | 57.53 | 53.66 | 1.06 | 1.86 | 4.69 | 5.67 | 24.47 |
| Black Krim | 68.96 | 52.58 | 51.86 | 1.02 | 1.94 | 4.41 | 4.52 | 29.13 |
| Pomodoro Fiorentino | 72.14 | 63.76 | 38.34 | 1.67 | 1.39 | 3.82 | 6.90 | 58.47 |
| Pomodoro Marmande | 92.08 | 63.14 | 44.99 | 1.40 | 1.82 | 4.25 | 7.62 | 99.81 |
| CV (%) | 15.82 | 7.22 | 5.34 | 27.12 | 28.73 | 22.43 | 37.40 | 23.89 |

CV: coefficient of variation.

TABLE 3: Epicarp, mesocarp, and endocarp color of fruits from eight tomato genotypes.

| Genotypes | Color | |
| --- | --- | --- |
| | Epicarp | Mesocarp and endocarp |
| Ellus | Rosy red | Rosy red |
| Black Mauri | Brownish red | Brownish red |
| Green Zebra | Yellow with green stripes | Green |
| Green Tomato | Light green | Green |
| Pitanga | Cherry red | Cherry red |
| Black Krim | Intense red | Rosy red |
| Pomodoro Fiorentino | Red | Red |
| Pomodoro Marmande | Rosy red | Rosy red |

Lycopene is the most abundant carotene in red tomato fruits [14]. The biosynthesis is associated with the change in fruit color from green to red when chloroplasts are transformed to chromoplasts [15]. Its abundance is associated with genetic control (especially by increasing the expression of hp, og$^c$, and CrtL genes) [8, 16, 17], fruit development time [18], water supply [19], and climatic conditions [8]. In addition, lycopene biosynthesis can be inhibited if high temperatures (above 30°C) occur in fruits, often due to a high incidence of solar radiation in the epicarp [8].

The genotypes Green Zebra and Green Tomato stood out from the others due to the yellowish and greenish epicarp, respectively, and the green pulp (Table 3). The pigments present in genotypes Green Zebra and Green Tomato are possibly xanthophylls (lutein and zeaxanthin), oxygenated derivatives of carotenoids. Xanthophylls are powerful antioxidants associated with a reduction in the risk of eye diseases (e.g., age-related macular degeneration and cataract) [20, 21]. In addition, high concentrations of lutein, zeaxanthin, and vitamin C in the plasma could reduce the incidence of chronic diseases (cancer, diabetes, cardiovascular disease, and stroke) by preventing leukocyte telomere shortening [22].

Several enzymatic complexes, especially the lycopene cyclase, mediate carotenogenesis, in tomato fruits [16, 17]. Wild species holding green (S. neorickii, S. peruvianum, and S. pennellii) and yellow fruit (S. cheesmaniae) have polymorphisms in the coding region of the gene that encodes the enzyme β lycopene cyclase [23]. Therefore, the inability of Green Zebra and Green Tomato fruits to accumulate lycopene is probably related with the mutation of genes associated with the lycopene cyclase.

3.1. Multivariate Analysis. The first and second components of the multivariate analysis of plant traits explained 73.38% of the total variance (Figure 1). The genotype Black Krim was represented in the first quadrant, Black Mauri in the second quadrant, and Ellus in the fourth quadrant. Such representation indicated that the traits number of fruits per plant at 80 DAT (Black Mauri), number of flowers per plant at 35 DAT (Black Krim), plant height at 35 DAT, and estimated production per plant and per hectare (Ellus) prevail in describing these genotypes. The genotypes Pomodoro Marmande, Pomodoro Fiorentino, Pitanga, and Green Zebra were located in the third quadrant and were characterized by having a low number of fruits per plant at both 35 DAT and 80 DAT.

The production of fruits per plant/area was positively correlated with plant height at 35 DAT. According to Piotto and Peres [24], the tomato plant only flourishes after issuing eight to twelve leaves. Therefore, genotypes that are more precocious with respect to vegetative growth may be more susceptible to hormonal and environmental stimuli, which favor flowering. Finally, this early vegetative maturity may provide greater accumulation of photoassimilates due to the increased competitiveness in capturing light, leading to a greater success in fruiting.

The first and second components of the multivariate analysis conducted for the fruit traits (Figure 2) explained 84.13% of the total variance. The occurrence of genotype Ellus in the second quadrant shows its prominence in regard to mesocarp thickness, fresh weight, and transverse diameter and confirms the importance of these traits in describing the

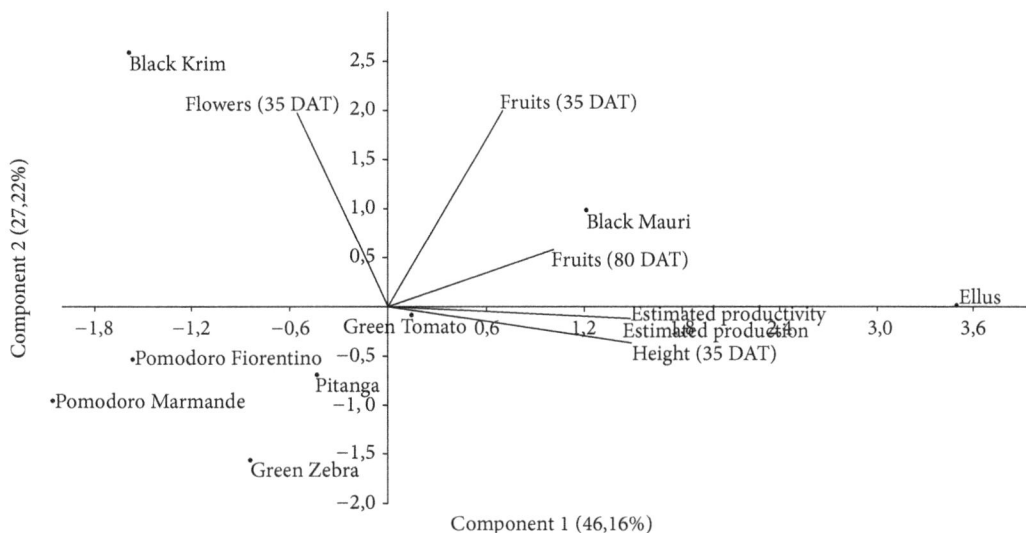

FIGURE 1: Biplot of plant variables (height, flowers, number of fruits, and production) and tomato genotypes relative to principal components 1 and 2.

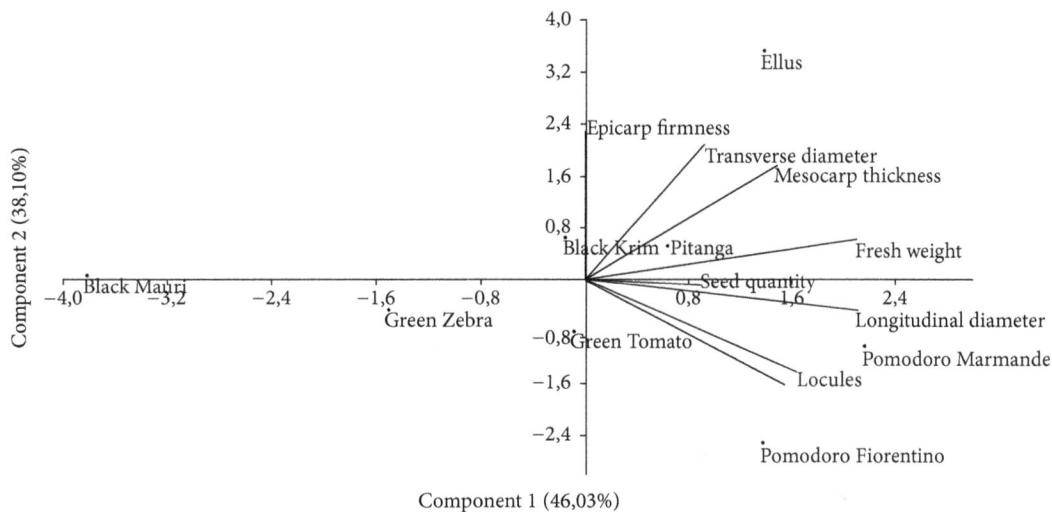

FIGURE 2: Biplot of fruit variables (fresh weight, longitudinal and transverse diameter, longitudinal and transverse diameter ratio, epicarp firmness, mesocarp thickness, locules, and number seeds) and tomato genotypes relative to principal components 1 and 2.

genotype Ellus. Meanwhile, low firmness of the epicarp was the best descriptor of fruit of the genotype Green Tomato.

The Italian genotypes Pomodoro Marmande and Fiorentino, located in the fourth quadrant, stood out from the others, especially by having a high number of locules and longitudinal/transverse diameter ratio. The locules are cavities within the fruit, derived from the ovary, where seed are submerged in placental mucilage. The number of locules is positively correlated with the number of seeds and fruit size in the fourth quadrant (Figure 2).

The genotype Black Mauri, represented in the first quadrant, differed mainly by the low mass and fruit size (Figure 2) and had twice the number of fruits as genotype Ellus (Table 1). Still, due to the low average mass of genotype Black Mauri (Table 2), production per plant and per area was approximately three times lower than for genotype Ellus.

Genotypes Black Krim and Pitanga had low numbers of seed per fruit.

Genotypes of specific growth, as Black Krim, are suitable for the production of fruits for industrial use. The combination of small size (Table 1), intense red peel (Table 3), and uniform maturation leads to an optimization of the mechanical harvesting and the processing of sauces and ketchups [8, 24].

Cluster analysis of plant traits enabled grouping the eight genotypes in only two groups (Figure 3). Group 1 comprised genotypes Green Tomato, Pitanga, Ellus, and Black Mauri and Group 2 genotypes Green Zebra, Pomodoro Fiorentino, Pomodoro Marmande, and Black Krim. The cluster of fruit traits (Figure 4) also indicated the existence of two groups: Group 1 (Green Tomato, Pomodoro Marmande, and Ellus) and Group 2 (Green Zebra, Pitanga, Black Krim, Pomodoro

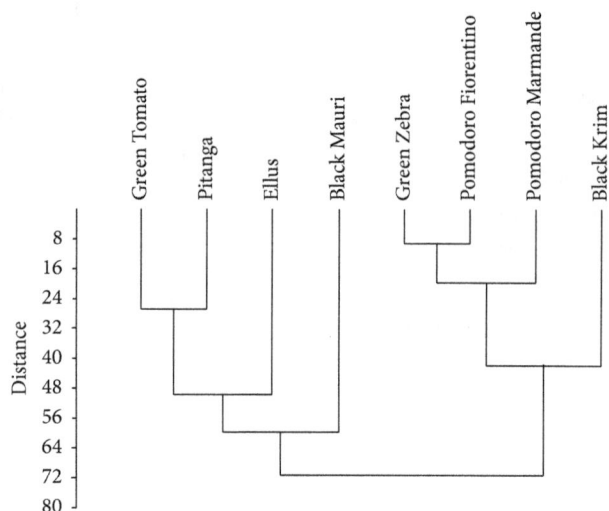

FIGURE 3: Similarity dendrogram of tomato genotypes associated with plant characteristics (height, flowers, fruits, and production).

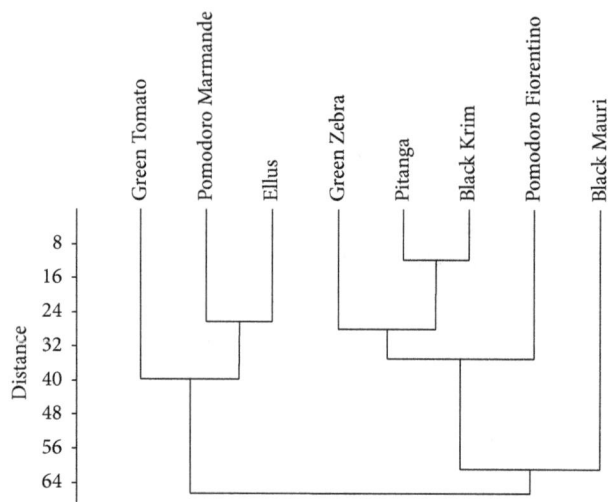

FIGURE 4: Similarity dendrogram of tomato genotypes associated with fruit characteristics (fresh mass weight, longitudinal and transverse diameter, relationship between longitudinal and transverse diameter, epicarp firmness, loci, and number of seeds).

Fiorentino, and Black Mauri). The intersection of the group pairs indicates a high level of similarity between genotypes Green Tomato and Ellus and genotypes Green Zebra, Pomodoro Fiorentino, and Black Krim.

## 4. Conclusion

The cluster and principal component analyses were efficient in characterizing plants and/or fruits of different tomato genotypes. Such efficiency enhances result interpretation and proposed inferences, with applied relevance for the producers.

The genotype Ellus has a combination of morphoagronomic plant and fruit traits superior to other genotypes. Such superior traits enable a high productivity.

## Conflict of Interests

The authors declare that there is no conflict of interests regarding the publication of this paper.

## References

[1] D. O. Ibitoye, P. E. Akin-Idowu, and O. T. Ademoyegun, "Agronomic and lycopene evaluation in tomato (*Lycopersicon lycopersicum* Mill.) as a function of genotype," *World Journal of Agricultural Sciences*, vol. 5, pp. 892–895, 2009.

[2] R. Ilahy, C. Hdider, M. S. Lenucci, I. Tlili, and G. Dalessandro, "Antioxidant activity and bioactive compound changes during fruit ripening of high-lycopene tomato cultivars," *Journal of Food Composition and Analysis*, vol. 24, no. 4-5, pp. 588–595, 2011.

[3] A. Raiola, M. M. Rigano, R. Calafiore, L. Frusciante, and A. Barone, "Enhancing the health-promoting effects of tomato fruit for biofortified food," *Mediators of Inflammation*, vol. 2014, Article ID 139873, 16 pages, 2014.

[4] P. M. Kris-Etherton, K. D. Hecker, A. Bonanome et al., "Bioactive compounds in foods: their role in the prevention of cardiovascular disease and cancer," *The American Journal of Medicine*, vol. 113, no. 9, supplement 2, pp. 71S–88S, 2002.

[5] M. Giovannetti, L. Avio, R. Barale et al., "Nutraceutical value and safety of tomato fruits produced by mycorrhizal plants," *British Journal of Nutrition*, vol. 107, no. 2, pp. 242–251, 2012.

[6] K. Zu, L. Mucci, B. A. Rosner et al., "Dietary lycopene, angiogenesis, and prostate cancer: a prospective study in the prostate-specific antigen era," *Journal of the National Cancer Institute*, vol. 106, no. 2, article djt430, 2014.

[7] Y. Li, V. L. W. Go, and F. H. Sarkar, "The role of nutraceuticals in pancreatic cancer prevention and therapy: targeting cellular signaling, MicroRNAs, and Epigenome," *Pancreas*, vol. 44, no. 1, pp. 1–10, 2015.

[8] S. Brandt, Z. Pék, É. Barna, A. Lugasi, and L. Helyes, "Lycopene content and colour of ripening tomatoes as affected by environmental conditions," *Journal of the Science of Food and Agriculture*, vol. 86, no. 4, pp. 568–572, 2006.

[9] Y. Bai and P. Lindhout, "Domestication and breeding of tomatoes: what have we gained and what can we gain in the future?" *Annals of Botany*, vol. 100, no. 5, pp. 1085–1094, 2007.

[10] J. M. Moita Neto and G. C. Moita, "Uma introdução à análise exploratória de dados multivariados," *Química Nova*, vol. 21, no. 4, pp. 467–469, 1998.

[11] J. F. Hair Junior, R. E. Anderson, R. L. Tatham et al., *Análise multivariada de dados*, Artmed, Porto Alegre, Brazil, 2005.

[12] M. T. A. Gusmão, S. A. L. Gusmão, and J. A. C. Araújo, "Produtividade de tomate tipo cereja cultivado em ambiente protegido e em diferentes substratos," *Horticultura Brasileira*, vol. 24, no. 4, pp. 431–436, 2006.

[13] Z. Pék and L. Helyes, "The effect of daily temperature on truss flowering rate of tomato," *Journal of the Science of Food and Agriculture*, vol. 84, no. 13, pp. 1671–1674, 2004.

[14] N. Lavi, Y. Tadmor, A. Meir et al., "Characterization of the intense pigment tomato genotype emphasizing targeted fruit

metabolites and chloroplast biogenesis," *Journal of Agricultural and Food Chemistry*, vol. 57, no. 11, pp. 4818–4826, 2009.

[15] Z. Pék, L. Helyes, and A. Lugasi, "Color changes and antioxidant content of vine and postharvest ripened tomato fruits," *HortScience*, vol. 45, no. 3, pp. 466–468, 2010.

[16] I. Pecker, R. Gabbay, F. X. Cunningham Jr., and J. Hirschberg, "Cloning and characterization of the cDNA for lycopene β-cyclase from tomato reveals decrease in its expression during fruit ripening," *Plant Molecular Biology*, vol. 30, no. 4, pp. 807–819, 1996.

[17] G. Ronen, M. Cohen, D. Zamir, and J. Hirschberg, "Regulation of carotenoid biosynthesis during tomato fruit development: expression of the gene for lycopene epsilon-cyclase is down-regulated during ripening and is elevated in the mutant Delta," *The Plant Journal*, vol. 17, no. 4, pp. 341–351, 1999.

[18] D. Znidarcic and T. Pozrl, "Comparative study of quality changes in tomato cv. 'Malike' (*Lycopersicon esculentum* Mill.) whilst stored at different temperatures," *Acta Agriculture Slovenica*, vol. 87, no. 2, pp. 235–243, 2006.

[19] L. Helyes, A. Lugasi, and Z. Pék, "Effect of irrigation on processing tomato yield and antioxidant components," *Turkish Journal of Agriculture and Forestry*, vol. 36, no. 6, pp. 702–709, 2012.

[20] J. L. Olea, J. A. Aragón, M. E. Zapata, and J. Tur, "Características de la población con ingesta baja en luteína y zeaxantina en pacientes con degeneración macular asociada a la edad variante húmeda," *Archivos de la Sociedad Española de Oftalmología*, vol. 87, no. 4, pp. 112–118, 2012.

[21] J. Karppi, J. A. Laukkanen, and S. Kurl, "Plasma lutein and zeaxanthin and the risk of age-related nuclear cataract among the elderly Finnish population," *British Journal of Nutrition*, vol. 108, no. 1, pp. 148–154, 2012.

[22] A. Sen, G. Marsche, P. Freudenberger et al., "Association between higher plasma lutein, zeaxanthin, and vitamin C concentrations and longer telomere length: results of the Austrian Stroke Prevention Study," *Journal of the American Geriatrics Society*, vol. 62, no. 2, pp. 222–229, 2014.

[23] A. H. Araújo, M. E. D. N. Fonseca, and L. S. Boiteux, "Nucleotide diversity of a major carotenoid biosynthetic pathway gene in wild and cultivated *Solanum* (Section *lycopersicon*) species," *Brazilian Journal of Plant Physiology*, vol. 19, no. 3, pp. 233–237, 2007.

[24] F. A. Piotto and L. E. P. Peres, "Base genética do hábito de crescimento e florescimento em tomateiro e sua importância na agricultura," *Ciência Rural*, vol. 42, no. 11, pp. 1941–1946, 2012.

# 13

# Origin, Domestication, and Dispersing of Pear (*Pyrus* spp.)

**G. J. Silva, Tatiane Medeiros Souza, Rosa Lía Barbieri, and Antonio Costa de Oliveira**

*Plant Genomics and Breeding Center, Federal University of Pelotas, 96001-970 Pelotas, RS, Brazil*

Correspondence should be addressed to Antonio Costa de Oliveira; acostol@terra.com.br

Academic Editor: Innocenzo Muzzalupo

The pear (*Pyrus communis* L.) is a typical fruit of temperate regions, having its origin and domestication at two different points, China and Asia Minor until the Middle East. It is the fifth most widely produced fruit in the world, being produced mainly in China, Europe, and the United States. Pear belongs to rosaceous family, being a close "cousin" of the apple, but with some particularities that make this fruit special with a delicate flavor. Thus, it deserves a special attention and a meticulous review of all the history involved, and the recent research devoted to it, because of the economic and cultural importance of this fruit in a range of countries and cultures. Therefore, the purpose of this literature review is to approach the history of the origin, domestication, and dispersal of pears, as well as reporting their botany, their current scenario in the world, and their breeding and conservation.

## 1. Introduction

Pear, a typical fruit of temperate climates, with delicate pleasant taste and smooth, has a wide acceptance throughout the world. By its shape, it inspires designers and architects. The fruit pleases generations; already in 1661, Jean-Baptiste de La Quintinie, lawyer and botanist, responsible for the gardens of the Versailles palace, passionate about the cultivation of pears, wrote in reports: "It must be confessed that, among all fruits in this place, nature does not show anything so beautiful nor so noble as this pear. It is pear that makes the greatest honor on the tables..."

The pear is mainly consumed *in natura*, pies, cakes, accompanying strong cheese or carpaccio, risotto, jams, and ice creams and is a great fruit to be consumed in diets because of its low caloric value. It has high nutritional value with reasonable amounts of vitamins A, B1, B2, B3, and C and minerals like sodium, potassium, phosphorus, calcium, magnesium, and iron. It has a lot of fiber, giving excellent results in the treatment of constipation and intestine inflammation. Many recommend pears to cure anomalies such as cystitis and kidney stones [1].

Belonging to the genus *Pyrus*, which originated in the Tertiary period, in Western China, the pear had its dispersion from northern Italy, Switzerland, former Yugoslavia, Germany, Greece, Moldova, and Ukraine to the East, in countries such as Iran, Uzbekistan, China, Japan, Korea, and Bhutan. Commercially, it is divided into two major groups: European and Asian pears. The first, with elongated and full-bodied texture, and the second, with sandy texture and rounded body, make this fruit the ninth in world production, being mainly a commodity in China [2–4].

## 2. Taxonomy, Origin, and Speciation

The name pear is derived from Latin, *pera* or *pira*, with some variants like in French as *poire*, in German as *peer*, and in Greece as *acras* as wild type and *apios* as cultivated pear.

It belongs to Equisetopsida C. Agardh class of vascular plants, Magnoliidae Novák ex Takht subclass, characterized by plants that have ribbed leaves and flowers. Belonging to the Rosales Bercht. & J.Presl order, and Rosaceae Juss family, with hermaphrodite flowers, polypetalae and perigynics stamens, the pear, of *Pyrus* L., gender is a fruit of big importance for the agriculture of latitude moderate countries, being cultivated on a large scale in China, Western Europe, and the United States [5–7].

The Maloideae subfamily, where the *Pyrus* gender belongs, has a basic chromosome number as $x = 17$, which is fair if compared with other species of Rosaceae, where $x = 7$ or $x = 9$. Of the three hypotheses that emerged from the 1920s to explain the event, the most accepted theory [8] suggests an allotetraploid or allopolyploid from the cross between two primitive forms of Rosaceae family, Prunoideae with $x = 8$ and Spiraeoideae with $x = 9$. This theory was based on the observation of a predominance of univalent (unpaired chromosomes) and not from multivalent chromosomes during meiosis. Subsequently, isozyme studies supported this theory [9]. Most cultivated pears are diploid ($2n = 34$), but there are a few polyploid cultivars of P. communis and Pyrus × bretschneideri. According to some authors [9], the speciation of Pyrus occurred without a change in chromosome number. It is believed that gender Pyrus originated during the Tertiary period (65 to 55 million years ago) in the mountainous regions of western China where a very large number of species of the gender Pomoideae and Prunoideae are concentrated. Taking into consideration the areas of distribution of the various genres of Pomoideae, it is likely that the common ancestor of these was widely distributed in that territory during the Cretaceous or Paleocene and prior to the Tertiary. Evidence suggests that pear dispersion and speciation followed the mountain ranges to both the east and the west [10, 11]. In this period, only few traces of leaves in some localities from eastern Europe and the Caucasus were found, as the village of Parschlug, Austria, and the Kakhetia mountains, where Pyrus theobroma fossils were found. Whereas in eastern Georgia, Horizon Akchagyl, Azerbaijan, and Turkey, Pyrus communis L. fossil leaves were also found. In postglacial records, traces of fruits were found in lacustrine deposits in Switzerland and Italy [12]. It is believed that the process of domestication followed what is currently seen in the Caucasus, where one can find many types of pear trees that grow abundantly [13].

There are two domestication centers and primary origin of the genus Pyrus: the first is located in China, the second located in Asia Minor to the Middle East, in the Caucasus mountains, and a third secondary center located in Central Asia [14, 15].

The number of cataloged species varies greatly according to the interpretation of each author, 20 to 75 species [16]. There are 23 wild species cataloged, all native to Europe, temperate Asia, and northern mountainous regions of Africa [7, 17, 18]. Pears are classified into three groups according to the number of carpels and fruit size: small fruits that have two carpels known as Asian pears, large fruits with five carpels, and fruits with three to four carpels that are hybrids of fruits mentioned above. Asian pears have a crisp texture, while the European pear has a buttery and juicy texture, with characteristic flavor and aroma. Pears are propagated by grafting, where the graft is adapted against stresses such as soil alkalinity, drought, cold. Species diversity is concentrated in western Eurasia to eastern Asia and especially in China (Table 1), but several species are mentioned by many authors, without a consensus, which hampers an organization, as many are hybrids between species, and in some cases, different regions use different names for the same cultivars [10, 19, 20]. In these two regions, two distinct groups of species, eastern and western, are formed

(Table 2). Studies indicate that there is a large genetic distance between these two groups [21]. The first is focusing on most cultivated pears, found in Europe, North Africa, Asia Minor, Iran, part of Soviet Central Asia, and Afghanistan. The second group includes species that are concentrated in East Asia, the Tien-Shan and Hindu Kush mountains, and Japan. In the latter, there is a very large group of cultivars in China and Japan [11, 22]. Currently there are several works that aim to estimate the genetic distance among the different cultivars, concentrated in gene banks and breeding programs.

Researchers at the University of Lleida (UDL-ETSIA) could estimate the genetic distance of 141 Spanish accessions of P. communis (past and current) through eight SSR markers. Thirteen well-known Spanish cultivars that represent their diversity were also used, but all thirteen were grouped into a single cluster, showing the narrow genetic base of cultivars P. communis in Spain, mainly caused by market demands [23].

Another study was conducted by a group of Chinese researchers, in which, through six SSR markers, it was possible to verify the genetic distance of 98 species of Pyrus, including 51 Pyrifolia, Japanese and Chinese Pyrus, 11 P. ussuriensis, 24 Chinese white pears, six wild types, two Korean species, two P. communis cultivars, and 2 unidentified types. The results showed the grouping of these cultivars in 10 groups, with 4 groups composed of white and sandy pears of Chinese and Japanese origin. The results showed that Japanese cultivars have as parents, Chinese sandy pear. Western cultivars formed separated and distant groups from the eastern pears [24].

Many studies have been conducted in the context to identify genetic variations and clustering of populations of cultivated pear in China, since the fruit is a commodity of great importance to this country, as a study of 233 landraces of P. pyrifolia, the "sandy pear," was able to determine the level of genetic diversity and relatedness of companies by 14 SSR markers [25].

In 2013, the pear genome sequencing was completed by combining the illumina sequencing technology and a BAC by BAC (bacterial artificial chromosome) strategy in an Asian pear named "Suli" [26]. This strategy minimized the limitation of the sequencing of a heterozygous genome. The results showed a frequency of 1.02% of SNPs and 53.1% of repeated sequences in the pear genome. It was verified that the genomic portion of pear and apple is very similar, and the major differences between them are the repeated sequences that are actively transposing.

The pear genome sequencing project concluded that the average density of genes is one per 12 kb in at least 42,812 gene loci, a similar number comparing to other plants, and that the pear and apple genome are almost equal in gene numbers. The project also showed that the lignin content found in pear is similar to that of poplar, indicating that this lignin content is involved in the stone cell formation [26]. From the genomic approaches used in this project, a better understanding of this fruit crop has been achieved, which will reflect on future improvements.

TABLE 1: *Pyrus* species and hybrids from Asia.

| Species | Site of origin | Crop |
|---|---|---|
| *Pyrus alnifolia* (S. and Z.) Franch. and Sav. | Russian Far East, China, Japan, Korea, Taiwan | * |
| *Pyrus armeniacifolia* T. T. Yu | China | * |
| *P. aucuparia* var. randaiensis Hayata | Taiwan | * |
| *Pyrus baccata* L. | Russia, Mongolia, China, Korea | * |
| *Pyrus baccata* var. *aurantiaca* Regel | Russia, Mongolia, China, Korea | * |
| *Pyrus baccata* var. *himalaica* Maxim. | China, Bhutan, India, Nepal | * |
| *Pyrus baccata* var. *mandshurica* Maxim. | Russia, China, Japan, Korea | * |
| *Pyrus betulifolia* Bunge | China, Laos | * |
| *Pyrus* × *bretschneideri* Rehder | China | * |
| *Pyrus calleryana* Decne. | China, Korea, Taiwan, Vietnam | USA, Canada |
| *Pyrus calleryana* var. *dimorphophylla* (Makino) Koidz. | Japan | * |
| *Pyrus calleryana* var. *fauriei* (C. K. Schneid.) Rehder | Korea | * |
| *Pyrus calleryana* var. *koehnei* (C. K. Schneid.) T. T. Yu | China | * |
| *Pyrus cathayensis* Hemsl. | China | * |
| *Pyrus delavayi* Franch. | China | * |
| *Pyrus discolor* Maxim. | China | * |
| *Pyrus doumeri* Bois | China, Taiwan, Laos, Vietnam | * |
| *Pyrus folgner* (C. K. Schneid.) Bean | China | * |
| *Pyrus foliolosa* Wall. | Burma, Bhutan, India, Nepal, China | * |
| *Pyrus glabra* Boiss. | Iran | * |
| *Pyrus gracilis* Siebold and Zucc. | Japan | * |
| *Pyrus harrowiana* Balf. f. and W. W. Sm. | China, India, Nepal, Burma | * |
| *Pyrus heterophylla* Regel and Schmalh. | Kyrgyzstan, Tajikistan, China | * |
| *Pyrus hondoensis* Nakai and Kikuchi | Japan | * |
| *Pyrus* × *hopeiensis* T. T. Yu | China | * |
| *Pyrus hupehensis* Pamp. | China, Taiwan | * |
| *Pyrus indica* Wall. | South Asia and Far East Asia | * |
| *Pyrus japonica* Thunb. | Japan | * |
| *Pyrus keissleri* (C. K. Schneid.) H. Lev. | China, Myanmar | * |
| *Pyrus kansuensis* Batalin | China | * |
| *Pyrus lanata* D. Don | Afghanistan, India, Nepal, Pakistan | * |
| *Pyrus matsumurana* Makino | Japan | * |
| *Pyrus nussia* Buch.-Ham. ex D. Don | Far East, South Asia | * |
| *Pyrus* × *phaeocarpa* Rehder | China | * |
| *Pyrus pohuashanensis* Hance | Russia, China, Korea | * |
| *Pyrus prattii* Hemsl. | China | * |
| *Pyrus prunifolia* Willd. | China | * |
| *Pyrus pseudopashia* T. T. Yu | China | * |
| *Pyrus pyrifolia* var. *pyrifolia* | China, Laos, Vietnam | * |
| *Pyrus ringo* Wenz. | China, Korea | * |
| *Pyrus ringo* var. *kaido* Wenz | China | * |
| *Pyrus scabrifolia* Franch. | China | * |
| *Pyrus scalaris* (Koehne) Bean | China | * |
| *Pyrus* × *serrulata* Rehder | China | * |
| *Pyrus sieboldii* Regel | China, Japan | * |
| *Pyrus sikkimensis* Hook. f. | China, Bhutan, India | * |
| *Pyrus sinensis* var. *maximowicziana* H. Lev. | Korea | * |
| *Pyrus* × *sinkiangensis* T. T. Yu | China | * |
| *Pyrus spectabilis* Aiton | China | * |
| *Pyrus taiwanensis* Iketani and H. Ohashi | Taiwan | * |

| Species | Site of origin | Crop |
|---|---|---|
| *Pyrus ussuriensis* Maxim. | Russia, China, Japan, Korea, Brazil | Brazil |
| *Pyrus* × *uyematsuana* Makino | Japan, Korea | * |
| *Pyrus vestita* Wall. ex G. Don | China, Bhutan, India, Nepal, Myanmar | * |
| *Pyrus vilmorinii* (C. K. Schneid.) Asch. and Graebn. | China | * |
| *Pyrus xerophila* T. T. Yu | China | * |
| *Pyrus yunnanensis* Franch. | China, Myanmar | * |
| *Pyrus zahlbruckneri* (C. K. Schneid.) Cardot | China | * |
| *Pyrus tschonoskii* Maxim. | Japan | * |
| *Pyrus cydonia* L. | Iran, Armenia, Azerbaijan, Russia, Turkmenistan | * |
| *Pyrus germanica* (L.) Hook. f. | Middle East and Northern Asia | * |
| *Pyrus korshinskyi* Litv. | Afghanistan, Tajikistan, Uzbekistan | * |
| *Pyrus kumaoni* Decne. | Middle East, Far East and South Asia | * |
| *Pyrus salicifolia* Pall. | Iran, Armenia, Turkey, Arzebaijão | * |
| *Pyrus trilobata* (Poir.) DC. | Israel, Lebanon, Turkey, Bulgaria, Greece | * |
| *Pyrus turkestanica* Franch. | Kyrgyzstan, Tajikistan, Turkmenistan, Afghanistan | * |

*The same origin.
Source: USDA (2012) [5].

## 3. Domestication and Breeding

Domestication has as a consequence the change in gene frequencies regarding to the original populations. A fully domesticated species is dependent upon man for its survival; in other words, it cannot reproduce in nature itself. The domestication of fruits began only around 6,000 years ago, through vegetative propagation, due to high rate of heterozygosity in them. As a consequence, self-fertility in pear and peach trees, hermaphroditism in grape, parthenocarpy, seedless fruits on banana, and absence of spines in some fruits [27] emerged. During this period, ancient Mediterranean fruits such as grape, olive, fig, and pomegranate have been domesticated. Even citrus, banana, apple, pear, quince, medlar, almond, apricot, cherry, peach, and plum were domesticated in Central and East Asia. Some fruits such as kiwi, blueberry, and pecan were domesticated only in the 19th and 20th centuries. The earliest mention of growing pears in Europe was made by Homer in ancient Greece, a little less than three thousand years ago, who wrote that "Pears are a gift of God" [2]. It was then that breeding started and also the history of the pear as a cultivated plant. Theophrastus (371–287 BC), another Greek, also made important reports on pear. The same distinguished the wild forms from cultivated ones and suggested that bred genotypes received a special name and other important observations on breeding in general [28]. A large contribution to pear cultivation was made by the Romans. Portius-Cato (235–150 BC) described the methods of propagation, grafting, and caring for fruit and also described six cultivars of pear. Another great writer of ancient Rome, Terentius Varro, dedicated some of his work to agriculture (116–27 BC), describing grafting methods and storage. Among the Roman historians, the most important of all was Pliny the Elder (23–79 AD), who described in detail almost all varieties of the season, in a manuscript with more than sixty editions. In summary, the ancient Romans

reported more than 40 cultivars existing in the 1st century BC and described methods of cultivation similar to the practiced currently [2, 28]. Little is known about introducing pear in France, but in the mid-800, the cultivation has developed very well on site, making the country in the sixteen and seventeen centuries the world's largest producer of the fruit. During the eighteenth century, Belgium developed numerous cultivars, including some that are important even today, as the varieties "Beurre Bosc," "Beurre d'Anjou," "Flemish Beauty," and "Winter Nelis" [2].

The pear improvement happened in Europe from two species: *Pyrus communis* and *P. nivalis*. The first, European Common pear, is completely barren and has in its gene pool an influence of other species such as *P. eleagrifolia*, *P. spinosa*, *P. nivalis*, and *P. syriaca* [29]. The second, used to make wine, has been of great importance in Britain and France for over 400 years. Most cultivars released in Europe were developed via open pollination and fruits were selected according to their softness and buttery aspect.

In Asia, the cultivation began over 2500 years ago, with the main species *Pyrus pyrifolia*, *Pyrus serotina*, and *Pyrus ussuriensis*. The result was reported in written Chinese (Shi Jing) and other books for at least 1500 years [30]. In Japan, pear seeds dating from the years 200–300 were found. During the Edo period in Japan (1603–1868) over 150 cultivars were documented; this time the pears were planted in the corners, like a talisman to avoid the "evil eye."

One of the main characteristics of Asian pears is the crispy, sweet, and juicy acid pulp. The pulp is characterized by having "stone cells" which are sclerenchyma cells that differ from fiber because they are very elongated. They also offer a sandy texture to the fruit [2]. The sizes vary from rounded as apples, these being the most cultivated, until pears to the top and bottom elongated bulbous pears, similar to the European pears. The fruits are very sensitive to physical damage, both at harvest and in the classification as storage and marketing.

TABLE 2: *Pyrus* species and hybrids originating in Europe and Southern Africa.

| Species | Geographic distribution-site of origin | Crop |
|---|---|---|
| *Pyrus aria* (L.) Ehrh. | Canary Islands, North Africa, All of Europe | * |
| *Pyrus aria* (L.) Ehrh. var. cretica Lindl. | North Africa, Middle East, Central Europe Oriental and Southern and Turkmenistan | * |
| *Pyrus aucuparia* var. dulcis (K.) A. and G. | All Europe | North America |
| *Pyrus boissieriana* Buhse | Azerbaijan, Turkmenistan, Iran | * |
| *Pyrus korshinskyi* Litv. subsp. bucharica (Litv.) B. K. | Former Soviet Union | * |
| *Pyrus bulgarica* Kuth. and Sachokia (*Pyrus* × nivalis Jacq.) | Western Europe, Central Eastern and Southern | * |
| *Pyrus caucasica* Fed. | Eastern Europe and Central Greece | * |
| *Pyrus chamaemespilus* (L.) Ehrh. | Western Europe, Central Eastern and Southern | * |
| *Pyrus communis* L. | All Europe | Eastern Europe Central, South and West, and South America |
| *P. communis* var. cordata (Desv.) H.f. | UK, Portugal, Spain, France | * |
| *P. communis* subsp gharbiana (T.) Maire | Algeria, Morocco | * |
| *P. communis* subsp. marmorensis (Trab.) Maire | Morocco | * |
| *P. communis* subsp. pyraster (L.) Ehrh. | Western Europe, Central Eastern, and Southern | * |
| *Pyrus* × *complexa* Rubtzov | Former Soviet Union | * |
| *Pyrus cossonii* Rehder | Algeria | * |
| *Pyrus crataegifolia* Savi | Turkey, Albania, Serbia, Greece, Italy, Macedonia | * |
| *Pyrus cuneifolia* Guss. | Central Eastern Europe, South and Central | * |
| *Pyrus decipiens* Bechst. | All Europe and North Africa | * |
| *Pyrus domestica* (L.) Sm. | Algeria, Cyprus, Eastern Europe Central West and Meridional | * |
| *Pyrus elaeagrifolia* Pall. | Turkey, Ukraine, Albania, Bulgaria, Greece, Romania | * |
| *Pyrus elaeagrifolia* subsp. kotschyana | Turkey | * |
| *Pyrus germanica* (L.) Hook. f. | Middle East, Eastern Europe, Central, Southern and Northern Asia | * |
| *Pyrus gharbiana* Trab. | Algeria, Morocco | * |
| *Pyrus intermedia* Ehrh. | All Europe | * |
| *Pyrus malus* subsp. paradisiaca (L.) Schubl. and G. Martens | Western, Eastern, and Central Europe and Greece | * |
| *Pyrus minima* Ley | UK | * |
| *Pyrus nebrodensis* Guss. | Italy - Sicily | * |
| *Pyrus pinnatifida* Ehrh. | All Europe | * |
| *Pyrus praemorsa* Guss | South of Italy, France | * |
| *Pyrus sachokiana* Kuth. | Georgia | * |
| *Pyrus spinosa* Forssk. | Central Eastern Europe, South, and Central | * |
| *Pyrus sudetica* Tausch | Western Europe, Central Eastern, and Southern | * |
| *Pyrus syriaca* Boiss. | Caucasus and Middle East Region | * |
| *Pyrus torminalis* (L.) Ehrh. | North Africa, Middle East, South Caucasus, whole Europe | * |
| *Pyrus trilobata* (Poir.) DC. | Turkey, Bulgaria, Greece, Israel, Lebanon | * |

*The same origin.
Source: USDA (2012) [5].

TABLE 3: *Pyrus* species and hybrids originating in the Americas.

| Species | Place of origin | Crop |
|---|---|---|
| *Pyrus americana* DC | Greenland, USA, Canada | * |
| *Pyrus angustifolia* Aiton | USA, Canada | * |
| *Pyrus arbutifolia* (L.) L. f. | USA | * |
| *Pyrus arbutifolia* (L.) L. f. var. nigra Willd. | USA | Northern and Eastern Europe Center |
| *Pyrus coronaria* L. | Canada, USA | * |
| *P. coronaria* var. ioensis Alph. Wood | USA | * |
| *Pyrus diversifolia* Bong. | USA, Canada | * |
| *Pyrus floribunda* Lindl. | USA, Canada | Korea, Russia, Sweden, Czech Republic, Slovakia, Germany, Latvia, Bulgaria |
| *Pyrus fusca* (Raf.) C. K. Schneid. | USA, Canada | * |
| *Pyrus sanguinea* Pursh | Canada, USA | * |

*The same origin.
Source: USDA (2012) [5].

Pear was introduced by the English and French settlers in the United States and Canada, and in 1629, there was record of its cultivation in New England [28]. Unlike Europe, which grew via grafting pears, pear in the United States was initially cultivated by seeds, which resulted in a much higher genetic variability than in Europe [31], resulting in a number of different varieties in America. Currently, many European pears are well established in North America; however, the U.S. genotypes cannot adapt to the climate and European soil (Table 3). In the United States, in the latter half of the nineteenth century, breeders have used the pear wild type (crosses between Asian and European pears) to their crosses, in order to obtain greater resistance to cold and "fire blight" disease caused by the bacterium *Erwinia amylovora* that is widely spread, though causing a large reduction in the quality of the fruit, which was repaired with successive backcrosses. The most notable difference between these junctions is undoubtedly the texture [31–33]. Wild type pears are used today as a rootstock because of their cold tolerance and adaptability to different environments [31].

## 4. Production and Economic Importance

A medium-sized fruit has about 58 calories, 6 grams of fiber, and 7.0 mg of vitamin C, besides being free of fat and sodium and possessing significant amounts of calcium, iron, magnesium, phosphorus, potassium, zinc, copper, manganese, and phytosterols [5]. Pears, because they are part of the family Rosaceae, have sorbitol as their main translocated sugar that is converted into glucose, fructose, and sucrose. The sugar content varies greatly among Japanese, Chinese, and European pears [34]. Japanese and Chinese pears are those with higher and lower sucrose content, respectively, and the European pears are those with a high content of fructose.

Pear is used mostly for fresh consumption or for the production of jams [35], being the ninth most important cultivated fruit in the world (Table 4). China is the world's largest producer (Asian pear) and the United States is the second largest producer, being the first producer of European

TABLE 4: World production of fruit crops in the years 2010 and 2012 in tonnes.

| Type of fruit | 2010 | 2012 |
|---|---|---|
| Watermelons | 101,342,555 | 105,372,341 |
| Banana | 105,726,175 | 101,992,743 |
| Apples | 70,581,492 | 76,378,738 |
| Orange | 69,045,495 | 68,223,759 |
| Grape | 67,460,130 | 67,067,129 |
| Melon | 31,495,365 | 31,925,787 |
| Fruit fresh nes | 29,414,585 | 31,447,977 |
| Tangerines | 23,867,076 | 27,060,756 |
| Pears | 22,705,619 | 23,580,845 |
| Pineapples | 20,377,660 | 23,333,886 |

FAO 2010 and 2012.

pear type. Together, the top ten producers occupy an area of 1.360.230 HA annually (Table 5).

The European pear (*P. communis*) is grown into five major regions: Europe, North America, South America, South Africa, and Oceania, while production of Asian pear (*P. pyrifolia*) is concentrated in Asia.

China's pear production has increased steadily during the 1980s and early 1990s because of the expensive planting. This rate of growth generated an amount of 7.74 million metric tons of fresh pear in that time. Data show that China produces more than twice the total world production, making the crop a commodity of great importance to this country [4].

## 5. Conclusion

Documentation of botanists and biologists over the last hundred years was of great importance to collect the available data in this review.

Undoubtedly, a fruit that produces around 24 billion tonnes per year is considered a great success in the world market. This success is mainly due to the wide commercial acceptance around the world, its nutritional importance, and

TABLE 5: World production of pear in 2012 (tonnes) and the area (ha) harvested in the ten most productive countries.

| Position | Country | Production (tonnes) | Area harvested (ha) |
|---|---|---|---|
| 1 | China | 16,266,000 | 1,136,700 |
| 2 | USA | 778,582 | 22,015 |
| 3 | Argentina | 700,000 | 26,500 |
| 4 | Italy | 645,540 | 35,195 |
| 5 | Turkey | 439,656 | 34,067 |
| 6 | Spain | 400,600 | 25,000 |
| 7 | Republic of Korea | 394,596 | 14,353 |
| 8 | India | 340,000 | 38,500 |
| 9 | South Africa | 338,584 | 13,000 |
| 10 | Japan | 299,000 | 14,900 |

FAO, 2012.

its adaptability in places with large planting conditions and marketing.

The recent advances achieved in the last year with the pear sequencing genome project will provide new opportunities for developing improved genotypes tolerant to biotic and abiotic stresses and also high quality fruits regarding nutritional and sugar content.

The understanding of the history of pear for agriculture is of paramount importance, since scientists and students could have a better grasp of the richness of this fruit crop and its trajectory associated to humankind.

## Conflict of Interests

The authors declare that there is no conflict of interests regarding the publication of this paper.

## References

[1] P. E. Gonsalves, "As frutas e seus benefícios," in *Frutas Que Curam*, vol. 1, pp. 131–166, 2002.

[2] R. E. C. Layne and H. A. Quamme, "Pears," in *Advances in Fruit Breeding*, J. Janick and J. N. Moore, Eds., pp. 38–70, Purdue University Press, West lafayette, Ind, USA, 1975.

[3] T. Shen, "Pears in China," *HortScience*, vol. 15, no. 1, pp. 13–17, 1980.

[4] Food and Agricultural organization of the United Nation FAO, http://faostat.fao.org/site/342/default.aspx.

[5] USDA, National Agriculture Library, "National Nutrient Database for Standard Reference," 2012, http://ndb.nal.usda.gov/.

[6] A. Dzhangaliev, T. Salova, and P. Turekhanova, "Pome fruits," in *The Wild Fruit and Nut Plantas of Kazakhstan*, J. Janick, Ed., Chapter 1, pp. 311–324, Horticultural Reviews, 2003.

[7] C. Fideghelli, "Origine ed evoluzione," in *Il Pero*, R. Angelini, Ed., vol. 1, Bayer/CropScience, Milano, Italy, 2007.

[8] K. Sax, "The origin and relationships of the pomoideae," *Journal of the Arnold Arboretum*, vol. 12, 1931.

[9] N. Weeden and R. C. Lamb, "Genetics and linkage analysis of 19 isozyme loci in apple," *Journal of the American Society of Horticultural Science*, vol. 112, pp. 865–872, 1987.

[10] Q. B. Zielinski and M. M. Thompson, "Speciation in Pyrus: chromosome number and meiotic behavior," *Botanical Gazette*, vol. 128, pp. 109–112, 1967.

[11] G. A. Rubzov, "Geographical distribution of the genus *Pyrus* and trends and factors in its evolution," *American Naturalist*, vol. 78, pp. 358–366, 1944.

[12] A. C. Zeven and P. M. Zhukovsky, *Dictionary of Cultivated Plants and Their Centres of Diversity*, vol. 2, Centre for Agricultural Publishing and Documentation, Wageningen, The Netherlands, 1975.

[13] N. I. Vavilov, "The origin, variation, immunity and breeding of cultivated plants," *Chronica Botanica*, vol. 13, 1951.

[14] P. M. Zukovskij, *Cultivated Plants and Their Wild Relatives*, Farnham Royal, London, UK, 1962, Translated by P.S. Hudson.

[15] N. I. Vavilov, *Origin and Geography of Cultivated Plants*, vol. 15, Cambridge University Press, Cambridge, UK, 1992.

[16] W. Erhardt, E. Götz, N. Bödeker, and S. Seybold, *Zander. Handwörterbuch der Pflanzennamen*, vol. 1, Eugen Ulmer Verlag, Stuttgart, Germany, 2002.

[17] M. N. Westwood, "Pear germplasm of the new national clonal repository: its evaluation and uses," *Acta Horticulturae*, vol. 124, pp. 57–65, 1982.

[18] R. L. Bell, "Pears," in *Fruit Breeding*, J. Janick and J. N. Moore, Eds., Tree and Tropical Fruits, pp. 441–514, John Wiley & Sons, 1996.

[19] J. S. Challice and M. N. Westwood, "Phenolic compounds of the genus *Pyrus*," *Phytochemistry*, vol. 11, pp. 37–44, 1972.

[20] A. Masseron, *El Nashi*, vol. 2, Mundi Prensa, Madrid, Spain, 1992.

[21] S. Teramoto, "DNA finger printing," in *Techniques on Gene Diagnosis and Breeding in Fruit Trees*, T. Hayashi, M. Omura, and N. S. Scott, Eds., pp. 74–76, Fruit Tree Research Station, Tsukuba, 1993.

[22] A. Rehder, *Manual of Cultivated Tress and Shrubs Hardy in North America Exclusive of the Subtropical and Warmer Temperate Regions*, Collier Macmillan, New York, NY, USA, 1940.

[23] C. Miranda, J. Urrestarazu, L. G. Santesteban, J. B. Royo, and V. Urbina, "Genetic diversity and structure in a collection of ancient Spanish pear cultivars assessed by microsatellite markers," *Journal of the American Society for Horticultural Science*, vol. 135, no. 5, pp. 428–437, 2010.

[24] L. Bao, K. Chen, D. Zhang, Y. Cao, T. Yamamoto, and Y. Teng, "Genetic diversity and similarity of pear (Pyrus L.) cultivars native to East Asia revealed by SSR (simple sequence repeat) markers," *Genetic Resources and Crop Evolution*, vol. 54, pp. 959–971, 2007.

[25] J. Zhengwang, T. Feiyan, H. Hongwen, H. Hongju, and C. Qiliang, "Assessment of genetic diversity of Chinese sand pear landraces (Pyrus pyrifolia Nakai) using simple sequence repeat markers," *HortScience*, vol. 44, no. 3, pp. 619–626, 2009.

[26] J. Wu, Z. Wang, Z. Shi et al., "The genome of the pear (Pyrus bretschneideri Rehd.)," *Genome Research*, vol. 23, no. 2, pp. 396–408, 2013.

[27] P. Spiegel-Roy, "Domestication of fruits trees," in *The Origin and Domestication of Cultivated Plants*, C. Barigozzi, Ed., pp. 201–211, Elsevier, Amsterdam, The Netherlands, 1986.

[28] U. P. Hedrick, *The Pears of New York*, New York State Agricultural Experiment Station, 1921.

[29] V. H. Heywood and D. Zohary, "A catalogue of the wild relatives of cultivated plants native to Europe," *Herbarium Mediterraneum Panormitanum*, vol. 5, pp. 375–415, 1997.

[30] T. Shen, "Pears in China," *HortiScience*, vol. 15, no. 1, pp. 13–17, 1980.

[31] R. Watkins, "Apple and pear," in *Evolution of Crop Plants*, N. W. Simmonds, Ed., vol. 1, pp. 247–250, Edinburgh School of Agriculture, Edinburgh, UK, 1986.

[32] A. Itai, "Pear," in *Genome Mapping and Molecular Breeding in Plants*, C. Kole, Ed., Chapter 6, pp. 157–170, Springer, Berlin, Germany, 2007.

[33] R. L. Bell and A. Itai, "Pyrus," in *Wild Crop Relatives: Genomic and Breeding Resources—Temperate Fruits*, C. Kole, Ed., Chapter 8, pp. 147–178, Springer, 2011.

[34] I. Kajiura, S. Yamaki, M. Omura, T. Akihama, and Y. Machida, "Improvement of sugar content and composition in fruits and classification of east asian pears by the principal component analysis of sugar compositions in fruits," *Japan Journal of Breeding*, vol. 29, no. 1, pp. 12–11, 1979.

[35] J. E. Jackson, "The growing of apples and pears," in *Biology of Apples and Pears*, Chapter 1, pp. 4–20, Cambridge University Press, Cambridge, UK, 2003.

# 14

# Weed Control in White Bean with Various Halosulfuron Tankmixes

**Nader Soltani,[1] Robert E. Nurse,[2] Christy Shropshire,[1] and Peter H. Sikkema[1]**

[1] *University of Guelph Ridgetown Campus, 120 Main Street East, Ridgetown, ON, Canada N0P 2C0*
[2] *Agriculture and Agri-Food Canada, Harrow, ON, Canada N0R 1G0*

Correspondence should be addressed to Nader Soltani; soltanin@uoguelph.ca

Academic Editor: Albino Maggio

Four field trials were conducted over a three-year period (2011–2013) in southwestern Ontario to evaluate the level of weed control provided by various halosulfuron tankmixes applied preplant incorporated (PPI) in white bean. Trifluralin, s-metolachlor, halosulfuron, and imazethapyr applied alone or in combination caused 4% or less visible injury 1 and 4 weeks after emergence (WAE) in white bean. Trifluralin, s-metolachlor, halosulfuron, and imazethapyr applied PPI provided 80–96%, 84–95%, 83–100%, and 75–92% control of redroot pigweed; 19–28%, 30–40%, 97–99%, and 73–84% control of common ragweed; 94–96%, 63–82%, 96–100%, and 96–100% control of common lambsquarters; 14-15%, 12–35%, 100%, and 96–97% control of wild mustard; and 96–97%, 95–97%, 53–56%, and 80–82% control of green foxtail, respectively. The two- and three-way tankmixes of halosulfuron with trifluralin, s-metolachlor, or imazethapyr provided 85–100% control of redroot pigweed, 90–98% control of common ragweed, 97–100% control of common lambsquarters, 100% control of wild mustard, and 93–98% control of green foxtail. Weed density, weed biomass and white bean seed yields reflected the level of visible weed control.

## 1. Introduction

Ontario is one of the main regions for the production of white (navy) bean (*Phaseolus vulgaris* L.) in North America. In 2012, white bean growers in Ontario planted 28,000 hectares and produced 68,000 MT of white bean with a farm-gate value of approximately $38 million [1]. White bean has short physical stature and therefore is more vulnerable to weed interference and subsequent yield losses if weeds are not adequately controlled [2–6]. Yield losses of 59% have been attributed to weed interference in dry beans in Ontario which is substantially greater than other field crops such as winter wheat (3%), spring cereals (12%), soybean (38%), and corn (52%) [7]. There is only one registered soil applied broadleaf herbicide, imazethapyr, for weed control in white bean in Ontario. Imazethapyr provides only marginal control of common lambsquarters (*Chenopodium album* L.) and common ragweed (*Ambrosia artemesiifolia* L.) and has a narrow margin of crop safety in dry bean [2, 6]. More research is needed to identify new herbicide options for annual broadleaf weed management in white bean.

Halosulfuron is a newly registered sulfonylurea herbicide that controls redroot pigweed (*Amaranthus retroflexus* L.), common lambsquarters, common ragweed, wild mustard (*Sinapis arvensis* L.), nutsedge species (*Cyperus spp.*), ladysthumb (*Polygonum persicaria* L.), velvetleaf (*Abutilon theophrasti* Medic.), and cocklebur (*Xanthium strumarium* L.), including glyphosate and triazine resistant biotypes [8]. However, halosulfuron is weak on controlling troublesome annual grasses such as *Setaria*, *Digitaria*, *Echinochloa*, and *Panicum* species.

Trifluralin is a dinitroaniline herbicide that controls several annual grasses including *Setaria*, *Digitaria*, *Echinochloa*, and *Panicum* spp. and some broadleaved weeds such as common lambsquarters and redroot pigweed, including acetolactate synthase and triazine resistant biotypes [7].

S-metolachlor is a chloroacetanilide herbicide that controls annual grasses such as green foxtail (*Setaria viridis* (L.) Beauv.), yellow foxtail (*Setaria glauca* (L.) Beauv.), giant foxtail (*Setaria faberii* Herrm.), smooth crabgrass (*Digitaria ischaemum* (Schreb) Muhl.), large crabgrass (*Digitaria sanguinalis* (L.) Scop.), barnyardgrass (*Echinochloa crusgalli* (L.)

TABLE 1: Percent visible injury 1 and 4 WAE, crop moisture at harvest, and yield of white bean treated with halosulfuron tankmixes applied PPI at Exeter (2011–2013) and Harrow, ON (2012). Means followed by the same letter within a column are not significantly different according to Fisher's Protected LSD at $P < 0.05$[a].

| Herbicide treatment | Rate (g ai ha$^{-1}$) | Injury 1 WAE | Injury 4 WAE | Moisture | Yield |
|---|---|---|---|---|---|
| | | | % | | T ha$^{-1}$ |
| Weedy check | | 0 | 0 a | 20.8 | 0.6 e |
| Weed-free check | | 0 | 0 a | 19.9 | 2.2 ab |
| trifluralin | 600 | 0 | 0 a | 20.3 | 1.2 d |
| s-Metolachlor | 1050 | 0 | 0 a | 20.4 | 1.4 cd |
| Halosulfuron | 35 | 0 | 0 a | 19.6 | 1.7 bc |
| Imazethapyr | 45 | 0 | 2 bc | 19.9 | 1.8 bc |
| Trifluralin + halosulfuron | 600 + 35 | 0 | 0 a | 19.4 | 2.2 ab |
| Trifluralin + imazethapyr | 600 + 45 | 0 | 2 bc | 19.5 | 2.0 ab |
| Trifluralin + halosulfuron + imazethapyr | 600 + 35 + 45 | 0 | 4 c | 19.5 | 2.1 ab |
| s-Metolachlor + halosulfuron | 1050 + 35 | 0 | 0 a | 19.6 | 2.2 ab |
| s-Metolachlor + imazethapyr | 1050 + 45 | 0 | 1 b | 19.8 | 2.0 ab |
| s-Metolachlor + halosulfuron + imazethapyr | 1050 + 35 + 45 | 0 | 3 bc | 19.9 | 2.0 ab |

[a]Abbreviations: WAE, weeks after crop emergence; PPI, preplant incorporated.

Beauv.), fall panicum (*Panicum dichotomiflorum* Michx.), and witchgrass (*Panicum capillare* L.) [7, 8] and some annual broadleaf weeds such as redroot pigweed and Eastern black nightshade (*Solanum ptycanthum* Dunal).

Tankmixing halosulfuron with trifluralin, S-metolachlor, and/or imazethapyr can provide Ontario white bean growers with a herbicide option that provides full-season control of annual grass and broadleaf weeds. Utilization of herbicide tankmixes will also reduce the selection pressure for herbicide resistant weeds.

There is little information on the crop safety and efficacy of halosulfuron tankmixed with trifluralin, s-metolachlor, and/or imazethapyr applied preplant incorporated (PPI) for the control of annual grass and broadleaf weeds under Ontario environmental conditions. The objective of this research was to evaluate the level of crop injury and weed control provided by the two- and three-way tankmixes of halosulfuron with trifluralin, s-metolachlor, and/or imazethapyr applied PPI in white bean.

## 2. Materials and Methods

Field studies were conducted in 2011, 2012, and 2013 at the Huron Research Station, University of Guelph, Exeter, Ontario, and in 2012 at the Agriculture and Agri-Food Canada Research Centre, Harrow, Ontario. The soil at Exeter was a Brookston clay loam (Orthic Humic Gleysol, mixed, mesic, and poorly drained) with 32% sand, 42% silt, 26% clay, 3.7% organic matter, and pH 7.8 in 2011; 41% sand, 35% silt, 24% clay, 3.2% organic matter, and pH 7.9 in 2012; and 29% sand, 44% silt, 27% clay, 3.6% organic matter, and pH 7.7 in 2013. The soil at Harrow was a Fox sandy loam (Brunisolic Gray Brown Luvisol) with 82.5% sand, 5% silt, 12.5% clay, 2.6% organic matter, and pH of 6.0. Seedbed preparation at all sites consisted of fall moldboard plowing followed by two passes with a field cultivator with rolling basket harrows in the spring.

The experiments were arranged in a randomized complete block design (RCBD) with four replications. There were twelve treatments as listed in Table 1. Plots consisted of four rows of white bean "T9905" planted at a rate of 250,000 seeds ha$^{-1}$ to a depth of 5 cm in rows that were spaced 0.75 m apart and were 8 or 10 m long in late May to early June of each year.

Herbicide treatments were made with a $CO_2$-pressurized backpack sprayer calibrated to deliver 200 L ha$^{-1}$ of spray solution at a pressure of 200/240 kPa using low drift nozzles (ULD120-02, Spraying Systems Co., P.O. Box 7900. Wheaton, IL). The boom was 1.5 m wide with four nozzles spaced 0.5 m apart. PPI herbicides were applied 1-2 days before planting and were immediately incorporated. Weed-free check plots were maintained with interrow cultivation and hand hoeing during the growing season.

White bean injury and weed control were visually estimated on a scale of 0 (no injury/control) to 100% (complete plant death). White bean injury was assessed 1 and 4 weeks after emergence (WAE). Weed control was assessed 4 and 8 WAE, respectively. Weed density and dry weight were evaluated 8 WAE by counting and cutting plants at the soil surface from two 0.5 m$^2$ quadrats per plot and separating by species. Plants were dried at 60 C to a constant moisture (at least for 48 hours) and then weighed. White bean was considered mature when 90% of the pods in the weed-free check had turned from green to a golden colour. Beans were harvested from the two center rows of each plot with a small plot combine, weight and seed moisture content were recorded, and seed yields were adjusted to 18% seed moisture content.

Data were analyzed as an RCBD using PROC MIXED in SAS 9.2. Herbicide treatment was considered a fixed effect, while environment (year-location combinations), the interaction between environment and herbicide treatment, and replicate nested within environment were considered random effects. Significance of the fixed effect was tested

TABLE 2: Percent control, density, and dry weight of redroot pigweed in white bean treated with halosulfuron tankmixes applied PPI at Exeter (2011-2012) and Harrow, ON (2012). Means followed by the same letter within a column are not significantly different according to Fisher's Protected LSD at $P < 0.05$[a].

| Herbicide treatment | Rate (g ai ha$^{-1}$) | Control | | Density | Dry weight |
| --- | --- | --- | --- | --- | --- |
| | | 4 WAE | 8 WAE | | |
| | | % | | # m$^{-2}$ | g m$^{-2}$ |
| Weedy check | | 0 | 0 | 6.0 c | 29.7 c |
| Weed-free check | | 100 | 100 | 0.0 a | 0.0 a |
| Trifluralin | 600 | 80 | 96 | 0.7 ab | 4.9 b |
| s-Metolachlor | 1050 | 84 | 95 | 0.6 a | 4.3 ab |
| Halosulfuron | 35 | 83 | 100 | 0.8 ab | 2.5 ab |
| Imazethapyr | 45 | 75 | 92 | 2.0 b | 3.9 ab |
| Trifluralin + halosulfuron | 600 + 35 | 85 | 100 | 0.6 a | 3.3 ab |
| Trifluralin + imazethapyr | 600 + 45 | 78 | 100 | 0.8 ab | 5.8 b |
| Trifluralin + halosulfuron + imazethapyr | 600 + 35 + 45 | 91 | 100 | 0.6 a | 3.5 ab |
| s-Metolachlor + halosulfuron | 1050 + 35 | 87 | 100 | 0.4 a | 2.0 ab |
| s-Metolachlor + imazethapyr | 1050 + 45 | 87 | 100 | 0.3 a | 1.6 ab |
| s-Metolachlor + halosulfuron + imazethapyr | 1050 + 35 + 45 | 91 | 100 | 0.1 a | 0.3 a |

[a]Abbreviations: WAE, weeks after crop emergence; PPI, preplant incorporated.

using $F$-test and random effects were tested using a $Z$-test of the variance estimate. The UNIVARIATE procedure was used to test data for normality and homogeneity of variance. For all ratings, any treatments assigned a value of zero (weedy and weed-free checks for crop injury; weedy check for weed control; weed-free check for weed density and dry weight) were excluded from the analysis. However, all values were compared independently to zero to evaluate treatment differences with the weedy and/or weed-free check. To satisfy the assumptions of the variance analyses, injury 1 and 4 WAE and weed control were arcsine square root transformed and weed density, dry weight, and crop moisture were log transformed. Treatment comparisons were made using Fisher's Protected LSD at a level of $P < 0.05$. Data compared on the transformed scale were converted back to the original scale for presentation of results.

## 3. Results and Discussion

*3.1. Crop Injury.* The analysis of variance (ANOVA) indicated no effect of years or treatment by year interaction for crop injury; therefore, data were pooled and averaged for environments.

Trifluralin, s-metolachlor, halosulfuron, and imazethapyr applied alone or in combination applied PPI caused 4% or less visible injury 1 and 4 weeks after emergence (WAE) in white bean (Table 1). This is similar to other studies that have shown up to 6% injury in white bean with halosulfuron applied alone or in tankmixes with other herbicides [9–11]. Silvey et al. [12] also found minimal and transient injury (5%) with halosulfuron in snap bean. In other studies, halosulfuron reduced yield of white bean as much as 9% and adzuki bean as much as 68% but had no effect on the seed yield of black, cranberry, kidney, otebo, pinto, and small red Mexican beans [9, 10]. In another study halosulfuron applied postemergence reduced seed yield of white bean as much as 21% [11].

However, other studies have shown significant injury with halosulfuron alone or in tankmixes with other herbicides in adzuki bean [10]. There was no difference in seed moisture content between treatments evaluated (Table 1).

All treatments increased white bean seed yield compared to the weedy check (Table 1). Weed interference with trifluralin, s-metolachlor, halosulfuron, and imazethapyr applied alone resulted in a decrease in white bean seed yield of 45, 36, 23, and 18% compared to the weed-free check, respectively (Table 1). In contrast, the level of weed control with trifluralin + halosulfuron, trifluralin + imazethapyr, trifluralin + halosulfuron + imazethapyr, s-metolachlor + halosulfuron, s-metolachlor + imazethapyr, and s-metolachlor + halosulfuron + imazethapyr at rates evaluated resulted in white bean seed yield that was equivalent to the weed-free check (Table 1).

*3.2. Weed Control.* The dominant weed species in this study were redroot pigweed, common ragweed, common lambsquarters, wild mustard, and green foxtail. The ANOVA indicated no effect of environments or treatment by environments for weed control; therefore, data were pooled and averaged for environments (Tables 2–6).

*3.2.1. Redroot Pigweed.* Trifluralin, s-metolachlor, halosulfuron, imazethapyr, trifluralin + halosulfuron, trifluralin + imazethapyr, trifluralin + halosulfuron + imazethapyr, s-metolachlor + halosulfuron, s-metolachlor + imazethapyr, and s-metolachlor + halosulfuron + imazethapyr applied PPI controlled redroot pigweed 80–96%, 84–95%, 83–100%, 75–92%, 85–100%, 78–100%, 91–100%, 87–100%, 87–100%, and 91–100%, (Table 2).

Trifluralin, s-metolachlor, halosulfuron, and imazethapyr applied PPI reduced redroot pigweed density 67–90% and biomass 84–92% (Table 2). Trifluralin + halosulfuron, trifluralin + imazethapyr, trifluralin + halosulfuron + imazethapyr,

TABLE 3: Percent control, density, and dry weight of common ragweed in white bean treated with halosulfuron tankmixes applied PPI at Exeter, ON (2012-2013). Means followed by the same letter within a column are not significantly different according to Fisher's Protected LSD at $P < 0.05^a$.

| Herbicide treatment | Rate (g ai ha$^{-1}$) | Control | | Density | Dry weight |
| --- | --- | --- | --- | --- | --- |
| | | 4 WAE | 8 WAE | | |
| | | % | | # m$^{-2}$ | g m$^{-2}$ |
| Weedy check | | 0 e | 0 e | 1.8 b | 5.2 bc |
| Weed-free check | | 100 a | 100 a | 0.0 a | 0.0 a |
| Trifluralin | 600 | 28 d | 19 e | 2.7 b | 6.6 c |
| s-Metolachlor | 1050 | 30 d | 40 d | 2.5 b | 18.6 c |
| Halosulfuron | 35 | 99 a | 97 ab | 0.1 a | 0.0 a |
| Imazethapyr | 45 | 84 c | 73 c | 1.2 ab | 1.7 ab |
| Trifluralin + halosulfuron | 600 + 35 | 96 ab | 90 abc | 0.3 a | 0.6 ab |
| Trifluralin + imazethapyr | 600 + 45 | 88 c | 78 bc | 1.0 ab | 2.3 ab |
| Trifluralin + halosulfuron + imazethapyr | 600 + 35 + 45 | 98 a | 96 ab | 0.0 a | 0.0 a |
| s-Metolachlor + halosulfuron | 1050 + 35 | 98 a | 93 abc | 0.2 a | 0.3 ab |
| s-Metolachlor + imazethapyr | 1050 + 45 | 90 bc | 78 bc | 0.9 ab | 1.2 ab |
| s-Metolachlor + halosulfuron + imazethapyr | 1050 + 35 + 45 | 98 a | 94 ab | 0.0 a | 0.0 a |

$^a$Abbreviations: WAE, weeks after crop emergence; PPI, preplant incorporated.

TABLE 4: Percent control, density, and dry weight of common lambsquarters in white bean treated with halosulfuron tankmixes applied PPI at Exeter (2011–2013) and Harrow, ON (2012). Means followed by the same letter within a column are not significantly different according to Fisher's Protected LSD at $P < 0.05^a$.

| Herbicide treatment | Rate (g ai ha$^{-1}$) | Control | | Density | Dry weight |
| --- | --- | --- | --- | --- | --- |
| | | 4 WAE | 8 WAE | | |
| | | % | | # m$^{-2}$ | g m$^{-2}$ |
| Weedy check | | 0 d | 0 d | 6.8 c | 9.3 b |
| Weed-free check | | 100 a | 100 a | 0.0 a | 0.0 a |
| Trifluralin | 600 | 94 bc | 96 b | 0.5 ab | 1.4 a |
| s-Metolachlor | 1050 | 82 c | 63 c | 3.9 c | 12.7 b |
| Halosulfuron | 35 | 96 ab | 100 a | 0.5 ab | 1.4 a |
| Imazethapyr | 45 | 96 ab | 100 a | 1.3 b | 2.2 a |
| Trifluralin + halosulfuron | 600 + 35 | 97 ab | 100 a | 0.4 ab | 1.7 a |
| Trifluralin + imazethapyr | 600 + 45 | 96 ab | 100 a | 0.4 ab | 1.1 a |
| Trifluralin + halosulfuron + imazethapyr | 600 + 35 + 45 | 99 ab | 100 a | 0.2 a | 0.5 a |
| s-Metolachlor + halosulfuron | 1050 + 35 | 99 ab | 100 a | 0.2 a | 1.6 a |
| s-Metolachlor + imazethapyr | 1050 + 45 | 98 ab | 100 a | 0.3 ab | 0.8 a |
| s-Metolachlor + halosulfuron + imazethapyr | 1050 + 35 + 45 | 98 ab | 100 a | 0.2 a | 0.9 a |

$^a$Abbreviations: WAE, weeks after crop emergence; PPI, preplant incorporated.

s-metolachlor + halosulfuron, s-metolachlor + imazethapyr, and s-metolachlor + halosulfuron + imazethapyr applied PPI reduced redroot pigweed density 87–98% and biomass 80–99% (Table 2).

In other studies, halosulfuron provided 80-81% control of redroot pigweed [11]. Halosulfuron plus fomesafen POST provided 89–92% control of redroot pigweed and the tankmix of halosulfuron plus bentazon plus fomesafen also provided 89–91% control of redroot pigweed [11]. Imazethapyr applied alone or in combination with other herbicides has been shown to reduce redroot pigweed density 69–100% [13].

*3.2.2. Common Ragweed.* Trifluralin, s-metolachlor, halosulfuron, imazethapyr, trifluralin + halosulfuron, trifluralin + imazethapyr, trifluralin + halosulfuron + imazethapyr, s-metolachlor + halosulfuron, s-metolachlor + imazethapyr, and s-metolachlor + halosulfuron + imazethapyr applied PPI controlled common ragweed 19–28%, 30–40%, 97–99%, 73–84%, 90–96%, 78–88%, 96–98%, 93–98%, 78–90%, and 94–98%, respectively (Table 3). The two- and three-way tankmixes of halosulfuron with trifluralin, s-metolachlor, and imazethapyr were most effective as they provided 90–98% control of common ragweed.

Trifluralin and s-metolachlor applied PPI had no effect on common ragweed density or biomass but halosulfuron reduced common ragweed density 94% and biomass 100% compared to the weedy check (Table 3). Trifluralin + halosulfuron, trifluralin + imazethapyr, trifluralin + halosulfuron + imazethapyr, s-metolachlor + halosulfuron, s-metolachlor + imazethapyr, and s-metolachlor + halosulfuron + imazethapyr applied PPI reduced common ragweed density 44–100% and biomass 56–100% (Table 3). Halosulfuron applied alone or in tankmix combination with trifluralin, s-metolachlor, and imazethapyr provided the most consistent density

TABLE 5: Percent control, density, and dry weight of wild mustard in white bean treated with halosulfuron tankmixes applied PPI at Exeter, ON (2011 and 2013). Means followed by the same letter within a column are not significantly different according to Fisher's Protected LSD at $P < 0.05$[a].

| Herbicide treatment | Rate (g ai ha$^{-1}$) | Control | | Density | Dry weight |
| --- | --- | --- | --- | --- | --- |
| | | 4 WAE | 8 WAE | | |
| | | % | | # m$^{-2}$ | g m$^{-2}$ |
| Weedy check | | 0 d | 0 d | 12.6 c | 95.1 c |
| Weed-free check | | 100 a | 100 a | 0.0 a | 0.0 a |
| Trifluralin | 600 | 15 c | 14 c | 12.8 c | 103.4 c |
| s-Metolachlor | 1050 | 35 b | 12 c | 13.4 c | 122.2 c |
| Halosulfuron | 35 | 100 a | 100 a | 0.1 ab | 0.1 a |
| Imazethapyr | 45 | 97 a | 96 b | 1.4 b | 0.7 ab |
| Trifluralin + halosulfuron | 600 + 35 | 100 a | 100 a | 0.3 ab | 0.3 ab |
| Trifluralin + imazethapyr | 600 + 45 | 99 a | 97 b | 1.6 b | 2.2 b |
| Trifluralin + halosulfuron + imazethapyr | 600 + 35 + 45 | 100 a | 100 a | 0.0 a | 0.0 a |
| s-Metolachlor + halosulfuron | 1050 + 35 | 100 a | 100 a | 0.0 a | 0.0 a |
| s-Metolachlor + imazethapyr | 1050 + 45 | 98 a | 98 b | 1.0 ab | 0.3 ab |
| s-Metolachlor + halosulfuron + imazethapyr | 1050 + 35 + 45 | 100 a | 100 a | 0.0 a | 0.0 a |

[a]Abbreviations: WAE, weeks after crop emergence; PPI, preplant incorporated.

TABLE 6: Percent control, density, and dry weight of green foxtail in white bean treated with halosulfuron tankmixes applied PPI at Exeter, ON (2011–2013). Means followed by the same letter within a column are not significantly different according to Fisher's Protected LSD at $P < 0.05$[a].

| Herbicide treatment | Rate (g ai ha$^{-1}$) | Control | | Density | Dry weight |
| --- | --- | --- | --- | --- | --- |
| | | 4 WAE | 8 WAE | | |
| | | % | | # m$^{-2}$ | g m$^{-2}$ |
| Weedy check | | 0 g | 0 f | 36.7 c | 38.2 d |
| Weed-free check | | 100 a | 100 a | 0.0 a | 0.0 a |
| Trifluralin | 600 | 97 bc | 96 bc | 1.7 a | 1.0 abc |
| s-Metolachlor | 1050 | 95 cd | 97 b | 2.6 a | 1.5 abc |
| Halosulfuron | 35 | 53 f | 56 e | 52.5 c | 28.2 d |
| Imazethapyr | 45 | 82 e | 80 d | 11.0 b | 4.5 c |
| Trifluralin + halosulfuron | 600 + 35 | 93 d | 93 c | 2.4 a | 2.9 bc |
| Trifluralin + imazethapyr | 600 + 45 | 96 bcd | 94 bc | 1.7 a | 0.8 abc |
| Trifluralin + halosulfuron + imazethapyr | 600 + 35 + 45 | 99 b | 97 b | 0.6 a | 0.2 ab |
| s-Metolachlor + halosulfuron | 1050 + 35 | 95 cd | 93 c | 1.5 a | 0.6 abc |
| s-Metolachlor + imazethapyr | 1050 + 45 | 97 bc | 94 bc | 1.1 a | 0.8 abc |
| s-Metolachlor + halosulfuron + imazethapyr | 1050 + 35 + 45 | 98 b | 96 bc | 1.2 a | 0.8 abc |

[a]Abbreviations: WAE, weeks after crop emergence; PPI, preplant incorporated.

reduction (83–100%) and biomass reduction (88–100%) of common ragweed among herbicide treatments evaluated (Table 3).

In other studies, halosulfuron provided 91–94% control of common ragweed. Halosulfuron plus fomesafen applied POST provided 93–95% control of common ragweed [11]. Also, the tankmix of halosulfuron plus bentazon plus fomesafen provided 97–98% control of common ragweed [11]. Imazethapyr and imazamox have been shown to control common ragweed 76 to 100% in dry bean [14–16].

*3.2.3. Common Lambsquarters.* S-metolachlor provided only 63–82% control of common lambsquarters but trifluralin, halosulfuron, imazethapyr, trifluralin + halosulfuron, trifluralin + imazethapyr, trifluralin + halosulfuron + imazethapyr,

s-metolachlor + halosulfuron, s-metolachlor + imazethapyr, and s-metolachlor + halosulfuron + imazethapyr controlled redroot pigweed 94–100% (Table 4). Similarly, s-metolachlor had no effect on density or biomass of common lambsquarters but other treatments evaluated reduced common lambsquarters density 81–97% and biomass 76–95% (Table 4). The two- and three-way tankmixes of halosulfuron with trifluralin, s-metolachlor, and imazethapyr provided 97–100% control of common lambsquarters.

In other studies, trifluralin applied PPI provided better control of common lambsquarters (83 versus 71%) compared with s-metolachlor [17]. Halosulfuron plus fomesafen POST provided 55–59% control of common lambsquarters [11]. The tankmix of halosulfuron plus bentazon plus fomesafen provided 79–82% control of common lambsquarters [11].

Imazethapyr applied alone or in combination with other herbicides has been shown to reduce common lambsquarters density 69–100% [13]. Imazethapyr and imazamox also have been shown to control common lambsquarters 76 to 100% in dry bean [14–16].

*3.2.4. Wild Mustard.* Trifluralin and s-metolachlor provided only 12–35% control of wild mustard but halosulfuron, imazethapyr, trifluralin + halosulfuron, trifluralin + imazethapyr, trifluralin + halosulfuron + imazethapyr, s-metolachlor + halosulfuron, s-metolachlor + imazethapyr, and s-metolachlor + halosulfuron + imazethapyr provided 96–100% control of wild mustard in white bean (Table 5). Halosulfuron applied alone or in two- and three-way tankmixes of halosulfuron with trifluralin, s-metolachlor, and imazethapyr provided 100% control of wild mustard (Table 5). Trifluralin and s-metolachlor had no effect on density or biomass of wild mustard but other treatments evaluated reduced wild mustard density 87–100% and biomass 98–100% (Table 5).

*3.2.5. Green Foxtail.* Trifluralin, s-metolachlor, halosulfuron, imazethapyr, trifluralin + halosulfuron, trifluralin + imazethapyr, trifluralin + halosulfuron + imazethapyr, s-metolachlor + halosulfuron, s-metolachlor + imazethapyr, and s-metolachlor + halosulfuron + imazethapyr applied PPI controlled green foxtail as much as 97, 97, 56, 82, 93, 96, 99, 95, 97, and 98%, respectively (Table 6). Halosulfuron applied PPI alone had no effect on density or biomass of green foxtail but other treatments evaluated reduced green foxtail density 70–98% and biomass 88–99% (Table 6). The two- and three-way tankmixes of halosulfuron with trifluralin, s-metolachlor, and imazethapyr provided 93–98% control of green foxtail. In other studies, the control of some grasses such as *E. crus-galli* increased from 58–96% to 98% when s-metolachlor or trifluralin was applied PPI in combination with broadleaf herbicides such as imazethapyr [2].

## 4. Conclusions

Trifluralin, s-metolachlor, halosulfuron, imazethapyr, trifluralin + halosulfuron, trifluralin + imazethapyr, trifluralin + halosulfuron + imazethapyr, s-metolachlor + halosulfuron, s-metolachlor + imazethapyr, and s-metolachlor + halosulfuron + imazethapyr applied PPI caused minimal and transient injury in white bean. Trifluralin applied PPI at 600 g ai ha$^{-1}$ provided excellent full season control of redroot pigweed, common lambsquarters, and green foxtail and poor control of common ragweed and wild mustard. S-metolachlor applied PPI at 1050 g ai ha$^{-1}$ provided excellent full season control of green foxtail, good control of redroot pigweed, fair control of common lambsquarters, and poor control of common ragweed and wild mustard. Halosulfuron applied PPI at 35 g ai ha$^{-1}$ provided excellent full season control of common lambsquarters, redroot pigweed, common ragweed, and wild mustard and fair control of green foxtail. Imazethapyr applied PPI at 45 g ai ha$^{-1}$ provided excellent full season control of redroot pigweed, common lambsquarters, and wild mustard and good control of common ragweed

and green foxtail. The two- and three-way tankmixes of halosulfuron with other herbicides evaluated provided good to excellent full season control of redroot pigweed, common ragweed, common lambsquarters, wild mustard, and green foxtail. Generally, white bean yields reflected the level of weed control. Based on these results, halosulfuron applied PPI in two-way or three-way tankmix with trifluralin, s-metolachlor, and/or imazethapyr can be used safely for the control of annual grass and broadleaf weeds in white bean.

## Conflict of Interests

The authors have no conflict of interests to report.

## Acknowledgment

Funding for this project was provided by the Ontario Bean Growers and the GF2 program of Agricultural Adaptation Council.

## References

[1] K. Kulasekera, "Estimated area, yield, production and farm value of specified field crops, Ontario, 2001–2013, (metric units)," 2014, http://www.omafra.gov.on.ca/english/stats/crops/estimate_metric.htm.

[2] N. R. Arnold, W. M. Murray, J. E. Gregory, and D. Smeal, "Weed control in pinto beans (*Phaseolus vulgaris*) with imazethapyr combinations," *Weed Technology*, vol. 7, pp. 361–364, 1993.

[3] T. A. Bauer, K. A. Renner, D. Penner, and J. D. Kelly, "Pinto bean (*Phaseolus vulgaris*) varietal tolerance to imazethapyr," *Weed Science*, vol. 43, no. 3, pp. 417–424, 1995.

[4] R. E. Blackshaw and G. Saindon, "Dry bean (*Phaseolus vulgaris*) tolerance to imazethapyr," *Canadian Journal of Plant Science*, vol. 76, no. 4, pp. 915–919, 1996.

[5] C. P. Urwin, R. G. Wilson, and D. A. Mortensen, "Response of dry edible bean (*Phaseolus vulgaris*) cultivars to four herbicides," *Weed Technology*, vol. 10, no. 3, pp. 512–518, 1996.

[6] R. G. Wilson and S. D. Miller, "Dry edible bean (*Phaseolus vulgaris*) responses to imazethapyr," *Weed Technology*, vol. 5, pp. 22–26, 1991.

[7] Ontario Ministry of Agriculture and Food and Rural Affairs (OMAFRA), *Guide to Weed Control*, Publication 75, Ontario Ministry of Agriculture and Food and Rural Affairs (OMAFRA), Toronto, Canada, 2013.

[8] S. A. Senseman, *Herbicide Handbook*, Weed Sci. Soc. Am., Champaign, Ill, USA, 9th edition, 2007.

[9] N. Soltani, C. Shropshire, and P. H. Sikkema, "Response of dry beans to halosulfuron applied postemergence," *Canadian Journal of Plant Science*, vol. 92, no. 4, pp. 723–728, 2012.

[10] C. L. Stewart, R. E. Nurse, C. Gillard, and P. H. Sikkema, "Tolerance of adzuki bean to preplant-incorporated, pre-emergence, and post-emergence herbicides in Ontario, Canada," *Weed Biology and Management*, vol. 10, no. 1, pp. 40–47, 2010.

[11] N. Soltani, R. E. Nurse, and P. H. Sikkema, "Weed management in white beans with postemergence herbicide tankmixes," *Canadian Journal of Plant Science*, vol. 93, no. 4, pp. 669–674, 2013.

[12] B. D. Silvey, W. E. Mitchem, A. W. Macrae, and D. W. Monks, "Snap bean (*Phaseolus vulgaris*) tolerance to halosulfuron PRE,

POST, or PRE followed by POST," *Weed Technology*, vol. 20, no. 4, pp. 873–876, 2006.

[13] D. Wall, "Bentazon tank-mixtures for improved redroot pigweed (*Amaranthus retroflexus*) and common lambsquarters (*Chenopodium album*) control in navy bean (*Phaseolus vulgaris*)," *Weed Technology*, vol. 9, no. 3, pp. 610–616, 1995.

[14] O. C. Burnside, W. H. Ahrens, B. J. Holder, M. J. Weins, M. M. Johnson, and E. A. Ristau, "Efficacy and economics of various mechanical plus chemical weed control systems in dry beans (*Phaseolus vulgaris*)," *Weed Technology*, vol. 8, no. 2, pp. 238–244, 1994.

[15] M. J. Vangessel and P. Westra, "Economics and efficacy of postemergence spurred anoda (*Anoda cristata*) control in pinto beans (*Phaseolus vulgaris*)," *Weed Technology*, vol. 11, no. 2, pp. 329–334, 1997.

[16] R. G. Wilson Jr., "Response of dry bean and weeds to fomesafen and fomesafen tank mixtures," *Weed Technology*, vol. 19, no. 1, pp. 201–206, 2005.

[17] P. H. Sikkema, N. Soltani, C. Shropshire, and T. Cowan, "Tolerance of white beans to postemergence broadleaf herbicides," *Weed Technology*, vol. 18, no. 4, pp. 893–901, 2004.

# 15

# Ecological Complexity and the Success of Fungal Biological Control Agents

**Guy R. Knudsen[1] and Louise-Marie C. Dandurand[2]**

[1] *Soil & Land Resources Division, University of Idaho, Moscow, ID 83844, USA*
[2] *Department of Plant, Soil, & Entomological Sciences, University of Idaho, Moscow, ID 83844, USA*

Correspondence should be addressed to Guy R. Knudsen; gknudsen@uidaho.edu

Academic Editor: Tibor Janda

Fungal biological control agents against plant pathogens, especially those in soil, operate within physically, biologically, and spatially complex systems by means of a variety of trophic and nontrophic interspecific interactions. However, the biocontrol agents themselves are also subject to the same types of interactions, which may reduce or in some cases enhance their efficacy against target plant pathogens. Characterization of these ecologically complex systems is challenging, but a number of tools are available to help unravel this complexity. Several of these tools are described here, including the use of molecular biology to generate biocontrol agents with useful marker genes and then to quantify these agents in natural systems, epifluorescence and confocal laser scanning microscopy to observe their presence and activity in situ, and spatial statistics and computer simulation modeling to evaluate and predict these activities in heterogeneous soil habitats.

## 1. Introduction

Reduction of chemical pesticide usage, including chemicals for control of soilborne plant pathogens, is widely recognized as a desirable goal for agriculture and forestry. Many traditional chemical means of disease control (e.g., methyl bromide and certain fungicides) are being or have been phased out due to economic considerations and/or mandate. Application of fungi and bacteria as microbial antagonists of plant pathogens offers prospects of environmentally benign pest control. A wide range of fungi have shown potential as agents for biological control of soilborne phytopathogens, and this review will focus on these fungal biocontrol agents.

Unlike gnotobiotic systems, soil is characterized by a multitude of both trophic (food webs) and nontrophic (e.g., mutualism, commensalism, neutralism, amensalism, antagonism, and competition) interspecific relationships [1]. The distinction between trophic and nontrophic relationships is somewhat arbitrary, since it focuses only on "what an organism eats and what eats it." Parasitism typically is categorized as a trophic interaction, since a parasite derives nutrition from its host, whereas mutualism is generally considered a nontrophic interaction. Nonetheless, in an alternate taxonomy of interspecific relationships, mutualism and parasitism are often considered to be part of a continuum [2, 3].

Biocontrol microbes that are applied to seeds or soil prior to planting may colonize the spermosphere and/or rhizosphere of seedlings and thus may be present at or near infection courts of soilborne pathogens. There, biocontrol agents may participate in a variety of trophic and nontrophic interaction mechanisms including production of antifungal compounds, hyperparasitism of pathogens, stimulation of host plant defenses, or competitive colonization of spermosphere and rhizosphere substrates. However, soils are both densely populated and highly competitive habitats, and poor competitive ability of introduced fungal agents in soil communities, leading to unsuccessful establishment and activity, has often been observed [4].

Understanding effects of biodiversity on soil ecosystem functions requires integration both within trophic levels (i.e., horizontal diversity, the number of species within a trophic level) and across trophic levels (vertical diversity, the number of trophic levels) [5, 6]. The significance of both horizontal diversity and vertical diversity, from an agronomic perspective, may vary with the system being considered. A number of soilborne fungi (e.g., some members of the genera *Fusarium*, *Rhizoctonia*, *Pythium*, *Sclerotinia*, and others) are pathogenic on roots of both crop plants and weeds. Thus, they may be viewed either as harmful pathogens or as beneficial biocontrol agents, depending on the agronomic significance of the plant host. For example, *Sclerotinia sclerotiorum* and *S. minor* are important crop pathogens and also have been reported as weed biocontrol agents: *S. sclerotiorum* against knapweed [7] and *S. minor* against dandelion [8].

When analysis is limited only to host and parasite, the system may be horizontally diverse (several potential pathogens) but not vertically diverse (only two trophic levels: host and parasite). Studies of *Sclerotinia* as a weed biocontrol agent might typically be less vertically diverse (two trophic levels: host and parasite) compared to attempts at biological control of phytopathogenic *Sclerotinia* with a fungal hyperparasite such as *Trichoderma* (three trophic levels: crop plant, parasite, and biocontrol hyperparasite). Screening a large number of potential antagonists to a plant pathogen is an exercise that focuses exclusively on horizontal diversity. However, failure to account for nontrophic interactions and for trophic vertical diversity is likely one reason why promising potential biocontrol agents often fail to live up to expectations in the field. As one example, fluorescent pseudomonads and other bacteria have been shown to inhibit and lyse germlings of *Trichoderma* spp. [9]. We observed that a soil pseudomonad, *Pseudomonas fluorescens* isolate 2–79, significantly inhibited radial growth and hyphal density of the biocontrol fungus *T. harzianum* in soil [10]. In another study, we isolated a fungivorous nematode (*Aphelenchoides* sp.) from field soil and found that the nematode significantly reduced hyphal growth and total population of *T. harzianum* under soil environmental conditions favorable to nematode activity [11]. Because hyphal growth and proliferation are an attribute of the biocontrol fungus that appears to correlate with its success as a biocontrol agent, consideration of such nontrophic and vertically diverse trophic interactions may provide insight into the "failure parameters" that often plague biocontrol efforts.

Although numerous studies have suggested that, for effective biocontrol activity, successful hyphal growth and proliferation of the agent in soil or the rhizosphere are critical, the presence of an indigenous fungal microbiota makes this difficult to evaluate. As noted by Martin and English [12], the structural, physical, and biological complexity of the soil environment in which pathogens interact with plant roots constrain disease control options, including biological control. As seeds germinate and roots subsequently elongate, the spatial and temporal availability of infection courts are constantly changing. Colonization dynamics of biocontrol agents on seeds and roots, in the presence of an indigenous rhizosphere microbial community, will determine how well these potential infection courts are protected. One of the long-term goals of our research has been to understand and quantify some of the biotic and abiotic parameters that determine the establishment and activity of fungal agents introduced into natural habitats.

Certain organisms have received considerable research attention as potential biocontrol agents; one of the most well studied of these is the fungal genus *Trichoderma*. *Trichoderma* spp. have received considerable attention as potential biological control agents against a wide range of soilborne plant pathogenic fungi [13, 14]. Members of the genus produce a number of fungal cell wall-degrading enzymes including chitinases, 1,3-glucanases, proteases, mannanases, and other hydrolases [15, 16]. *Trichoderma* spp. are highly ecologically successful fungi and have been used increasingly in commercial agriculture, especially in the greenhouse industry where a significant proportion of soilborne disease control products are based on *T. harzianum* [15].

Isolates of *Trichoderma harzianum* have been reported as antagonists of several soilborne plant pathogens. An Idaho isolate of *T. harzianum*, ThzID1, controlled Fusarium wilt of Douglas-fir seedlings [17] and colonized sclerotia of *Sclerotinia sclerotiorum* under laboratory and field conditions [10, 18, 19]. Experimental biocontrol of the important pea pathogens *Pythium ultimum* and *Aphanomyces euteiches* f. sp. *pisi* has been obtained with agents including several bacterial species and also members of the genus *Trichoderma* [9, 18, 20–24]. Seed treatment of pea and other crops with *T. harzianum* controlled *Pythium* seed rot and/or preemergence damping-off [25–27].

*Trichoderma* spp. also have shown considerable potential for bioremediation activity in soil, especially for hydrocarbon pollutants, and many of the same ecological constraints on efficacy may apply in bioremediation systems. *Trichoderma* spp. and especially *T. harzianum* can degrade a number of soil pollutants; for example, Katayama and Matsumura [28] reported on the ability of *T. harzianum* to degrade organochlorine pesticides, particularly endosulfan. Ermisch and Rehm [29] described degradation of polycyclic aromatic hydrocarbons, including anthracene, by *T. harzianum* in soil. Because of the wealth of background information on *Trichoderma* spp. as both biocontrol and bioremediation agents, the genus is an excellent model for the investigation of factors influencing survival and proliferation of recombinant fungi introduced into the soil.

## 2. Complex Soil Microbial Interactions: New Tools and Old Challenges

*2.1. Molecular and Microscopic Tools.* Several factors constrain the observation, quantification, and evaluation of biocontrol fungi in soil. Although various methods have been used to study the occurrence and distribution of *Trichoderma* in natural soils [30–32], relatively few methods have allowed quantitative evaluation of population dynamics and proliferation. For example, in efforts to quantify the influence of temperature, soil matric potential, nutrient source, and antagonistic bacteria, on the hyphal growth and biocontrol efficacy of pelletized *Trichoderma harzianum*, it was not possible to differentiate the hyphal growth of this fungal agent from indigenous *Trichoderma* strains in natural soils [33, 34].

The use of dilution plating for numerical estimation of fungal population does not differentiate among the different propagules (hyphal fragments, conidia, and chlamydospores) that may generate colonies when plated on agar and thus is an unreliable estimate of fungal biomass and active physiological status [35]. The use of mutant strains resistant to specific fungicides may partially overcome problems related to non-specific recovery [36], but this method does not allow for in situ monitoring of growth dynamics and survival structures of introduced *Trichoderma* strains or differentiation of introduced *Trichoderma* strains from indigenous strains.

More recently, genetic engineering of biocontrol agents with reporter or marker genes has provided useful tools for detection and monitoring of introduced biocontrol agents in natural environments [31, 37]. The selectable hygromycin B (*hygB*) phosphotransferase gene, encoding for resistance to this antibiotic, has been used to monitor survival of biocontrol agents in the rhizosphere or on the phylloplane [37, 38]. The β-glucuronidase (GUS) reporter gene also is a promising tool for ecological studies of biocontrol agents [31]. Advantages of this reporter gene system include the low background activity of GUS in fungi and plants, the relative ease and sensitivity of detection [38], and the apparent lack of influence of GUS expression on biocontrol efficacy [39]. However, some GUS activity may be present in unsterile systems or natural soils. For example, *Aspergillus niger* has some indigenous GUS activity [39]. Therefore, for study of growth patterns of an introduced fungus in natural ecosystems, this reporter gene system may be less useful. The green fluorescent protein gene (GFP) of the jellyfish *Aequorea victoria* also has been developed as a reporter for gene expression [40]. Expression of cloned GFP has been reported in several organisms [41, 42]. GFP was shown to be a useful tool for studying host-fungal pathogen interactions *in vivo* [43] and has been used to assess colonization and dispersion of *Aureobasidium pullulans* in the phyllosphere [44]. GFP requires only UV or blue light and oxygen to induce green fluorescence. An exogenous substrate, which GUS requires, is not needed for the detection system to function, thus avoiding problems related to cell permeability and substrate uptake [42].

In our lab, we were able to use a stable transformant of *T. harzianum* expressing both GFP and GUS phenotypes [45] to evaluate aspects of the ecology of this organism in soil. The fungus was cotransformed with GFP, GUS, and hygromycin B (*hygB*) resistance, using PEG-mediated transformation. One cotransformant (strain ThzID1-M3) has been phenotypically stable with respect to GFP expression for several years, despite repeated subculturing without selection pressure. Morphology of ThzID1-M3 is similar to that of the wild-type strain. In nonsterile soil, growth, sporulation, and the ability of the cotransformant to colonize sclerotia of *Sclerotinia sclerotiorum* were compared with the wild-type strain. Wild-type and transformant strains both increased colonization of sclerotia above background levels of colonization by indigenous *Trichoderma* spp. in untreated controls. There were no significant differences in colonization levels between wild-type and cotransformant strains; however, presence of the GFP and GUS marker genes in the cotransformant strain allowed differentiation of introduced *Trichoderma* from indigenous strains. Formation of greenfluorescing conidiophores and conidia was observed within the first three days of incubation in soil, followed by formation of terminal and intercalary chlamydospores and subsequent disintegration of older hyphal segments. Addition of X-Gluc substrate to recovered glass slides confirmed the activity of GUS as well as GFP in the cotransformant in soil; the green color of cotransformant hyphae was clearly visible under UV epifluorescence, while indigenous fungi in the same samples were barely visible microscopically. In one study [11], interactions between ThzID1-M3 and the fungivorous nematode *Aphelenchoides* sp. were investigated in untreated field soil using epifluorescence microscopy; the GFP phenotype allowed for effective identification and quantification of hyphae of ThzID1-M3 in nonsterile soil. We have also used this method to evaluate effects of indigenous soil microflora on biomass and biocontrol efficacy of ThzID1-M3 in soil [46]. Combining epifluorescence with computer image analysis allowed quantification of mycelial biomass, over time, of the GFP-transformed ThzID1-M3 following its introduction into nonsterile soil [47].

Real-time PCR amplification of Trichoderma spp. gene sequences allowed quantification of sclerotia of *S. sclerotiorum* by *Trichoderma* spp. in nonsterile soil [48]. The fact that *T. harzianum* strain ThzID1-M3 is phenotypically different (in that it expresses GFP) from other soil fungi (including other *Trichoderma* strains) and that also the corresponding novel gene is well characterized provided another avenue of ecological investigation, since the GFP gene (which is of jellyfish origin) provided a unique sequence to design a polymerase chain reaction (PCR) primer/probe set to exclusively amplify that sequence from ThzID1-M3 [49]. We also compared quantitative real-time PCR with confocal microscopy to evaluate colonization of sclerotia by ThzID1-M3 and concluded that microscopy combined with image analysis provides better information on the spatial and temporal dynamics of sclerotial colonization, while real-time PCR can provide a more precise assessment of the total extent of colonization over time, including biomass estimates for the biocontrol fungus, and can more easily be used to sample entire sclerotia [50]. Real-time PCR primers were designed for the pea pathogen *Fusarium solani* f. sp. *pisi*, allowing for quantitative assessment of competitive interactions between the biocontrol fungus (*T. harzianum* ThzID1-M3) and the pathogen *F. solani* in the pea rhizosphere [51]. Thus, GFP activity of the transformant is a useful tool for nondestructive monitoring of hyphal proliferation and competitive ability of the fungus in a natural soil, and such transformants may provide useful models to investigate the ecological fitness of introduced fungi in soil.

Sometimes, scanning confocal laser microscopy (SLCM, or "confocal") can provide several advantages over conventional light or standard epifluorescence microscopy for visualization of microbes, such as those in colonized plant debris in soil or inside target pathogen structures (e.g., sclerotia). A confocal microscope combines fluorescence microscopy with electronic image analysis to obtain three-dimensional images. Confocal microscopy has proven to be a powerful tool for examining the structure, organization, and physiology of microbial cells on surfaces, among other uses [50, 52, 53]. The shallow depth of field (as little as 0.5–1.5 $\mu$m) of a confocal microscope allows information to be collected from well-defined optical sections rather than from most of the specimen as in conventional microscopy. Thus, out-of-focus fluorescence is eliminated, resulting in increased contrast and clarity. Effectively, the sample can be optically sectioned, and stacks of optical sections taken at successive focal planes (i.e., a "Z-series") can be reconstructed to produce a focused view of the sample.

*2.2. Spatial Complexity of Soil Biocontrol Systems.* Biological organisms and their controlling variables rarely are distributed in a random or in a uniform way, since the environment is spatially structured by various energy inputs that result in patchy structures or gradients [54]. The rhizoplane is a good example of this, since energy input is largely due to root exudates, and certain zones of roots produce more exudate than others [55]. Although only 1-2% of a root system may be colonized by microbes, space can be a limiting factor [56, 57]. Sites may be preferentially colonized by some rhizoplane microbes and thus may no longer be available to others [58]. As a result, microbial distributions are not always random or uniform. The tendency for rhizosphere microbial population to conform to lognormal or similar frequency distributions has been noted [59–61], although there has been less attention to mechanisms of population development that might lead to such distributions.

Successful manipulation of rhizoplane microflora to enhance native or introduce beneficial microorganisms depends on knowledge of their ecological associations over time and space [62, 63]. Mandeel and Baker [64] observed that potential infection courts on the rhizoplane can be protected by an agent that actively competes for these sites and that this mechanism has more potential impact on biocontrol efficiency than does rhizosphere nutrient competition. They suggested that efficiency values are influenced by spatial relationships, especially relative proximities of the pathogen to penetration sites compared to proximity of the biocontrol agent. We would further add that these spatial arrangements and their interactions evolve over time, so that understanding them is critical to effective biocontrol. Additionally, most previous work in this area has been done using gnotobiotic systems (e.g., [65–68]), so effects of the indigenous microbial community on the distributions of the pathogen and the inoculated biocontrol agent are not yet well understood.

Spatial variability of exudates from seeds and roots may influence sites of colonization of biocontrol agents as well as pathogens. Chemotaxis of zoospores and growth of mycelium towards roots are highly regulated by root exudates [69–71]. Zoospores are an important infectious propagule for *Pythium* spp. The greatest accumulation of zoospores has been reported to be at approximately 2.5 mm behind the root tip (zone of cell elongation), where a major portion of diffusible compounds are exuded [71]. However, we found that *Pythium ultimum* var. *sporangiiferum* encysted throughout the root region and that the extent of zoospore aggregation

was density dependent [66]. *Pythium ultimum* var. *sporangiiferum* represents a promising choice as a model organism to investigate spatial-temporal interactions between biocontrol agents and pathogens, since if access to spatially important entry sites for pathogens is blocked by biocontrol agents or other microbes, then infection and disease may be reduced. For example, Dandurand and Menge found that prior colonization of citrus roots by *Fusarium solani* reduced numbers of encysted zoospores of *Phytophthora parasitica* and *P. citrophthora* and reduced *P. parasitica* population [70, 72].

Deacon and Donaldson [73] described zoospores as "homing agents" or "site-selection agents," because their motility is linked to receptor functions for detecting environmental signals. They described the zoospore homing response as a sequence requiring two factors: a chemotactic stimulus and a suitable surface on which zoospores can orient [73]. Further, they pointed out that zoospores can precisely locate root tips, wounds, or even individual root cells, so that understanding the homing response and factors that may modify it is central to understanding zoosporic fungi and for attempts to control them. Spatial statistics may provide an effective tool for understanding this phenomenon and its ecological significance.

*2.3. Geostatistical Analysis as a Tool to Evaluate Spatial Complexity.* Nonpathogenic rhizoplane colonizers (biocontrol agents) at or near infection courts may be well positioned to modify the zoospore encystment process and subsequent root infection. For example, the frequency distribution of cucumber root sections without encysted zoospores of *Pythium aphanidermatum* was higher for roots treated with biocontrol bacteria than for untreated roots [74]. However, we suggest that frequency distribution analyses alone are not adequate to describe spatially structured biological phenomena [75–77], in part, because they assume independence of observations. The existence of spatial structure implies that the assumption of independence is not met, because any ecological phenomenon located at a given sampling point may have an influence on other points close by or some distance away [75].

What information can spatial statistics (e.g., geostatistics) provide that simple observation or distributional methods (e.g., frequency distributions and dispersion indices) do not? The variable success of biocontrol agents in controlling diseases may, in part, be caused by a lack of understanding of the spatial partitioning of resources in the rhizoplane. Spatial analysis of biocontrol agents may increase our predictive ability for effective biocontrol agents. For example, colonization patterns of bacteria were reported by Fukui et al. [20]; although it was observed that two strains colonized various parts of sugar beet seeds, a quantitative analysis of the spatial patterns of the two strains was not made, and conclusions derived from spatial patterns of the two bacteria based on observation only are difficult to interpret. A quantitative analysis of spatial patterns is particularly important to determine whether biocontrol agents change the spatial patterns of pathogens in predictable and consistent ways. Spatial statistical analysis provides a mechanism to explore processes that generate different patterns of organisms over time and helps to determine the sensitivity of spatial pattern to variations in these processes. Spatial analysis is defined here as any analysis that quantitatively evaluates variations or changes based on spatial orientation within a defined area or volume.

Frequency distribution methods are commonly based on mean/variance ratios, which do not provide reliable interpretations of spatial structure, since information on the location of each sample site is ignored. Although such indices are useful for estimation of population means, they do not maintain the spatial integrity of samples, making spatial analysis impossible [76–79]. Unlike frequency analysis, spatial analysis requires that the spatial integrity of observations be maintained; that is, spatial coordinates are recorded for each sample point. One method for spatial analysis, geostatistics, provides a quantitative assessment of spatial distributions that maintains the spatial integrity of data and is able to analyze the degree of association (autocorrelation) based on direction and distance between samples [80, 81]. Although geostatistics evolved primarily through mining applications, it has proven highly applicable to biological systems. For example, geostatistics has been used to evaluate insect spatial distributions [82, 83] and a spatial simulation model [77, 84], plant disease patterns [76, 85], and spatial patterns of zoospores and bacteria on roots [66, 67, 86]. Some advantages of geostatistics may be summarized as follows: geostatistics is independent of the relationship between the mean and variance; geostatistics maintains the spatial integrity of locations of samples and uses the variation between points to evaluate spatial dependence; geostatistics assesses spatial dependence quantitatively and can be used to compare spatial dependence at different points in time or at the same point in time under different conditions.

Although geostatistical analysis cannot say specifically what the mechanism of biocontrol activity of *T. harzianum* is, it can help give credence to hypotheses about mechanisms. As an example, Mandeel and Baker [64] observed that potential infection courts on the rhizoplane can be protected by an agent that actively competes for these sites, and they suggested that this mechanism has more potential impact on biocontrol efficiency than does rhizosphere nutrient competition. Mandeel and Baker further observed that biocontrol efficiency values are influenced by spatial relationships, especially relative proximities of the pathogen to penetration sites compared to proximity of the biocontrol agent [64]. If space is indeed a mechanism in biocontrol efficacy, then proliferation of the biocontrol agent would be expected to generate a change in observed spatial patterns of the pathogen population, as our preliminary results have indicated. However, these preliminary results were obtained in a gnotobiotic system. Of course, a recurring question in biological control research is why is biocontrol efficacy so variable in natural soils? Again, spatial statistical analysis can provide quantitative answers about whether spatial attributes of a biocontrol agent will change or not change in the presence of indigenous microbes. Effective development of spatially rigorous analysis techniques will provide a necessary

framework for evaluating the effects of mechanisms that are studied at the genetic and biochemical level.

*2.4. Simulation Modeling of Biocontrol Agent Performance.* Fungi in soil perform beneficial roles that include biological control of soilborne plant pathogens. However, relatively little predictive information is available about the growth and activity of fungal hyphae in soil habitats. A number of fungal growth models deal primarily with aspects of physiology and morphogenesis (e.g., [87, 88]), and these may not be directly applicable to predict fungal proliferation in natural habitats; other models deal with colony pattern formation in 2D and 3D space. Knudsen & Stack [89] introduced the idea of a simulation model for hyphal growth of a fungal hyperparasite through soil and use of the model to predict the incidence of hyperparasitism of sclerotia of certain soilborne plant pathogens. Lejeune and Baron [90] and Lejeune et al. [91] simulated the 3D growth of the filamentous fungus *Trichoderma reesei*, based on properties of mycelial growth (total hyphal length and total number of tips). Cross and Kenerley [92], using a combination of the Ratkowsky and Arrhenius equations, modeled colony growth of *T. virens* at different temperatures. Other models of fungal growth [93, 94] are useful for studying fungal physiology and morphogenesis but do not have an explicit spatial component (i.e., they assume spatial homogeneity) and thus may not be applicable to predict colony expansion in a heterogeneous environment.

Models developed for fungal growth in homogeneous environments, such as on agar surfaces or in culture media, may be less applicable to predict colony expansion in a heterogeneous environment. One potential way to deal with problems of spatial heterogeneity is the development of an "individual-based" model that tracks spatial attributes of the organism of interest as well as those of the local environment. With an individual-based model, information about individual organisms and/or habitat patches is retained in a database, and the model can then aggregate attributes of individuals to generate dynamics of the larger population. Thus, individual-based models represent an inductive approach to spatial population modeling in that population statistics are derived from collective attributes of individuals [95]. For example, we have used this approach to predict epizootics caused by the entomopathogen *Beauveria bassiana* in population of Russian wheat aphid on uniform versus nonuniform host plant distributions [77, 84].

There are important reasons why spatially homogeneous models of microbial growth kinetics and interactions are inadequate for some purposes, especially for analysis of mechanisms that generate different spatial dynamics as the population develops. If a biocontrol system is made more complex with patchily distributed agents and/or target pathogens, the spatial dimensions of the system become central to effective analysis and comprehensive understanding. Although description of spatially complex biological processes may sometimes most effectively be accomplished by modeling individuals, the concept of an "individual" is ambiguous for clonal organisms such as fungi. Numerically, physiologically, and often genetically, a single fungal colony is best described as a network of individual hyphal segments. Two advantages of this approach are as follows: first, the spatial structure of the fungal colony (hyphal network) can be explicitly reconstructed at any point in time; second, the approach has the flexibility to accommodate environments that are heterogeneous with respect to the spatial distribution of physical and biotic factors, and where phenotypic heterogeneity among individuals is present.

The objective of our "Fungmod" study [96] was to develop a modeling framework, in the form of an individual-based computer simulation model, that is suitable for exploration of the spatial dynamics of hyphal growth of a biocontrol fungus in a three-dimensional soil habitat. In that model, records of spatial location and branching hierarchy are maintained for individual hyphal nodes or for one or more fungal colonies. Thus, each hyphal segment is spatially referenced (coordinates in three dimensions) in terms of its ending node position, with its starting location referenced as the ending position of its parent node. In this way, the entire spatial structure of the fungal colony can be explicitly reconstructed from the data set at any point in time. The model proceeds at 1-day time steps and allows prediction of colony diameter, fungal biomass, and the spatial distribution of hyphae over time. The model was validated against data derived from image analysis of hyphal biomass accumulation in soil.

Although simulation models have been used extensively for arthropod biocontrol systems (e.g., [97]), relatively few models have been used for plant disease biocontrol studies. Knudsen and Hudler [98] described a simulation model to predict population dynamics of antibiotic-producing strains of *Pseudomonas fluorescens* on conifer foliage and subsequent effects on conidial germination of the fungal pathogen *Gremmeniella abietina*. Spurr and Knudsen [99] and Knudsen and Spurr [100] described how population models for antagonistic phylloplane bacteria might be incorporated into predictive models for foliar fungal diseases. However, those studies did not address spatial aspects of growth and proliferation of the putative control agents. For control of sclerotia-forming pathogens in soil, the ability of the biocontrol agent to effectively explore the three-dimensional soil space via hyphal growth probably is more important than spatially localized proliferation, for example, via sporulation. During the initial stage of colony development in soil, hyphae of biocontrol fungi will obtain their energy from the formulation matrix and/or from substrata located within the soil volume in which they are growing. The extent of mycoparasitism thus would depend on the density and spatial arrangement both of target sclerotia and of hyphae of the biocontrol agent. Quantification of patterns and pattern changes in hyphal growth should enhance prediction of the likelihood of successful mycoparasitic events.

*2.5. Trophic Complexity Influences the Success of Fungal Biocontrol Agents.* A wide range of biotic factors, in addition to abiotic factors (e.g., [101]), influence growth and proliferation of fungi introduced into natural soils. However, there has been relatively little investigation of quantitative effects of the indigenous soil biota on growth and establishment of fungi introduced into natural soils as potential biocontrol agents. Especially, interactions between introduced fungi and soil microanimals have largely been neglected in studies relating to the potential proliferation and ecological impact of introduced fungal agents. Nematodes are the most abundant soilinhabiting animals; there are many described species (more than 80,000), and population densities typically are high in the top 5 cm of soils, with substantial seasonal differences in agricultural soils [102, 103]. Identified nematode trophic groups include fungivores, bacterivores, omnivores, predators, and plant parasites [104]. Fungivorous nematodes have stylets and feed on mycelia of many different species of soil fungi, including beneficial fungi as well as plant pathogens [105].

Numerous species of fungivores have been found in soils [106]. The most common genera found in agricultural soils are *Aphelenchoides*, *Aphelenchus*, *Tylenchus*, and *Ditylenchus* [106, 107]; some species within this group are plant parasites. Fungivores within this group of nematodes usually are found at lower densities than bacterivores or plant parasites [107]. However, if suitable fungi as food are available, population of fungivores may rapidly increase. Hofman and S'jacob [107] reported that numbers of the mycophagous nematodes *Aphelenchoides* sp., *Aphelenchus avenae*, and unidentified microbivorous nematodes increased severalfold within a few days on dying roots of flax that had been precolonized by *Rhizoctonia solani*. Many plant ectoparasitic nematodes also may be fungivorous, as the need arises [106]. For example, *Ditylenchus destructor*, which is parasitic on carrot, potato, alfalfa, radish, and sweet potato, also can be cultured on a variety of filamentous fungi [107].

Fungivorous nematodes may be a significant biotic constraint on activity of biocontrol fungi in the field, and experiments suggest that their presence introduces a layer of trophic complexity that needs to be factored into attempts to predict the fitness of introduced biocontrol fungi in soil. In our laboratory, we isolated a fungivorous nematode (*Aphelenchoides* sp.) from field soil and found that population of the nematode significantly reduced hyphal growth and total population of *T. harzianum* under some environmental conditions, in nonsterile field soil [11]. Addition of *Aphelenchoides* resulted in lower numbers of sclerotia colonized by *T. harzianum* compared to untreated controls [11]. Because the nematode feeds (via stylet) on hyphae of the fungus but is unable to feed on conidia or chlamydospores, active nematode population appears to have a disproportionate effect on fungal population that is actively growing (i.e., with a comparatively higher ratio of hyphae : spores) compared to population of *Trichoderma* that is primarily inactive. In a typical biocontrol scenario, the introduced fungus is applied to soil, often with a food base, from which it initiates hyphal growth (e.g., [10, 19, 34]). Meanwhile, the bulk of the indigenous *Trichoderma* population may be relatively inactive. Thus, if conditions are favorable for nematode activity, nematode feeding effectively will be selected against the introduced organism, and its apparent relative fitness compared to indigenous *Trichoderma* will be less. Perhaps ironically, soil environmental conditions that are favorable for nematode activity (adequate soil moisture and moderate temperature) are those that might otherwise be considered "ideal" for introduction of the fungal agent. The nematode x fungus interaction just described represents the first level of trophic complexity that may be a determinant of the (apparent) fitness of a recombinant fungus released into the environment.

Another aspect of trophic complexity may be implicated if hyphae of the biocontrol agent are able to contact and colonize the target substratum, that is, sclerotia of *S. sclerotiorum*, since colonized sclerotia potentially represent a refuge where the agent is protected from nematode activity and thus can persist. Similarly, because *Trichoderma* spp. are effective colonizers of plant debris, thecolonization of dead plant material may also provide a refuge from nematode activity. Refuge theory has played a prominent role in a large number of recent ecological studies, although to our knowledge it has not been addressed in microbial ecology. Hawkins et al. [108] applied principles of refuge theory to the biological control of insect pests using parasitoids and suggested that both theoretical and experimental results predict that success of biological control efforts (in insect pest/parasitoid systems) is inversely related to the proportion of insects protected from parasitoid attack. In general, parasitism rates increased as refuge level decreased [109]. However, it was also pointed out that research on refuge theory in entomology has largely been restricted to correlative studies and theory and that theoretical work has advanced more rapidly than experimental tests of the models' assumptions in part due to the difficulty of manipulating the feeding biology and refuges of individual insects in natural systems [109]. In the *Aphelenchoides x Trichoderma x Sclerotinia* system, however, the roles are quite different: one hypothesized refuge for *T. harzianum* is sclerotia of *S. sclerotiorum*, the target organism itself. Ability of the biocontrol agent to rapidly colonize this refuge may simultaneously effectuate biocontrol of the target sclerotia and also escape from feeding activity of fungivorous nematodes (however, recent experiments suggest that the nematode may also be able to enter sclerotia and feed there; R. Garcia De la Cruz & G. R. Knudsen, unpublished). One aspect that these very different biological control systems (insect parasitoids and their hosts versus fungal hyperparasites and their hosts) have in common is, as described by Hawkins et al. [108], that refuges "...provide a general mechanism for interpreting ecological patterns at both the community level (their species diversity) and population level (their dynamics)." Again, it is anticipated that study of this phenomenon will be helpful in linking ecological theory with applied biotechnology risk assessment considerations.

## 3. Conclusions

As Spurr and Knudsen [99] noted, the search for a "silver bullet," a biological control organism that can simply be applied as if it were a chemical pesticide, is unlikely to succeed in an ecologically complex milieu. Ecological systems, as described by Levin [110], are the prototypical complex adaptive systems (CAS) "*in which macroscopic system properties such as trophic structure, diversity-productivity relationships, and patterns of nutrient flux emerge from interactions among components,*

*and may feed back to influence the subsequent development of those interactions.*" In recent years, the study of complex systems has contributed significantly to important research areas in ecology, namely, food (trophic) webs, patch dynamics and spatial structure, and population fluctuations [111–113]. We believe that each of these areas will continue to prove highly applicable and provide many research opportunities for biological control of soilborne plant pathogens.

## Conflict of Interests

The research reported here was supported by the public sector, and the authors have no conflict of interests to report.

## References

[1] V. Vasas and F. Jordán, "Topological keystone species in ecological interaction networks: considering link quality and nontrophic effects," *Ecological Modelling*, vol. 196, no. 3-4, pp. 365–378, 2006.

[2] E. A. Herre, N. Knowlton, U. G. Mueller, and S. A. Rehner, "The evolution of mutualisms: exploring the paths between conflict and cooperation," *Trends in Ecology and Evolution*, vol. 14, no. 2, pp. 49–53, 1999.

[3] C. Neuhauser and J. E. Fargione, "A mutualism-parasitism continuum model and its application to plant-mycorrhizae interactions," *Ecological Modelling*, vol. 177, no. 3-4, pp. 337–352, 2004.

[4] J. A. van Veen, L. S. van Overbeek, and J. D. van Elsas, "Fate and activity of microorganisms introduced into soil," *Microbiology and Molecular Biology Reviews*, vol. 61, no. 2, pp. 121–135, 1997.

[5] J. E. Duffy, B. J. Cardinale, K. E. France, P. B. McIntyre, E. Thébault, and M. Loreau, "The functional role of biodiversity in ecosystems: incorporating trophic complexity," *Ecology Letters*, vol. 10, no. 6, pp. 522–538, 2007.

[6] D. S. Srivastava and T. Bell, "Reducing horizontal and vertical diversity in a foodweb triggers extinctions and impacts functions," *Ecology Letters*, vol. 12, no. 10, pp. 1016–1028, 2009.

[7] J. S. Jacobs, R. L. Sheley, and B. D. Maxwell, "Effect of *Sclerotinia sclerotiorum* on the interference between bluebunch wheatgrass (*Agropyron spicatum*) and spotted knapweed (*Centaurea maculosa*)," *Weed Technology*, vol. 10, no. 1, pp. 13–21, 1996.

[8] M. H. Abu-Dieyeh and A. K. Watson, "Efficacy of *Sclerotinia minor* for dandelion control: effect of dandelion accession, age and grass competition," *Weed Research*, vol. 47, no. 1, pp. 63–72, 2007.

[9] J. P. Hubbard, G. E. Harman, and Y. Hadar, "Effects of soilborne *Pseudomonas* spp. on the biological control agent, *Trichoderma harzianum*, on pea seeds," *Phytopathology*, vol. 73, pp. 655–659, 1983.

[10] L. Bin, G. R. Knudsen, and D. J. Eschen, "Influence of an antagonistic strain of *Pseudomonas fluorescens* on growth and ability of *Trichoderma harzianum* to colonize sclerotia of *Sclerotinia sclerotiorum* in soil," *Phytopathology*, vol. 81, pp. 994–1000, 1991.

[11] Y.-S. Bae and G. R. Knudsen, "Influence of a fungus-feeding nematode on growth and biocontrol efficacy of *Trichoderma harzianum*," *Phytopathology*, vol. 91, no. 3, pp. 301–306, 2001.

[12] F. N. Martin and J. T. English, "Population genetics of soilborne fungal plant pathogens," *Phytopathology*, vol. 87, no. 4, pp. 446–447, 1997.

[13] G. E. Harman, C. R. Howell, A. Viterbo, I. Chet, and M. Lorito, "*Trichoderma* species—opportunistic, avirulent plant symbionts," *Nature Reviews Microbiology*, vol. 2, no. 1, pp. 43–56, 2004.

[14] M. Vermaa, S. K. Brara, R. D. Tyagia, R. Y. Surampallib, and J. R. Valéroa, "Antagonistic fungi, *Trichoderma* spp.: panoply of biological control," *Biochemical Engineering Journal*, vol. 37, no. 1, pp. 1–20, 2007.

[15] G. E. Harman, "Myths and dogmas of biocontrol: changes in perceptions derived from research on *Trichoderma harzianum* T-22," *Plant Disease*, vol. 84, no. 4, pp. 377–393, 2000.

[16] M. C. Limón, M. R. Chacón, R. Mejías et al., "Increased antifungal and chitinase specific activities of *Trichoderma harzianum* CECT 2413 by addition of a cellulose binding domain," *Applied Microbiology and Biotechnology*, vol. 64, no. 5, pp. 675–685, 2004.

[17] M. R. Mousseaux, R. K. Dumroese, R. L. James, D. L. Wenny, and G. R. Knudsen, "Efficacy of *Trichoderma harzianum* as a biological control of *Fusarium oxysporum* in container-grown Douglas-fir seedlings," *New Forests*, vol. 15, no. 1, pp. 11–21, 1998.

[18] L.-M. Dandurand, R. D. Mosher, and G. R. Knudsen, "Combined effects of Brassica napus seed meal and *Trichoderma harzianum* on two soilborne plant pathogens," *Canadian Journal of Microbiology*, vol. 46, no. 11, pp. 1051–1057, 2000.

[19] G. R. Knudsen, D. J. Eschen, L. M. Dandurand, and L. Bin, "Potential for control of *Sclerotinia sclerotiorum* through colonization of sclerotia by *Trichoderma harzianum*," *Plant Disease*, vol. 75, pp. 466–470, 1991.

[20] R. Fukui, E. I. Poinar, P. H. Bauer et al., "Spatial colonization patterns and interaction of bacteria on inoculated sugar beet seed," *Phytopathology*, vol. 84, no. 11, pp. 1338–1345, 1994.

[21] G. E. Harman, R. Petzoldt, A. Comis, and J. Chen, "Interactions between *Trichoderma harzianum* strain T22 and maize inbred line Mo17 and effects of these interactions on diseases caused by *Pythiuin ultimum* and *Colletotrichum graminicola*," *Phytopathology*, vol. 94, no. 2, pp. 147–153, 2004.

[22] W. Mao, R. D. Lumsden, J. A. Lewis, and P. K. Hebbar, "Seed treatment using pre-infiltration and biocontrol agents to reduce damping-off of corn caused by species of *Pythium* and *Fusarium*," *Plant Disease*, vol. 82, no. 3, pp. 294–299, 1998.

[23] A. Muthukumar, A. Eswaran, S. Nakkeeran, and G. Sangeetha, "Efficacy of plant extracts and biocontrol agents against *Pythium aphanidermatum* inciting chilli damping-off," *Crop Protection*, vol. 29, no. 12, pp. 1483–1488, 2010.

[24] T. C. Paulitz, "Effect of *Pseudomonas putida* on the stimulation of *Pythium ultimum* by seed volatiles of pea and soybean," *Phytopathology*, vol. 81, pp. 1282–1287, 1991.

[25] G. E. Harman, "Myths and dogmas of biocontrol: changes in perceptions derived from research on *Trichoderma harzianum* T-22," *Plant Disease*, vol. 84, no. 4, pp. 377–393, 2000.

[26] N. Benhamou and I. Chet, "Cellular and molecular mechanisms involved in the interaction between *Trichoderma harzianum* and *Pythium ultimum*," *Applied and Environmental Microbiology*, vol. 63, pp. 2095–2099, 1997.

[27] C. R. Howell, "Mechanisms employed by *Trichoderma* species in the biological control of plant diseases: the history and evolution of current concepts," *Plant Disease*, vol. 87, no. 1, pp. 4–10, 2003.

[28] A. Katayama and F. Matsumura, "Degradation of organochlorine pesticides, particularly endosulfan, by *Trichoderma harzianum*," *Environmental Toxicology and Chemistry*, vol. 12, no. 6, pp. 1059–1065, 1993.

[29] O. Ermisch and H. J. Rehm, "Degradation of polycyclic aromatic hydrocarbons by the immobilized mold *Trichoderma harzianum* in soil," in *Proceedings of the DECHEMA Annual Meeting on Biotechnology and 58th Meeting of the European Biotechnology Federations, and Joint Meeting of Society for Industry Microbiology (SIM '89)*, pp. 780–781, 1989.

[30] I. Grishkan, E. Nevo, S. P. Wasser, and A. Beharav, "Adaptive spatiotemporal distribution of soil microfungi in "Evolution Canyon" II, Lower Nahal Keziv, western Upper Galilee, Israel," *Biological Journal of the Linnean Society*, vol. 78, no. 4, pp. 527–539, 2003.

[31] H. Green and D. F. Jensen, "A tool for monitoring *Trichoderma harzianum*: II. The use of a GUS transformant for ecological studies in the rhizosphere," *Phytopathology*, vol. 85, no. 11, pp. 1436–1440, 1995.

[32] C.-L. Zhang, I. S. Druzhinina, C. P. Kubicek, and T. Xu, "*Trichoderma* biodiversity in China: evidence for a North to South distribution of species in East Asia," *FEMS Microbiology Letters*, vol. 251, no. 2, pp. 251–257, 2005.

[33] G. R. Knudsen and L. Bin, "Effects of temperature, soil moisture and wheat bran on growth of *Trichoderma harzianum* from alginate pellets," *Phytopathology*, vol. 80, pp. 724–727, 1990.

[34] G. R. Knudsen, D. J. Eschen, L. M. Dandurand, and Z. G. Wang, "Method to enhance growth and sporulation of pelletized biocontrol fungi," *Applied and Environmental Microbiology*, vol. 57, no. 10, pp. 2864–2867, 1991.

[35] R. D. Lumsden, J. P. Carter, J. M. Whipps, and J. M. Lynch, "Comparison of biomass and viable propagule measurements in the antagonism of *Trichoderma harzianum* against *Pythium ultimum*," *Soil Biology and Biochemistry*, vol. 22, no. 2, pp. 187–194, 1990.

[36] J. Rigot and F. Matsumura, "Assessment of the rhizosphere competency and pentachlorophenol-metabolizing activity of a pesticide-degrading strain of *Trichoderma harzianum* introduced into the root zone of corn seedlings," *Journal of Environmental Science and Health B: Pesticides, Food Contaminants, and Agricultural Wastes*, vol. 37, no. 3, pp. 201–210, 2002.

[37] C.-T. Lo, E. B. Nelson, C. K. Hayes, and G. E. Harman, "Ecological studies of transformed *Trichoderma harzianum* strain 1295-22 in the rhizosphere and on the phylloplane of creeping bentgrass," *Phytopathology*, vol. 88, no. 2, pp. 129–136, 1998.

[38] I. N. Roberts, R. P. Oliver, P. J. Punt, and C. A. M. J. J. Van den Hondel, "Expression of the *Escherichia coli* β-glucuronidase gene in industrial and phytopathogenic filamentous fungi," *Current Genetics*, vol. 15, no. 3, pp. 177–180, 1989.

[39] C. Thrane, M. Lubeck, H. Green et al., "A tool for monitoring *Trichoderma harzianum*: I. Transformation with the GUS gene by protoplast technology," *Phytopathology*, vol. 85, no. 11, pp. 1428–1435, 1995.

[40] M. Chalfie, Y. Tu, G. Euskirchen, W. W. Ward, and D. C. Prasher, "Green fluorescent protein as a marker for gene expression," *Science*, vol. 263, no. 5148, pp. 802–805, 1994.

[41] B. P. Cormack, G. Bertram, M. Egerton, N. A. R. Gow, S. Falkow, and A. J. P. Brown, "Yeast-enhanced green fluorescent protein (yGFP): a reporter of gene expression in *Candida albicans*," *Microbiology*, vol. 143, no. 2, pp. 303–311, 1997.

[42] J. Sheen, H. S. Hwang Seongbin, Y. Niwa, H. Kobayashi, and D. W. Galbraith, "Green-fluorescent protein as a new vital marker in plant cells," *Plant Journal*, vol. 8, no. 5, pp. 777–784, 1995.

[43] T. Spellig, A. Bottin, and R. Kahmann, "Green fluorescent protein (GFP) as a new vital marker in the phytopathogenic

fungus *Ustilago maydis*," *Molecular and General Genetics*, vol. 252, no. 5, pp. 503–509, 1996.

[44] A. J. Vanden Wymelenberg, D. Cullen, R. N. Spear, B. Schoenike, and J. H. Andrews, "Expression of green fluorescent protein in *Aureobasidium pullulans* and quantification of the fungus on leaf surfaces," *Bioechniques*, vol. 23, no. 4, pp. 686–690, 1997.

[45] Y. S. Bae and G. R. Knudsen, "Cotransformation of *Trichoderma harzianum* with β-glucuronidase and green fluorescent protein genes provides a useful tool for monitoring fungal growth and activity in natural soils," *Applied and Environmental Microbiology*, vol. 66, no. 2, pp. 810–815, 2000.

[46] Y. S. Bae and G. R. Knudsen, "Soil microbial biomass influence on growth and biocontrol efficacy of *Trichoderma harzianum*," *Biological Control*, vol. 32, no. 2, pp. 236–242, 2005.

[47] K. A. Orr and G. R. Knudsen, "Use of green fluorescent protein and image analysis to quantify proliferation of *Trichoderma harzianum* in nonsterile soil," *Phytopathology*, vol. 94, no. 12, pp. 1383–1389, 2004.

[48] T. G. Kim and G. R. Knudsen, "Quantitative real-time PCR effectively detects and quantifies colonization of sclerotia of *Sclerotinia sclerotiorum* by *Trichoderma* spp," *Applied Soil Ecology*, vol. 40, no. 1, pp. 100–108, 2008.

[49] T. G. Kim and G. R. Knudsen, "Colonization of *Sclerotinia sclerotiorum* sclerotia by a biocontrol isolate of *Trichoderma harzianum*, and effects on myceliogenic germination," *Biocontrol Science and Technology*, vol. 19, no. 10, pp. 1081–1085, 2009.

[50] T. G. Kim and G. R. Knudsen, "Comparison of real-time PCR and microscopy to evaluate sclerotial colonisation by a biocontrol fungus," *Fungal Biology*, vol. 115, no. 4-5, pp. 317–325, 2011.

[51] T. G. Kim and G. R. Knudsen, "Relationship between the biocontrol fungus *Trichoderma harzianum* and the phytopathogenic fungus *Fusarium solani* f.sp. *pisi*," *Applied Soil Ecology*, vol. 68, pp. 57–60, 2013.

[52] C.-T. Huang, F. P. Yu, G. A. McFeters, and P. S. Stewart, "Nonuniform spatial patterns of respiratory activity within biofilms during disinfection," *Applied and Environmental Microbiology*, vol. 61, no. 6, pp. 2252–2256, 1995.

[53] J. S. McLean, O. N. Ona, and P. D. Majors, "Correlated biofilm imaging, transport and metabolism measurements via combined nuclear magnetic resonance and confocal microscopy," *ISME Journal*, vol. 2, no. 2, pp. 121–131, 2008.

[54] P. Legendre and M. J. Fortin, "Spatial pattern and ecological analysis," *Vegetatio*, vol. 80, no. 2, pp. 107–138, 1989.

[55] A. D. Rovira, "Zones of exudation along plant roots and spatial distribution of microorganisms in the rhizosphere," *Pesticide Science*, vol. 4, no. 3, pp. 361–366, 1973.

[56] S. Compant, C. Clément, and A. Sessitsch, "Plant growth-promoting bacteria in the rhizo- and endosphere of plants: their role, colonization, mechanisms involved and prospects for utilization," *Soil Biology and Biochemistry*, vol. 42, no. 5, pp. 669–678, 2010.

[57] A. Ramette and J. M. Tiedje, "Multiscale responses of microbial life to spatial distance and environmental heterogeneity in a patchy ecosystem," *Proceedings of the National Academy of Sciences of the United States of America*, vol. 104, no. 8, pp. 2761–2766, 2007.

[58] A. J. H. Davey, C. P. Doncaster, and O. D. Jones, "Distinguishing between interference and exploitation competition for shelter in a mobile fish population," *Environmental Modeling and Assessment*, vol. 14, no. 5, pp. 555–562, 2009.

[59] E. Limpert, W. A. Stahel, and M. Abbt, "Log-normal distributions across the sciences: keys and clues," *BioScience*, vol. 51, no. 5, pp. 341–352, 2001.

[60] J. E. Loper, T. V. Suslow, and M. N. Schroth, "Lognormal distribution of bacterial populations in the rhizosphere," *Phytopathology*, vol. 74, pp. 1454–1460, 1984.

[61] G. Wieland, R. Neumann, and H. Backhaus, "Variation of microbial communities in soil, rhizosphere, and rhizoplane in response to crop species, soil type, and crop development," *Applied and Environmental Microbiology*, vol. 67, no. 12, pp. 5849–5854, 2001.

[62] J. M. Raaijmakers, T. C. Paulitz, C. Steinberg, C. Alabouvette, and Y. Moënne-Loccoz, "The rhizosphere: a playground and battlefield for soilborne pathogens and beneficial microorganisms," *Plant and Soil*, vol. 321, no. 1-2, pp. 341–361, 2009.

[63] M. E. Stanghellini and S. L. Rasmussen, "Root prints: a technique for the determination of the in situ spatial distribution of bacteria on the rhizoplane of field-grown plants," *Phytopathology*, vol. 79, pp. 1131–1134, 1989.

[64] Q. Mandeel and R. Baker, "Mechanisms involved in biological control of Fusarium wilt on cucumber with strains of nonpathogenic *Fusarium oxysporum*," *Phytopathology*, vol. 81, pp. 462–469, 1991.

[65] T. F. C. Chin A Woeng, W. de Priester, A. J. van der Bij, and B. J. Lugtenberg, "Description of the colonization of a gnotobiotic tomato rhizosphere by *Pseudomonas fluorescens* biocontrol strain WCS365, using scanning electron microscopy," *Molecular Plant Microbe Interactions*, vol. 10, no. 1, pp. 79–86, 1997.

[66] L. M. Dandurand, G. R. Knudsen, and D. J. Schotzko, "Quantification of *Pythium ultimum* var. *sporangiiferum* zoospore encystment patterns using geostatistics," *Phytopathology*, vol. 85, no. 2, pp. 186–190, 1995.

[67] L. M. Dandurand, D. J. Schotzko, and G. R. Knudsen, "Spatial patterns of rhizoplane populations of *Pseudomonas fluorescens*," *Applied and Environmental Microbiology*, vol. 63, pp. 3211–3217, 1997.

[68] M. Simons, A. J. van der Bij, I. Brand, L. A. de Weger, C. A. Wijffelman, and B. J. J. Lugtenberg, "Gnotobiotic system for studying rhizosphere colonization by plant growth-promoting *Pseudomonas* bacteria," *Molecular Plant Microbe Interactions*, vol. 9, pp. 600–607, 1996.

[69] C. A. Walker and P. van West, "Zoospore development in the oomycetes," *Fungal Biology Reviews*, vol. 21, no. 1, pp. 10–18, 2007.

[70] L. M. Dandurand and J. A. Menge, "Influence of *Fusarium solani* on chemotaxis of zoospores of *Phytophthora parasitica* and *Phytophthora citrophthora* and on distribution of 14C in citrus tissues and root exudate," *Soil Biology and Biochemistry*, vol. 26, no. 1, pp. 75–79, 1994.

[71] R. T. Mitchell and J. W. Deacon, "Differential (host-specific) accumulation of zoospores of *Pythium* on roots of graminaceous and non-graminaceous plants," *New Phytologist*, vol. 102, pp. 113–122, 1986.

[72] L. M. Dandurand and J. A. Menge, "Influence of *Fusarium solani* on citrus root rot caused by *Phytophthora parasitica* and *Phytophthora citrophthora*," *Plant and Soil*, vol. 144, no. 1, pp. 13–21, 1992.

[73] J. W. Deacon and S. P. Donaldson, "Molecular recognition in the homing responses of zoosporic fungi, with special reference to *Pythium* and *Phytophthora*," *Mycological Research*, vol. 97, no. 10, pp. 1153–1171, 1993.

[74] T. Zhou and T. C. Paulitz, "In vitro and in vivo effects of *Pseudomonas* spp. on *Pythium aphanidermatum*: zoospore behavior in exudates and on the rhizoplane of bacteria-treated cucumber roots," *Phytopathology*, vol. 83, pp. 872–876, 1993.

[75] P. Legendre, "Spatial autocorrelation: trouble or new paradigm?" *Ecology*, vol. 74, no. 6, pp. 1659–1673, 1993.

[76] P. C. Nicot, D. I. Rouse, and B. S. Yandell, "Comparison of statistical methods for studying spatial patterns of soilborne plant pathogens in the field," *Phytopathology*, vol. 74, pp. 1399–1402, 1984.

[77] D. J. Schotzko and G. R. Knudsen, "Use of geostatistics to evaluate a spatial simulation of Russian wheat aphid (Homoptera: Aphididae) movement behavior on preferred and nonpreferred hosts," *Environmental Entomology*, vol. 21, no. 6, pp. 1271–1282, 1992.

[78] P. A. Jumars, D. Thistle, and M. L. Jones, "Detecting two-dimensional spatial structure in biological data," *Oecologia*, vol. 28, no. 2, pp. 109–123, 1977.

[79] A. J. Sawyer, "Inconstancy of Taylor's b: simulated sampling with different quadrat sizes and spatial distributions," *Researches on Population Ecology*, vol. 31, no. 1, pp. 11–24, 1989.

[80] E. H. Isaaks and R. H. Srivastava, *Applied Geostatistics*, Oxford University Press, Oxford, UK, 1989.

[81] B. B. Trangmar, R. S. Yost, and G. Uehara, "Application of geostatistics to spatial studies of soil properties," *Advances in Agronomy*, vol. 38, pp. 45–94, 1985.

[82] W. P. Kemp, T. M. Kalaris, and W. F. Quimby, "Rangeland grasshopper (Orthoptera: Acrididae) spatial variability: macroscale population assessment," *Journal of Economic Entomology*, vol. 82, pp. 1270–1276, 1989.

[83] D. J. Schotzko and C. M. Smith, "Effect of host plant on the plant to plant distribution of the Russian wheat aphid (Homoptera: Aphididae)," *Journal of Economic Entomology*, vol. 84, pp. 1725–1734, 1991.

[84] G. R. Knudsen and D. J. Schotzko, "Spatial simulation of epizootics caused by Beauveria bassiana in Russian wheat aphid populations," *Biological Control*, vol. 16, no. 3, pp. 318–326, 1999.

[85] D. A. Johnson, J. R. Alldredge, J. R. Allen, and R. Allwine, "Spatial pattern of downy mildew in hop yards during severe and mild diesease epidemics," *Phytopathology*, vol. 81, pp. 1369–1374, 1991.

[86] G. R. Knudsen and L. M. Dandurand, "Analysis of spatial patterns of rhizoplane colonization," in *The Spatial Distribution of Microbes in the Environment*, R. B. Franklin and A. L. Mills, Eds., pp. 109–133, Springer, New York, NY, USA, 2007.

[87] L. Edelstein, "The propagation of fungal colonies: a model for tissue growth," *Journal of Theoretical Biology*, vol. 98, no. 4, pp. 679–701, 1982.

[88] L. Edelstein, Y. Hadar, and I. Chet, "A model for fungal colony growth applied to *Sclerotium rolfsii*," *Journal of General Microbiology*, vol. 129, no. 6, pp. 1873–1881, 1983.

[89] G. R. Knudsen and J. P. Stack, "Modeling growth and dispersal of fungi in natural environments," in *Handbook of Applied Mycology, Vol. I: Soil and Plants*, D. K. Arora, K. G. Mukerji, B. Rai, and G. R. Knudsen, Eds., pp. 625–645, Marcel Dekker, New York, NY, USA, 1991.

[90] R. Lejeune and G. V. Baron, "Simulation of growth of a flamentous fungus in 3 dimensions," *Biotechnology and Bioengineering*, vol. 53, no. 2, pp. 139–150, 1997.

[91] R. Lejeune, J. Nielsen, and G. V. Baron, "Morphology of *Trichoderma reesei* QM 9414 in submerged cultures," *Biotechnology and Bioengineering*, vol. 47, no. 5, pp. 609–615, 1995.

[92] D. Cross and C. M. Kenerley, "Modelling the growth of *Trichoderma virens* with limited sampling of digital images," *Journal of Applied Microbiology*, vol. 97, no. 3, pp. 486–494, 2004.

[93] A. P. J. Trinci, "Regulation of hyphal branching and hyphal orientation," in *The Ecology and Physiology of Fungal Mycelium*, D. H. Jennings and A. D. M. Rayner, Eds., pp. 23–52, Cambridge University Press, Cambridge, UK, 1984.

[94] J. G. H. Wessels, "A steady-state model for apical wall growth in fungi," *Acta Botanica Neerlandica*, vol. 37, pp. 3–16, 1988.

[95] M. Huston, D. DeAngelis, and W. Post, "New computer models unify ecological theory," *BioScience*, vol. 38, pp. 682–691, 1988.

[96] G. R. Knudsen, J. P. Stack, S. O. Schuhmann, K. Orr, and C. LaPaglia, "Individual-based approach to modeling hyphal growth of a biocontrol fungus in soil," *Phytopathology*, vol. 96, no. 10, pp. 1108–1115, 2006.

[97] A. A. de Souza, S. G. F. Martins, and M. S. Zacarias, "Computer simulation applied to the biological control of the insect *Aphis gossypii* for the parasitoid *Lysiphlebus testaceipes*," *Ecological Modelling*, vol. 220, no. 6, pp. 756–763, 2009.

[98] G. R. Knudsen and G. W. Hudler, "Use of a computer simulation model to evaluate a plant disease biocontrol agent," *Ecological Modelling*, vol. 35, no. 1-2, pp. 45–62, 1987.

[99] H. W. Spurr Jr. and G. R. Knudsen, "Biological control of leaf diseases with bacteria," in *Biological Control on the Phylloplane*, C. E. Windels and S. E. Lindow, Eds., pp. 45–62, American Phytopathological Society, St. Paul, Minn, USA, 1985.

[100] G. R. Knudsen and H. W. Spurr Jr., "Field persistence and efficacy of five bacterial preparations for control of peanut leaf spot," *Plant Disease*, vol. 71, no. 5, pp. 442–445, 1987.

[101] B. K. Duffy, B. H. Ownley, and D. M. Weller, "Soil chemical and physical properties associated with suppression of take-all of wheat by *Trichoderma koningii*," *Phytopathology*, vol. 87, no. 11, pp. 1118–1124, 1997.

[102] E. A. Paul and F. E. Clark, "Components of the soil biota," in *Soil Microbiology and Biochemistry*, pp. 69–107, Academic Press, New York, NY, USA, 2nd edition, 1996.

[103] G. W. Yeates, D. A. Wardle, and R. N. Watson, "Responses of soil nematode populations, community structure, diversity and temporal variability to agricultural intensification over a seven-year period," *Soil Biology and Biochemistry*, vol. 31, no. 12, pp. 1721–1733, 1999.

[104] W. L. Nicholas, *The Biology of Free-Living Nematodes*, Clarendon, Oxford, UK, 2nd edition, 1984.

[105] G. L. Barnes, C. C. Russell, W. D. Foster, and R. W. McNew, "Aphelenchus avenae, a potential biological control agent for root rot fungi," *Plant Disease*, vol. 65, pp. 423–424, 1981.

[106] D. W. Freckman and E. P. Caswell, "The ecology of nematodes in agroecosystems," *Annual Review of Phytopathology*, vol. 23, pp. 275–296, 1985.

[107] T. W. Hofman and J. J. S'jacob, "Distribution and dynamics of mycophagous and microbivorous nematodes in potato fields and their relationship to some food sources," *Annals of Applied Biology*, vol. 115, no. 2, pp. 291–298, 1989.

[108] B. A. Hawkins, M. B. Thomas, and M. E. Hochberg, "Refuge theory and biological control," *Science*, vol. 262, no. 5138, pp. 1429–1432, 1993.

[109] E. E. Porter and B. A. Hawkins, "Coexistence of specialist parasitoids with host refuges in the laboratory and the dynamics of spatial heterogeneity in attack rate," *Oikos*, vol. 100, no. 2, pp. 232–240, 2003.

[110] S. A. Levin, "Ecosystems and the biosphere as complex adaptive systems," *Ecosystems*, vol. 1, no. 5, pp. 431–436, 1998.

[111] M. Anand, A. Gonzalez, F. Guichard, J. Kolasa, and L. Parrott, "Ecological systems as complex systems: challenges for an emerging science," *Diversity*, vol. 2, no. 3, pp. 395–410, 2010.

[112] J. Jabiol, B. G. McKie, A. Bruder, C. Bernadet, M. O. Gessner, and E. Chauvet, "Trophic complexity enhances ecosystem functioning in an aquatic detritus-based model system," *Journal of Animal Ecology*, vol. 82, no. 5, pp. 1042–1051, 2013.

[113] R. E. Ulanowicz, R. D. Holt, and M. Barfield, "Limits on ecosystem trophic complexity: insights from ecological network analysis," *Ecology Letters*, vol. 17, no. 2, pp. 127–136, 2014.

# 16

# Mineralization Rates of Soil Forms of Nitrogen, Phosphorus, and Potassium as Affected by Organomineral Fertilizer in Sandy Loam

**Ayeni Leye Samuel and Adeleye Omotayo Ebenezer**

*Department of Agricultural Science, Adeyemi College of Education, PMB 520, Ondo State, Ondo City, Nigeria*

Correspondence should be addressed to Ayeni Leye Samuel; leye_sam@yahoo.com

Academic Editor: Othmane Merah

Farmers tend to use organomineral fertilizers as a result of inadequacies embedded in the sole use of organic and mineral fertilizers. A laboratory incubation study to determine the rate of the forms of N, P, and K released by organomineral fertilizer was conducted at Adeyemi College of Education, Ondo, southwest Nigeria, in 2013. Organomineral fertilizer (OMF) at the rates of 0, 0.125, 0.25, 0.5, and 1.0 g/100 g soil to represent 0, 2.5, 5, 10, and 20 t ha$^{-1}$ OMF, respectively, was incubated for ninety days. The treatments were replicated three times and arranged in a completely randomized design. The determined forms of N were total N, $NH_4$–N, and $NO_3$–N; the forms of P were total P, solution P, and available P while the forms of K were total K, solution K, and exchangeable K. Organomineral fertilizer significantly increased N, $NH_4$–N, $NO_3$–N, total P, solution P, exchangeable P, solution K, and exchangeable K at all rates with different values. The rate of ammonification of N was higher than the rate of nitrification of $NH_4$ + N to $NO_3$ + N especially at 10 and 20 t ha$^{-1}$ OMF. Application of 5 and 10 t ha$^{-1}$ OMF could be used to increase soil forms of N, P, and K.

## 1. Introduction

The aim of a farmer is to have bumper harvest and to get it depending on soil productivity. This leads to the philosophy that emphasises the need to apply fertilizers when economic yield is likely to occur. The essential nutrient elements are N, P, K, Ca, Mg, and S. Among these nutrient elements, N ranks first in plant requirement and phosphorus (P) ranks second followed by potassium (K). N, P, and K are referred to essential nutrients because nearly all plants use them for growth and development. Their deficiencies cannot be corrected by another element and are needed by the plants before they can complete their vegetative and reproductive cycles. Many plants require them throughout their life cycles. The nutrients are expected to be present in the soil in proper form and in proper balance. Their misuse can pose threats to water quality and can also cause nutrient antagonism which may result in low crop yield.

Farmers are in the habit of improving soil fertility through the addition of mineral fertilizers and organic manures, but the sole use of either fertilizer has not solved the problem of nutrient deficiencies. This necessitates research into the use of combined organic and inorganic fertilizers [1]. Organomineral fertilizers are produced by fortifying agro wastes with mineral fertilizer. The interest in organomineral fertilizers arose from the demerits of organic and mineral fertilizers such as high cost, scarcity of inorganic fertilizer, and their deleterious effect on soil physical properties as well as bulkiness and low nutrient content in the case of organic manures [2]. Organomineral fertilizers combine the attributes of mineral and organic manures. They are known to contain considerable amount of N, P, and K which are the three major nutrient elements needed by crop to complete both vegetative and reproductive cycles [3]. Integrated application of agro wastes and household refuse have been found to increase the productive capacity of soils [4].

Forms of N present in soils include organic N, $NO_3$–N, $NH_4$–N, $NO_2$, $N_2O$, NO, and $NH_3$ gas. Apart from $NO_3$–N and $NH_4$–N, other forms of N such as $N_2O$, NO, and $NH_3$ can escape into atmosphere through denitrification and volatilization. Other sources of N loss from the soil include erosion, crop removal, and leaching. In crop production, attention is focused on $NO_3$–N and $NH_4$–N. This is because plants absorb N in form of $NO_3$–N or $NH_4$–N. The N in form of $NO_3$–N is very soluble in water and highly mobile and can easily leach into the soil beyond the reach of plant root. Ammonium nitrogen is less subject to losses from soil by leaching and denitrification. Absorption of $NH_4^+$ by root reduces $Ca^{2+}$, $Mg^{2+}$, and $K^+$ uptake while increases absorption of $H_2PO_3^-$, $SO_4^{2-}$, and $Cl^-$ [5].

Forms of soil phosphorus are total P, organic soil P, and inorganic P. These forms of phosphorus can be found in solid, liquid phase, or in the interface between solid and liquid. Phosphorus is absorbed by plants in form of orthophosphate ions ($H_2PO_4^-$ or $HPO_4^-$). Plants absorb $H_2PO_4^-$ more than $HPO_4^{2-}$. Availability of high Ca or Al in the soil may hinder P uptake by plants. The problem associated with long-term applications of phosphatic fertilizers is the increase of P levels exceeding crop requirements [6].

Potassium is not commonly deficient in tropical soils as nitrogen and phosphorus. This is because it is adsorbed by the soil particles and this is the reason why it is not easily lost. Potassium is common in most soils, but its availability depends on the amount of exchangeable K present in the soil.

Ayeni [1] found that the integrated soil nutrition management is more feasible in maintaining nutrients status as well as increasing crop production than single application of mineral or organic fertilizers and it advocated the use of combined organic wastes and mineral fertilizers. Understanding the rate by which OMF released plant nutrients to the soil will be a guide to the fertilizer recommendation to avoid its over- or underutilization by farmers. Most arable crops complete their vegetative and reproductive cycles within three months of planting. Thus, predicting the amount of plant available N, P, and K produced by agrowaste amended soils is necessary for proper plant nutrition and protection of ground and surface water quality [7]. The objective of this study was to determine the effect of organomineral fertilizer on the forms of N, P, and K released into Alfisol in southwestern Nigeria within 90 days.

## 2. Materials and Methods

Laboratory incubation study was conducted in sandy loam soil to determine the rate of forms of N, P, and K released from organomineral fertilizer. Surface soil sample 0–20 cm was randomly collected and bulked from a farmer's field in Ondo in southwestern Nigeria. The soil samples were transferred to Adeyemi College of Education, Department of Agricultural Science, Laboratory for the incubation study.

The soil sample was air-dried and allowed to pass through 2 mm sieve. 100 grams of sieved soil sample was weighed into a well-labelled cup according to the treatments. The soil samples used for the experiment were preincubated for one week at 60% water holding capacity prior to the treatments application. The treatment applied was the OMF produced by the Ondo State Government under the Waste to Wealth Management Programme in Nigeria. The treatments used were 0, 0.125, 0.25, 0.5, and 1.0 g/100 g soil to represent 0, 2.5, 5, 10, and 20 t ha$^{-1}$ OMF, respectively. Equal volume of distilled water was added to each soil sample weekly for three (3) months. The pots were covered with perforated asbestos to give room for exchange of air. At the end of the 90 days, the incubated soil samples were air-dried and chemically analysed.

The pH of the soil samples was determined using glass electrode photometer in 1 : 2 soil-water ratio. Organic matter content of the sample was determined with the use of Walkley-Black wet oxidation method.

Total nitrogen (N) was determined by the normal microkjedahl method. Nitrate nitrogen and ammonium nitrogen were extracted with 1 M KCl solution and determined using steam distillation technique [8]. Ammonium nitrogen and nitrate nitrogen contents were determined with the use of spectrophotometer and flow injection analyzer method, respectively. Organic carbon was determined by Walkley and Black [9] method. Available phosphorus (P) was extracted by Bray-1 method and determined colourimetrically. The Murphy and Riley [10] procedure was also used to determine solution P (i.e., addition of a mixed reagent containing ammonium molybdate, antimony potassium tartrate, ascorbic acid, and sulphuric acid to a neutral). The absorbance of the blue solution was measured at a wavelength of 882 nm. Digestion method was used to extract total P, that is, sulphuric acid-potassium sulphate-copper sulphate digestion procedure, and determined by AAS.

Potassium content was fractionated into water soluble K, exchangeable K, and total K. Water soluble K was extracted in 1 : 2 soil-water suspension after shaking for 2 hours with mechanical shaker and allowed to stand for 16 hours according to Mclean [11]. Exchangeable K was extracted from soil with 1 N $NH_4OAc$ buffered at pH 7 while total K was obtained by digestion according to Pratt method [12]. The K extracts were determined by flame photometer.

Exchangeable bases were extracted with neutral ammonium acetate. Ten (10) grams of soil sample was put into conical flask. 100 mL of the extracting solution ($NH_4O$ : Ac) was added. It was shaken for an hour, filtered, and made up to 100 mL with $NH_4OAC$. Exchangeable bases Ca, Mg, and Na were determined from the filtrate by atomic absorption spectrophotometer (AAS).

*2.1. Statistical Analysis.* Statistical analysis was carried out using ANOVA and LSD was used to separate the mean and presented in form of graph.

## 3. Result and Discussion

According to the manufacturer label, the organomineral fertilizer used for the experiment contained 3.5, 2.5, and 4% for N, P, and K, respectively. The total N of the soil used for the experiment was lower than 0.15% showing that the soil

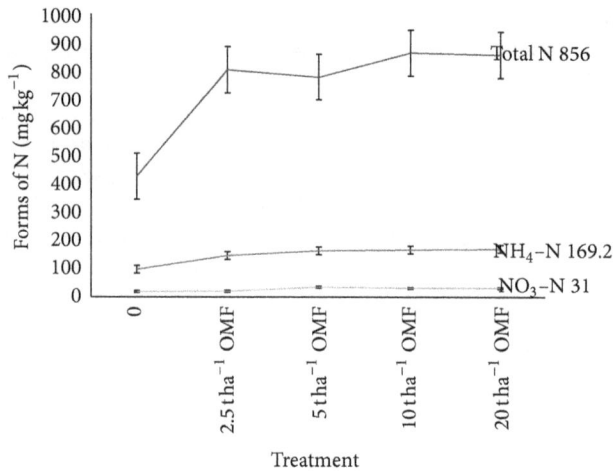

FIGURE 1: Effect of organomineral fertilizer on forms of N. Total N Lsd (0.05) = 11.20, $NO_3$-N = 4.02, and $NH_4$-N = 5.11.

TABLE 1: Percentage increase in forms of nitrogen.

| Treatment | Total N | $NO_3$-N | $NH_4$-N |
|---|---|---|---|
| 0 | — | — | — |
| 2.5 t ha$^{-1}$ OMF | 50.4 | 5 | 49 |
| 5 t ha$^{-1}$ OMF | 92 | 84 | 66 |
| 10 t ha$^{-1}$ OMF | 113 | 63 | 70 |
| 20 t ha$^{-1}$ OMF | 111 | 63 | 74 |

TABLE 2: Percentage increase in forms of phosphorus.

| Treatment | Total P | Solution P | Available P |
|---|---|---|---|
| 0 | — | — | — |
| 2.5 t ha$^{-1}$ OMF | 10 | 83 | 87 |
| 5 t ha$^{-1}$ OMF | 9 | 137 | 137 |
| 10 t ha$^{-1}$ OMF | 11 | 191 | 189 |
| 20 t ha$^{-1}$ OMF | 17 | 239 | 242 |

TABLE 3: Percentage increase in forms of potassium.

| Treatment | Exch. K | Solution K | Total K |
|---|---|---|---|
| 0 | — | — | — |
| 2.5 t ha$^{-1}$ OMF | −0.06 | 118 | 48 |
| 5 t ha$^{-1}$ OMF | −0.11 | 71 | 120 |
| 10 t ha$^{-1}$ OMF | 6.35 | 82 | 131 |
| 20 t ha$^{-1}$ OMF | 10 | 205 | 186 |

was deficient in N and thereby needed N fertilization. The soil contained 20 ppm $NNO_3$-N (medium), 7.8 ppm available P (low), 2.14 C mol kg$^{-1}$ (medium), and K 1.64 C mol kg$^{-1}$ according to Sobulo and Osiname [13], Bationo et al. [14], Sanchez et al. [15], Akinrinde and Obigbesan [16], and Agboola and Corey [17]. The soil textural class was sandy loam with 73.6, 8.6, and 17.8% sand, silt, and clay, respectively.

In Figure 1, total N, $NO_3$-N, and $NH_4$-N increased as the level of OMF increased indicating that OMF as low as 2.5 t ha$^{-1}$ is beneficial as a source of N to the soil. Addition of organomineral fertilizer was expected to add N, P, and K to the soil since analysis showed that it contains N, P, and K. Organomineral fertilizer applied at 10 and 20 t ha$^{-1}$ recorded the highest increase in total N, $NH_4$-N, and $NO_3$-N (Figure 1). Though total N does not indicate plant available N and is not the sum of $NH_4$-N and $NO_3$-N, it is a tool for determining the amount of organic N that can mineralize to inorganic form of N in future. The amount of total N present in the soil samples treated with OMF at all rates was significantly high (Figure 1). It was expected that the rate of conversion of $NH_4$-N to $NO_3$-N should be rapid [18], but the reverse was the case in the soil samples treated with 2.5, 10, and 20 t ha$^{-1}$ OMF. The soil conditions might have favoured ammonifying bacteria than the nitrifying bacteria. The higher increase in $NH_4$-N than in $NO_3$-N recorded in this experiment signified that the rate of ammonification was higher than the rate of nitrification. It was noted that application of 5, 10, and 20 t ha$^{-1}$ OMF raised the level of $NO_3$-N to 30 mg that is recommended as the critical level for optimum crop production in southwest Nigeria. It was also noted that the $NH_4^+$ released by OMF within the period of the incubation was too high compared with $NO_3$-N as clearly shown by the percentage increase in $NO_3$-N and $NH_4$-N (Table 1). The percentage rate of conversion of $NH_4$-N to $NO_3$-N was the highest in 5 t ha$^{-1}$ OMF showing that mineralization rate was the highest in 5 t ha$^{-1}$ OMF. The N present in 5 t ha$^{-1}$ OMF would be easily released for the immediate crop. The high value of total N compared with $NH_4$-N and $NO_3$-N recorded

in all the treatments showed that not all the N present in soil samples treated with OMF was mineralized during the period of incubation. The remaining N would likely be useful for crop uptake in the nearest future.

There was no much change in soil pH when OMF was added to the soil samples. Soil organic matter decreases soil pH during nitrification as a result of the release of hydrogen ion [19]. Table 4 shows that addition of OMF at all rates significantly increased ($P < 0.05$) soil Ca compared with control which might have obliterated the acidic effect of the mineral fertilizer blended with organic manure. This also might have reduced nitrification. It could also be argued that the presence of high amount of $K^+$ might have reduced $NH^+$ fixation. Samuel et al. [5] stated that the presence of $K^+$ often restricts $NH_4^+$ since $K^+$ can also fill fixation sites.

Ammonium tolerance limits are narrowed with excessive levels producing toxic reactions. Caution must be exercised in applying large dose of OMF, that is, 20 t ha$^{-1}$, to avoid excessive ammonification which may retard crop growth, restricts uptake of $K^+$, and produce symptoms of K deficiency. Saturating soils with a ready supply of $NO^{3+}$ can also cause denitrification; that is, conversion of nitrate to various forms of N can be lost to the atmosphere.

Compared with control, all the treatments significantly ($P < 0.05$) increased soil total P (Figure 2(a)). Available P increased as the level of OMF increased. Water P increased as the level of OMF increased up to 10 t ha$^{-1}$ and decreased at 20 t ha$^{-1}$. Compared with total P, the amount of P mineralized to solution P and available P was low showing that the high

TABLE 4: Effect of organomineral fertilizer on soil chemical properties.

| Treatment | pH | OC | Ca | Mg | Na | EA | ECEC | %BS |
|---|---|---|---|---|---|---|---|---|
| 0 | 6.33[a] | 1.90[a] | 3.94[d] | 1.83[b] | 1.11[c] | 1.42[b] | 8.30[c] | 82.89[a] |
| 2.5 t ha$^{-1}$ OMF | 6.24[ab] | 1.77[a] | 4.00[c] | 0.99[c] | 1.52[b] | 1.82[a] | 8.33[c] | 78.15[a] |
| 5 t ha$^{-1}$ OMF | 6.15[b] | 1.73[a] | 6.01[b] | 2.55[a] | 1.48[b] | 1.64[ab] | 11.68[b] | 85.96[a] |
| 10 t ha$^{-1}$ OMF | 6.14[b] | 1.89[a] | 6.58[ab] | 2.68[a] | 1.83[a] | 1.75[a] | 12.84[ab] | 86.37[a] |
| 20 t ha$^{-1}$ OMF | 6.98[a] | 1.79[a] | 7.75[a] | 2.67[a] | 2.40[b] | 1.61[b] | 14.43[a] | 88.84[a] |

Treatments with the same letter are not significantly different at 5% level using DMRT.

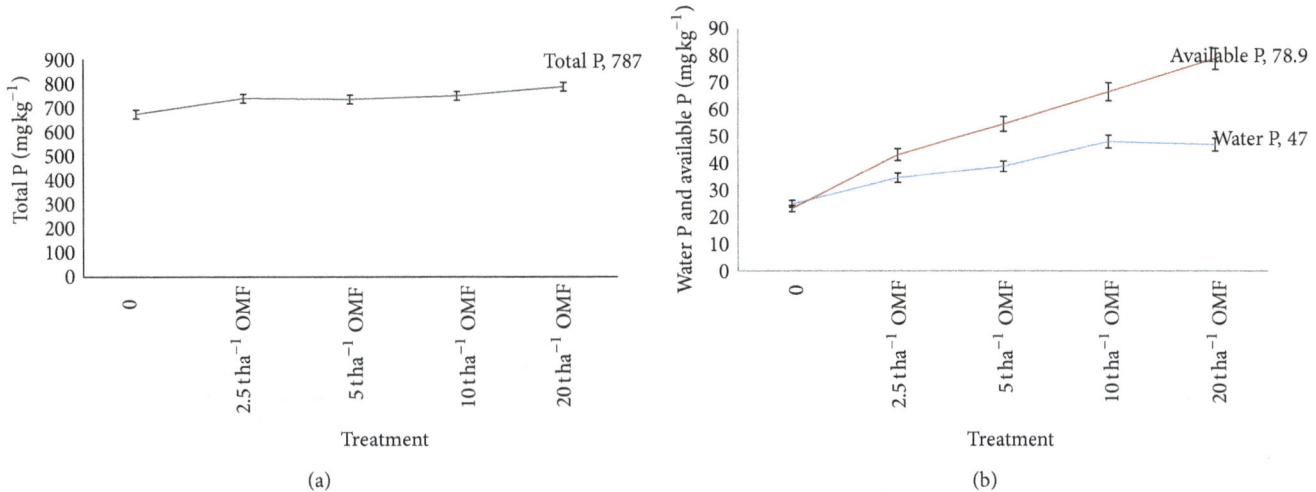

(a)

(b)

FIGURE 2: (a) Effect of organomineral fertilizer on total P. Lsd (0.05) = 11. (b) Effect of organomineral fertilizer on water P and available P. Lsd (0.05) water P = 6.03, avail. P = 8.07.

amount of P released to the soil by OMF might have been fixed (Figure 2).

Orthophosphates are readily absorbed by plants. Available P is the form of P mostly needed by crop. In this research, application of OMF as low as 2.5 t ha$^{-1}$ released more than 30 mg kg$^{-1}$ that is recommended as the critical level for optimum arable crop production in southwestern Nigeria within 90 days of incubation. The pH of the soil after the addition of OMF ranged within the level at which P and K could be easily mineralized. The favourable soil pH, moisture, and fine texture of the soil were expected to enhance the release of the orthophosphates and K to the soil. These conditions are favourable for fungi and bacteria that aid in the solubilization of P from inorganic to organic form. This experiment showed that organomineral fertilizer as low as 2.5 t ha$^{-1}$ can increase soil available P to the level that can increase crop production (Table 2). Addition of large dose of OMF to increase soil P might lead to excess P which could lead to nutrient imbalance especially when the native P is at medium or high level. Excess P application is known to move from soil into surface waters and lead to excessive growth of vegetation which can damage aquatic ecosystem [18].

Total K and available K increased as the level of OMF increased, but water K did not follow a definite pattern, though 20 t ha$^{-1}$ OMF recorded the highest value (Figure 3).

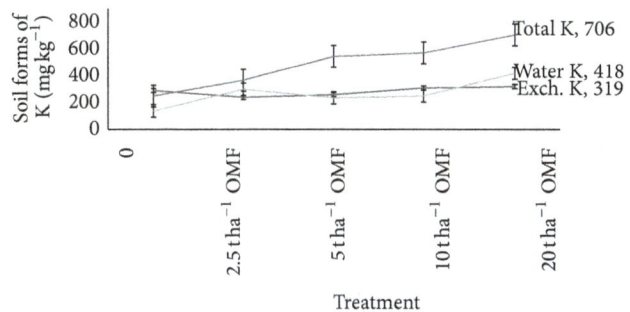

FIGURE 3: Effect of organomineral fertilizer on soil forms of K. Total K Lsd (0.05) = 10.05, water K = 10.66, and exch. K = 12.05.

Compared with control, application of OMF at all rates significantly increased all forms of K under study. All the treatments increased the available K above critical level.

Compared with control, 2.5 and 20 t ha$^{-1}$ OMF significantly increased ($P < 0.050$) soil pH while 5 and 10 t ha$^{-1}$ reduced the soil pH but still favoured N, P, and K mineralization (Table 3). This shows inconsistency of organomineral fertilizers on soil pH which might indirectly affect nutrients mineralization. OMF applied at the highest level recorded the highest pH among the treatments. This is in line with the assertion of Ojeniyi, [20] that increase in organic manure is dependent on the amount of organic manure present in the soil. The pH favoured the mineralization of N, P, and K as recorded in Figures 1, 2, and 3. Researches have shown that some organic manures contain reasonable amount of cations especially Ca that can reduce soil acidity [21]. There was no significant increase in soil OC and base saturation. Ano and Agwu [22] observed similar decrease in soil organic carbon when an acid soil was incubated with organic materials and attributed the observation to the increase in microbial respiration stimulated by the added manure which might have caused temporary immobilization. Relative to control, all the treatments significantly increased soil Ca, Mg, and Na and exchangeable cation capacity (ECEC) (Table 4). Application of 2.5 and $10\,t\,ha^{-1}$ OMF significantly increased exchange acidity compared with control.

## 4. Conclusion

Laboratory incubation study was conducted in sandy loam soil to determine the rate of forms of N, P, and K released from organomineral fertilizer. Organomineral fertilizer especially at 2.5, 5, 10, and $20\,t\,ha^{-1}$ increased soil forms of N, P, and K to sufficient levels that can boost food production. Further research is needed on the field to substantiate the finding as environmental conditions might have interfered with the incubation study.

## Conflict of Interests

The authors declare that there is no conflict of interests regarding the publication of this paper.

## Acknowledgments

The authors are grateful to Tertiary Education Trust Fund (TETFUND) and the Management of Adeyemi College of Education, Ondo, for providing the fund for conducting this research.

## References

[1] L. S. Ayeni, "Effect of combined cocoa pod ash and NPK fertilizer on soil properties, nutrient uptake and yield of maize (*Zea mays*)," *Journal of American Science*, vol. 6, no. 3, pp. 79–84, 2010.

[2] E. A. Makinde, O. K. Oluwa, A. O. Oke, and P. O. Duyile, "Effects of organic, organomineral and NPK fertilizer treatments on fresh and dry matter yield of *Amaranthus cruentus* L on soil types in Lagos, Nigeria," *New York Science Journal*, vol. 3, no. 4, pp. 12–17, 2010.

[3] J. O. Olaniyi and W. B. Akanbi, "Effect of organo mineral and inorganic fertilizers on the yield quality of fluted Pumpkin (*Telfaria occidentalis* hook.F.)," *African Crop Science Conference Proceedings*, vol. 8, pp. 347–350, 2007.

[4] P. K. Ghosh, A. K. Tripathi, K. K. Bandyopadhyay, and M. C. Manna, "Assessment of nutrient competition and nutrient requirement in soybean/sorghum intercropping system," *European Journal of Agronomy*, vol. 31, no. 1, pp. 43–50, 2009.

[5] L. T. Samuel, W. H. Tisdale, D. B. James, and L. H. John, *Soil Fertility and Fertilizer*, 2003.

[6] P. B. DeLaune, P. A. Moore Jr., and J. L. Lemunyon, "Effect of chemical and microbial amendment on phosphorus runoff from composted poultry litter," *Journal of Environmental Quality*, vol. 35, no. 4, pp. 1291–1296, 2006.

[7] J. Hirzel, P. Undurraga, and I. Walter, "Nitrogen mineralization and released nutrients in a volcanic soil amended with poultry litter," *Chilean Journal of Agricultural Research*, vol. 70, no. 1, pp. 113–121, 2010.

[8] J. M. Bremner, "Inorganic forms of nitrogen," in *Methods of Soil Analysis. Part 2*, C. A. Black et al., Ed., vol. 9 of *Agronomy Monograph*, pp. 1179–1237, ASA, Madison, Wis, USA, 1965.

[9] A. C. Walkley and T. A. Black, "Estimation of soil organic carbon by chromic acid titration method," *Soil Science*, vol. 47, pp. 29–38, 1935.

[10] J. Murphy and J. P. Riley, "A modified single solution method for the determination of phosphate in natural waters," *Analytica Chimica Acta*, vol. 27, pp. 31–36, 1962.

[11] A. J. Mclean, "Potassium supplying power of some Canadian soils," *Canadian Journal of Soil Science*, vol. 41, pp. 196–206, 1961.

[12] P. F. Pratt, "Potassium," in *Methods of Soil Analysis. Agronomy 9*, C. A. Black, Ed., pp. 1022–1030, American Society of Agronomy, 1965.

[13] R. A. Sobulo and A. Osiname, "Soil and fertilizer use in southwestern Nigeria," Research Bulletin 11, IAR&T, Ibadan, Nigeria, 1981.

[14] A. Bationo, J. Kihara, B. Vanlauwe, B. Waswa, and J. Kimetu, "Soil organic carbon dynamics, functions and management in West African agro-ecosystems," *Agricultural Systems*, vol. 94, no. 1, pp. 13–25, 2007.

[15] P. A. Pearson Sanchez, K. D. Shepherd, M. J. Soule, F. M. Place, A. U. Mukwunye, and R. J. Bursch, "Soil fertility replenishment in Africa: an investment in natural resource capital," in *Replenishing Soil Fertility in Africa*, R. J. Bursch, P. A. Sanchez, and F. Calhoon, Eds., pp. 1–46, Soil Science Society of America, Madison, Wis, USA, 1997.

[16] E. A. Akinrinde and G. O. Obigbesan, "Evaluation of the fertility status of selected soils for crop production in five ecological zones of Nigeria," in *Proceedings of the 26th Annual Conference of Soil Science Society of Nigeria*, pp. 279–288, Ibadan, Nigeria, 2000.

[17] A. A. Agboola and R. B. Corey, "The relationship between soil pH, organic matter, available P, exchangeable K, Ca, Mg and nine elements in the maize tissue," *Soil Science*, vol. 115, no. 5, pp. 367–375, 1973.

[18] D. Horneck, A. D. M. Sullivan, J. S. Owen, and J. M. Hart, *Soil Interpretation Guide*, Oregon State University Extension Service, 2011.

[19] R. J. Haynes and M. S. Mokolobate, "Amelioration of Al toxicity and P deficiency in acid soils by additions of organic residues: a critical review of the phenomenon and the mechanisms involved," *Nutrient Cycling in Agroecosystems*, vol. 59, no. 1, pp. 47–63, 2001.

[20] S. O. Ojeniyi, "Effect of goat manure on soil nutrient and yield of okra in a rain forest area on Nigeria," *Applied Tropical Agriculture*, vol. 5, pp. 20–23, 2000.

[21] O. O. Babalola, "Beneficial bacteria of agricultural importance," *Biotechnology Letters*, vol. 32, no. 11, pp. 1559–1570, 2010.

[22] A. O. Ano and J. A. Agwu, "Effect of animal manure on selected soil chemical properties," *Nigerian Journal of Soil Science*, vol. 13, pp. 14–19, 2005.

# 17

# Efficacy of Cow Urine as Plant Growth Enhancer and Antifungal Agent

**Savita Jandaik,[1] Preeti Thakur,[2] and Vikas Kumar[2]**

[1]Shoolini University of Biotechnology and Management Sciences, Solan, Himachal Pradesh 173212, India
[2]Shoolini Institute of Life Sciences and Business Management, Solan, Himachal Pradesh 173212, India

Correspondence should be addressed to Savita Jandaik; drsavitajandaik@gmail.com

Academic Editor: Tibor Janda

The present study was conducted to determine antifungal activity of three different concentrations (5, 10, and 15%) of cow urine against three fungal pathogens (*Fusarium oxysporum*, *Rhizoctonia solani*, and *Sclerotium rolfsii*) isolated from infected plants of Methi and Bhindi that showed symptoms of damping off and wilting disease by poison food technique. The extent of growth of test fungi in plates poisoned with cow urine was lesser when compared with the control plates. Among these concentrations cow urine at 15% concentration was most effective. When the three fungal organisms were compared, maximum growth suppression was observed in *Fusarium oxysporum* (78.57%) at 15% concentration of cow urine followed by *Rhizoctonia solani* (78.37%) and *Sclerotium rolfsii* (73.84%). Finally we concluded that the cow urine has antifungal activities and the inhibitory activity can be used in the control of fungi. The nutritional effect of cow urine on plant growth was also tested with *Trigonella foenum-graecum* (Methi) and *Abelmoschus esculentus* (Bhindi) plants and the chlorophyll and protein content was also estimated.

## 1. Introduction

Vegetable plants suffer from diseases caused by various kinds of pathogens such as bacteria, fungi, viruses, nematodes, and mycoplasma. Among these, fungi are considered as most aggressive pathogens causing qualitative and quantitative damage. Fungal pathogens, namely, *Fusarium oxysporum*, *Rhizoctonia solani*, and *Sclerotium rolfsii*, are associated with damping off and wilting of Methi and Bhindi (Okra). The plant diseases have significant role in agriculture in terms of reduction of yield and economy. One of the most widely used strategies to control plant diseases is the use of chemical agents. However, overuse and abuse of these chemical agents resulted in certain hazardous effects. These chemicals suffer from drawbacks such as high cost, toxicity to nontarget organisms, residual problem, and development of resistance in pathogens. This situation triggered interest in searching alternates for disease control. Natural products, in particular from plants, can be the potential candidates which can be used against phytopathogenic fungi. The use of these agents is risk-free when compared to synthetic chemicals.

In ancient Ayurveda cow urine has been greatly mentioned for its pharmacological importance. Okra contains nutrients that may confer a number of health advantages, including a decreased risk of several serious medical problems. Methi is one of the oldest medicinal herbs; ongoing research in India and abroad is currently uncovering new possibilities for its potential role in the treatment of diabetes and high cholesterol levels associated with coronary heart disease, both of which plague many industrial societies [1]. Cow urine is one of the ingredients of "Panchagavya" (urine, dung, milk, curd, and ghee) which is capable of treating many diseases as it has several medicinal properties [2] and it is the best remedy to cure fungal and bacterial diseases. It has an excellent germicidal power, antibiotics and antimicrobial activity. Therefore, cow urine can kill varieties of germs and it also boosts immunity [3]. Cow urine contains many beneficial elements, that is, chemical properties, potentialities, and constituents which help in removing all the ill effects and imbalances of body caused by infectious agents. Cow urine contains 95% water, 2.5% urea, and the remaining 2.5% a mixture of salts, hormones, enzymes, and minerals [4]. It has been considered that cow urine is very useful in agricultural operations as a biofertilizer and biopesticide [5] as it can kill number of pesticide and herbicide resistant bacteria, viruses, and fungi. Cow urine in combination with plant extracts is used to prepare disinfectant which is biodegradable and ecofriendly with good antibacterial action [6]. Majority of people in India use cow urine to get rid of various diseases due to its therapeutic values. Cow urine has several biological activities such as antioxidant, antidiabetic, antitumor, antiprotozoal, and molluscicidal [7–9].

## 2. Materials and Methods

*2.1. Collection of Cow Urine.* Fresh cow urine was collected in a sterile container from a local variety of cow. The urine was filtered through Whatman No. 1 filter paper to get rid of debris and precipitated material and was stored in airtight container at 4°C before use.

*2.1.1. Isolation of Fungal Pathogens.* *Fusarium oxysporum*, *Rhizoctonia solani*, and *Sclerotium rolfsii* were isolated from infected plants of Methi and Bhindi (Okra) that showed symptoms of damping off and wilting.

*2.2. Antifungal Activity.* The three concentrations (5%, 10%, and 15% v/v) of cow urine were prepared. 10 mL of different concentrations of cow urine was amended in 10 mL of potato dextrose agar medium and mixed thoroughly by stirring. Control was maintained in which distilled water was used instead of cow urine. The medium was autoclaved and poured into sterilized Petri plates and left. The fungal discs of 5 mm diameter were taken from actively growing cultures by using cork borer and the discs were transferred aseptically on PDA plates poisoned with cow urine. Plates were incubated at 28 ± 2°C temperature in incubator for 7 days. After 7 days plates were observed and colony diameters were measured with the help of ruler [10]. The percent of inhibition was calculated using the following formula given by [11]:

Percent inhibition of mycelial growth (%)

$$I = \frac{100 (C - T)}{C}, \tag{1}$$

where $I$ is inhibition percentage, $C$ is colony diameter in control plates, and $T$ is colony diameter in poisoned plates.

*2.3. Effect of Cow Urine on Plant Growth*

*2.3.1. Collection of Seeds.* The seeds of *Trigonella foenum-graecum* (Methi) and *Abelmoschus esculentus* (Bhindi) were purchased from the local market of Solan, Himachal Pradesh.

*2.3.2. Pot Culture Experiment.* The pot culture study was conducted to find out the effect of various concentrations of cow urine on growth of Methi and Bhindi plants. The seeds were soaked in water over night and then 5 seeds were sown in different pots filled with sterile garden soil. The garden soil was sterilized in an autoclave at 15 lbs pressure

for half an hour. The pH of the soil was adjusted to 7. Each pot was irrigated twice a day with different concentrations (1%, 2%, 3%, 4%, and 5% (v/v)) of cow urine. In control pots, the seeds were irrigated with tap water instead of cow urine. When the plants grew randomly 3 seedlings from each treatment were uprooted without disturbing the root system and different parameters such as plant height, shoot and root length, number of leaves and branches, and leaf length and breadth were measured after 25 days to observe the plant growth.

*2.3.3. Estimation of Protein.* 10 mg of coomassie brilliant blue G250 was mixed with 10 mL of 88% phosphoric acid and 45 mL of absolute alcohol. Then the mixture was diluted to 100 mL with distilled water. 1 gm of fresh germinated seedlings (test seedlings) was ground in 20 mL of distilled water. It was filtered and filtrate was made up to 20 mL. Then 0.1 mL of filtrated solution was added with 0.9 mL of water to which 2 mL of coomassie blue was added. The absorbance was read at 595 nm. Same procedure was repeated for the seedlings that were treated with water as control [12].

*2.3.4. Estimation of Carbohydrate.* 2 gm of anthrone was diluted in one liter of sulphuric acid and stored in dark bottle and labeled as anthrone reagent. 5 gm of plant sample was collected from control plant and test plants separately and ground in 2 mL of 80% acetone. The homogenized solution was filtered. 1 mL of filtered solution was added with 5 mL of anthrone reagent. The solution was heated in water bath for 5 minutes. The OD was taken for the above mentioned sample(s) at 600 nm. Standard of glucose was prepared by dissolving 100 mg of glucose in 100 mL water [13]. The concentration of carbohydrate was calculated using the formula:

Concentration of test sample

$$= \frac{\text{Absorbance of Test}}{\text{Absorbance of Std.}} \times \text{Concentration of Std.} \tag{2}$$

*2.3.5. Estimation of Chlorophyll.* Fresh leaves were collected from control plant and test plant separately and 1 gm of leaves was weighed. The leaves were cut into small pieces and homogenized in a mortar and pestle with excess of acetone and then filtered using Whatman No. 1 filter paper. The filtrate was collected and made up to 100 mL with acetone. 5 mL of extract was transferred into 50 mL volumetric flask and diluted to 50 mL with 80% acetone. Absorbance was read at 645 nm and 663 nm using spectrophotometer [14]. The quantity of chlorophyll a, chlorophyll b, and total chlorophyll was calculated using the following formula:

Chlorophyll a (mg/g)

$$= 12.7 (A_{663}) - 2.69 (A_{645}) \times \frac{V}{1000 \times w},$$

TABLE 1: Antifungal activity of cow urine by poison food technique.

| S. number | Fungal pathogen | Concentrations (%) | Colony diameter (mm) | Percent of inhibition (%) |
|-----------|-----------------|--------------------|-----------------------|----------------------------|
| 1 | *Fusarium oxysporum* | 0 (control) | 84 | 0.0 |
| | | 5 | 38 | 54.76 |
| | | 10 | 22 | 73.80 |
| | | 15 | 18 | 78.57 |
| 2 | *Rhizoctonia solani* | 0 (control) | 74 | 0.0 |
| | | 5 | 38 | 48.60 |
| | | 10 | 30 | 59.45 |
| | | 15 | 16 | 78.37 |
| 3 | *Sclerotium rolfsii* | 0 (control) | 65 | 0.0 |
| | | 5 | 29 | 55.38 |
| | | 10 | 24 | 63.07 |
| | | 15 | 17 | 73.84 |

Chlorophyll b (mg/g)

$$= 22.9 \left(A_{645}\right) - 4.68 \left(A_{663}\right) \times \frac{V}{1000 \times w},$$

Total chlorophyll (mg/g)

$$= 20.2 \left(A_{645}\right) + 8.02 \left(A_{663}\right) \times \frac{V}{1000 \times w}, \tag{3}$$

where $A$ is optical density, $V$ is final volume of 80% acetone (mL), and $w$ is dry weight of sample taken (g).

*2.4. Statistical Analysis.* The experiment was performed in triplicate. The results were represented as mean ± standard deviation (SD) to facilitate the comparison of the data.

## 3. Results

In the present studies three fungal pathogens, namely, *Fusarium oxysporium* var. trifoli, *Rhizoctonia solani* Kuhn (Corticium Vagum B. & C.), and *Sclerotium rolfsii* Sacc. were isolated from the diseased *Trigonella foenum-graecum* (Methi) and *Abelmoschus esculentus* (Bhindi) plants. Data presented in Table 1 depicts that all concentrations (5, 10, and 15%) of cow urine were effective against the growth of the fungus which exhibited the significant inhibition in the growth of fungal plant pathogens. With increase in concentration of cow urine there was corresponding increase in the inhibition of vegetative growth of the fungal pathogens. The diameter of the fungal colonies in poisoned plates was lesser when compared to control plates and it indicates the antifungal effect of cow urine. Maximum inhibition was shown against *Fusarium oxysporum* (78.57%) followed by *Rhizoctonia solani* (78.37%) and *Sclerotium rolfsii* (73.84%), whereas minimum inhibition was recorded with 5 percent concentration of cow urine in *Rhizoctonia solani* (48.60%) followed by *Fusarium oxysporum* (54.76%) and *Sclerotium rolfsii* (55.38%) (Figures 1(a)–1(c)).

Pot culture studies were carried out to find out the effects of cow urine spray on the phenotypic characters of *Trigonella foenum-graecum* (Methi) (Figure 2) and *Abelmoschus esculentus* (Bhindi) after 25 days (Figure 3). Parameters such as plant height, shoot length and root length, number of leaves, and leaf length and breadth were observed in the experimental and control plants. It is clear from the results (Tables 2 and 3) that plant height of Methi increased with increase in concentration of cow urine and duration of time. Maximum plant height of Methi was 14.30 ± 0.40 cm with maximum concentration, that is, 5% of cow urine. Plant height of Bhindi plants consistently increased to a maximum of 13.97 ± 0.50 cm in the plants treated with 5% concentration of cow urine; this was followed by the 12.03 ± 0.42 cm in 4% concentration of cow urine. The mean height of Methi plants was 9.00 ± 0.46 cm in control plants. Shoot length of Methi was 7.27 ± 0.25, 7.97 ± 0.25, 9.17 ± 0.31, 9.67 ± 0.25, and 10 ± 0.20 cm when sprayed with 1, 2, 3, 4, and 5% concentration of cow urine, respectively. Shoot length in control pot was 6.8 ± 0.30 cm. Maximum shoot length and root length of Bhindi plants were 7.83 ± 0.15 and 6.20 ± 0.36 cm with the plants sprayed with 5% concentration of cow urine followed by 4% concentration.

The root length of Methi was maximum 4.13 ± 0.35 cm with maximum concentration (5%) of cow urine. However, root length increased in all the experimental plants as compared to control. Increase in root length supports the fact that the application of cow urine influences growth by increasing the mitotic index.

The protein content found in seedlings sprayed with cow urine showed more protein irrespective of the concentration as compared to the control (Table 4). The maximum protein content was found in the seedlings of Methi treated with 5% cow urine, that is, 174.97 ± 0.50 mg/mL, whereas the control had only 36.50 ± 0.46 mg/mL of protein and the seedlings of Bhindi treated with 5% cow urine contain 164.40 ± 0.32 mg/mL of protein content while the control of Bhindi had only 38.80 ± 0.46 mg/mL protein content. Carbohydrate content in 1, 2, 3, 4, and 5% cow's urine irrigated plant sample of Methi was 144.09 ± 0.17 mg/mL, 162.03 ± 0.56 mg/mL,

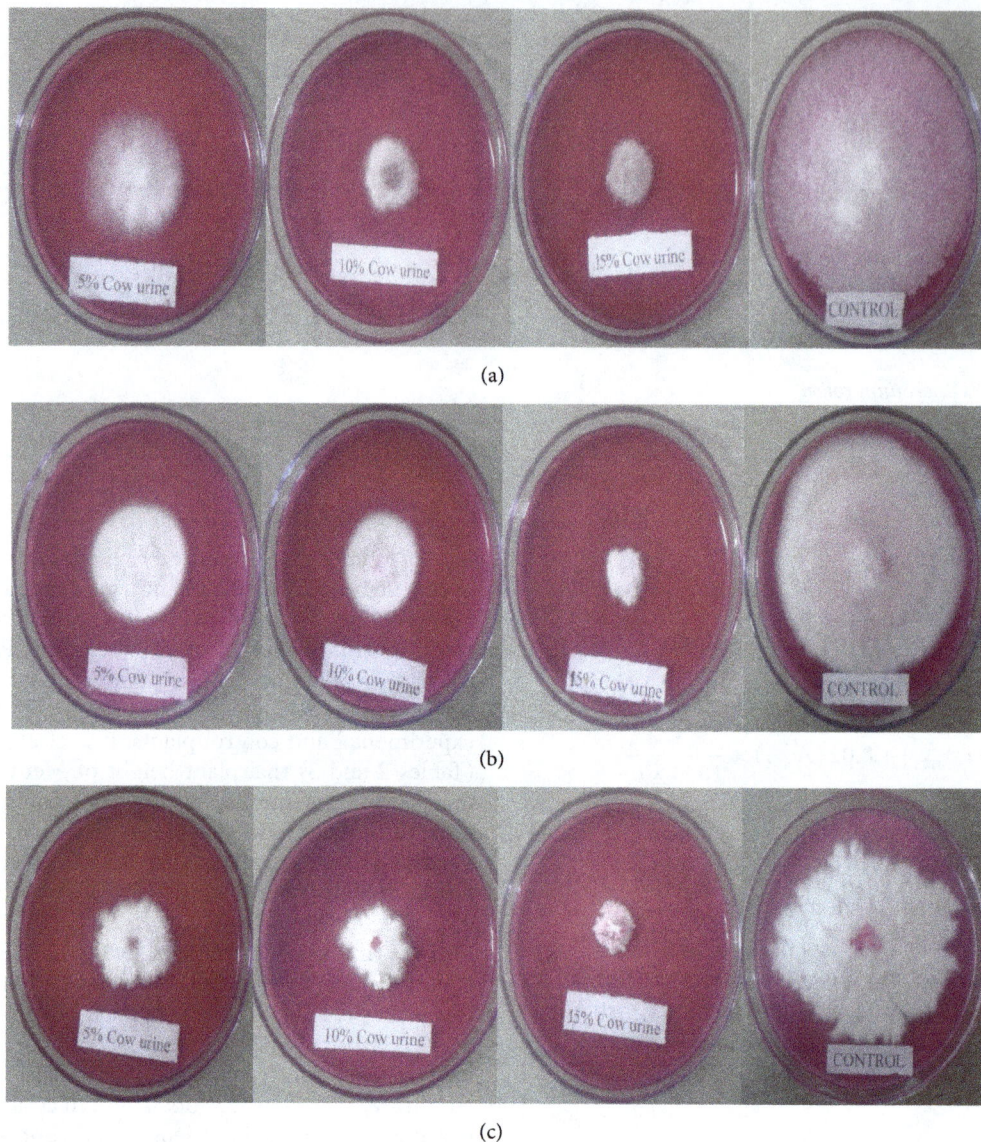

(a)

(b)

(c)

FIGURE 1: (a) Effect of different concentrations (5%, 10%, and 15%) of cow urine on *Fusarium oxysporum*. (b) Effect of different concentrations (5%, 10%, and 15%) of cow urine on *Rhizoctonia solani*. (c) Effect of different concentrations (5%, 10%, and 15%) of cow urine on *Sclerotium rolfsii*.

TABLE 2: Effect of cow urine on exomorphological characters of *Trigonella foenum-graecum* (Methi) by pot culture experiment after 25 days.

| Conc. of cow urine (%) | Plant height (cm) | Root length (cm) | Shoot length (cm) | Number of leaves | Number of branches | Leaf length (cm) | Leaf breadth (cm) |
|---|---|---|---|---|---|---|---|
| 1 | 10.23 ± 0.75 | 2.97 ± 0.50 | 7.27 ± 0.25 | 7.00 ± 1.00 | 4.67 ± 0.58 | 0.90 ± 0.26 | 0.57 ± 0.15 |
| 2 | 11.17 ± 0.65 | 3.20 ± 0.40 | 7.97 ± 0.25 | 9.33 ± 0.58 | 6.00 ± 1.00 | 1.13 ± 0.25 | 0.73 ± 0.12 |
| 3 | 12.70 ± 0.66 | 3.50 ± 0.36 | 9.17 ± 0.31 | 11.00 ± 1.00 | 6.67 ± 0.58 | 1.57 ± 0.25 | 1.23 ± 0.35 |
| 4 | 13.43 ± 0.75 | 3.77 ± 0.31 | 9.67 ± 0.25 | 12.33 ± 0.58 | 8.00 ± 0.00 | 1.83 ± 0.12 | 1.53 ± 0.38 |
| 5 | 14.30 ± 0.40 | 4.13 ± 0.35 | 10.00 ± 0.20 | 14.00 ± 1.00 | 8.33 ± 0.58 | 1.97 ± 0.15 | 1.73 ± 0.21 |
| Control | 9.00 ± 0.46 | 2.13 ± 0.25 | 6.80 ± 0.30 | 6.67 ± 0.58 | 4.00 ± 0.00 | 0.90 ± 0.20 | 0.40 ± 0.10 |
| CD(0.05) | 0.096 | 0.056 | 0.031 | 0.290 | 0.267 | 0.066 | 0.101 |

TABLE 3: Effect of cow urine on exomorphological characters of *Abelmoschus esculentus* (Bhindi) by pot culture experiment after 25 days.

| Conc. of cow urine (%) | Plant height (cm) | Root length (cm) | Shoot length (cm) | Number of leaves | Number of branches | Leaf length (cm) | Leaf breadth (cm) |
|---|---|---|---|---|---|---|---|
| 1 | 8.33 ± 0.51 | 1.87 ± 0.29 | 6.47 ± 0.25 | 1.67 ± 0.58 | 1.33 ± 0.58 | 1.63 ± 0.21 | 1.43 ± 0.31 |
| 2 | 10.23 ± 0.57 | 3.27 ± 0.35 | 6.97 ± 0.23 | 2.00 ± 0.00 | 2.00 ± 0.00 | 1.80 ± 0.26 | 1.47 ± 0.31 |
| 3 | 11.27 ± 0.50 | 3.93 ± 0.25 | 7.33 ± 0.25 | 2.00 ± 1.00 | 2.67 ± 0.58 | 2.17 ± 0.25 | 1.90 ± 0.20 |
| 4 | 12.03 ± 0.42 | 4.50 ± 0.17 | 7.50 ± 0.30 | 2.67 ± 0.58 | 3.00 ± 0.00 | 2.40 ± 0.36 | 2.17 ± 0.25 |
| 5 | 13.97 ± 0.50 | 6.20 ± 0.36 | 7.83 ± 0.15 | 3.00 ± 0.00 | 3.00 ± 0.00 | 2.77 ± 0.25 | 2.40 ± 0.26 |
| Control | 6.50 ± 0.46 | 1.67 ± 0.31 | 4.80 ± 0.20 | 1.33 ± 0.58 | 1.67 ± 0.58 | 1.23 ± 0.31 | 0.97 ± 0.25 |
| CD(0.05) | 0.068 | 0.069 | 0.046 | 0.367 | 0.242 | 0.035 | 0.038 |

FIGURE 2: Effect of different concentrations (1%, 2%, 3%, 4%, and 5%) of cow urine on *Trigonella foenum-graecum* (Methi) after 25 days.

FIGURE 3: Effect of different concentrations (1%, 2%, 3%, 4%, and 5%) of cow urine on *Abelmoschus esculentus* (Bhindi) after 25 days.

TABLE 4: Protein estimation of *Trigonella foenum-graecum* (Methi) and *Abelmoschus esculentus* (Bhindi) sprayed with different concentrations of cow urine.

| S. number | Conc. of cow urine (v/v%) | Protein (mg/mL) | |
|---|---|---|---|
| | | Methi | Bhindi |
| 1 | 1 | 42.33 ± 0.51 | 44.03 ± 0.11 |
| 2 | 2 | 50.23 ± 0.57 | 62.33 ± 0.17 |
| 3 | 3 | 71.27 ± 0.50 | 97.07 ± 0.59 |
| 4 | 4 | 103.03 ± 0.42 | 123.93 ± 0.12 |
| 5 | 5 | 174.97 ± 0.50 | 164.40 ± 0.32 |
| 6 | Control | 36.50 ± 0.46 | 38.80 ± 0.46 |
| | CD | 0.168 ± 0.11 | 0.271 ± 0.12 |

TABLE 5: Carbohydrate estimation of *Trigonella foenum-graecum* (Methi) and *Abelmoschus esculentus* (Bhindi) sprayed with different concentrations of cow urine.

| S. number | Conc. of cow urine (%) | Carbohydrate (mg/mL) | |
|---|---|---|---|
| | | Methi | Bhindi |
| 1 | 1 | 144.09 ± 0.17 | 145.07 ± 0.17 |
| 2 | 2 | 162.03 ± 0.56 | 162.03 ± 0.56 |
| 3 | 3 | 187.07 ± 0.23 | 177.07 ± 0.23 |
| 4 | 4 | 188.93 ± 0.78 | 180.09 ± 0.78 |
| 5 | 5 | 194.40 ± 0.27 | 184.60 ± 0.12 |
| 6 | Control | 138.80 ± 0.89 | 134.20 ± 0.90 |

187.07 ± 0.23 mg/mL, 188.93 ± 0.78 mg/mL, and 194.40 ± 0.27 mg/mL, respectively, and the carbohydrate content in 1, 2, 3, 4, and 5% cow's urine irrigated plant sample of Bhindi was 44.03 ± 0.11 mg/mL, 62.33 ± 0.17 mg/mL, 97.07 ± 0.59 mg/mL, 123.93 ± 0.12 mg/mL, and 164.40 ± 0.32 mg/mL, respectively (Table 5). However, the carbohydrate content in control plants of Methi and Bhindi was 138.80 ± 0.89 mg/mL and 134.20 ± 0.90 mg/mL, respectively.

The chlorophyll content of Methi and Bhindi (Okra) is shown in Table 6. The total chlorophyll content of *Trigonella foenum-graecum* (Methi) was recorded more in plant leaves irrigated with 5% cow urine. It had 0.972 ± 3 mg/g of chlorophyll a, 0.897 ± 96 mg/g of chlorophyll b, and 1.869 ± 21 mg/g of total chlorophyll. Control plants had 0.437 ± 71 mg/g of total chlorophyll. The total chlorophyll content in *Abelmoschus esculentus* (Bhindi) leaves was maximum with 5% cow urine irrigation (2.246 ± 28 mg/g) followed by 4% (1.599 ± 88 mg/g), 3% (0.803 ± 98 mg/g), 2% (0.690 ± 44 mg/g), and 1% (0.544 ± 2 mg/g). The control of Bhindi plants had 0.146 ± 66 mg/g of chlorophyll a, 0.025 ± 8 mg/g of chlorophyll b, and 0.171 ± 91 mg/g of total chlorophyll.

## 4. Discussion

This study revealed that the cow urine at different concentrations had considerable effect on the vegetative growth of *R. stolonifer*, *Sclerotium rolfsii*, and *F. oxysporum*. It is clear from the results that 15 percent concentration of cow

TABLE 6: Chlorophyll estimation of *Trigonella foenum-graecum* (Methi) and *Abelmoschus esculentus* (Bhindi) sprayed with different concentrations of cow urine.

| S. number | Conc. of cow urine | Chlorophyll (mg/g) | | | | | |
|---|---|---|---|---|---|---|---|
| | | Methi | | | Bhindi | | |
| | | Chlorophyll a | Chlorophyll b | Total chlorophyll | Chlorophyll a | Chlorophyll b | Total chlorophyll |
| 1 | 1 | 0.407 ± 17 | 0.327 ± 12 | 0.733 ± 09 | 0.368 ± 27 | 0.176 ± 01 | 0.544 ± 02 |
| 2 | 2 | 0.451 ± 09 | 0.365 ± 80 | 0.816 ± 38 | 0.449 ± 81 | 0.241 ± 07 | 0.690 ± 44 |
| 3 | 3 | 0.563 ± 87 | 0.466 ± 63 | 1.029 ± 92 | 0.502 ± 43 | 0.301 ± 30 | 0.803 ± 98 |
| 4 | 4 | 0.922 ± 23 | 0.738 ± 07 | 1.660 ± 02 | 0.996 ± 18 | 0.603 ± 08 | 1.599 ± 88 |
| 5 | 5 | 0.972 ± 03 | 0.897 ± 96 | 1.869 ± 21 | 1.404 ± 92 | 0.842 ± 02 | 2.246 ± 28 |
| 6 | Control | 0.248 ± 28 | 0.189 ± 10 | 0.437 ± 71 | 0.146 ± 66 | 0.025 ± 08 | 0.171 ± 91 |

urine showed maximum inhibition in growth of all the three fungal pathogens as compared to control. Inhibitory activity of cow urine against fungal pathogens have been reported by different workers [8, 15, 16]. Pot culture studies revealed that increase in cow urine concentrations increased the performance of all phenotypic characters of Methi and Bhindi. Present studies are in accordance with the findings of [17] Oliveira et al. 2009 who reported that the increase in cow urine concentrations increased the performance of all lettuce characteristics like fresh and dry leaf mass, fresh and dry stem mass, stem length, fresh root mass, fresh head mass, and commercial yield. The work of [18] Tharmaraj, 2011, reported that growth substances in panchagavya help to bring rapid changes in phenotypes of plants and also improve the growth and productivity. The protein and carbohydrate content found in seedlings sprayed with cow urine was more irrespective of the concentration as compared to the control. The biochemical contents (carbohydrates, protein, and amino acids) in *Abelmoschus esculentus* (L.) Moench and *Vigna mungo* increased with 3% concentration of panchagavya spray [8, 19]. It is evident from the results that the chlorophyll content of Methi and Bhindi plants increased with increase in concentration of cow urine. Similar findings with panchagavya spray were observed in *Arachis hypogaea* [20] and *Abelmoschus esculentus* (L.) Moench [8].

## 5. Conclusion

It was revealed from the study that cow urine caused inhibition in growth of all the three fungal pathogens used in the present studies. This demonstrated fungitoxic potential of cow urine against the three pathogenic fungi. The biochemical contents of both the plants increased when sprayed with cow urine. Therefore the use of cow urine provides better alternative to synthetic chemicals which are expensive and pose potential danger to the farmers, marketers, consumers, and environment. The cow urine can be used as biopesticide.

## Conflict of Interests

The authors declare that they have no conflict of interests.

## Acknowledgment

Authors are thankful to Department of Microbiology, Shoolini Institute of Life Sciences and Business Management, for offering facilities to carry out this work.

## References

[1] A. Sathasivam, M. Muthuselvam, and R. Rajendran, "Antimicrobial activities of cow urine distillate against some clinical pathogens," *Global Journal of Pharmacology*, vol. 4, no. 1, pp. 41–44, 2010.

[2] M. L. Pathak and A. Kumar, "Cow praising and importance of Panchyagavya as medicine," *Sachitra Ayurveda*, vol. 5, pp. 56–59, 2003.

[3] R. Chauhan, B. P. Singh, and L. K. Singhal, "Immunomodulation with kamdhenu ark in mice," *Journal of Immunology and Immunopathology*, vol. 71, pp. 89–92, 2001.

[4] H. Bhadauria, "Cow urine- a magical therapy. Vishwa Ayurveda Parishad," *International Journal of Cow Science*, vol. 1, pp. 32–36, 2002.

[5] K. Dharma, R. Rajesh, R. S. Chauhan, and T. Simmi, "Panchgavya (Cowpathy): an overview," *International Journal of Cow Science*, vol. 1, no. 1, pp. 1–15, 2005.

[6] S. A. Mandavgane, A. K. Rambhal, and N. K. Mude, "Development of cow urine based disinfectant," *Natural Product Radiance*, vol. 4, pp. 410–412, 2005.

[7] K. Krishnamurthi, D. Dutta, S. D. Sivanesan, and T. Chakrabarti, "Protective effect of distillate and redistillate of cow's urine in human polymorphonuclear leukocytes challenged with established genotoxic chemicals," *Biomedical and Environmental Sciences*, vol. 17, no. 3, pp. 247–256, 2004.

[8] M. Rajesh and K. Jayakumar, "Changes in morphological, biochemical and yield parameters of *Abelmoschus esculents* (L.) Moench due to panchagavya spray," *International Journal of Modern Plant & Animal Sciences*, vol. 1, no. 2, pp. 82–95, 2013.

[9] K. N. Rakesh, N. Dileep, N. A. S. Nawaz, S. Junaid, and P. T. R. Kekuda, "Antifungal activity of cow urine against fungal pathogens causing rhizome rot of ginger," *Environment and Ecology*, vol. 31, no. 3, pp. 1241-1244, 2013.

[10] O. D. Dhingra and A. C. Kapoor, "Nutritive value of mango seed kernel," *Journal of the Science of Food and Agriculture*, vol. 36, no. 8, pp. 752–756, 1985.

[11] J. M. Vincent, "Distortion of fungal hyphæ in the presence of certain inhibitors," *Nature*, vol. 159, no. 4051, p. 850, 1947.

[12] M. M. Bradford, "A rapid and sensitive method for the quantitation of microgram quantities of protein utilizing the principle of protein-dye binding," *Analytical Biochemistry*, vol. 72, no. 1-2, pp. 248–254, 1976.

[13] J. E. Hedge and B. T. Hofreiter, "Determination of total carbohydrates by anthrone," in *Carbohydrate Chemistry*, R. L. Whistler and J. N. Be Miller, Eds., p. 420, Academic Press, New York, NY, USA, 1962.

[14] D. I. Arnon, "Copper enzymes in isolated chloroplasts. Polyphenoloxidase in *Beta vulgaris*," *Plant Physiology*, vol. 24, no. 1, pp. 1–15, 1949.

[15] A. B. Basak, M. W. Lee, and T. S. Lee, "Inhibitive activity of cow urine and cow dung against *Sclerotinia sclerotiorum* of cucumber," *Mycobiology*, vol. 30, no. 3, pp. 175–179, 2002.

[16] K. N. Rakesh, N. Dileep, S. Junaid, K. T. R. Prashith, K. S. Vinayaka, and N. A. S. Noor, "Inhibitory effect of cow urine extracts of selected plants against pathogens causing rhizome rot of ginger," *Science, Technology and Arts Research Journal*, vol. 2, no. 2, pp. 92–96, 2013.

[17] N. L. Oliveira, M. Puiatti, R. H. S. Santos, P. R. Cecon, and P. H. R. Rodrigues, "Soil and leaf fertilization of lettuce crop with cow urine," *Horticultura Brasileirae*, vol. 27, pp. 102–536, 2009.

[18] K. Tharmaraj, P. Ganesh, R. Suresh Kumar, A. Anandan, and K. Kolanjinathan, "A critical review on Panchagavya—a boon plant growth," *International Journal of Pharmaceutical and Biological Archive*, vol. 2, no. 6, pp. 1611–1614, 2011.

[19] S. L. Chopra and J. S. Kunwar, "Effect of panchagavya on growth of *Oryza sativa*," in *Analytical Agricultural Chemistry*, pp. 180–244, Kalyan Publishers, New Delhi, India, 1976.

[20] A. Subramaniyan, "Effect of Panchagavya on *Escherichia coli* in procured milk," *Indian Veterinary Journal*, vol. 82, pp. 799–800, 2005.

# 18

# Rhizospheric Microflora Escalating Aroma Constituents and Yield Attributes in *Ocimum tenuiflorum* (L.) cv. CIM-Ayu

**Shilpi Khare Saikia and Rakesh Pandey**

*Microbial Technology and Nematology Department, CSIR-Central Institute of Medicinal and Aromatic Plants, P.O. CIMAP, Lucknow 226015, India*

Correspondence should be addressed to Rakesh Pandey; r.pandey@cimap.res.in

Academic Editor: Gábor Kocsy

The exploration of rhizospheric microbial flora for crop yield enhancement is well established. Rhizospheric microbes influence the plant physiology by imparting several beneficial effects, namely, Nitrogen fixation, increased nutrient uptake, and secondary metabolites production on their host plants. The present study investigates the response of *Bacillus megaterium* ATCC No. 13525, *Pseudomonas fluorescens* ATCC No. 14581, and *Trichoderma viride* MTCC No. 167 in alone and combined treatments for their effect on growth and yield parameters in a commercially important *Ocimum tenuiflorum* L. cv. CIM-Ayu. The plant is therapeutically important for its essential oil constituents, namely, eugenol, $\beta$-caryophyllene, and various monoterpenes. The combination treatments, T7 (*B. megaterium* + *P. fluorescens*) and T8 (*B. megaterium* + *P. fluorescens* + *T. viride*), showed maximum enhancement (27.27%) of percentage essential oil as compared to untreated control. Nutrient uptake especially $N_2$ content was significantly increased (43%) with the treatment T8 (*B. megaterium* + *P. fluorescens* + *T. viride*). Amongst major essential oil constituents, eugenol content was maximally increased by 58.5% as compared to 42.9% (control) indicating a cumulative role of microbial inoculants for crop yield boost-up.

## 1. Introduction

The members of genus *Ocimum* (Gk. ozo = smell) from family Lamiaceae are medicinally important plants owing to their therapeutic potentials as antiseptics, antioxidants, antistressors, antipyretics, antimicrobials, and insecticidal [1]. *Ocimum tenuiflorum* is an imperative species from the genus, widely cultivated for its high essential oil yields (0.5 to 0.7%), rich in eugenol, methyl chavicol, and linalool content [2]. The commercial importance of essential oil depends on its constituents and hence methods are required to enhance these valuable phytomolecules.

Rhizospheric microorganisms are recognized as an economic and sustainable input for increasing the productivity of several agricultural, horticultural, forestry, and medicinal crops [3–5]. Numerous microbes such as *Pseudomonas, Azospirillum, Azotobacter, Klebsiella, Enterobacter, Alcaligenes, Arthrobacter, Burkholderia, Bacillus,* and *Serratia* have been reported to enhance the plant growth [6, 7]. These microorganisms coexist in the rhizosphere, which is a thin

soil layer immediately surrounding plant roots. Rhizospheric microbial wealth is widely acclaimed in agricultural practices for enhancement of crop yield attributes as they share their environment with the host plants and thus exhibit better adaptation [8]. Rhizospheric microbes play significant role/s in improving the growth and yield of host plant [9] by imparting several beneficial effects, namely, $N_2$ fixation, increased nutrient uptake, siderophores, and secondary metabolite/s production [8, 10]. This interest is linked to environmental concerns for reduced use of chemicals as well as an appreciation for utilization of biological and organics in agriculture.

Therefore, a proper understanding of microbial species associated with the host plants is essential in improving the quality and yield in a desired crop. In this regard, the present experiment was designed to study the influence of selected rhizospheric microbial inoculants (previously proven as effective biocontrol agents), namely, *Bacillus megaterium* ATCC No. 13525, *Pseudomonas fluorescens* ATCC No. 14581, and *Trichoderma viride* MTCC No. 167 under greenhouse conditions on the agronomical attributes of *O. tenuiflorum* cv. CIM-Ayu. The study also validates the role of microbial inoculants for plant growth promotion and alterations in aromatic oil symphony due to them.

## 2. Materials and Methods

*2.1. Plant Materials and Growth Conditions.* Seeds of high yielding cultivar of *O. tenuiflorum* "CIM-Ayu," obtained from the National Gene Bank for Medicinal and Aromatic Plants at the Central Institute of Medicinal and Aromatic Plants (CSIR-CIMAP), Lucknow, India, were surface-sterilized by soaking in 10% (v/v) sodium hypochlorite solution for 5 minutes, washed with distilled water, and soaked for 4 h. After sterilization and soaking, healthy looking uniform sized seeds were sown in content plug plates filled with sterilized soil in a greenhouse under natural light conditions, a daytime temperature of about 28°C and relative humidity of 65–70%. Twenty-one days after sowing, four healthy leafed stage seedlings were transplanted into 7.0 kg soil capacity clay pots containing a mixture of autoclaved soil (76% sand, 8% silt, and 16% clay, pH 7.7) and composted farm manure in 5:1 ratio. Soil mineralizable nitrogen, phosphorus, and potassium content were estimated [11–13]. Plants were irrigated manually at alternate days to ensure adequate soil moisture throughout the experimentation.

*2.2. Microbial Inoculants (Culture and Maintenance).* The rhizospheric microbes *Bacillus megaterium* (ATCC-13525), *Pseudomonas fluorescens* (ATCC No. 14581), and *T. viride* (MTCC-167) are continuously maintained in the Microbial Technology and Nematology Department, CSIR-CIMAP, Lucknow. These microbes have a proven role in biocontrol experiments [14–16] and thus were selected for observing plant growth promotion effects. The fungus *T. viride* was cultured using sand maize media while the bacterial strains were cultured in Luria broth. The fungal seed culture was incubated at $30 \pm 1°C$ for 96 h. After mass-multiplication, the mycelial mat with conidia was homogenized and suspended in 500 mL of 0.1 M phosphate buffer ($K_2HPO_4$; $KH_2PO_4$) adjusting the colony forming units (CFU) at $1.2 \times 10^6$ per mL. The bacterial cultures were incubated at $28 \pm 1°C$ for 48 h under shaking conditions (200 rpm). After proper multiplication, the cultures were centrifuged at 6000 g for 10 min. The supernatants were discarded and the pellets containing bacterial cells were suspended in 500 mL of 0.9% saline, adjusting the CFU as $2.5 \times 10^8$ per $mL^{-1}$ for *B. megaterium* and $1.8 \times 10^8$ for *P. fluorescens*. Bacterial cell suspensions were inoculated as 10 mL of $10^8$ CFU/pot whereas the fungal seed culture was inoculated as 10 mL of $10^6$ CFU/pot. Nine microbial treatments were considered (T1 = uninoculated control, T2 = *B. megaterium*, T3 = *P. fluorescens*, T4 = *T. viride*, T5 = *B. megaterium* + *T. viride*, T6 = *P. fluorescens* + *T. viride*, T7 = *B. megaterium* + *P. fluorescens*, and T8 = *B. megaterium* + *P. fluorescens* + *T. viride*). Nitrogenous fertilizers were added to all the treatments according to the soil test. Shoot height, root length, shoots and root weight (fresh and dry), and N, P, and K content were determined after 90 days of microbial inoculation, at the harvest of the crop.

*2.3. Chemical Composition of Essential Oil (GC-Analysis).* Essential oil from the aerial biomass was hydrodistilled using Clevenger apparatus for 4 h. The oil was dried over anhydrous sodium sulphate ($Na_2SO_4$) and then analysed for gas chromatography (GC) on an Agilent Perkin Elmer GC Instrument (Model) fitted with FID (flame ionization detector) and electronic integrator with (30 m × 32 mm i.d., 0.25 $\mu$m film thickness) fused-silica capillary column. Nitrogen (at 0.4 mL/min) was used as a carrier gas. Samples were injected in the split mode at a ratio of 1:10–1:100. The injector was kept at 250°C and the transfer line at 280°C. The column was maintained at 50°C for 2 min and then programmed to 220°C at 5°C/min and held for 10 min at 300°C. The relative quantities of individual components of essential oil were calculated based on software computed GC peak area percentage without applying correction for FID response factor. The identification of the compounds was performed by comparing their retention indices (RI), determined with reference to a homologous series of n-alkanes (C9–C24, Polyscience Corp.) under identical experimental conditions in both polar and nonpolar columns, coinjection with standards. The relative quantities of the individual components were calculated based on computer calculated GC peak areas without correction for flame ionization detection response factors.

*2.4. Nutrient Uptake Analysis.* Soil mineralizable nitrogen, phosphorus, and potassium content were estimated in various treatments for assessing growth promotion effects. Initial soil samples were analyzed for available N ($kg\,ha^{-1}$), available P (Olsen's $P_2O_5$, $kg\,ha^{-1}$), and available K (exchangeable $K_2O$, $kg\,ha^{-1}$) following Jackson [17]. The estimations were made through FOSS application notes AN5222, ENISO 11732:1997, and AN5249 ISO 6878: 1996 with slight modifications in flame photometer 128, Cistronic model.

*2.5. Statistical Analysis.* The plant treatment was conducted in a random complete block design with three replicates per treatment. The data were subjected to analysis of variance (ANOVA) by using Duncan's multiple test, and treatment means were compared using least significant difference at $P \leq 0.05$. Least significant difference (LSD) was calculated at 5% probability level ($P = 0.05$) for comparing the significance of difference between any two treatment means [18]. The statistical analyses were performed using ASSISTAT 2012 software (version 7.6, Brazil). When the main effects were significant, differences among factor levels were tested for significance using the mean sum of square statement at $P = 0.05$.

## 3. Results

*3.1. Growth Promoting Parameters.* The results reveal that the microbial treatments not only enhanced the growth parameters in *O. tenuiflorum* but also significantly modulated aromatic oil's quality. The physical parameters of soil indicate good soil health with a neutral pH (Table 1). Electrical conductivity is the ability of a material to transmit an electrical current and is expressed as milli-Siemens per meter (mS/m). In the present experiment, the pH and conductivity of the

TABLE 1: Physical parameters of the soil.

| Physical parameters | Values |
|---|---|
| Electrical conductance (dsm. sup.$^{-1}$) | 135 $\mu$S/cm |
| N | 182.74 Kg/h |
| K | 84 Kg/h |
| P | 52 Kg/h |
| Na | 205 Kg/h |
| Organic carbon | 0.68% |
| pH | 7.53 |

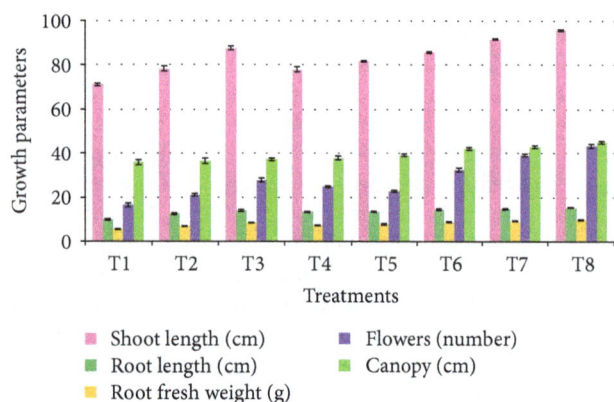

FIGURE 1: Growth parameters as influenced by the microbial treatments in *Ocimum tenuiflorum* cv. CIM-Ayu. a. Standard error of means (SEM) is indicated by the error bars.

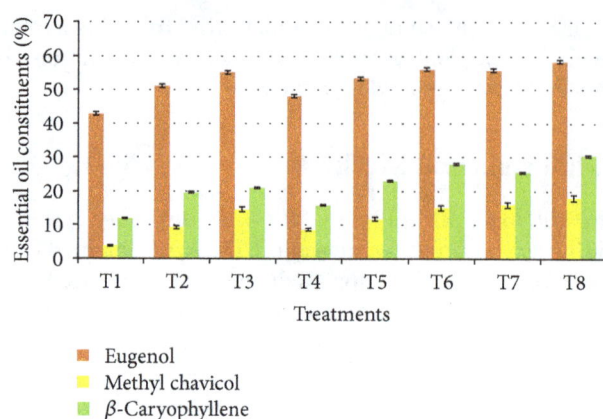

FIGURE 2: The major essential oil constituents influenced by different microbial treatments in *Ocimum tenuiflorum* cv. CIM-Ayu.

initial soil sample were 7.53 and 135 $\mu$Scm$^{-1}$, respectively. According to USDA criteria, EC was lower than 400 $\mu$Sm$^{-1}$ indicating neutral stress of soil against the existing flora signifying for the good quality of the soil. The attributes like shoot height and weight, root length and weight, and so forth are not considered as an important character determining essential oil extraction but are momentous for growth promotion decision. The traits are also desirable as the flowering tops and young shoots of *O. tenuiflorum* are also used for herbal tea preparations. All of inoculation treatments increased shoot length which were significantly different ($P < 0.05$) when compared with uninoculated control. All single and coinoculation treatments enhanced shoot length ranging from 88.33 cm to 95.67 cm (Figure 1). In the single inoculation, *P. fluorescens* (T3) treatment was observed with maximum effects as the plants depicted a 16.67 cm increase in shoot length over the control plants. The treatments had a lower variation coefficient of 1.52 (Table 2(a)) depicting the significant impact of all the treatments in enhanced growth. *P. fluorescens* intervention also resulted in the maximum increase in root length (13.8 cm) in alone treatments with an overall mean root length augmentation of 5.6 cm (9.7 to 15.3 cm). Fresh root weight was enhanced chiefly due to the combined treatment of *P. fluorescens* and *B. megaterium* (9.2 g). In the alone treatments, *T. viride* treatment depicted an augmentation of 2 cm over the control plants while the maximum shift of approximately 10 cm was observed in the combined treatment (T8) of all the bioinoculants, reflecting

the growth promotional aspect of the organic treatments (Figure 1).

*3.2. Leaf Yield and Essential Oil Content (%w/v).* Leaf yield that is leaf fresh weight per plant is an important character in determining the aromatic oil yield. The microbial inoculants can vary the constituents of essential oil and thus determine its quality. Observations for leaf biomass depicted a maximum of 49.17 g increase (T8) against the control (T1) plants. In the alone treatments, the highest leaf and oil yield were recorded in *P. fluorescens treatment* with an overall increase of 15.6% and 11%, respectively, against the control plants (Table 2(a)). Essential oil content showed a minimum of 21% enhancement in the combined treatments (T7 and T8) of microbial inoculants. Table 2(b) depicts a quick glance of ANOVA analysis for biomass and percentage essential oil with highly significant results as is indicated by the $P$ values. Means were compared using the least significant difference (LSD) at probability level $P = 0.05$ and revealed significant mean differences against the control plants. CV% values for leaf yield and percentage essential oil were 1.1 and 1.88, respectively, which showed greater variation among various treatments.

*3.3. Identification of Essential Oil Components.* To identify the mixture components, calculated retention indices (RI) were used. Retention indices of "neutral" components were compared with those reported (Adams 1995). When not less than three RI values are given for the same compound in different sources they were randomized. The results of Head Space analysis indicated that essential oil is rich in mono- and sesquiterpenoids. The major essential oil components as quantified through gas chromatography showed much variability. In alone treatments, the maximum increase (12.34%) in eugenol content was recorded in *P. fluorescens* treatment followed by *B. megaterium* treatment, showing 8.33% increase over the control plants (Figure 2). The combined treatment of *P. fluorescens* and *T. viride* enhanced eugenol content up to 56.22% as compared to 42.86% in control plants.

TABLE 2: (a) Relation between plant biomass and percentage essential oil in *O. tenuiflorum* cv. CIM-Ayu. (b) ANOVA table for biomass and percentage essential oil.

(a)

| Treatments | Shoot length (cm) | Root length (cm) | Root fresh weight (g) | Leaf yield (g) | Essential oil yield (%) |
|---|---|---|---|---|---|
| T1 | $71.00^f$ | $9.67^e$ | $5.33^f$ | $160.83^f$ | $0.56^e$ |
| T2 | $78.33^e$ | $12.33^d$ | $6.67^e$ | $178.33^e$ | $0.61^d$ |
| T3 | $87.67^c$ | $13.83^{bc}$ | $8.33^c$ | $185.83^d$ | $0.67^c$ |
| T4 | $78.00^e$ | $13.17^{cd}$ | $7.167^{de}$ | $181.50^e$ | $0.60^d$ |
| T5 | $81.67^d$ | $13.33^c$ | $7.67^d$ | $187.17^d$ | $0.67^c$ |
| T6 | $85.67^c$ | $14.33^b$ | $8.67^b$ | $196.50^c$ | $0.71^b$ |
| T7 | $91.67^b$ | $14.50^{ab}$ | $9.17^{ab}$ | $203.50^b$ | $0.77^a$ |
| T8 | $95.67^a$ | $15.33^a$ | $9.67^a$ | $210.00^a$ | $0.77^a$ |
| LSD ($P = 0.05$) | 2.21 | 0.88 | 0.59 | 3.59 | 0.02 |
| CV% | 1.52 | 3.83 | 4.32 | 1.10 | 1.88 |

[a]Mean in each column followed by the same letters does not differ significantly ($P < 0.05$) according to Duncan's multiple range test.
[b]Values followed by different small letters within columns are significantly different ($P < 0.05$) according to the LSD test.

(b)

| Source of variation | Degree of freedom (df) | Shoot length (cm) | Root length (cm) | Root fresh weight (g) | Leaf yield (g) | Essential oil yield (%) |
|---|---|---|---|---|---|---|
| Treatment | 7 | 1358.96** | 63.24** | 42.50** | 5026.79** | 0.09** |
| Error | 16 | 26.00 | 4.17 | 1.83 | 68.67 | 0.003 |
| Total | 23 | 1384.96 | 67.41 | 44.33 | 5095.46 | 0.09 |

**Significance at a level of 5% of probability ($P < 0.05$).

Other essential oil constituents responsible for the characteristic aroma of *O. tenuiflorum*, namely, $\beta$-caryophyllene and methyl chavicol, were also appreciably enhanced by microbial treatments. A quick insight into major essential oil constituent for three best treatments is depicted through their chromatograms (Figure 3). The peaks for eugenol, methyl chavicol, and $\beta$-caryophyllene were identified at retention index (RI) of 26.6, 28.2, and 29.5 minutes, respectively.

*3.4. Nutrient Uptake.* Nitrogen uptake was drastically enhanced in all the treatments with *B. megaterium* showing the maximum uptake of 2.83% followed by *P. fluorescens* as 2.62% in the alone treatments. The combined treatment of *B. megaterium*, *P. fluorescens*, and *T. viride* showed maximum uptake (2.93%) for nitrogen which was much higher than the control (1.67%) values. An enhanced uptake for nitrogen in all the treatments clearly shows that these microbial inoculants have an unambiguous role in growth promotion of plants (Table 3(a)). The phosphorus uptake results were quite contrasting to $N_2$ uptake as the phosphorus content declined in most of the alone treatments. However, the combined treatments showed a slight rise in phosphorus uptake with the highest phosphorus content recorded in the plants inoculated with the consortia of *B. megaterium*, *P. fluorescens*, and *T. viride*. An increased potassium uptake was also little influenced by the microbial treatments as is evident from the least significant difference and *P* values (Table 3(b)). These inoculants also depicted an augmented uptake when combined with *T. viride*.

TABLE 3: (a) Effect of rhizospheric microbes on nutrient uptake in *O. tenuiflorum* cv. CIM-Ayu. (b) ANOVA analysis for nutrient uptake in *O. tenuiflorum* cv. CIM-Ayu.

(a)

| Treatments | Nutrient uptake (%) | | |
|---|---|---|---|
| | N | P | K |
| T1 | $1.67^g$ | $0.32^d$ | $2.27^a$ |
| T2 | $2.83^b$ | $0.28^{ef}$ | $2.30^a$ |
| T3 | $2.62^c$ | $0.36^c$ | $1.62^d$ |
| T4 | $2.23^d$ | $0.31^{de}$ | $1.93^c$ |
| T5 | $2.13^e$ | $0.25^f$ | $1.85^c$ |
| T6 | $2.07^f$ | $0.33^d$ | $2.08^b$ |
| T7 | $2.11^e$ | $0.41^b$ | $2.24^a$ |
| T8 | $2.93^a$ | $0.44^a$ | $2.33^a$ |
| LSD ($P = 0.05$) | 0.04 | 0.03 | 0.11 |
| CV% | 0.98 | 5.50 | 2.96 |

[a]Mean in each column followed by same letters does not differ significantly ($P < 0.05$) according to Duncan's multiple range test.

(b)

| Source of variation | Degree of freedom (df) | N% | P% | K% |
|---|---|---|---|---|
| Treatment | 7 | 3.99** | 0.085** | 1.37** |
| Error | 16 | 0.01 | 0.01 | 0.06 |
| Total | 23 | 3.99 | 0.09 | 1.43 |

**Significance at a level of 5% of probability ($P < 0.01$).

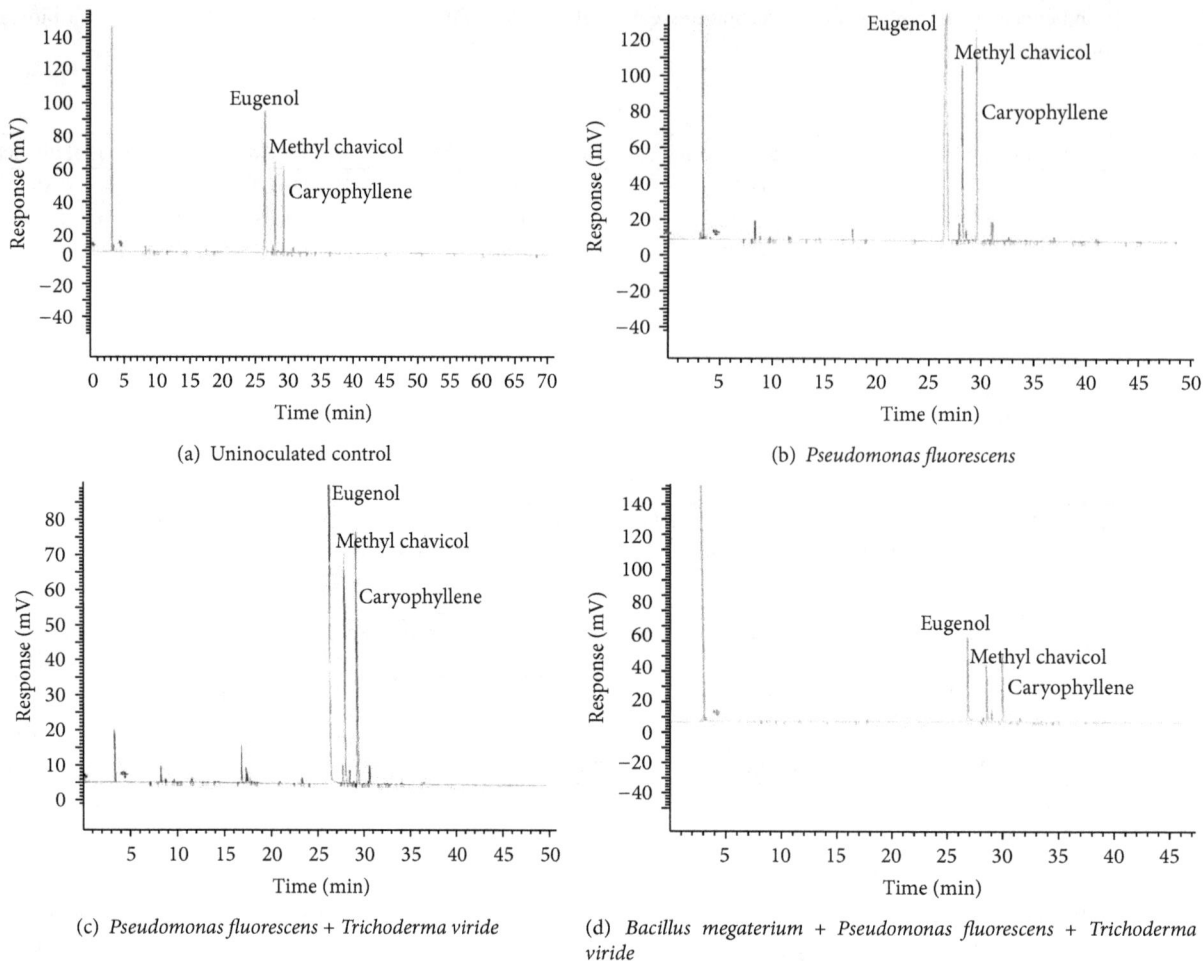

(a) Uninoculated control

(b) *Pseudomonas fluorescens*

(c) *Pseudomonas fluorescens + Trichoderma viride*

(d) *Bacillus megaterium + Pseudomonas fluorescens + Trichoderma viride*

FIGURE 3: GC chromatograms of selected microbial treatments depicting major essential oil constituents.

## 4. Discussion

Medicinal plants constitute a large segment of the flora, which provide raw materials for pharmaceutical, cosmetic, and fragrance industries. Noteworthy increases in growth and yield of vital crops can be the outcome of plant growth promoting rhizobacteria (PGPR) inoculation. PGPRs actively colonize plant roots and increase plant growth and yield [19]. In the present study, growth parameters were significantly enhanced by the microbial treatments with an improved oil quality and quantity.

Glick et al. [20] considered plant growth promoting rhizobacteria as a better alternative to the established chemical strategies for facilitating plant growth. Rhizospheric microbes influence plant growth through the release of several antibiotics such as celastramycins A-B [21], kakadumycins [22], and demethylnovobiocins [23]. Furthermore PGPRs are also known to secrete a number of bioactive metabolites, for example, Taxol, maytansinoids. PGPRs facilitate the production of plant growth promotion compounds, such as auxins, cytokinins, and gibberellins, or by producing siderophore to bind $Fe^{3+}$ from the environment and help to improve nutrient uptake, supply of plant nutrients (nitrogen,

phosphate, and other mineral nutrients), or suppression of stress ethylene production by 1-aminocyclopropane-1-carboxylate (ACC) deaminase activity [24].

The present experimentation evaluates the effect of plant growth promoting rhizobacteria (PGPR) on nutrient uptake, growth, and yield attributes in Basil (*O. tenuiflorum*). The estimated impacts of microbial treatments on plant growth indices, mineral content, and essential oils in combined treatments validate the consortia inoculation to be a better option for enhancing growth parameters and yield attributes. Nitrogen and phosphorus are the two major plant nutrients responsible for influencing vegetative and reproductive phase of the plant growth, respectively. The colonization of root by PGPRs stimulates increased nutrient uptake at the root interface which results in better absorption of water and nutrients from the soil [25]. In the present study, nitrogen content in the biomass of the inoculated plants was much higher and it was significantly high in *B. megaterium* treatments. *Bacillus* spp. being a free living $N_2$-fixer supplements more nitrogen to biomass during the growth. In addition to the improvement of plant growth, these microbes also improve the soil fertility by aggregation and adding nutrients to soil. Similarly better response of *O. basilicum* plants inoculated with consortia of

*Glomus fasciculatum, P. fluorescens,* and *B. megaterium* has been reported [26]. Thus it is suggested that microbial treatment improves the plant growth, biomass, and yield by supplementing plant nutrients and producing growth hormones.

Increased nutrient uptake by plants inoculated with PGPRs has been attributed to the production of plant growth regulators at the root interface, which stimulated root development and resulted in better absorption of water and nutrients from the soil as PGPRs are able to solubilise "unavailable" forms of minerals by excreting organic acids which dissolve or chelate mineral ions to soluble form, thereby improving the competitiveness and external stress responses of the plant. The results in our experiment indicated a higher response with all the microbial treatments especially in combined microbial treatments. Essential oils are commonly extracted from aromatic crops through steam distillation. Since the method requires huge amounts of biomass, the microbial treatment proves to be beneficial for improved biomass yield. The GC-analysis of the oil having a combined treatment of *B. megaterium, P. fluorescens,* and *T. viride* revealed a major compound (58.5%) with a Kovat's index of 1356. The compound was identified as eugenol, a monoterpene having antimicrobial, insecticidal, antihelminthic, and nematicidal properties. Methyl chavicol also possesses antifungal and antibacterial activities.

## 5. Conclusions

The results reveal that microbial association not only enhanced the growth parameters in *O. tenuiflorum* but also significantly modulated aromatic oil's quality. The observations explained appreciable correlations between functional rhizospheric microbes and growth promoting traits, namely, plant height, shoot and root fresh weight, N, P, and K uptake, leaf yield and essential oil content, and major oil constituents. The study also emphasises that *P. fluorescens* greatly alters the growth parameters in *O. tenuiflorum*. With this experiment as primary step, further study is needed to develop biofertilizers consortium for commercially grown medicinal plants.

## Conflict of Interests

The authors declare that there is no conflict of interests regarding the publication of this paper.

## Acknowledgments

The authors wish to thank the Director, CSIR-CIMAP, Lucknow (India), for providing necessary facilities and Dr. D. Saikia for his editorial suggestions.

## References

[1] A. A. Farooqi and B. S. Sreeramu, *Cultivation of Medicinal and Aromatic Crops,* University Press, Hyderabad, India, 2001.

[2] R. K. Lal, S. P. S. Khanuja, A. K. Agnihotri et al., "High essential oil and eugenol yielding cultivar of *Ocimum sanctum* "CIM-AYU"," US Patent 2005/0091705 A1, 2005.

[3] F. A. Atta and O. A. O. Saad, "Biofertilizers as potential alternative of chemical fertilizer for *Cathranthus roseus,*" *Journal of Agricultural Science,* vol. 26, pp. 7193–7208, 2001.

[4] M. M. Wagenaar and J. Clardy, "Dicerandrols, new antibiotic and cytotoxic dimers produced by the fungus *Phomopsis longicolla* isolated from an endangered mint," *Journal of Natural Products,* vol. 64, no. 8, pp. 1006–1009, 2001.

[5] S. M. Nadeem, I. Hussain, M. Naveed, H. N. Ashgar, Z. A. Zahir, and M. Arshad, "Performance of plant growth promoting rhizobacteria containing ACC-deaminase activity for improving growth of maize under salt-stressed conditions," *Pakistan Journal of Agricultural Sciences,* vol. 43, pp. 114–121, 2006.

[6] J. W. Kloepper, R. Lifshitz, and R. M. Zablotowicz, "Free-living bacterial inocula for enhancing crop productivity," *Trends in Biotechnology,* vol. 7, no. 2, pp. 39–44, 1989.

[7] B. Giri, P. H. Giang, R. Kumari, R. Prasad, and A. Varma, "Microbial diversity in soils," in *Roles in Genesis and Functions,* F. Buscot and S. Varma, Eds., pp. 195–212, Springer, Heidelberg, Germany, 2005.

[8] J. Barriuso, B. Ramos Solano, C. Santamaría, A. Daza, and F. J. Gutiérrez Mañero, "Effect of inoculation with putative plant growth-promoting rhizobacteria isolated from Pinus spp. on Pinus pinea growth, mycorrhization and rhizosphere microbial communities," *Journal of Applied Microbiology,* vol. 105, no. 5, pp. 1298–1309, 2008.

[9] J. Vacheron, G. Desbrosses, M. L. Bouffaud et al., "Plant growth promoting rhizobacteria and root system functioning," *Frontiers in Plant Science,* vol. 4, p. 356, 2013.

[10] J. W. Kloepper, C. Ryu, and S. Zhang, "Induced systemic resistance and promotion of plant growth by *Bacillus* spp.," *Phytopathology,* vol. 94, no. 11, pp. 1259–1266, 2004.

[11] L. Pessoa, M. McKenna, E. Gutierrez, and L. G. Ungerleider, "Neural processing of emotional faces requires attention," *Proceedings of the National Academy of Sciences of the United States of America,* vol. 99, no. 17, pp. 11458–11463, 2002.

[12] R. Harisaranraj, S. S. Babu, and K. Suresh, "Callus induction and plant regeneration of *Vigna mungo* (L.) Hepper via half seed explant," *Ethnobotanical Leaflets,* vol. 12, pp. 577–585, 2008.

[13] B. V. Subbiah and G. L. Asija, "A rapid procedure for the estimation of available nitrogen in soils," *Current Science,* vol. 25, pp. 259–260, 1956.

[14] R. Pandey, A. Kalra, M. L. Gupta, and P. Sharma, "Phytonematodes: Major pest of MAPs," in *Proceedings of the 1st National Interactive Meet on Medicinal and Aromatic Plants,* S. Mathur, Ed., pp. 188–197, CIMAP, Lucknow, India, 2003.

[15] S. K. Saikia, S. Tiwari, and R. Pandey, "Rhizospheric innovations for growth enhancement and *Meloidogyne incognita* management in *Mentha arvensis* cv. Kosi," *International Journal of Environmental Science and Technology,* vol. 3, no. 1, pp. 26–34, 2012.

[16] R. Pandey, A. Gupta, H. N. Singh, and A. Kalra, "Phytonematode management through Bacteria: an Underground Battle for existence," in *Recent Advances. Biopesticides Biotechnological Applications,* J. K. Johri and N. B. R. I. Lucknow, Eds., pp. 1–26, 2009.

[17] M. L. Jackson, *Soil Chemical Analysis,* Prentice Hall of India, New Delhi, India, 1973.

[18] G. W. Snedecor and W. G. Cochran, *Statistical Methods,* Iowa State University Press, 8th edition, 1989.

[19] S. Shawky, R. Z. El-shennawy, and A. M. Shady, "Biological control of *Meloidogyne javaniica* on tomato plants with isolated

bioagent in Egypt," *The Journal of Agricultural Science, Mansoura University*, vol. 37, pp. 6049–6063, 2006.

[20] B. R. Glick, B. Todorovic, J. Czarny, Z. Cheng, J. Duan, and B. McConkey, "Promotion of plant growth by bacterial ACC deaminase," *Critical Reviews in Plant Sciences*, vol. 26, no. 5-6, pp. 227–242, 2007.

[21] C. Pullen, P. Schmitz, K. Meurer et al., "New and bioactive compounds from *Streptomyces* strains residing in the wood of Celastraceae," *Planta*, vol. 216, no. 1, pp. 162–167, 2002.

[22] U. Castillo, J. K. Harper, G. A. Strobel et al., "Kakadumycins, novel antibiotics from Streptomyces sp. NRRL 30566, an endophyte of Grevillea pteridifolia," *FEMS Microbiology Letters*, vol. 224, no. 2, pp. 183–190, 2003.

[23] Y. Igarashi, "Screening of novel bioactive compounds from plant-associated actinomycetes," *Actinomycetologica*, vol. 18, pp. 63–66, 2004.

[24] V. Kannan and R. Sureendar, "Synergistic effect of beneficial rhizosphere microflora in biocontrol and plant growth promotion," *Journal of Basic Microbiology*, vol. 49, no. 2, pp. 158–164, 2009.

[25] P. Nimnoi, N. Pongsilp, and S. Lumyong, "Endophytic actinomycetes isolated from *Aquilaria crassna* Pierre ex Lec and screening of plant growth promoters production," *World Journal of Microbiology and Biotechnology*, vol. 26, no. 2, pp. 193–203, 2010.

[26] B. A. Iwalokun, G. O. Gbenle, T. A. Adewole, S. I. Smith, K. A. Akinsinde, and E. O. Omonigbehin, "Effects of *Ocimum gratissimum* L. essential oil at subinhibitory concentrations on virulent and multidrug-resistant *Shigella* strains from Lagos, Nigeria," *Acta Pathologica, Microbiologica et Immunologica Scandinavica*, vol. 111, no. 4, pp. 477–482, 2003.

# 19

# Response of Short Duration Tropical Legumes and Maize to Water Stress: A Glasshouse Study

**Hossain Sohrawardy[1] and Md. Lokman Hossain[2]**

[1]Department of Crop Production Systems in the Tropics, Faculty of Agricultural Sciences, Georg-August-University, 37077 Göttingen, Germany
[2]Global Change Ecology (International Elite Graduate Programme), Faculty of Biology, Chemistry and Geosciences, University of Bayreuth, 95440 Bayreuth, Germany

Correspondence should be addressed to Md. Lokman Hossain; lokmanbbd@gmail.com

Academic Editor: Mumtaz Cheema

The study was conducted as a pot experiment in the tropical glasshouse to evaluate the response of grain legumes (*Phaseolus vulgaris*, *Vigna unguiculata*, and *Lablab purpureus*) in comparison to maize (*Zea mays*) and estimate their potential and performance. Two experiments were established using completely randomized design. Physiological measurements (stomatal conductance, photosynthetic activities, and transpiration rates) were measured using LCpro instrument. Scholander bomb was used for the measurement of plant cell water potential. Significant difference was observed in different plant species with increase of different water regimes. Among the legumes, *L. purpureus* showed better response in water stressed conditions. At the beginning, in dry watered treatment the photosynthetic rate was below 0 $\mu$mol m$^{-2}$ s$^{-1}$ and in fully watered condition it was 48 $\mu$mol m$^{-2}$ s$^{-1}$. In dry treatment, total dry weight was 10 g/pot and in fully watered condition it was near to 20 g/pot in *P. vulgaris*. The study concludes that water stress condition should be taken into consideration for such type of crop cultivation in arid and semiarid regions.

## 1. Introduction

Water is the most common limiting factor to primary productivity in arid and semiarid areas [1]. Areas with higher precipitation in a region have more above-ground primary productivity [2, 3]. Water-use efficiency is often equated with drought resistance and the improvement of crop yield under stress conditions [4]. Water has an influence on grain yield and yield contributing parameters. Canopy development is sensitive to water deficit [5]. Water stress is one of the most important factors for legumes production. It affects not only the production of the grains but also the whole process of growth of all organs of the plants and metabolism [6]. Plant growth and development are affected by water stress affecting physiological and biochemical processes, for example, ion uptake, photosynthesis, respiration and translocation [7–11]. Environmental stresses such as drought, salinity, heat, and cold represent a significant constraint to meet the world food demand, for example, Islamic Republic of Iran, especially under low precipitation (often <250 mm per annum). Furthermore, the uneven temporal and spatial distributions have diode agronomists to pick the foremost effective irrigation strategies or drought tolerant cultivars [12].

Grain legumes are a major source of protein in arid and semiarid regions of the world and play a key role in the economies of arid and semiarid regions [13]. Multipurpose (dual purpose) legumes provide fodder, forage, green manure, seeds, and leaves that are used in human food [14]. Cowpea (*Vigna unguiculata* L.) and lablab (*Lablab purpureus* L.) widely cultivated in the tropics have multipurpose uses [15, 16]. *L. purpureus* has been comparatively neglected in analysis and development and consequently our information of its genetic diversity is proscribed [16, 17].

Legumes fix atmospheric nitrogen and are important to restore soil fertility. Legumes also play important role in breakdown of disease cycles, improve physical structure of soil, encourage microorganisms, and release available phosphorus to the soil [18].

Morphological and physiological understanding of legumes could be useful for selecting varieties to obtain better yield under drought conditions [19, 20]. The response of plants to water stress varies considerably and depends upon the intensity and duration of stress as well as of genotype and growth stages of the plant [10, 21]. Understanding plant response to drought is important and essential to develop drought tolerant crops [22, 23].

### 1.1. Grain Legumes

#### 1.1.1. Common Bean (Phaseolus vulgaris L.).
P. vulgaris one of the most important legume species in the developing countries and three quarters of the world production (8.5 million tonnes) is cultivated annually [24]. A major constraint to bean production in many developing countries is drought, which affects 73% of the area planted to beans in Latin America [25]. More than 300 million people get protein, carbohydrates, and minerals from grain legumes such as V. unguiculata in these regions. P. vulgaris is the second important source of calories after Z. mays in many areas of the Sub-Saharan Africa [26]. P. vulgaris is primarily a food in sub-Saharan Africa. Furthermore, millions of small-scale farmers in Latin America and Africa rely on the production and sale of bean as an important source of household income.

#### 1.1.2. Cowpea (Vigna unguiculata L.).
V. unguiculata is also one of the most important grain legumes in the diet of humans living in the tropics and subtropics. V. unguiculata is also a major crop in West and Central Africa for livelihood of millions of people. They use it as food and fodder and also for cash income. V. unguiculata grains contain protein which is not expensive and both rural people and urban consumers use it as a major protein source. V. unguiculata grain contains about 25% protein and 64% carbohydrate [27]. V. unguiculata is a legume crop which is grown in hot-dry tropics and subtropics. It plays a substantial role by serving as grain and vegetables crop for the rural people in the east, west, south, and central part of Africa [28]. It is truly a dual purpose grain legume crop, providing food for man and feed for livestock and serving as a valuable and dependable revenue-generating commodity for small holder farmers and grain traders [29, 30]. V. unguiculata is also a valuable component of farming systems where soil fertility is limiting because V. unguiculata has a high rate of nitrogen fixation [31] and forms effective symbiosis with mycorrhizae [32] and has the ability to better tolerate a wide range of soil pH compared to other grain legumes [33]. V. unguiculata is utilized in crop rotation and it improves soil fertility for yield of the cereal grains [34–36]. In the developing countries where soil infertility is high, rainfall is limiting, and most of V. unguiculata is grown without the use of fertilizers and plant protection measures (i.e., pesticides or herbicides), and a wide variety of biotic and abiotic constraints also limit growth and severely limit yield [37, 38].

#### 1.1.3. Lablab (Lablab purpureus L.).
L. purpureus is fast growing multipurpose legume which is grown in tropical region.

Grain and forage types exist. It is suitable for ley and forage in Australia. L. purpureus originated from Africa but now it is cultivated in many tropical countries. It is grown as a grain legume crop in India and Africa. It is also adapted to grow in a wide range of rainfall and temperatures [39]. L. purpureus yields 4000 kg dry matter (DM)/ha of above-ground biomass in Australia under dry-land condition [40, 41]. Furthermore L. purpureus improves nitrogen (N) fertility of the soil [42, 43]. L. purpureus germination generally occurs within 5 days after planting. The growth period ranges between 75 and 300 days. Fruiting of some cultivars can be within 60–65 days of sowing and continues for 90–100 days. L. purpureus is adapted to drought condition [44] and known to be even more drought tolerant than P. vulgaris or V. unguiculata [45].

#### 1.1.4. Maize (Zea mays L.).
Z. mays is one of the important cereal crop species after wheat and rice. It grows throughout a wide range of climates [46]. Z. mays shows susceptibility to drought in every stage but these three stages are more crucial for drought stress, that is, early growth stage, flowering stage, and mid-to-late grain filling stage [47]. Although Z. mays makes efficient use of water, it is considered more susceptible to water stress than other crops because of its unusual floral structure with separate male and female floral organs and the near-synchronous development of florets on a single ear borne on each. Low wet dry spore grains reduce the silk receptiveness and spore viability. Z. mays has completely different responses to water deficit consistent with development stages [48]. Drought stress is particularly damaging to grain yield if it occurs early in the growing season at flowering and during mid-to-late grain filling [49]. During the seedling stage, water stress damages secondary root growth and development. Stem growth and development after floral initiation show rapidly in adequate amount of water. Plants will be shorter and will decrease individual and cumulative area in water stress condition [50]. Ten to fourteen days before and after flowering of the Z. mays is the critical time for water stress. During the flowering stage grain yield reduces two to three times than any other stages of the growing period of Z. mays [51]. Z. mays grain yield is correlated with kernel number per plant during the water stress at flowering and grain filling stages [52], highlighting the importance of adequate water supplies during flowering.

The study was conducted to evaluate the drought tolerance of grain legumes compared to Z. mays by measuring agronomic (biomass production, leaf area, and yield) and physiological parameters (stomata conductance and photosynthetic activity) and to identify phenotypic variability of three commonly grown grain legume species in semi-arid Africa compared with the common staple cereal crop Z. mays in Africa.

## 2. Materials and Methods

2.1. Materials. In experiments, common bean (P. vulgaris, KATX 56), cowpea (V. unguiculata, M 66), lablab (L. purpureus, Q 6880B), and maize (Z. mays, Ronaldinio KWS)

FIGURE 1: Seed of 4 species used in the experiment.

FIGURE 3: Components used for soil mix.

FIGURE 2: Soil used for the pot experiment.

were used in a glasshouse pot experiment (Figure 1). Figure 2 shows the soil used for the pot experiment.

*2.2. Methods.* Two glasshouse pot experiments were conducted as in the tropical glasshouse at Department of Crop Production Systems in the Tropics, Georg August University Göttingen, Göttingen, Germany. Experiment 1 was set using CRD with 3 different water regimes as treatments and replicated four times. Four species (*P. vulgaris, V. unguiculata, L. purpureus,* and *Z. mays*) were used in this experiment. Treatments were applied as (1) fully watered (watering three times in a week), (2) moderately watered (watering two times in a week), and (3) partially watered (watering one time in a week). Experiment 2 was set using CRD with 5 different water regimes as treatments and replicated three times. Three species (*P. vulgaris, V. unguiculata,* and *L. purpureus*) were used in this experiment. At first every pot was watered depending on weather condition. Treatments were applied as (1) fully watered (watering three times in a week), (2) moderately watered (watering two times in a week), (3) partially watered (watering one time in a week), (4) nearly dry (watering one time in every two weeks), and (5) dry (watering one time in every three weeks). Pot with a diameter of 23.5 cm and a depth of 19 cm with an internal capacity of approximately 8 lt. was used for both experiments. Pots were filled with substrate, clay, compost, and Schoninger sand (Figure 3). Soil pH (6.5) and bulk density ($0.7 \, gm/cm^3$) were measured before pot filling. Seeds were soaked in

distilled water for 24 hours and planted into pots directly. Three seeds were planted in each pot at a depth of 4.0 cm approximately. The pots were watered to field capacity by repeated watering. After germination pots were thinned to one plant per pot, pots were subsequently watered according to the above described treatments from 3 weeks after planting. Soil volumetric water content was measured by soil moisture meter (TDR) in a week once before watering and after watering. Data were taken from 12 cm deep of the soil. Physiological measurements (stomatal conductance, photosynthetic activities, and transpiration) were taken using LCpro-SD, ADC, BioScientific Ltd. Data were analysed using Statistica software version 10 and MS excel 2007. Treatment effects and interactions were considered statistically significant at ≤0.05 level of probability.

## 3. Results

*3.1. Cell Water Potential in Experiment 1.* The water potential was significantly different for the different plant species tested in the experiment during the vegetative stage. It can be assumed that higher water potential of the cells had higher water stress. For the partially watered treatment water potential was the highest for all tested plant species. But the intensity of the responsiveness differed among plant species. *Z. mays* seemed to be stressed the most (high plant cell water potential for the partially watered treatment) followed by *P. vulgaris,* whereas the cell water potential remained stable for the legumes *V. unguiculata* and *L. purpureus.* Plant cell water potential of *L. purpureus* and *V. unguiculata* was 5 and 6 bar in partially watered condition. It is a good sign for drought tolerance. On the other hand, water potential of *P. vulgaris* and *Z. mays* was 11 and 17 bar, respectively, in fully watered condition (Figure 4).

*3.2. Dry Weight of P. vulgaris, V. unguiculata, L. purpureus, and Z. mays in Experiment 1.* Higher water regimes showed higher dry weight in all legumes and they also showed significant difference between different water levels. But interaction between different plants and water regimes did not show significant difference. Total dry weight was about 43 g/pot in fully watered condition in common bean, whereas *V. unguiculata* and *L. purpureus* had 40 g/pot dry weight.

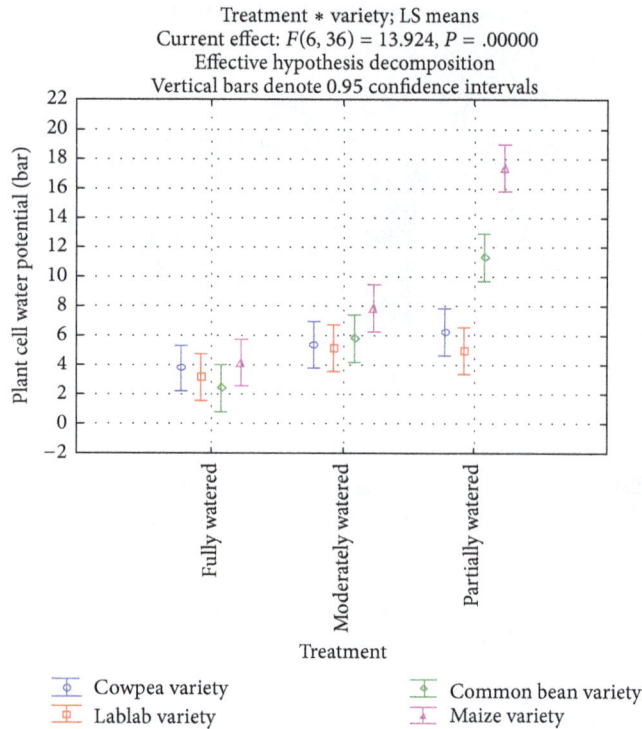

FIGURE 4: Plant cell water potential of *P. vulgaris*, *V. unguiculata*, *L. purpureus*, and *Z. mays* under different treatments in experiment 1.

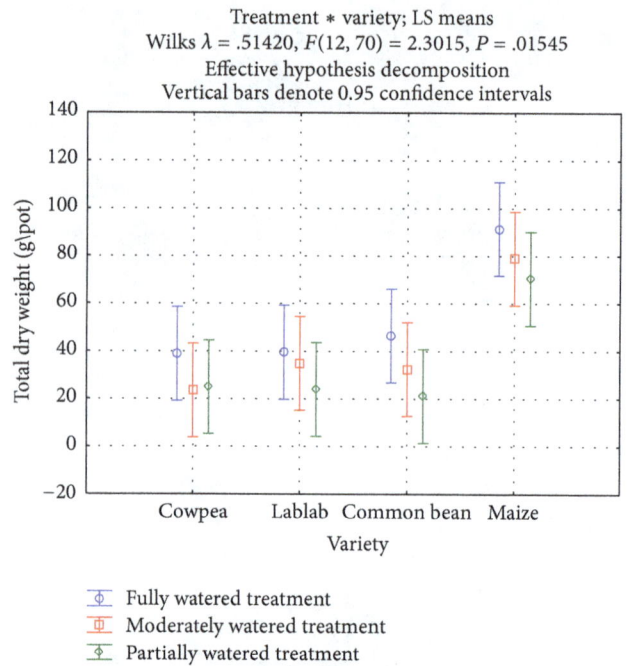

FIGURE 5: Total dry weight in different plants according to different watering treatments.

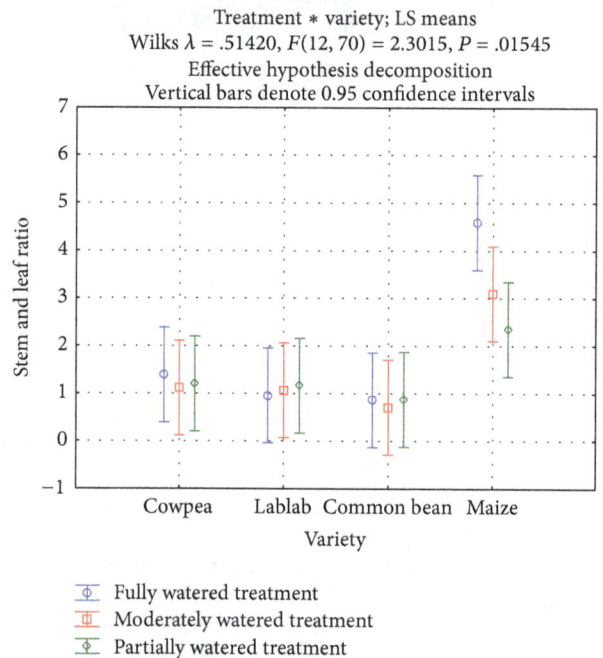

FIGURE 6: Stem-leaf ratio in different legumes according to different watering treatments.

Besides, partially watered condition *V. unguiculata* showed the highest value than other legumes, but *L. purpureus* showed the highest value in moderately watered treatment. This means that *V. unguiculata* and *L. purpureus* had low water use efficiency than *P. vulgaris* and partially watered condition *V. unguiculata* showed better yield than other legumes. On the other hand, *L. purpureus* had about 35 g/pot dry weight in partially watered treatment. Water stress affects total dry in *P. vulgaris*. In partially watered treatment, total dry weight was 20 g/pot in *P. vulgaris*. *P. vulgaris* showed above 40 g/pot total dry weight fully watered condition (Figure 5).

*3.3. Stem-Leaf Ratio of V. unguiculata, L. purpureus, P. vulgaris, and Z. mays in Experiment 1.* Different water regimes showed significant difference and different species also showed significant difference but interaction between variety and water regimes did not show significant difference. *L. purpureus* showed lowest stem and leaf ratio in fully watered condition in comparison to the moderately and partially watered treatments. This means that *L. purpureus* has low water use efficiency. On the other hand, non-legume crop, *Z. mays,* has more water use efficiency and water level was increased and hence, stem and leaf ratio showed higher value. In fully watered conditions, *Z. mays* had stem and leaf ratio was about 4.6 (Figure 6) than other watering treatments.

*3.4. Soil Moisture Content in Different Treatments of Four Species.* Differences between crops in the same watering regime were often small except fully watered condition. In case of dry, nearly dry, and moderately watered conditions, water consumption of *L. purpureus*, *V. unguiculata*, and *P. vulgaris* wasmuch more similar, but in fully watered conditions, *P. vulgaris* uptake more water than other crops. *V. unguiculata* and *L. purpureus* consumed less water in fully watered condition (Figure 7).

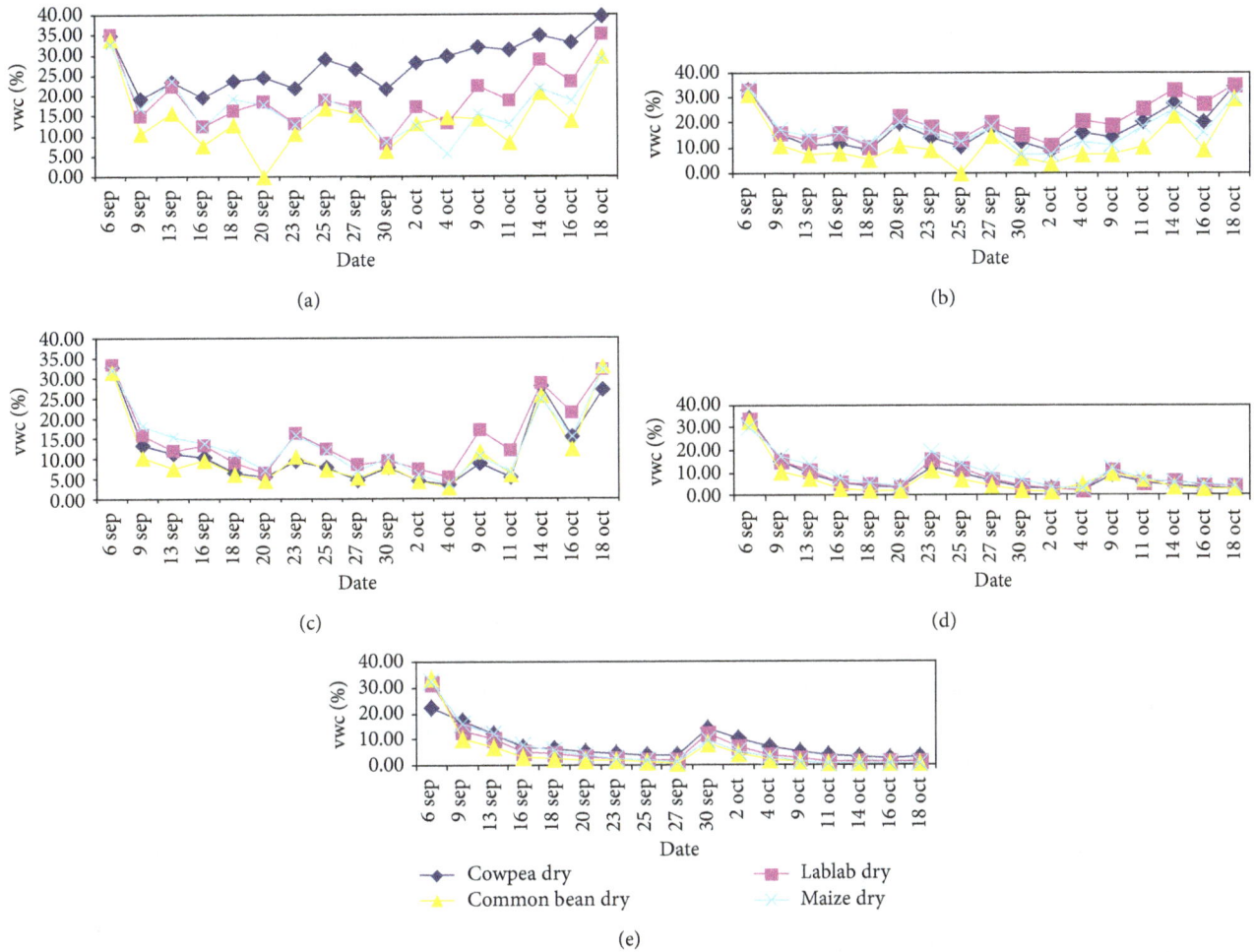

FIGURE 7: Soil moisture content of different water level: (a) fully watered (b) partially watered (c) moderately watered, (d) near to dry, and (e) dry.

### 3.5. Photosynthesis of V. unguiculata, L. purpureus, and P. vulgaris in Experiment 2.

Photosynthesis showed significant difference in variety and treatment interaction between before and after watering. Before watering, *V. unguiculata* showed the highest photosynthesis rate compared to the other plant species tested in dry condition. *L. purpureus* showed gradually increase of photosynthesis before and after watering (Figures 8 and 9). Photosynthesis was little bit increased in *P. vulgaris* and decreased in *L. purpureus* and *V. unguiculata* through watering. This also indicated that *L. purpureus* had lower water use efficiency than *P. vulgaris*.

### 3.6. Stomatal Conductance of V. unguiculata, L. purpureus, and P. vulgaris in Experiment 2.

Stomatal conductance showed significant difference in different water regimes in legumes. In *P. vulgaris*, stomatal conductance was increased when the water level increased. *P. vulgaris* showed more water use efficiency and stomatal conductance was gradually increased with the increase of water regimes. *L. purpureus* and *V. unguiculata* did not show too much variation; only *L. purpureus* in dry treatment suddenly showed very high stomatal conductance. *L. purpureus* and *V. unguiculata* showed less

water use efficiency than *P. vulgaris*. After watering, it showed that *P. vulgaris* showed higher stomatal conductance in nearly dry, moderately watered, and partially watered treatments. In *L. purpureus*, stomatal conductance was increased in case of nearly dry and moderately watered conditions but decreased little bit in fully watered and partially watered conditions. After watering, *V. unguiculata* showed water use efficiency in nearly dry and moderately watered conditions, but there was no change in fully watered and partially watered conditions (Figures 10 and 11).

### 3.7. Transpiration of V. unguiculata, L. purpureus, and P. vulgaris in Experiment 2.

Water regimes showed significant difference in transpiration before and after water use in different plants. In *P. vulgaris*, transpiration increased with the increase of water level except in dry treatment. After watering, evaporation increased slightly with the increase of water level in nearly dry and moderately watered conditions in *L. purpureus*, but transpiration was increased in partially and fully watered condition. In *V. unguiculata*, transpiration was increased in nearly dry and moderately watered

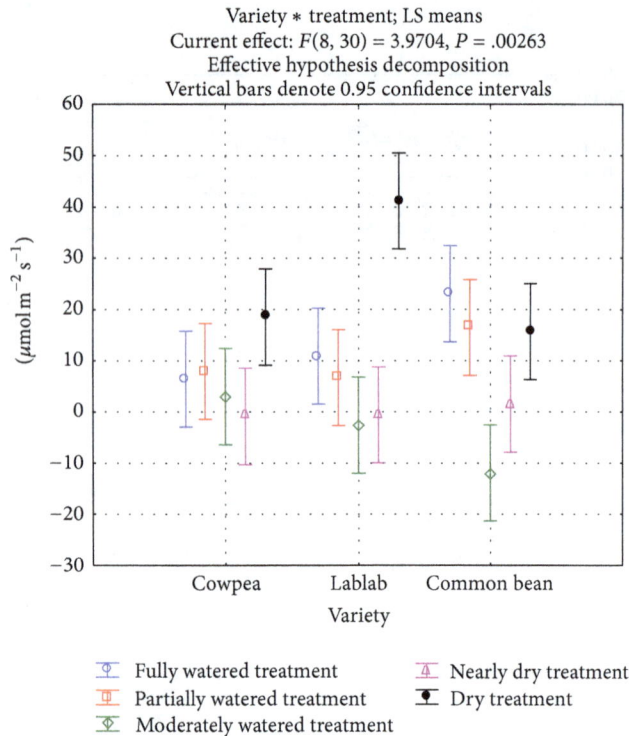

Variety * treatment; LS means
Current effect: $F(8, 30) = 3.9704$, $P = .00263$
Effective hypothesis decomposition
Vertical bars denote 0.95 confidence intervals

Fully watered treatment
Partially watered treatment
Moderately watered treatment
Nearly dry treatment
Dry treatment

FIGURE 8: Photosynthesis rate on 55 DAP.

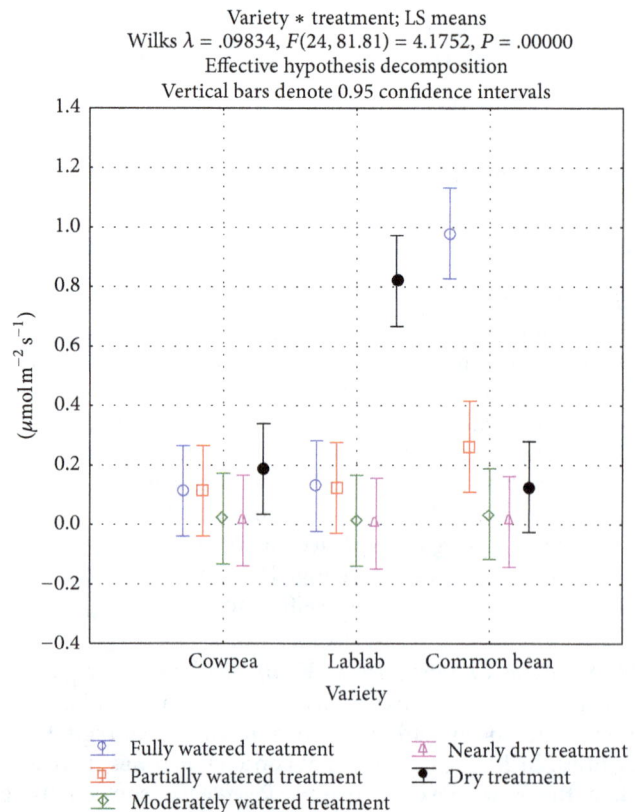

Variety * treatment; LS means
Current effect: $F(6, 24) = 4.3627$, $P = .00409$
Effective hypothesis decomposition
Vertical bars denote 0.95 confidence intervals

Fully watered treatment
Partially watered treatment
Moderately watered treatment
Nearly dry treatment

FIGURE 9: Photosynthesis rate on 58 DAP.

conditions but decreased in partially and fully watered conditions on 55 DAP (Figures 12 and 13).

### 3.8. Photosynthesis of *V. unguiculata, L. purpureus, and P. vulgaris on 55 DAP and 58 DAP in Experiment 2.*

Photosynthesis rate showed significant differences between different varieties and watering treatments on 55 DAP and 58 DAP. On 58 DAP, all varieties showed the highest photosynthesis rate in fully watered condition. On 55 DAP, *P. vulgaris* showed that increase of water level gradually increased photosynthetic rate. At the beginning, in dry watered treatment the photosynthetic rate was below 0 $\mu$mol m$^{-2}$ s$^{-1}$ and finally in fully watered condition it was 48 $\mu$mol m$^{-2}$ s$^{-1}$. Water level was increased followed by the increase of photosynthetic rate in fully watered condition. *V. unguiculata* and *L. purpureus* showed response to water level very well, but water use efficiency was lower in case of *V. unguiculata* and *L. purpureusthan* compared to *P. vulgaris* (Figures 14 and 15).

### 3.9. Stomatal Conductance of *V. unguiculata, L. purpureus, and P. vulgaris on 55 DAP and 58 DAP in Experiment 2.*

On 55 DAP, stomatal conductance showed significant difference in different legumes at different water regimes, but on 58 DAP stomatal conductance did not show significant difference. Stomatal conductance was gradually increased in *V. unguiculata* except for dry water treatments on 55 DAP (Figures 16 and 17).

### 3.10. Transpiration of *V. unguiculata, L. purpureus, and P. vulgaris on 55 DAP and 58 DAP in Experiment 2.* Transpiration

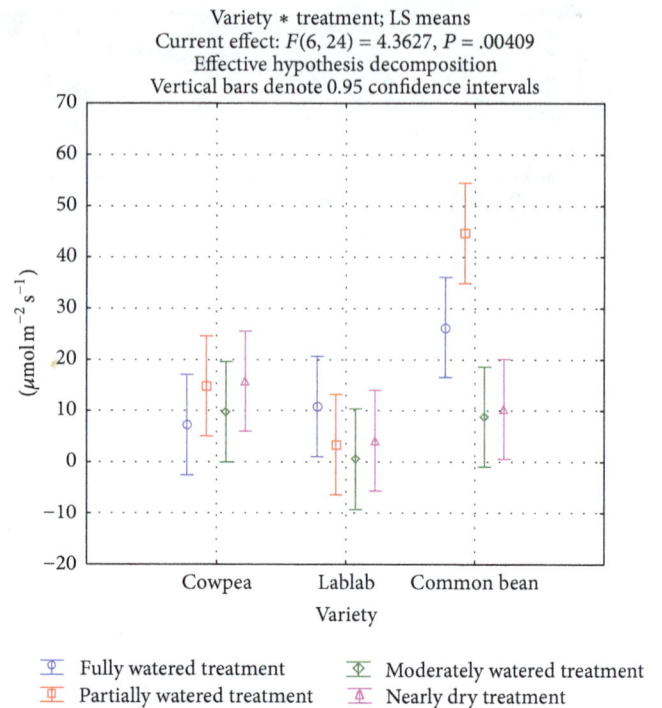

Variety * treatment; LS means
Wilks $\lambda = .09834$, $F(24, 81.81) = 4.1752$, $P = .00000$
Effective hypothesis decomposition
Vertical bars denote 0.95 confidence intervals

Fully watered treatment
Partially watered treatment
Moderately watered treatment
Nearly dry treatment
Dry treatment

FIGURE 10: Stomatal conductance on 55 DAP.

showed significant differences with different water treatments on 55 DAP and 58 DAP in different plants. *P. vulgaris* showed

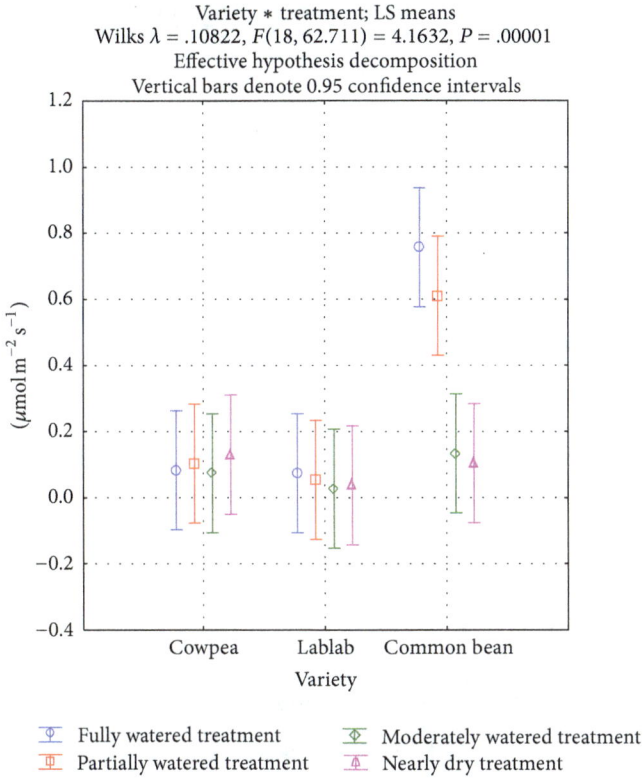

Variety * treatment; LS means
Wilks λ = .10822, F(18, 62.711) = 4.1632, P = .00001
Effective hypothesis decomposition
Vertical bars denote 0.95 confidence intervals

Fully watered treatment    Moderately watered treatment
Partially watered treatment    Nearly dry treatment

FIGURE 11: Stomatal conductance on 58 DAP.

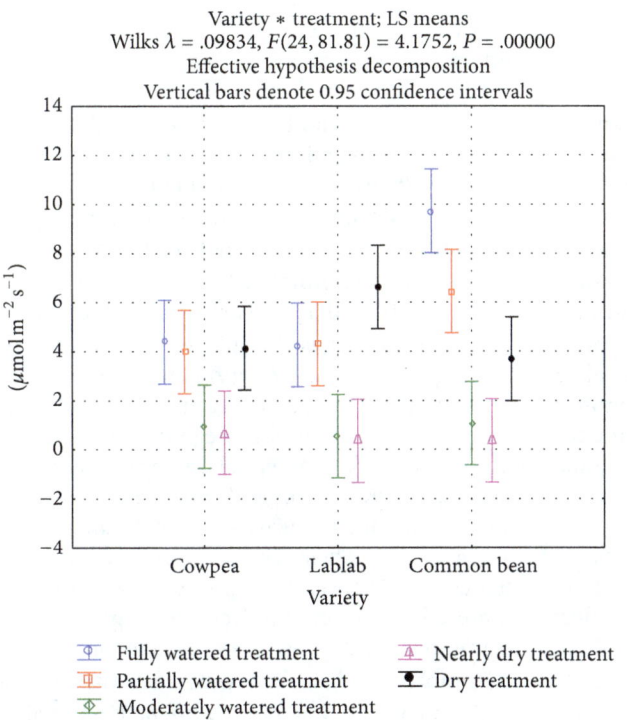

Variety * treatment; LS means
Wilks λ = .10822, F(18, 62.711) = 4.1632, P = .00001
Effective hypothesis decomposition
Vertical bars denote 0.95 confidence intervals

Fully watered treatment    Moderately watered treatment
Partially watered treatment    Nearly dry treatment

FIGURE 13: Transpiration on 58 DAP.

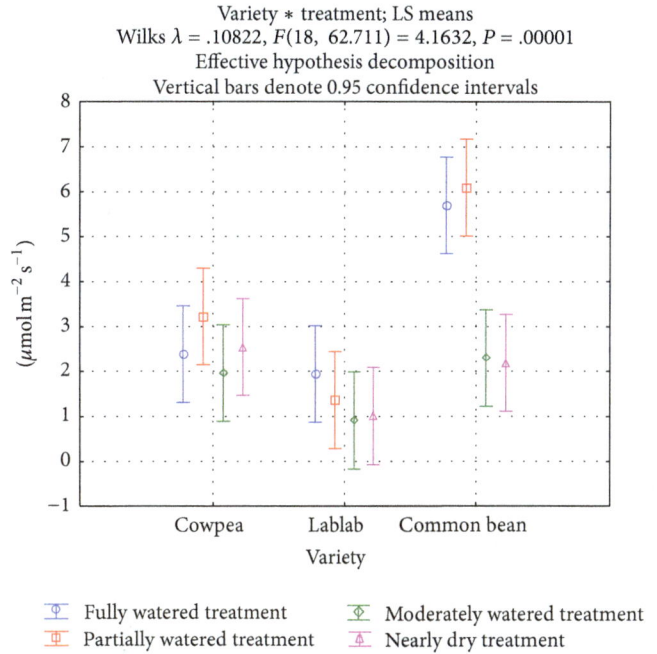

Variety * treatment; LS means
Wilks λ = .09834, F(24, 81.81) = 4.1752, P = .00000
Effective hypothesis decomposition
Vertical bars denote 0.95 confidence intervals

Fully watered treatment    Nearly dry treatment
Partially watered treatment    Dry treatment
Moderately watered treatment

FIGURE 12: Transpiration on 55 DAP.

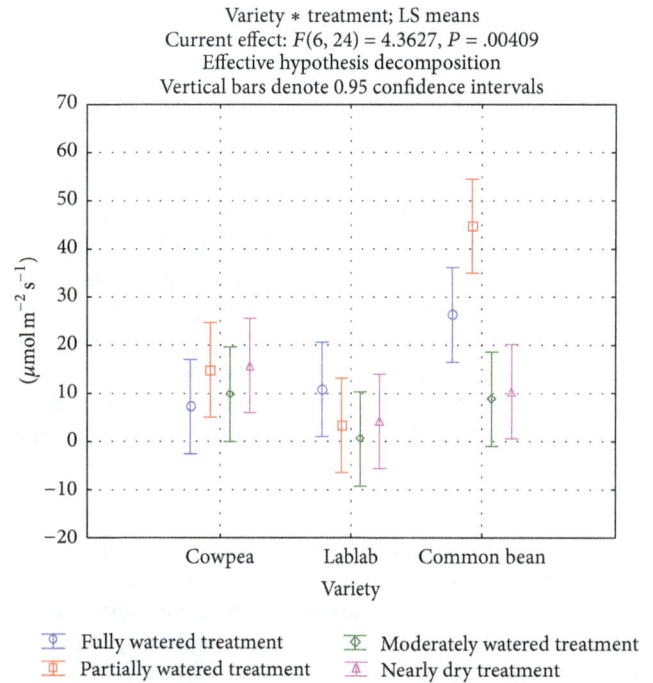

Variety * treatment; LS means
Current effect: F(6, 24) = 4.3627, P = .00409
Effective hypothesis decomposition
Vertical bars denote 0.95 confidence intervals

Fully watered treatment    Moderately watered treatment
Partially watered treatment    Nearly dry treatment

FIGURE 14: Photosynthesis on 55 DAP.

the gradual increases of transpiration with increasing water levels on 55 DAP. *L. purpureus* and *V. unguiculata* showed higher transpiration in fully watered condition on 55 DAP, but other treatments showed very little variation. *P. vulgaris* had more water use efficiency on 55 DAP. *P. vulgaris* also showed gradual increase of evaporation rate in different water regimes on 58 DAP. On 58 DAP, transpiration rate was increased after watering in *P. vulgaris* (Figures 18 and 19).

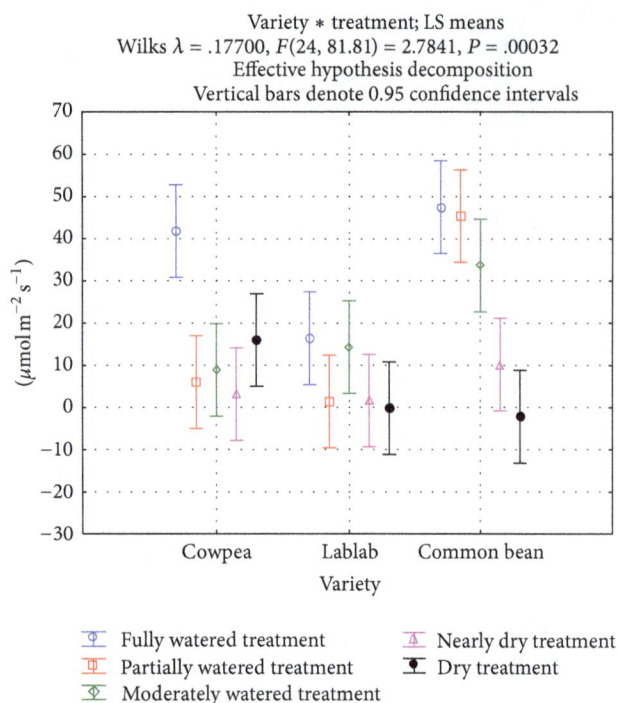

FIGURE 15: Photosynthesis rate on 58 DAP.

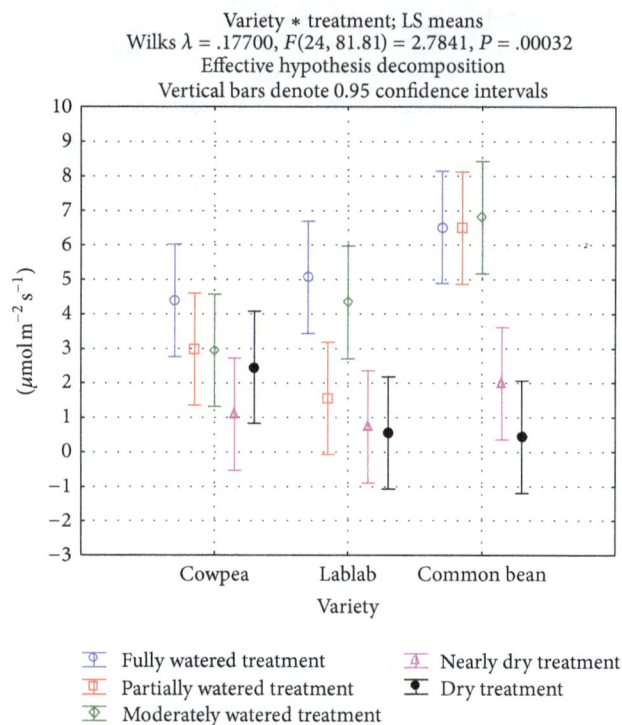

FIGURE 16: Stomatal conductance on 55 DAP.

*3.11. Stem-Leaf Ratio of V. unguiculata, L. purpureus, and P. vulgaris in Experiment 2.* Different water regimes and species followed significant difference and also the interaction between species and different water level showed significant difference. *P. vulgaris* showed the lowest stem leaf ratio in comparison to the other legume species in different water regimes. *V. unguiculata* had the highest stem leaf ratio compared to the other crops in fully watered condition. There was a clear effect of the watering treatment on the leaf stem ratio. The leaf stem ratio was increased for all plant species with increased water availability. Different plant species had different water use efficiency. But it was also clear that availability of water was responsible for increased stem and leaf ratio (Figure 20).

*3.12. Total Dry Weight of V. unguiculata, L. purpureus, and P. vulgaris in Experiment 2.* The effect of different water treatments on plant biomass production and dry weight was different among the three different legume species. For *P. vulgaris*, dry weight was decreased with the increased water stress. In dry treatment, total dry weight was 10 g/pot and in fully watered condition it was near to 20 g/pot in *P. vulgaris*. Total dry weight was increased double in *P. vulgaris* when water level was higher. For *L. purpureus* the effect of different water treatments on dry weight was not pronounced and remained almost constant. Surprisingly, *V. unguiculata* showed the highest dry weight in the moderately watered plants. Moderately watered condition showed the highest dry weight compared to the other treatments (Figure 21).

# 4. Discussions

*L. purpureus* showed gradually increased photosynthetic rate with response to different water level. In *P. vulgaris*, stomatal conductance was increased when the water level was increased and in fully watered condition it showed almost $1\,\mu\text{mol}\,\text{m}^{-2}\,\text{s}^{-1}$. *L. purpureus* and *V. unguiculata* did not show too much variation. Stomatal conductance was highest in *P. vulgaris*. *L. purpureus*, *V. unguiculata*, and *P. vulgaris* showed $0.0\,\mu\text{mol}\,\text{m}^{-2}\,\text{s}^{-1}$ stomatal conductance in nearly dry watering treatment, but in fully watered treatment stomatal conductance uplifted to $0.1\,\mu\text{mol}\,\text{m}^{-2}\,\text{s}^{-1}$ in *V. unguiculata* and *L. purpureus* each, and *P. vulgaris* gave value near to $1.0\,\mu\text{mol}\,\text{m}^{-2}\,\text{s}^{-1}$. The finding is also in agreement with Vogel et al. [53]. Among the legumes, *L. purpureus* showed lower photosynthetic rate than the other legumes. *P. vulgaris* had more water use efficiency than other legumes and stomatal conductance was gradually increased in regard to higher water regimes. *L. purpureus* and *V. unguiculata* did not show too much variation. *L. purpureus* and *V. unguiculata* followed less water use efficiency. In *L. purpureus*, stomatal conductance was increased in case of nearly dry and moderately watered conditions but decreased slightly in fully and partially watered conditions. The stomatal conductance was above $0.7\,\mu\text{mol}\,\text{m}^{-2}\,\text{s}^{-1}$ in fully watered treatment and $0.1\,\mu\text{mol}\,\text{m}^{-2}\,\text{s}^{-1}$ in partially watered treatment. On the other hand, *V. unguiculata* and *P. vulgaris* did not show too much change in fully watered treatment. Photosynthetic rate in *P. vulgaris* was increased with the increase of water level on 55 DAP. In dry watered treatment the photosynthetic rate was below $0\,\mu\text{mol}\,\text{m}^{-2}\,\text{s}^{-1}$, but in fully watered condition

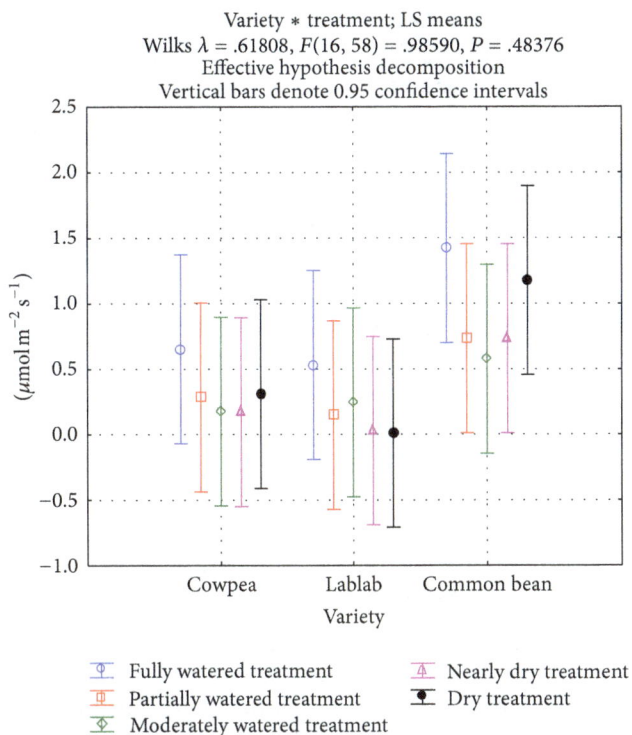

FIGURE 17: Stomatal conductance on 58 DAP.

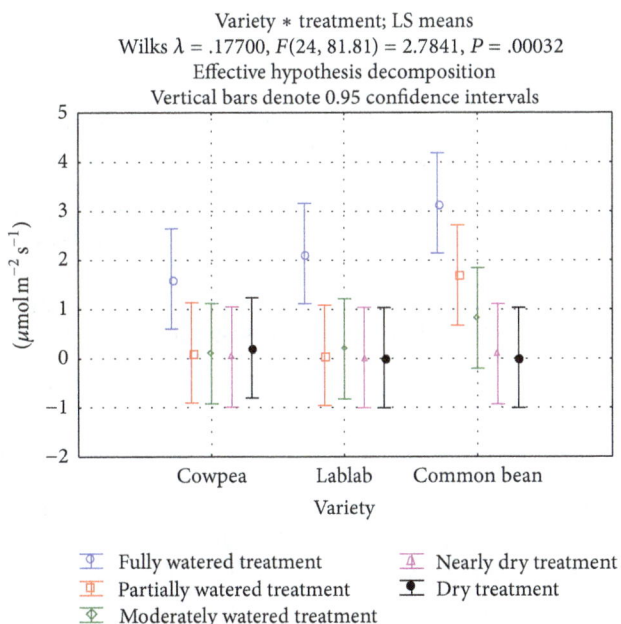

FIGURE 19: Transpiration on 58 DAP.

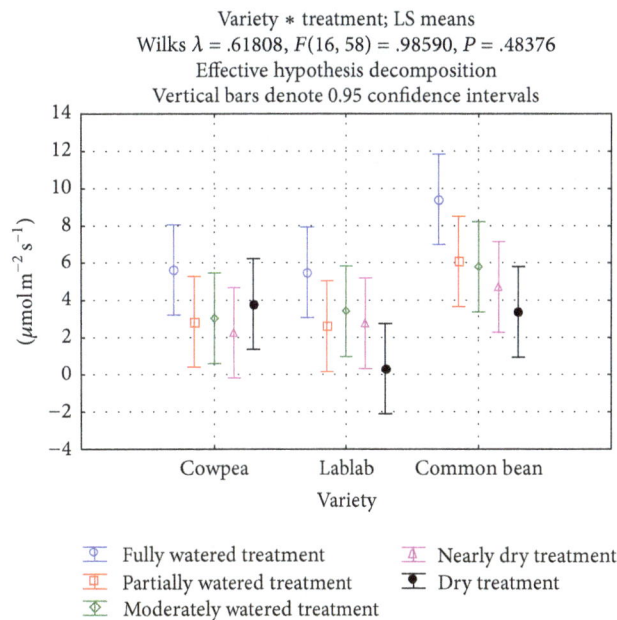

FIGURE 18: Transpiration on 55 DAP.

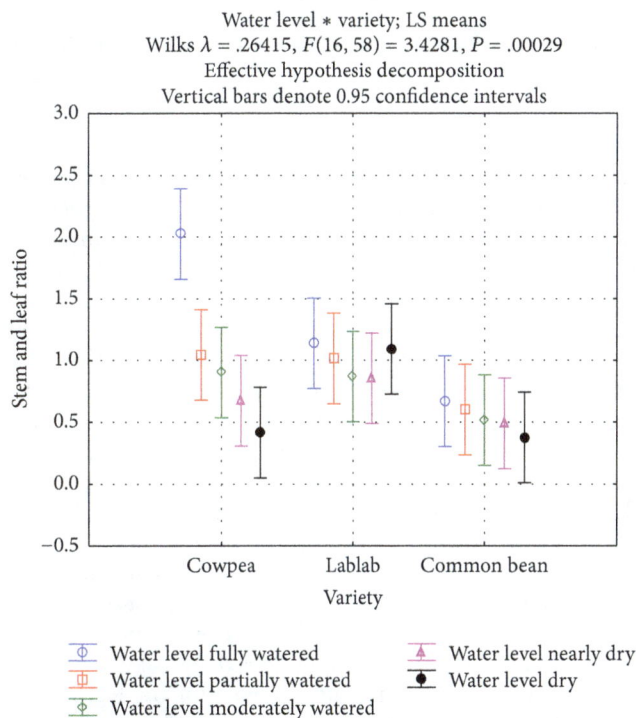

FIGURE 20: Stem-leaf ratio in different plants according to different watering treatments.

it was $48\,\mu\mathrm{mol\,m^{-2}\,s^{-1}}$. Moreover, *V. unguiculata* and *L. purpureus* also showed positive response to water level but lower water use efficiency than *P. vulgaris*. *P. vulgaris* showed the gradual increases of transpiration with increasing water levels on 55 DAP. *L. purpureus* and *V. unguiculata* showed higher transpiration in fully watered conditions on 55 DAP. After watering the plants, *P. vulgaris* also showed gradually

increases of transpiration rate in different water regimes on 58 DAP. It is proved that after watering transpiration rate was raised in all legumes, *P. vulgaris* showed clear gradual increase of transpiration than the other legumes. Higher water regimes showed higher dry weight in all legumes and they also showed significant difference between different water levels. Total dry weight in common bean was more (42 g/pot) in fully watered

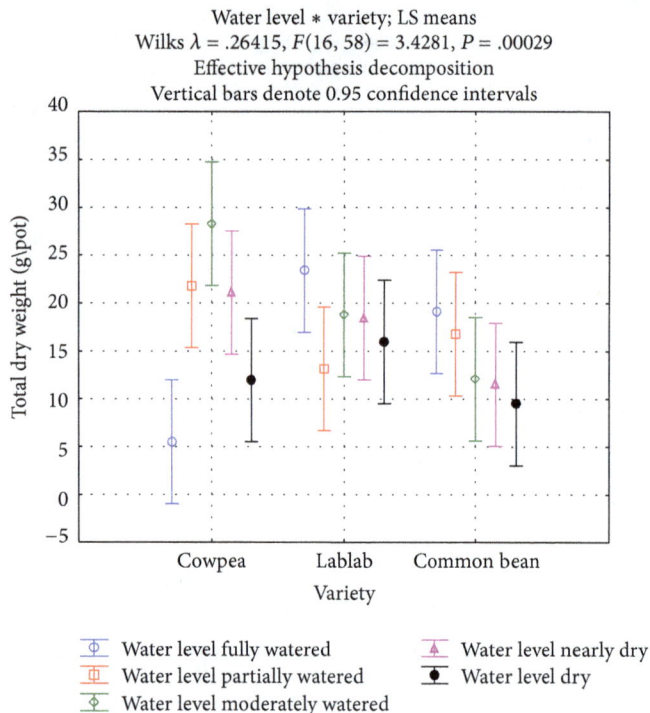

Water level * variety; LS means
Wilks $\lambda = .26415$, $F(16, 58) = 3.4281$, $P = .00029$
Effective hypothesis decomposition
Vertical bars denote 0.95 confidence intervals

○ Water level fully watered
□ Water level partially watered
◇ Water level moderately watered
△ Water level nearly dry
● Water level dry

FIGURE 21: Total dry weight in different plants according to different watering treatments.

condition. *V. unguiculata* and *L. purpureus* had 40 g/pot dry weight. Besides, in partially watered condition, *V. unguiculata* showed the highest value compared to the other legumes. On the other hand, *L. purpureus* had about 35 g/pot dry weight. In partially watered treatment, total dry weight was 20 g/pot in *P. vulgaris*. Yield was reduced in *P. vulgaris* due to drought treatments. The water potential was significantly different for the different plant species tested in the experiment during the vegetative stage. It can be assumed that higher water potential of the cells belongs to higher water stress. For the partially water treatment water potential was highest for all tested plant species. But the intensity of the responsiveness differed among plant species. *Z. mays* needed adequate amount of water for its growth and development. Different legumes also showed different responses to water stress. *Z. mays* seemed to be stressed the most (high plant cell water potential for the partially water treatment) followed by beans, whereas the cell water potential remained stable for the legumes *V. unguiculata* and *L. purpureus*. Plant cell water potential of *L. purpureus* and *V. unguiculata* was 5 and 6 bar in partially watered condition. It is a good sign for drought tolerance.

## 5. Conclusion

Water is the most limiting factor for agricultural production in particular in the semiarid areas of the world. Legumes are of great importance for food security particularly in the developing world. Furthermore, they have the ability to improve soil fertility in resource constraint small holder farming systems. Drought is the most limiting environmental

factor for the productivity of crops in primarily rain-fed based farming systems of semiarid areas. Plant developed totally different methods to avoid or minimize drought stress. The study examines responsiveness of three legumes and *Z. mays* to water stress. Significant difference was observed in different plant species with increase of different water regimes. *L. purpureus* showed better response in water stress compared to the other species. *P. vulgaris* showed a negative effect on dry weight with increased water stress. Therefore, this study concludes that selection of crops in water stress condition is highly necessary to improve the crop production and ensure the agricultural and environmental sustainability.

## Conflict of Interests

The authors declare that there is no conflict of interests regarding the publication of this paper.

## Acknowledgments

The authors extend their profound thanks to Professor Dr. Anthony Whitbread, PD Martin Worbes, Anne Sennhenn (Ph.D. student), and colleagues of the Department of Crop Sciences, Crop Production Systems in the Tropics, Georg August University Göttingen, Germany, for their assistance during the research work.

## References

[1] J. H. Brown, T. J. Valone, and C. G. Curtin, "Reorganization of an arid ecosystem in response to recent climate change," *Proceedings of the National Academy of Sciences of the United States of America*, vol. 94, no. 18, pp. 9729–9733, 1997.

[2] M. L. Rosenzweig, "Net primary productivity of terrestrial communities: prediction from climatological data," *The American Naturalist*, vol. 102, no. 923, pp. 67–74, 1968.

[3] O. E. Sala, W. J. Parton, L. A. Joyce, and W. K. Lauenroth, "Primary production of the central grassland region of the United States," *Ecology*, vol. 69, no. 1, pp. 40–45, 1988.

[4] A. Blum, "Drought resistance, water-use efficiency, and yield potential—are they compatible, dissonant, or mutually exclusive?" *Australian Journal of Agricultural Research*, vol. 56, no. 11, pp. 1159–1168, 2005.

[5] T. C. Hsiao, "The soil-plant-atmosphere continuum in relation to drought and crop production," in *Drought Resistance in Crops with Emphasis on Rice*, pp. 39–52, IRRI, Los Banos, Philippines, 1982.

[6] M. J. Zimmermann, M. Rocha, and T. Yolanda, *Bean Crop: Factors Affecting Productivity for the Presence of Potash and Phosphate*, 1988.

[7] C. A. Jaleel, R. Gopi, B. Sankar, M. Gomathinayagam, and R. Panneerselvam, "Differential responses in water use efficiency in two varieties of *Catharanthus roseus* under drought stress," *Comptes Rendus Biologies*, vol. 331, no. 1, pp. 42–47, 2008.

[8] C. A. Jaleel, P. Manivannan, G. M. A. Lakshmanan, M. Gomathinayagam, and R. Panneerselvam, "Alterations in morphological parameters and photosynthetic pigment responses of *Catharanthus roseus* under soil water deficits," *Colloids and Surfaces B: Biointerfaces*, vol. 61, no. 2, pp. 298–303, 2008.

[9] C. A. Jaleel, R. Gopi, P. Manivannan, M. Gomathinayagam, R. Sridharan, and R. Panneerselvam, "Antioxidant potential and indole alkaloid profile variations with water deficits along different parts of two varieties of *Catharanthus roseus*," *Colloids and Surfaces B: Biointerfaces*, vol. 62, no. 2, pp. 312–318, 2008.

[10] C. A. Jaleel, R. Gopi, and R. Panneerselvam, "Growth and photosynthetic pigments responses of two varieties of *Catharanthus roseus* to triadimefon treatment," *Comptes Rendus—Biologies*, vol. 331, no. 4, pp. 272–277, 2008.

[11] M. Farooq, S. M. A. Basra, A. Wahid, Z. A. Cheema, M. A. Cheema, and A. Khaliq, "Physiological role of exogenously applied glycinebetaine to improve drought tolerance in fine grain aromatic rice (*Oryza sativa* L.)," *Journal of Agronomy and Crop Science*, vol. 194, no. 5, pp. 325–333, 2008.

[12] A. Soltani and A. Faraji, *Soil Water and Plant Relationship*, Mashhad University, Mashhad, Iran, 2007.

[13] J. Singh and A. Patal, "Water statues, gaseous exchange, proline accumulation and yield of mung bean in response to water stress," *Annual of Biology Ludhiana*, vol. 12, pp. 77–81, 1996.

[14] A. I. Moalafi, J. A. N. Asiwe, and S. M. Funnah, "Germplasm evaluation and enhancement for the development of cowpea (*Vigna unguiculata* L.) dual-purpose $F_2$ genotypes," *African Journal of Agricultural Research*, vol. 5, no. 7, pp. 573–579, 2010.

[15] B. B. Singh, H. A. Ajeigbe, S. A. Tarawali, S. Fernandez-Rivera, and M. Abubakar, "Improving the production and utilization of cowpea as food and fodder," *Field Crops Research*, vol. 84, no. 1-2, pp. 169–177, 2003.

[16] B. C. Pengelly and B. L. Maass, "*Lablab purpureus* (L.) Sweet—diversity, potential use and determination of a core collection of this multi-purpose tropical legume," *Genetic Resources and Crop Evolution*, vol. 48, no. 3, pp. 261–272, 2001.

[17] B. L. Maass, R. H. Jamnadass, J. Hanson, and B. C. Pengelly, "Determining sources of diversity in cultivated and wild Lablab purpureus related to provenance of germplasm by using amplified fragment length polymorphism," *Genetic Resources and Crop Evolution*, vol. 52, no. 6, pp. 683–695, 2005.

[18] K. Hoshikawa, "Significance of legume crops in improving the productivity and stability of cropping systems," in *Phosphorus Nutrition of Grain Legumes in the Semi Arid Tropices*, C. Johannesen, K. K. Lee, and K. L. Sahrawat, Eds., pp. 173–181, International Crops Research Institute for the Semi-Arid Tropics, Paatancheru, India, 1991.

[19] N. H. Nam, Y. S. Chauhan, and C. Johansen, "Effect of timing of drought stress on growth and grain yield of extra-short-duration pigeonpea lines," *Journal of Agricultural Science*, vol. 136, no. 2, pp. 179–189, 2001.

[20] J. P. Martínez, H. Silva, J. F. Ledent, and M. Pinto, "Effect of drought stress on the osmotic adjustment, cell wall elasticity and cell volume of six cultivars of common beans (*Phaseolus vulgaris* L.)," *European Journal of Agronomy*, vol. 26, no. 1, pp. 30–38, 2007.

[21] M. M. Chaves, J. S. Pereira, J. Maroco et al., "How plants cope with water stress in the field. Photosynthesis and growth," *Annals of Botany*, vol. 89, pp. 907–916, 2002.

[22] A. R. Reddy, K. V. Chaitanya, and M. Vivekanandan, "Drought-induced responses of photosynthesis and antioxidant metabolism in higher plants," *Journal of Plant Physiology*, vol. 161, no. 11, pp. 1189–1202, 2004.

[23] C.-X. Zhao, L.-Y. Guo, C. A. Jaleel, H.-B. Shao, and H.-B. Yang, "Prospectives for applying molecular and genetic methodology to improve wheat cultivars in drought environments," *Comptes Rendus: Biologies*, vol. 331, no. 8, pp. 579–586, 2008.

[24] D. Pachico, "Trends in world *P. vulgaris* production," H. F. Schwartz and M. A. Pastor Corrales, Eds., pp. 1–8, CIAT, Cali, Colombia, 1989.

[25] A. V. Schoonhoven and O. Voysest, "Common beans in Latin America and their constraints," in *Bean Production Problems in the Tropics*, H. F. Schwartz and M. A. PastorCorrales, Eds., pp. 33–57, CIAT, Cali, Colombia, 1989.

[26] S. Beebe and B. McClafferty, *Biofortified Beans*, HarvestPlus, CIAT, Cali, Colombia, 2006, http://www.research4development .info/PDF/Outputs/Misc_Crop/beans.pdf.

[27] R. Bressani, "Nutritive value of cowpea," in *Cowpea: Research, Production and Utilization*, S. R. Singh and K. O. Rachie, Eds., pp. 353–360, John Wiley & Sons, New York, NY, USA, 1985.

[28] G. Lemma, W. Worku, and A. Woldemichael, "Moisture and planting density interactions affect productivity in cowpea (*Vigna unguiculata*)," *Journal of Agronomy*, vol. 8, no. 4, pp. 117–123, 2009.

[29] B. B. Singh, "Recent genetic studies in cowpea," in *Challenges and Opportunities for Enhancing Sustainable Cowpea Production*, C. A. Fatokun, S. A. Tarawali, B. B. Singh, P. M. Kormawa, and M. Tamo, Eds., pp. 3–13, International Institute of Tropical Agriculture, Ibadan, Nigeria, 2002.

[30] A. S. Langyintuo, J. Lowenberg-DeBoer, M. Faye et al., "Cowpea supply and demand in West and Central Africa," *Field Crops Research*, vol. 82, no. 2-3, pp. 215–231, 2003.

[31] H. O. A. Elowad and A. E. Hall, "Influences of early and late nitrogen fertilization on yield and nitrogen fixation of cowpea under well-watered and dry field conditions," *Field Crops Research*, vol. 15, no. 3-4, pp. 229–244, 1987.

[32] M. B. Kwapata and A. E. Hall, "Effects of moisture regime and phosphorus on mycorrhizal infection, nutrient uptake, and growth of cowpeas (*Vigna unguiculata* (L.) Walp.)," *Field Crops Research*, vol. 12, pp. 241–250, 1985.

[33] R. L. Fery, "The cowpea: production, utilization, and research in the United States," *Horticulture Review*, vol. 12, pp. 197–222, 1990.

[34] R. J. Carsky, B. Vanlauwe, and O. Lyasse, "Cowpea rotation as a resource management technology for cereal-based systems in the Savannas of West Africa," in *Challenges and Opportunities for Enhancing Sustainable Cowpea Production*, C. A. Fatokun, S. A. Tarawali, B. B. Singh, P. M. Kormawa, and M. Tamo, Eds., pp. 252–266, International Institute of Tropical Agriculture, Ibadan, Nigeria, 2002.

[35] S. A. Tarawali, B. B. Singh, S. C. Gupta et al., "Cowpea as a key factor for a new approach to integrated crop-livestock systems research in the dry savannas of West Africa," in *Challenges and Opportunities for Enhancing Sustainable Cowpea Production*, C. A. Fatokun, S. A. Tarawali, B. B. Singh, P. M. Kormawa, and M. Tamo, Eds., pp. 233–251, International Institute of Tropical Agriculture, Ibadan, Nigeria, 2002.

[36] N. Sanginga, K. E. Dashiell, J. Diels et al., "Sustainable resource management coupled to resilient germplasm to provide new intensive cereal-grain-legume-livestock systems in the dry savanna," *Agriculture, Ecosystems & Environment*, vol. 100, no. 2-3, pp. 305–314, 2003.

[37] B. B. Singh, "Cowpea [*Vigna unguiculata* (L.) Walp," in *Genetic Resources, Chromosome Engineering and Crop Improvement*, R. J. Singh and P. P. Jauhar, Eds., vol. 1, pp. 117–162, CRC Press, Boca Raton, Fla, USA, 2005.

[38] M. P. Timko, J. D. Ehlers, and P. A. Roberts, "Cowpea," in *Genome Mapping and Molecular Breeding in Plants*, C. Kole,

Ed., vol. 3 of *Pulses, Sugar and Tuber Crops*, pp. 49–67, Springer, Berlin, Germany, 2007.

[39] A. M. Murphy and P. E. Colucci, "A tropical forage solution to poor quality ruminant diets: a review of *Lablab purpureus*," *Livestock Research for Rural Development*, vol. 11, no. 2, 1999, http://ftp.sunet.se/wmirror/www.cipav.org.co/lrrd/lrrd11/2/colu112.htm.

[40] D. G. Cameron, "Tropical and subtropical pasture legumes," *Queensland Agricultural Journal*, vol. 114, no. 2, pp. 110–113, 1988.

[41] R. L. Clem, "Animal production from legume-based ley pastures in southeastern Queensland," in *Tropical Legumes for Sustainable Farming Systems in Southern Africa and Australia*, A. M. Whitbread and B. C. Pengelly, Eds., pp. 136–144, Australian Centre for International Agricultural Research, Canberra, Australia, 2004.

[42] R. D. Armstrong, B. J. Kuskopf, G. Millar, A. M. Whitbread, and J. Standley, "Changes in soil chemical and physical properties following legumes and opportunity cropping on a cracking clay soil," *Australian Journal of Experimental Agriculture*, vol. 39, no. 4, pp. 445–456, 1999.

[43] R. D. Armstrong, K. McCosker, S. B. Johnson et al., "Legume and opportunity cropping systems in central Queensland. 1. Legume growth, nitrogen fixation, and water use," *Australian Journal of Agricultural Research*, vol. 50, no. 6, pp. 909–924, 1999.

[44] P. M. Maundu, G. W. Ngugi, and C. H. S. Kabuye, *Traditional Food Plants of Kenya*, National Museums of Kenya, English Press, Nairobi, Kenya, 1999.

[45] C. V. Piper and W. J. Morse, *The Bonavist, Lablab or Hyacinth Bean*, Bulletin of the U.S. Department of Agriculture no. 318, USDA, Washington, DC, USA, 1915.

[46] *FAO Production Yearbook 2002*, 56:83, 2002.

[47] P. Zaidi, *Drought Tolerance in Maize: Theoretical Considerations & Practical Implications*, CIMMYT-Maize Program, Mexico DF, Mexico, 2002.

[48] R. Çakir, "Effect of water stress at different development stages on vegetative and reproductive growth of corn," *Field Crops Research*, vol. 89, no. 1, pp. 1–16, 2004.

[49] P. W. Heisey and G. O. Edmeades, "Maize production in drought-stressed environments: technical options and research resource allocation," World Maize Facts and Trends 1997/1998, 1999.

[50] R. C. Muchow, "Comparative productivity of maize, sorghum and pearl millet in a semi-arid tropical environment II. Effect of water deficits," *Field Crops Research*, vol. 20, no. 3, pp. 207–219, 1989.

[51] R. F. Grant, B. S. Jackson, J. R. Kiniry, and G. F. Arkin, "Water deficit timing effects on yield components in Maize," *Agronomy Journal*, vol. 81, no. 1, pp. 61–65, 1989.

[52] J. Bolaños and G. O. Edmeades, "The importance of the anthesis-silking interval in breeding for drought tolerance in tropical maize," *Field Crops Research*, vol. 48, no. 1, pp. 65–80, 1996.

[53] J. C. Vogel, A. Fuls, and A. Danin, "Geographical and environmental distribution of $C_3$ and $C_4$ grasses in the Sinai, Negev, and Judean deserts," *Oecologia*, vol. 70, no. 2, pp. 258–265, 1986.

# 20

# Forecasting Rice Productivity and Production of Odisha, India, Using Autoregressive Integrated Moving Average Models

**Rahul Tripathi, A. K. Nayak, R. Raja, Mohammad Shahid, Anjani Kumar,
Sangita Mohanty, B. B. Panda, B. Lal, and Priyanka Gautam**

*Crop Production Division, Central Rice Research Institute, Cuttack, Odisha 753006, India*

Correspondence should be addressed to Rahul Tripathi; rahulcrri@gmail.com

Academic Editor: Christos Tsadilas

Forecasting of rice area, production, and productivity of Odisha was made from the historical data of 1950-51 to 2008-09 by using univariate autoregressive integrated moving average (ARIMA) models and was compared with the forecasted all Indian data. The autoregressive ($p$) and moving average ($q$) parameters were identified based on the significant spikes in the plots of partial autocorrelation function (PACF) and autocorrelation function (ACF) of the different time series. ARIMA (2, 1, 0) model was found suitable for all Indian rice productivity and production, whereas ARIMA (1, 1, 1) was best fitted for forecasting of rice productivity and production in Odisha. Prediction was made for the immediate next three years, that is, 2007-08, 2008-09, and 2009-10, using the best fitted ARIMA models based on minimum value of the selection criterion, that is, Akaike information criteria (AIC) and Schwarz-Bayesian information criteria (SBC). The performances of models were validated by comparing with percentage deviation from the actual values and mean absolute percent error (MAPE), which was found to be 0.61 and 2.99% for the area under rice in Odisha and India, respectively. Similarly for prediction of rice production and productivity in Odisha and India, the MAPE was found to be less than 6%.

## 1. Introduction

Rice is one of the most important cereal crops of India occupying an area of 41.92 million hectare with an annual production of 89.09 million tonnes with an average productivity of 2.13 t ha$^{-1}$ (2009-10) (http://www.agricoop.nic.in/). It plays a vital role in the national food security and would continue to remain so because of its wider adaptability to grow under diverse ecosystems. Rice contributes 40.8% of total food grain and remains the principal source of livelihood for more than 58% of the population. With the stabilization of area under rice at around 42 million hectare, plateauing, and/or declining productivity trend, especially in the Northern and Southern zones and shrinking natural resource bases, the only opportunities for sustaining the current level of sufficiency are seen in the vast underexploited potential of rainfed Eastern India [1].

A proper trend analysis and forecast of production of such an important crop in the potential Eastern Region is having significance on many accounts. Critical analysis of production and productivity is a prerequisite for proper knowledge base on the ecology and appropriate research/development efforts for harvesting maximum possible potential. Trend analysis has been attempted for crops like papaya and garlic by several authors [2–4]. An unexpected decrease in production reduces marketable surplus and income of the farmers and leads to price rise. Similarly, an increase in production can lead to a sharp decrease in prices and has adverse effect on farmers' incomes. Impact on price of an essential commodity has a significant role in determining the inflation rate, wages, salaries, and various policies in an economy. The proper forecast would pave way for appropriate surplus and deficit management to stabilize the price and ensure profits for the farmers.

Several techniques like simulation modelling and remote sensing are largely being used for forecasting of the crop yield and acreage. But sometimes, forecasting is needed much before the crop harvest or even before the crop planting. This

TABLE 1: Descriptive statistics and Mann-Kendall trend analysis test for the time series data of rice cultivation in Odisha and India from 1950-51 to 2006-07.

| Parameters | Mean | Median | Standard deviation | Kurtosis | Skewness | Mann-Kendall trend ($Z$ value) | Sen's slope ($Q$) |
|---|---|---|---|---|---|---|---|
| All India | | | | | | | |
| Area (million hectare) | 38.63 | 39.42 | 4.33 | −0.83 | −0.46 | 9.59 | 0.26 |
| Productivity (kg ha$^{-1}$) | 1351.90 | 1234.55 | 435.12 | −1.28 | 0.32 | 9.64 | 25.32 |
| Production (million tonnes) | 53.94 | 48.74 | 22.42 | −1.27 | 0.31 | 9.81 | 1.31 |
| Odisha | | | | | | | |
| Area (million hectare) | 4.29 | 4.38 | 0.25 | −0.62 | −0.65 | 5.29 | 0.01 |
| Productivity (kg ha$^{-1}$) | 990.35 | 967.00 | 305.89 | −0.76 | 0.29 | 6.79 | 16.39 |
| Production (million tonnes) | 4.30 | 4.17 | 1.47 | −0.76 | 0.24 | 6.98 | 0.08 |

can be achieved only by modeling the past data and getting the predictions. Autoregressive integrated moving average (ARIMA) has been used for model building based on the past data and predictions are made. ARIMA models have been developed to forecast the cultivable area, production, and productivity of various crops of Tamil Nadu [5, 6] and wheat production in Pakistan [7] and Canada [8]. Univariate forecasting of state level agricultural production was also made by various authors using ARIMA models [9–12].

Keeping the above requirement in view, the present study was carried out to (i) analyze the trends of production, productivity, and area under rice in Odisha, an Eastern Indian state, and compare with all Indian scenarios and (ii) forecast and validate the rice area, production, and productivity using ARIMA models.

## 2. Materials and Methods

*2.1. Data Collection.* The data on cultivable area, production, and productivity of the rice in Odisha was collected from the Annual Report on "Orissa agriculture and statistics" on Agricultural Statistics published by the Directorate of Agriculture and Food Production, Government of Odisha, Bhubaneswar, India. The same data for India was obtained from Directorate of Economics and Statistics, Department of Agriculture and Cooperation, India. The data pertaining to the agricultural years 1950-1951 to 2006-2007 was used for the model building and forecasting. The data of 2007-08, 2008-09, and 2009-10 was used for validation of the model.

*2.2. Trend Analysis.* The time series data pertaining to rice area, productivity, and production in Odisha as well as India were analyzed using the Mann-Kendall trend test for assessing the trend present in the data. Initially, this test was used by Mann [13] and Kendall [14] and subsequently derived the test statistic distribution [15, 16]. This hypothesis test is a nonparametric, rank-based method for evaluating the presence of trends in time series data. The data are ranked according to time and then each data point is successively treated as a reference data point and is compared to all data points that follow in time. Compared with parametric statistical tests, nonparametric tests are thought to be more

suitable for nonnormally distributed data [17]. Since the time series data used in the study is mostly nonnormally distributed as evident from the skewness and kurtosis values given in Table 1, the nonparametric tests were used in the study.

The Mann-Kendall test statistic is given by

$$S = \sum_{i=1}^{n-1} \sum_{j=i+1}^{n} \text{sgn}\left(x_j - x_i\right), \qquad (1)$$

where $x_i$ and $x_j$ are the sequential data values, $n$ is the data set record length, and

$$\text{sgn}\left(\theta\right) = \begin{cases} +1 & \text{if } \theta > 0 \\ 0 & \text{if } \theta = 1 \\ -1 & \text{if } \theta < 1. \end{cases} \qquad (2)$$

The Mann-Kendall test has two parameters that are of importance to the trend detection. These parameters are the significance level that indicates the trend's strength and the slope magnitude estimate which indicates the direction as well as the magnitude of the trend.

For independent, identically distributed random variables with no tied data values, we have $E(S) = 0$;

$$\text{Var}(S) = \frac{n(n-1)(2n+5)}{18}. \qquad (3)$$

When some data value are tied, the correction to Var($S$) is

$$\text{Var}(S) = \frac{n(n-1)(2n+5) - \sum_{i=1}^{n} t_i(i)(i-1)(2i+5)}{18}, \qquad (4)$$

where $t_i$ denotes the number of ties of extent $i$. For $n$ larger than 10, the test statistic

$$Z_s = \begin{cases} \dfrac{S-1}{[\text{var}(S)]^{0.5}} & \text{for } S > 0 \\ 0 & \text{for } S = 0 \\ \dfrac{S+1}{[\text{var}(S)]^{0.5}} & \text{for } S < 0; \end{cases} \qquad (5)$$

$Z_s$ follows the standard normal distribution [14]. The magnitude of trend slopes can be also calculated (Sen, 1968). Sen's estimate for slope is associated with the Mann-Kendall test as follows:

$$\beta = \text{Median}\left(\frac{x_j - x_i}{j - i}\right), \quad \forall j > i, \qquad (6)$$

where $x_j$ and $x_i$ are considered data values at time $j$ and $i$ ($j > i$), correspondingly. The median of these $N$ values of $\beta_i$ is represented as Sen's estimator of slope which is given as

$$Q_i = \begin{cases} \beta_{(N+1)/2} & \text{when } N \text{ is odd} \\ \frac{1}{2}\left(\beta_{N/2} + \beta_{(N+2)/2}\right) & \text{when } N \text{ is even.} \end{cases} \qquad (7)$$

A positive value of $Q$ indicates an upward trend, whereas a negative value represents a downward trend.

### 2.3. ARIMA Model.
The ARIMA model analyzes and forecasts equally spaced univariate time series data. An ARIMA model predicts a value in a response time series as a linear combination of its own past values. The ARIMA approach was first popularized by Box and Jenkins [18], and ARIMA models are often referred to as Box-Jenkins models. In this study, the analysis performed by ARIMA is divided into three stages [19].

### 2.4. Notation for Pure ARIMA Models.
Consider

$$W_t = \mu + \frac{\theta(B)}{\Phi(B)} a_t, \qquad (8)$$

where $t$ indexes time, $W_t$ is the response series $Y_t$ or a difference of the response series, $\mu$ is the mean term, $B$ is the backshift operator, that is, $BX_t = X_{t-1}$, $\Phi(B)$ is the autoregressive operator, represented as a polynomial in the backshift operator: $\Phi(B) = 1 - \Phi_1 B - \cdots \Phi_p B^p$, $\theta B$ is the moving average operator, represented as a polynomial in the backshift operator $\theta(B) = 1 - \theta_1 B - \cdots \theta_p B^p$, and $a_t$ is the independent disturbance, also called the random error. For simple differencing, $W_t = (1 - B)^d Y_t$, where $d$ is the order of differencing.

*Identification Stage.* The stationary check of time series data was performed, which revealed that rice area, production, and productivity for India as well as for Odisha were nonstationary except for the area under rice in Odisha. The nonstationery time series data were made stationary by first order differencing and best fit ARIMA models were developed using the data from 1951 to 2007 and used to forecast the cultivable area, production, and productivity of rice for Odisha and India for the next three years, that is, 2007-2008, 2008-2009, and 2009-10.

Candidate ARIMA models were identified by finding the initial values for the orders of nonseasonal parameters "*p*" and "*q*." They were obtained by looking for significant spikes in autocorrelation and partial autocorrelation functions. At the identification stage, one or more models were tentatively

chosen which seem to provide statistically adequate representations of the available data. Then precise estimates of parameters of the model were obtained by least squares.

*Estimation Stage.* ARIMA models are fitted and accuracy of the model was tested on the basis of diagnostics statistics.

*Diagnostic Checking.* The best model was selected based on the following diagnostics.

*(i) Low Akaike Information Criteria (AIC).* AIC [20] is estimated by AIC = $(-2 \log L + 2m)$, where $m = p + q$ and $L$ is the likelihood function.

Sometimes, SBC [21] is also used and estimated by SBC = $\log \sigma^2 + (m \log n)/n$.

*(ii) Insignificance of Autocorrelations for Residuals.* If a model is an adequate representation of a time series, it should capture all the correlation in the series, and the white noise residuals should be independent of each other.

*(iii) Significance of the Parameters.* Significance tests for parameter estimates indicate whether some terms in the model might be unnecessary.

*Forecasting Stage.* Future values of the time series are forecasted.

### 2.5. Model Evaluation.
The mean absolute percent error (MAPE) as defined below was used as a measure of accuracy of the models:

$$\text{MAPE} = 100 * \left(\frac{\sum_{i=1}^{n}\left(|Y_F - Y|/Y\right)}{n}\right) \qquad (9)$$

$Y_F$ is forecasted variable, $Y$ is actual variable, and $n$ is number of variables.

SAS 9.2 software (SAS Institute, Inc., Cary, NC) was used for time series analysis and developing ARIMA models and forecasting.

## 3. Results and Discussion

*3.1. Trend Analysis.* Descriptive statistics for the time series data of rice area, production, and productivity for both Odisha and India is given in Table 1. The time series data is plotted in Figure 1. The time series data for rice area, production, and productivity are nonnormal which can be assessed from their probability density plot and values of skewness and kurtosis. Hence nonparametric Mann-Kendall test for trend analysis was performed to test the significance of trend. As evident from the values of Mann-Kendall's $Z$ statistics and Sen's slope estimate ($Q$), the time series data for all the parameters selected for analysis showed significant and positive trend. The Mann-Kendall $Z$ value as well as magnitude of slope indicated that the rate of increase was less for area, production, and productivity in Odisha as compared to all Indian scenarios.

The trend analysis of long term time series data (1950-51 to 2006-07) for the area under rice was found to be positive

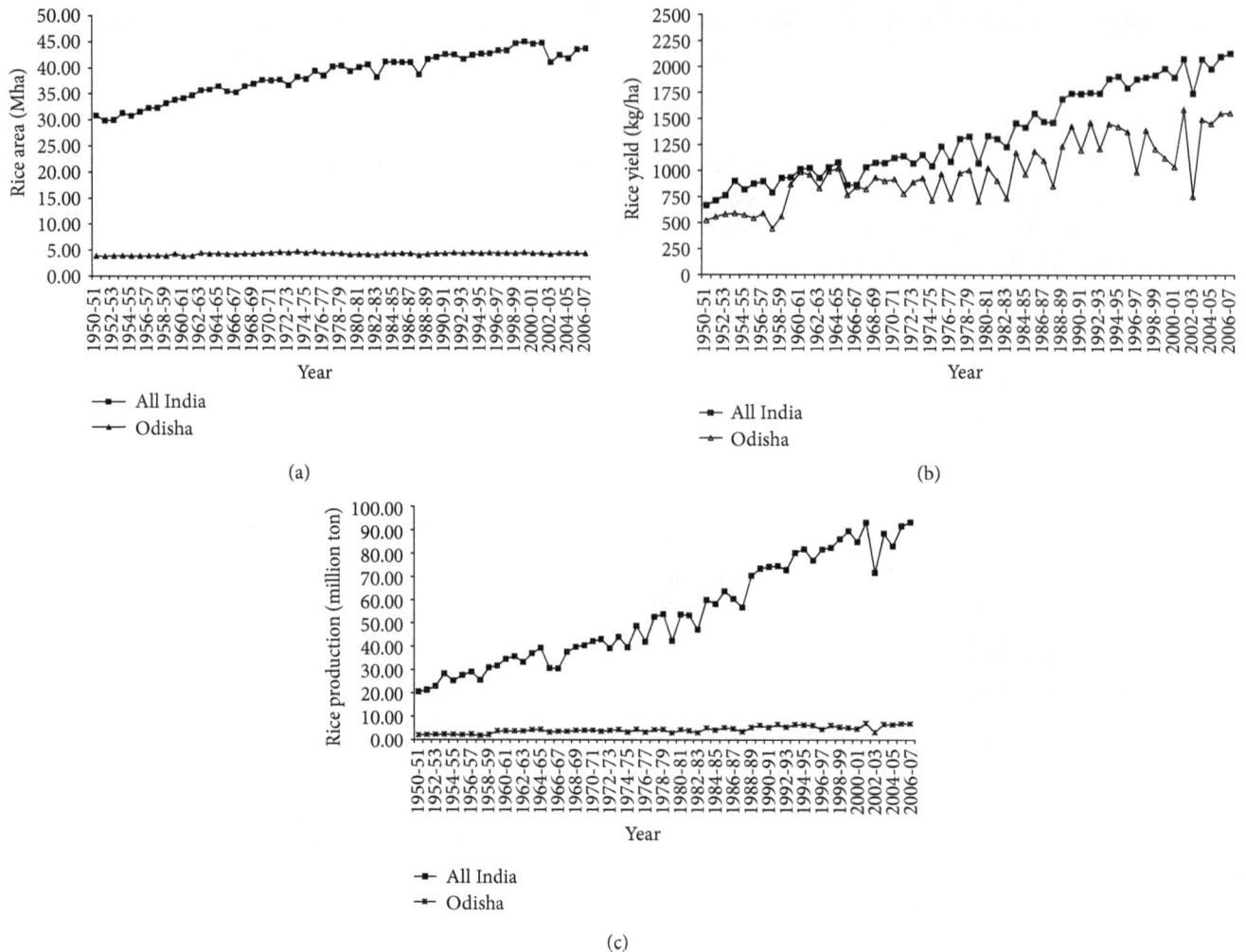

FIGURE 1: Trend of area, productivity, and production of rice in Odisha and India. (a) Area, million hectare; (b) productivity, kg ha$^{-1}$, and (c) production, million tonnes.

with a $Q$ value of 0.01 and 0.26 for both Odisha and India, respectively. The low $Q$ value can be explained by the fact that the area under rice remained more or less constant for the last 10 years due to competition from urbanization and industrialization. Area under rice in India was 43.45 million hectare and 43.81 million hectare, respectively, for the years 1997-98 and 2006-07, while during the same period area under rice in Odisha reduced to 4.45 million hectare from 4.50 million hectare. It is evident that there is plateauing in the area under rice in the last decade and the only option available to increase the rice production is vertical expansion.

Trend analysis also showed a considerable increase in all Indian average productivity of rice from 668 kg ha$^{-1}$ in 1950-51 to 2131 kg ha$^{-1}$ in 2006-07 and during the same period, rice productivity in Odisha increased from 520 kg ha$^{-1}$ to 1557 kg ha$^{-1}$. The rate of increase of productivity in Odisha is less than all Indian average as evident from Sen's slope estimate of 25.32 kg ha$^{-1}$ year$^{-1}$ and 16.39 kg ha$^{-1}$ year$^{-1}$ for India and Odisha, respectively, indicating an untapped growth potential for rice in Odisha. In order to tap this

potential Government of India has launched a programme "Bringing Green Revolution in Eastern India" since 2010-11.

*3.2. Building ARIMA Models.* The autoregressive ($p$) and moving average ($q$) parameters were identified based on the significant spikes in the plots of PACF and ACF of the different time series. While identifying the best fit ARIMA models, appropriate values of $p$, $d$, and $q$ were chosen corresponding to minimum value of the selection criterion, that is, AIC and SBC. The appropriate best fit models for rice area, production, and productivity of Odisha and India along with AIC and SBC are given in Table 2. The estimates of the autoregressive and moving average parameters along with the constant term are presented in Table 3. It is clear from the "$t$" value that all the parameters estimates were significant which is an essential criteria for the ARIMA models. It is evident from Figure 2(a) that ACF of area under rice for India has a significant spike at lag 1 and PACF declines gradually (Figure 3(a)), which indicated a moving average model of first order.

TABLE 2: Autoregressive integrated moving average (ARIMA) models fitted for time series data on rice area, productivity, and corresponding selection criterion, that is, Akaike information criteria (AIC) and Schwarz-Bayesian information criteria (SBC).

| Parameters | ARIMA model | AIC | SBC |
|---|---|---|---|
| All India | | | |
| Area, million hectare | ARIMA (0, 1, 1) | 165.33 | 169.38 |
| Productivity, kg ha$^{-1}$ | ARIMA (2, 1, 0) | 674.76 | 680.84 |
| Production, million tonnes | ARIMA (2, 1, 0) | 339.82 | 343.87 |
| Odisha | | | |
| Area, million hectare | ARIMA (2, 0, 0) | −46.67 | −40.54 |
| Productivity, kg ha$^{-1}$ | ARIMA (1, 1, 1) | 746.61 | 752.69 |
| Production, million tones | ARIMA (1, 1, 1) | 144.27 | 150.34 |

TABLE 3: Final estimates of parameters of autoregressive integrated moving average (ARIMA) models fitted for time series data on rice area, productivity, and production for Odisha and India.

| Parameters | ARIMA parameter | Estimate | Standard error | $t$-value |
|---|---|---|---|---|
| India | | | | |
| Area, million hectare | MU | 0.24 | 0.07 | 3.46* |
| | MA1 | 0.52 | 0.12 | 4.46* |
| Productivity, kg ha$^{-1}$ | MU | 25.14 | 3.99 | 6.31* |
| | AR1 | 25.67 | 6.14 | 4.18* |
| | AR2 | −0.79 | 0.13 | −6.10* |
| Production, million tones | MU | −0.36 | 0.13 | −2.74* |
| | AR1 | −0.76 | 0.13 | −5.76* |
| | AR2 | −0.30 | 0.13 | −2.22* |
| Odisha | | | | |
| Area, million hectare | MU | 3.85 | 0.14 | 27.53* |
| | AR1 | 0.50 | 0.12 | 4.18* |
| | AR2 | 0.49 | 0.12 | 4.06* |
| Productivity, kg ha$^{-1}$ | MU | 17.50 | 8.38 | 2.09* |
| | MA1 | 0.56 | 0.16 | 3.54* |
| | AR1 | −0.33 | 0.18 | −1.87* |
| Production, million tones | MU | 0.085 | 0.04 | 2.17* |
| | MA1 | 0.54 | 0.16 | 3.41* |
| | AR1 | −0.38 | 0.17 | −2.25* |

*Significant at $P = 0.05$.

Similarly significant spikes at lag 2 for the PACF of rice productivity and production of India indicate a second order autoregressive model of ARIMA $(2, 1, 0)$, which was found to be a best fit model. Significant spike at lag 2 of PACF (Figure 3(d)) and gradually declining ACF (Figure 2(d)) for area under rice for Odisha indicated a pure autoregressive model of order 2 and ARIMA $(2, 0, 0)$ and was found to be best fitted. Significant spike at lag 1 in Figures 2(e) and 2(f) and Figures 3(e) and 3(f), for both ACF and PACF, indicated a first order autoregressive as well as moving average model for both productivity and production of Odisha. The ACF and PACF were plotted for residuals of the fitted model and were lying within the limits, which showed that ARIMA model fitted well.

*3.3. Forecast Using ARIMA Models.* The observed and predicted values for rice area, production, and productivity along with percentage of deviation are presented in Table 4. The forecasted values of cultivable area of rice, for the years 2007-08, 2008-09, and 2009-10 for Odisha, were 4.46, 4.45, and 4.44 million hectare with the deviation of −0.22, 0, and −1.60%. Negative value in % deviation showed that predicted values were higher than the actual values. Similarly the forecasted values for cultivable area for all Indian average for the years 2007-08, 2008-09, and 2009-10 were 43.88, 44.12, and 44.35 million hectare with deviation from the actual of 0.07%, 3.12%, and −5.80%, respectively.

The forecasted values of productivity of rice for 2007-08, 2008-09, and 2009-10 for Odisha were 1503.89, 1544.61, and 1544.45 kg ha$^{-1}$ with deviation of 12.56 and −1.02 and 1.93%, respectively. The higher deviation of 12.56% in rice productivity of Odisha was due to jump in rice productivity in the year 2007-08 than the average productivity (Table 1). The forecasted values of rice productivity in India for the years 2007-08, 2008-09, and 2009-10 were 2121.24, 2173.59,

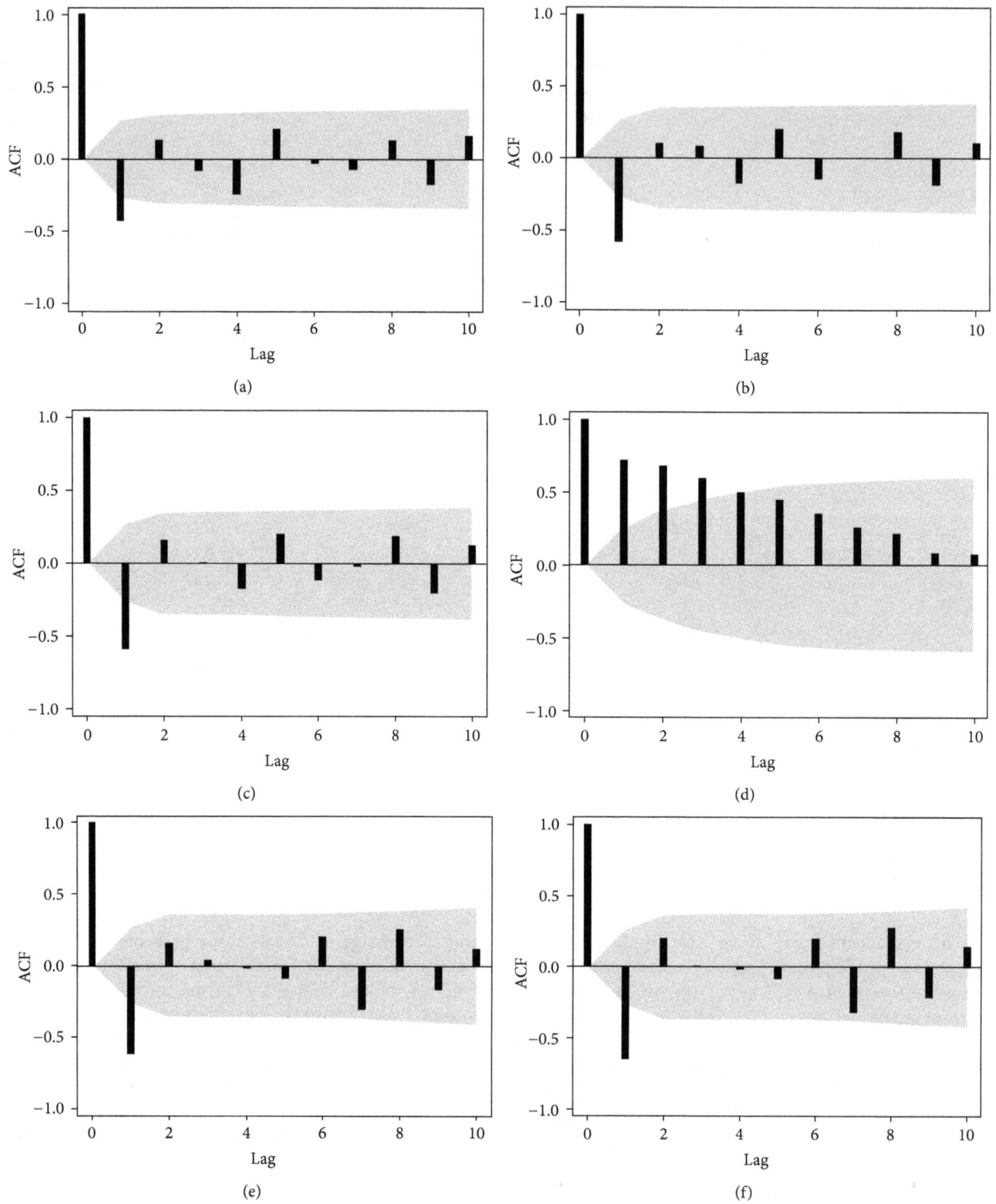

FIGURE 2: Autocorrelation function (ACF) for area, productivity, and production in Odisha and India. (a) Area (million hectare) in India; (b) productivity (kg ha$^{-1}$) in India; (c) production (million ton) in India; (d) area (million hectare) in Odisha; (e) productivity (kg ha$^{-1}$) in Odisha; and (f) production (million ton) in Odisha (shaded areas in the plots are significant at 5% level).

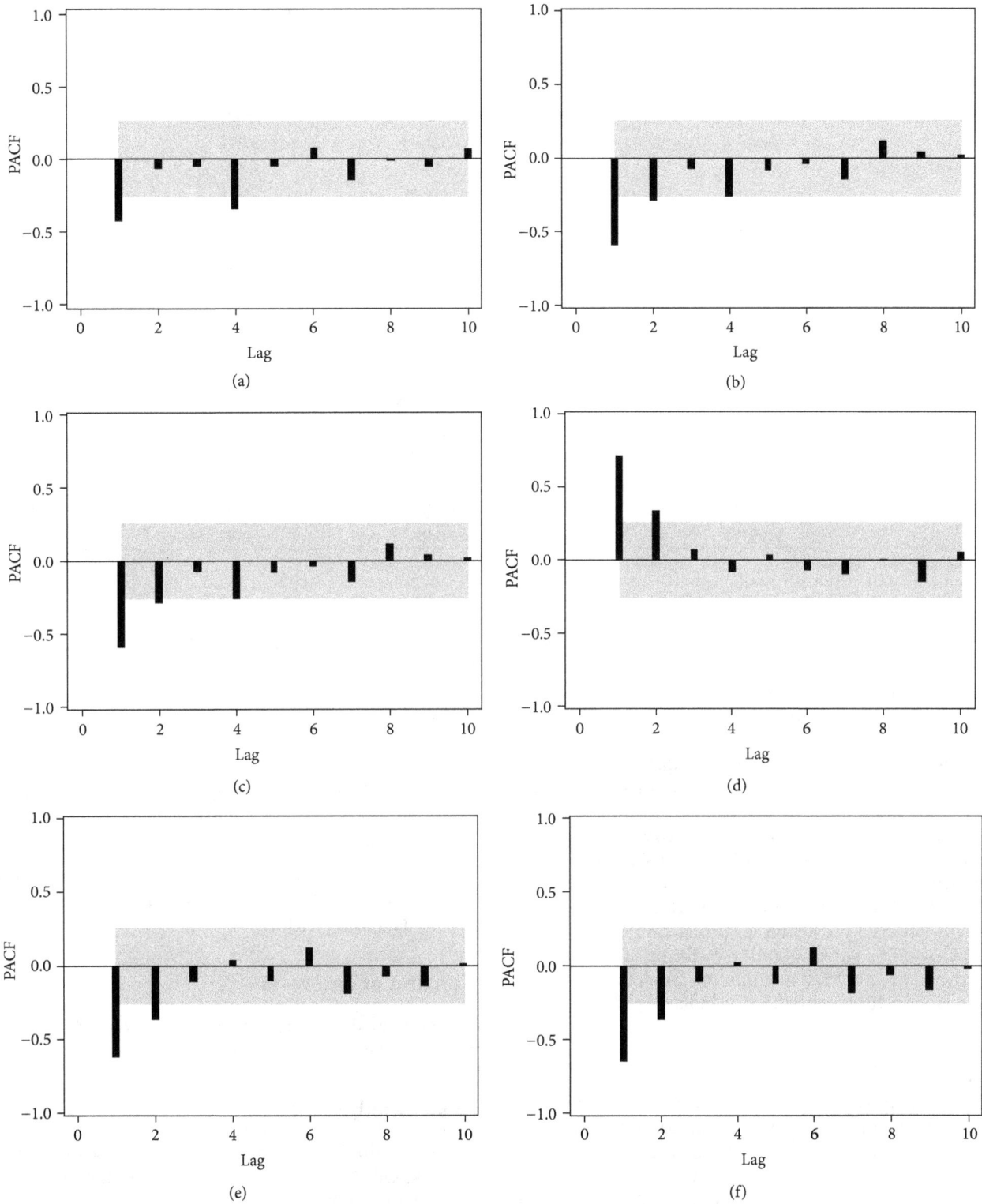

FIGURE 3: Partial autocorrelation function (PACF) for area, productivity, and production in Odisha and India. (a) Area (million hectare) in India; (b) productivity (kg ha$^{-1}$) in India; (c) production (million ton) in India; (d) area (million hectare) in Odisha; (e) productivity (kg ha$^{-1}$) in Odisha; and (f) production (million ton) in Odisha (shaded areas in the plots are significant at 5% level).

TABLE 4: Performance of autoregressive integrated moving average (ARIMA) models for rice area, productivity, and production for Odisha and India.

| Parameters | Years of prediction | Actual values | Forecasted values | % error in prediction (±) | MAPE in prediction |
|---|---|---|---|---|---|
| India | | | | | |
| | 2007-08 | 43.91 | 43.88 | 0.07 | |
| Area, million hectare | 2008-09 | 45.54 | 44.12 | 3.12 | 2.99 |
| | 2009-10 | 41.92 | 44.35 | −5.80 | |
| | 2007-08 | 2202 | 2121.24 | 3.67 | |
| Productivity, kg ha$^{-1}$ | 2008-09 | 2178 | 2173.59 | 0.20 | 2.32 |
| | 2009-10 | 2125 | 2190.89 | −3.10 | |
| | 2007-08 | 96.69 | 92.22 | 4.62 | |
| Production, million tones | 2008-09 | 99.18 | 95.26 | 3.95 | 5.41 |
| | 2009-10 | 89.09 | 95.92 | −7.67 | |
| Odisha | | | | | |
| | 2007-08 | 4.45 | 4.46 | −0.22 | |
| Area, million hectare | 2008-09 | 4.45 | 4.45 | 0.00 | 0.61 |
| | 2009-10 | 4.37 | 4.44 | −1.60 | |
| | 2007-08 | 1720 | 1503.89 | 12.56 | |
| Productivity, kg ha$^{-1}$ | 2008-09 | 1529 | 1544.61 | −1.02 | 5.17 |
| | 2009-10 | 1585 | 1554.45 | 1.93 | |
| | 2007-08 | 7.66 | 6.77 | 11.62 | |
| Production, million tones | 2008-09 | 6.81 | 6.95 | −2.06 | 4.90 |
| | 2009-10 | 6.92 | 6.99 | −1.01 | |

MAPE: mean absolute percent error.

and 2190.89 kg ha$^{-1}$ with 3.67, 0.20, and −3.10% deviations in prediction, respectively (Table 4).

The total forecasted production of rice in Odisha was 6.77, 6.95, and 6.99 million tonnes for the years 2007-08, 2008-09, and 2009-10 with prediction deviation of 11.62, −2.06, and −1.01%, respectively. This was due to high average productivity in 2007-08. Similarly the average production deviation for India was 4.62, 3.95, and −7.67%, respectively. The % error in prediction for area under rice varied from 0.07 to −5.80 and −0.22 to −1.60 for India and Odisha, respectively. The % deviation in prediction for rice productivity was 0.20 to 3.67 and −1.02 to 12.56% for India and Odisha, respectively. The % deviation in prediction for production of rice varied from 3.95 to −7.67 and −1.01 to 11.62% for India and Odisha, respectively. The MAPE was within 6% for all the forecasted parameters for Odisha as well as for India.

## 4. Conclusions

The trend analysis of the rice data showed an increasing productivity and production trend for both Odisha and India; the rate of increase was less in Odisha than all Indian average. This may be attributed to underexploitation of the potential of the state due to low input in agricultural operations and other biotic and abiotic factors. To bridge the gap between existing and potential productivity, rice varieties suitable to different ecologies can be introduced in farmer's field along with the nutrient and agronomic management practices. Based on the forecasting and validation results, it may be concluded that ARIMA model could be successfully used for forecasting rice

area, production, and productivity of Odisha as well as India for the immediate subsequent years.

## Highlights

(i) Trend analysis of rice area, production, and productivity of Odisha vis a vis India from the historical data of 1950-51 to 2008-09 is done.

(ii) Forecasting of rice area, production, and productivity of Odisha vis a vis India was made from the historical data using ARIMA models.

## Conflict of Interests

The authors declare that there is no conflict of interests regarding the publication of this paper.

## Acknowledgment

Authors thank Director of Central Rice Research Institute, Cuttack, Odisha (India), for providing all the help for performing this study.

## References

[1] E. A. Siddiq, "Bridging the rice yield gap in india," in *Bridging the Rice Yield Gap in the Asia -Pacific Region*, P. K. Minas, J. D. Frank, and J. H. Edward, Eds., pp. 84–111, Food and Agriculture

Organization of the United Nations Regional Office for Asia and the Pacific Bangkok, 2000.

[2] P. K. Sen, "Estimates of the regression coefficient based on Kendall's tau," *Journal of American Statistical Association*, vol. 39, pp. 1379–1389, 1968.

[3] S. C. Srivastava, U. C. Sharma, B. K. Singh, and H. S. Yadava, "A profile of garlic production in India: facts, trends and opportunities," *International Journal of Agriculture, Environment and Biotechnology*, vol. 5, no. 4, pp. 477–482, 2012.

[4] M. Mahesh and B. C. Jain, "Compound growth rate (CGR) of area, production and productivity of papaya in Raipur district of Chhattisgarh," *International Journal of Agriculture, Environment and Biotechnology*, vol. 6, no. 1, pp. 139–143, 2013.

[5] D. Balanagammal, C. R. Ranganathan, and R. Sundaresan, "Forecasting of agricultural scenario in Tamil Nadu—a time series analysis," *Journal of the Indian Society of Agricultural Statistics*, vol. 53, no. 3, pp. 273–286, 2000.

[6] P. Balasubramanian and P. Dhanavanthan, "Seasonal modeling and forecasting of crop production," *Statistics and Applications*, vol. 4, no. 2, pp. 107–118, 2002.

[7] N. Saeed, A. Saeed, M. Zakria, and T. M. Bajwa, "Forecasting of wheat production in Pakistan using ARIMA models," *International Journal of Agricultural Biology*, vol. 2, no. 4, pp. 352–353, 2000.

[8] V. K. Boken, "Forecasting spring wheat yield using time series analysis: a case study for the Canadian prairies," *Agronomy Journal*, vol. 92, no. 6, pp. 1047–1053, 2000.

[9] R. Indira and A. Datta, "Univariate forecasting of state-level agricultural production," *Economic and Political Weekly*, vol. 38, no. 18, pp. 1800–1803, 2003.

[10] K. P. Chandran and Prajneshu,, "Nonparametric regression with jump points methodology for describing country's oilseed yield data," *Journal of the Indian Society of Agricultural Statistics*, vol. 59, no. 2, pp. 126–130, 2005.

[11] K. K. Suresh and S. R. K. Priya, "Forecasting sugarcane yield of tamilnadu using ARIMA models," *Sugar Tech*, vol. 13, no. 1, pp. 23–26, 2011.

[12] Sarika, M. A. Iquebal, and C. Chattopadhyay, "Modelling and forecasting of pigeonpea (Cajanus cajan) production using autoregressive integrated moving average methodology," *Indian Journal of Agricultural Sciences*, vol. 81, no. 6, pp. 520–523, 2011.

[13] H. B. Mann, "Nonparametric tests against trend," *Econometrica*, vol. 13, pp. 245–259, 1945.

[14] M. G. Kendall, *Rank Correlation Measures*, Charles Griffin, London, UK, 1975.

[15] R. M. Hirsch and J. R. Slack, "Nonparametric trend test for seasonal data with serial dependence," *Water Resources Research*, vol. 20, no. 6, pp. 727–732, 1984.

[16] T. Y. Gan, "Hydroclimatic trends and possible climatic warming in the Canadian Prairies," *Water Resources Research*, vol. 34, no. 11, pp. 3009–3015, 1998.

[17] L. Dou, M. Huang, and Y. Hong, "Statistical assessment of the impact of conservation measures on streamflow responses in a watershed of the Loess Plateau, China," *Water Resources Management*, vol. 23, no. 10, pp. 1935–1949, 2009.

[18] G. E. Box and G. M. Jenkins, *Time Series Analysis. Forecasting and Control*, Holden-Day, San Francisco, Calif, USA, 1970.

[19] E. P. Box and G. M. Jenkins, *Time Series Analysis: Forecasting and Control*, Prentice-Hall, Englewood Cliffs, NY, USA, 1976.

[20] H. Akaike, "A new look at the statistical model identification," *IEEE Transactions on Automatic Control*, vol. 19, no. 6, pp. 716–723, 1974.

[21] A. Hirotsugu, "Likelihood and the Bayes procedure," in *Bayesian Statistics*, J. M. Bernardo, M. H. DeGroot, D. V. Lindley et al., Eds., pp. 143–166, University Press, Valencia, Spain, 1980.

# 21

# Effect of Fertilizer Types on the Growth and Yield of *Amaranthus caudatus* in Ilorin, Southern Guinea, Savanna Zone of Nigeria

**Olowoake Adebayo Abayomi and Ojo James Adebayo**

*Department of Crop Production, Kwara State University, PMB 1530, Ilorin, Nigeria*

Correspondence should be addressed to Olowoake Adebayo Abayomi; aolowoake@yahoo.com

Academic Editor: Albino Maggio

Field experiment was carried out at the Teaching and Research Farm of Kwara State University, Malete, Ilorin, to evaluate the effect of compost, organomineral, and inorganic fertilizers on the growth and yield of *Amaranthus caudatus* as well as its residual effects. *Amaranthus* was grown with compost Grade B (unamended compost), organomineral fertilizer Grade A (compost amended with mineral fertilizer), and NPK 15-15-15 and no fertilizer (control). All the treatments except control were applied at the rate of 100 kg N/ha. The results indicated that the *Amaranthus* yield of 18.9 t/ha produced from Grade A was significantly ($P < 0.05$) higher than 17.6 t/ha obtained from NPK fertilizer. Residual effect of *Amaranthus* growth parameters such as plant height, number of leaves, and yield values obtained from Grade A was also significantly ($P < 0.05$) higher than that of NPK, compost, and control values. Thus, organomineral fertilizer could be used in cultivation of *Amaranthus caudatus* in Ilorin and in similar type of soil in similar agroecology.

## 1. Introduction

Vegetable production in Africa is as old as peasant farming though its cultivation is still at the household level with very few farmers producing on a commercial level. This could be due to the fact that crops such as cereals, roots, and tubers and body-building crops like legumes are given much attention. Cereals and tubers form the bulk of food consumed in the tropics but they are deficient in minerals and vitamins compared to the body requirement to guarantee good healthy living [1].

*Amaranths* species is a leafy vegetable in the tropical region of world. It forms a high percentage of the daily intake of leafy vegetables [2]. *Amaranthus caudatus* is grown for its leaves and is among the highly prized leaf vegetables in Nigeria, due to their high nutritional and commercial significance.

There is an increasing awareness of value of leafy vegetable in contributing to balanced diet, particularly in area where animal protein is deficient. Leafy vegetables contribute significantly to the amount of carotene, vitamin c, protein, and minerals particularly calcium [3]. To grow vegetable amaranth it is necessary to know the effect of sources of nitrogen fertilization on its yield because nitrogen was found to be the primary limiting factors of *Amaranths* production [4]. Most Nigerian soils have low nitrogen and the low nitrogen status is usually supplemented with N fertilizer, and the importance of this source has increased over the year. However, the problem with the usage of chemical fertilizer is that while it can lead to high crop yield, it results into pollution of ground water after crop harvest [5]. Another major limitation to the usage of chemical fertilizers is due to the adverse effects they have on plant quality and disease susceptibility. A continual dependence on chemical fertilizers may be accompanied by a fall in organic matter content, increased soil acidity, degradation of soil physical properties and increased rate of erosion due to instability of soil aggregates [6, 7]. One of the ways to maintain or improve the soil fertility is by maintaining its organic matter. This is possible through the use of organic sources of fertilizer.

Research has shown that organic based fertilizers are less leached into ground water than the chemical fertilizer [8]. As a result of this fact, the use of organic based fertilizer has found favour in boosting crop production in Nigeria, because it is cheap and less likely to pollute the ground water as much as chemical fertilizer. It improves soil fertility status as well as increasing the income of farmers via increase in yield.

*Amaranthus*, like a number of other vegetables, requires soil with a high organic content and with adequate nutrient reserve for optimum yield. Hence organomineral fertilizer appears to be reliable organic source of N with a relatively large amount of N needed for growing *Amaranths*. Many researchers have reported that complimentary use of organic fertilizers is able to give the desired higher sustainable crop yields than sole use of inorganic fertilizer [1, 9, 10]. Therefore, the objective of this study is to determine the effect of inorganic, organic, and organomineral fertilizers on the growth and yield of *Amaranthus caudatus* as well as its residual effect.

## 2. Materials and Methods

The experimental site was located at the experimental plot of Kwara State University, Malete (08°42′48.5″N and 004°26′17.9″E), Ilorin, Nigeria, which lies in the southern guinea savanna belt of Nigeria. The annual rainfall in the area is about 1200 mm and temperature varies between 33°C and 34°C during the year, with a distinct dry season from December to March. The Kwara State University land area forms part of the South Western sector of Nigerian basement complex, a zone of basement reactivation and plutonism during the Pan-African orogeny [11]. The soil used for the trial has been cultivated in previous years; there was no record of fertilizer usage. The soil was well drained, of gentle slope. Twenty core soil samples were collected randomly from 0 to 15 cm depth in the site using soil auger thoroughly mixed and the bulk sample taken to the laboratory, air dried, and sieved to pass through a 2 mm screen for soil physical and chemical analysis. The soil sample was analyzed for soil texture, pH, organic carbon, total N, extractable P, exchangeable levels of Ca, Mg, Na and K, and cation exchange capacity. Soil texture was determined by the Bouyoucos hydrometer method [12]. Soil pH was measured electrometrically in a 1 : 2.5 soil-water suspension [13]. Organic carbon by Walkey-Black method [12]. Total nitrogen was determined by the Micro-Kjeldahl method [14], whereas extractable P was determined by Bray 1 method [15]. Exchangeable levels of Ca, Mg, K, and Na were determined by the atomic absorption spectrophotometer following the procedures outlined by Wilde et al. [16]. Micronutrients were extracted with 0.1 EDTA and determined using atomic absorption spectrophotometer.

*2.1. Experimental Layout.* The seeds of *Amaranthus caudatus* were sown on 23rd January, 2012, on a prepared nursery beds, watered regularly using a watering can and checked for seedling emergence. Transplanting of amaranths seedlings into their respective plots in the field took place two weeks after sowing on 6th February, 2012. The site was manually cleared and 12 raised beds were made to conserve the soil and

TABLE 1: Proximate analysis for organomineral fertilizer Grades A and B.

| Nutrient element | Concentration | |
| --- | --- | --- |
| | Grade A | Grade B |
| N (g kg$^{-1}$) | 50.9 | 10.2 |
| P (g kg$^{-1}$) | 44.0 | 7.6 |
| K (g kg$^{-1}$) | 10.8 | 20.9 |
| Mg (g kg$^{-1}$) | 1.9 | 2.4 |
| Ca (g kg$^{-1}$) | 27.7 | 23.4 |
| Na (g kg$^{-1}$) | 3.5 | 2.9 |
| Fe (mg kg$^{-1}$) | 7152.3 | 8915.4 |
| Zn (mg kg$^{-1}$) | 1.5 | 1.9 |
| Mn (mg kg$^{-1}$) | 93.3 | 106.7 |
| Cu (mg kg$^{-1}$) | 14.9 | 16.9 |

Source: Aleshinloye Fertilizer Company, Ibadan, Nigeria.

its nutrient availability. The experimental plot was divided into three blocks each containing four beds. Each bed size was 2 m × 1 m with 1 m alley between plots and blocks. Seedlings were transplanted at 2 weeks after transplanting on beds at spacing 50 cm × 20 cm. The treatments consisted of three fertilizer types: Grade A (compost amended with mineral fertilizer), Grade B (un-amended compost), NPK 15-15-15 and control (no soil additive). Grades A and B are commercial products of Aleshinloye Fertilizer Plant, Ibadan, Oyo State, Nigeria. The results of analyses of the fertilizer are summarized in Table 1. The fertilizers were applied by ring method, 5 cm radius and about 2 cm deep around the *Amaranthus* plant at the rate of 100 kg N/ha [6]. Grades A and B fertilizers were applied a week before transplanting while inorganic fertilizer was applied a day to transplanting. Each vegetable bed contained thirty plants out of which five were randomly tagged for data collection. Collection of data commenced from 2 weeks after transplanting and was done weekly till the fourth weeks. First harvesting was carried out on 5th March, 2012. The experiment was repeated immediately after harvesting without any fertilizer application at the second planting in order to evaluate the residual effects of the fertilizers. The harvesting of vegetables grown in the pot without treatment was done on 16th April, 2012 (6 weeks after planting).

The data taken include plant height, stem girth, number of leaves per plant, fresh root weight, and yield. The data collected were subjected to analysis of variance (ANOVA) and treatment means were separated by Duncan multiple range test (DMRT).

## 3. Results and Discussion

Table 2 shows the result of the physical and chemical analysis of the soil studied. The soil was clay and slightly acidic. The values of total nitrogen, available phosphorus, and potassium of the experimental soil were below the critical values of the soil of Guinea Savanna [17, 18]; this may be due to low soil organic matter. The low soil contents for the major nutrients

TABLE 2: Physicochemical properties of the experimental soil.

| Parameters | Soil test value |
|---|---|
| pH | 6.4 |
| Org. C (g kg$^{-1}$) | 2.0 |
| Total N (g kg$^{-1}$) | 2.44 |
| P Mehlich (m gkg$^{-1}$) | 9 |
| Exchangeable bases (cmol kg$^{-1}$) | |
| K | 1.2 |
| Mg | 3.2 |
| Na | 1.9 |
| Ca | 5.1 |
| Extractable micronutrients (cmol kg$^{-1}$) | |
| Fe | 53 |
| Zn | 13 |
| Mn | 49 |
| Cu | 2.6 |
| Textural class (%) | |
| Sand | 56.8 |
| Silt | 10.0 |
| Clay | 33.2 |
| Textural class | Clay |

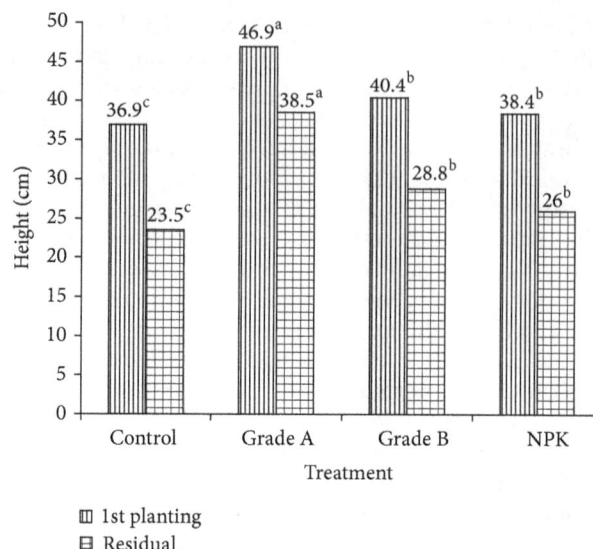

FIGURE 1: Plant height of *Amaranthus caudatus* as influenced by application of organic fertilizers and NPK at 4 weeks after transplanting. Means having the same letter along the columns indicate no significant difference using Duncan's multiple range test at 5% probability level.

signify the need for improvement of *Amaranthus caudatus* performance.

Plant heights as affected by organic amendments and NPK are shown in Figure 1. There were significant differences ($P < 0.05$) in plant height values obtained from the treatment of Grade B and NPK during the first and residual growing period. Results showed that at the end of 4 WAT Grade A, Grade B, and NPK plant height values differed significantly ($P < 0.05$) from control. The highest plant height 46.9 cm and 38.5 cm was obtained from organomineral Grade A both at first and residual planting, respectively.

The highest plant height obtained from Grade A may be probably due to favourable nutrient mineralization of this fertilizer as a result of the influence of the mineral component on the organic content of the compost [10, 19]. The control plants produced the shortest plants as they had to rely on the native soil fertility which from the result of chemical analysis was deficient in nutrients. Figure 2 shows that Grade A significantly enhanced the production of leaves and maintained the trend observed in plant height at both first and residual planting. Changes in the number of leaves are bound to affect the overall performance of *Amaranthus* as the leaves serve as photosynthetic organ of the plant [20].

There were no significant differences ($P < 0.05$) on number of leaves for plot treated with Grade B and NPK. The higher number of leaves produced from organomineral Grade A over the NPK throughout the growing period could be due to sustaining release of nutrients from the former over the latter [1]. Thus, the increase in number of leaves under amended, unamended, and NPK fertilizers application, reconfirmed the role of fertilizer in promoting vegetative growth in leafy vegetables [21].

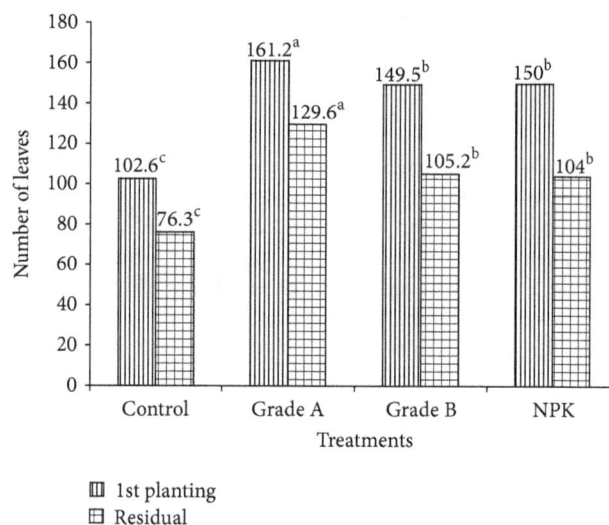

FIGURE 2: Number of leaves of *Amaranthus caudatus* as influenced by application of organic fertilizers and NPK at 4 weeks after transplanting. Means having the same letter along the columns indicate no significant difference using Duncan's multiple range test at 5% probability level.

Table 3 shows the response of stem girth and yield parameters of *Amaranthus caudatus* to application of organic fertilizer and NPK. All the fertilizers applied were found to increase the stem girth, root weight, and yield of *Amaranths* when compared with control. At 4 weeks after transplanting (WAT) stem girth of *Amaranths* with Grades A and B and NPK were significantly ($P < 0.05$) higher than stem girth of

TABLE 3: Effect of fertilizer types on the stem girth and yield parameters of *Amaranthus caudatus* at 4 weeks after transplanting.

| Treatment | Stem girth (cm) | Fresh root weight (g/plant) | Yield (t/ha) |
|---|---|---|---|
| 1st planting | | | |
| Control 0 | $6.0^c$ | $27.4^d$ | $12.1^c$ |
| Grade A | $7.5^a$ | $39.2^a$ | $18.9^a$ |
| Grade B | $6.9^b$ | $34.7^b$ | $17.6^b$ |
| NPK | $7.0^b$ | $31.0^c$ | $17.6^b$ |
| Residual effect | | | |
| Control | $4.4^c$ | $7.8^d$ | $3.6^c$ |
| Grade A | $5.9^a$ | $30^a$ | $7.5^a$ |
| Grade B | $5.0^b$ | $18.3^b$ | $5.0^b$ |
| NPK | $5.1^b$ | $15.3^c$ | $5.3^b$ |

Means having the same letter along the columns indicate no significant difference using Duncan's multiple range test at 5% probability level.

control both at first and residual planting. However, Grade A produced the highest stem girth of 7.5 cm and 5.9 cm which was 17% and 25% greater than the control.

With respect to the effects of different fertilizers on fresh roots and shoot yield at first and residual planting, all the treatments differed significantly ($P < 0.05$) from the control. The fresh roots weight of both first and residual planting were in the order Grade A > Grade B > NPK > control. The increase in root weight suggested that root growth parameters were differentially affected by application of fertilizer types.

However, the trend for shoot yield of *Amaranthus* was in order of Grade A > NPK > Grade B > control. Amaranth yield was least without application of fertilizer. This confirmed the findings of [22], who reported the application of organic, mineral and organomineral fertilizer for enhancement of *Amaranthus cruentus* yield. There were no significant difference ($P < 0.05$) in the yield of *Amaranthus* with plot treated with NPK and Grade B at both plantings. Table 3 also showed that organomineral Grade A had the highest yield of 18.9 t/ha compared to other obtained from fertilizer treatments. Residual effect of *Amaranthus* yield values obtained from organomineral Grade A was also significantly ($P < 0.05$) higher than that of NPK, compost, and control values. This was similar to the works of Akanni et al. [23], Ayeni [24], and Ogunlade et al. [1] who reported that the combinations of organic and mineral fertilizer perform better on the yield of tomato, maize, and *Solanum macrocarpon* than when each of them is solely used.

## 4. Conclusion

This study showed that Grade A fertilizer at 100 kg N/ha gave the best performances in all the *Amaranthus caudatus* growth parameters. Given its superior responses, organomineral Grade A fertilizer could be a very attractive fertilizer alternative particularly for annual crops with short growth cycle such as *Amaranthus*. It is therefore reasonable to recommend the use of organomineral Grade A fertilizer in the cultivation

of *Amaranthus caudatus* in Ilorin and similar type of soil in similar agroecology.

## Conflict of Interests

The authors declare that there is no conflict of interests regarding the publication of this paper.

## Acknowledgment

The authors are grateful to Centre for Community Development Unit (CCD), Kwara State University, Malete, Ilorin, Nigeria, for providing the fund for conducting this research.

## References

[1] M. O. Ogunlade, E. A. Adeyemi, D. O. Ogunleti, and P. S. Ibiyomi, "Effect of cocoa pod husk, urea fortified cocoa pod husk and NPK fertilizers on the growth and yield of *Solanum macrocarpon* cultivation," *International Journal of Organic Agriculture Research and Development*, vol. 3, pp. 1–8, 2011.

[2] M. C. Palada and L. C. Chang, "Suggested cultural practices for vegetable amaranths," International Cooperators' Guide 03-552, Asian Vegetable Research and Development Centre, 2003.

[3] H. D. Tindale, *Vegetable in the Tropics*, Macmillan, New York, NY, USA, 2nd edition, 1983.

[4] A. Pospišil, M. Pospišil, B. Varga, and Z. Svečnjak, "Grain yield and protein concentration of two amaranth species (*Amaranthus* spp.) as influenced by the nitrogen fertilization," *European Journal of Agronomy*, vol. 25, no. 3, pp. 250–253, 2006.

[5] W. B. Gordon, D. A. Whitney, and R. J. Raney, "Nitrogen management in furrow irrigated, ridge tilled corn," *Journal of Production Agriculture*, vol. 6, pp. 213–217, 1993.

[6] O. Adeoluwa and O. O. Adeogun, "Evaluation of feather as organic fertilizers on Amaranthus (*Amaranthus caudatus*)," in *Proceedings of the 1st Technical Workshop on Organic Agriculture Conference*, pp. 16–19, Ladoke Akintola University of Technology, Ogbomoso, Nigeria, 2010.

[7] A. A. Olowoake and G. O. Adeoye, "Comparative efficacy of NPK fertilizer and composted organic residues on growth, nutrient absorption and dry matter accumulation in maize," *International Journal of Organic Agriculture Research and Development*, vol. 2, pp. 43–53, 2010.

[8] M. K. C. Sridhar and G. O. Adeoye, "Organomineral fertilizer from urban wastes," *The Nigerian Field*, vol. 68, pp. 91–111, 2003.

[9] W. B. Akanbi, A. O. Togun, J. A. Adediran, and E. A. O. Ilupeju, "Growth, dry matter and fruit yields components of okra under organic and inorganic sources of nutrients," *American-Eurasian Journal of Sustainable Agriculture*, vol. 4, no. 1, pp. 1–13, 2010.

[10] G. O. Adeoye, M. K. C. Sridhar, O. O. Adeoluwa, M. Oyekunle, E. A. Makinde, and A. A. Olowoake, "Comparative evaluation of organo-mineral fertilizer (OMF) and mineral fertilizer (NPK) on yield and quality of maize (*Zea mays* (L) Moench)," *Nigerian Journal of Soil Science*, vol. 18, pp. 141–147, 2008.

[11] J. O. Olaniyan, *An evaluation of the soil map of Nigeria for land use planning in Kwara State [Ph.D. thesis]*, University of Ibadan, Ibadan, Nigeria, 2003.

[12] A. S. R. Juo, *Selected Methods for Soil and Plant Analysis*, Manual Series no. 1, IITA, Ibadan, Nigeria, 2nd edition, 1978.

[13] E. O. McLean, "Soil pH and Lime requirement," *Agronomy*, vol. 9, pp. 199–223, 1982.

[14] J. M. Bremner and C. S. Mulvaney, "Nitrogen total," in *Methods of Soil Analysis Part 2*, A. L. Page and R. H. Miller, Eds., vol. 9 of *Agronomy Monograph*, pp. 595–624, ASA, SSSA, Madison, Wis, USA, 1982.

[15] R. H. Bray and L. T. Kurtz, "Determination of total organic and available forms of phosphorus in soils," *Soil Science*, vol. 39, pp. 39–45, 1945.

[16] S. A. Wilde, R. B. Corey, V. G. Iver, and G. K. Voigt, *Soil and Plant Analysis for Tree Culture*, Oxford and IBH Publishing, New Delhi, India, 1979.

[17] M. O. Aduloju, "Acid extractable micronutrients (Mn and Zn) in selected soils of vegetables producing areas of Kwara State, Nigeria," *Nigeria Journal of Horticultural Science*, vol. 9, pp. 116–199, 2004.

[18] O. J. Ayodele, "Soil fertility management for the production of fruits and vegetables in South Western Nigeria," *Acta Horticulture*, vol. 128, pp. 237–242, 1983.

[19] L. J. Sikora and N. K. Enkiri, "Efficiency of compost-fertilizer blends compared with fertilizer alone," *Soil Science*, vol. 165, no. 5, pp. 444–451, 2000.

[20] K. E. Law-Ogbomo and S. O. Ajayi, "Growth and yield performance of *Amaranthus cruentus* influenced by planting density and poultry manure application," *Notulae Botanicae Horti Agrobotanici Cluj-Napoca*, vol. 37, no. 2, pp. 195–199, 2009.

[21] H. Tijani-Eniola, O. W. F. Nwagwu, and O. P. Aiyelari, "Response of *Celosia argentea* L. to different nitrogen sources and frequency of harvest," in *Proceeding of the 18th HORTSON Conference*, vol. 28, pp. 151–160, AR/ABU, Zaria, Nigeria, May 2000.

[22] A. A. Olowoake, "Influence of organic, mineral and organomineral fertilizers on growth, yield, and soil properties in grain amaranth (*Amaranthus cruentus* L.)," *Journal of Organics*, vol. 1, no. 1, pp. 39–47, 2014.

[23] D. I. Akanni, S. O. Ojeniyi, and M. A. Awodun, "Soil properties, growth yield and nutrient content of Maize, Pepper and Amaranthus as influenced by organic and organomineral fertilizer," *Journal of Agricultural Science and Technology*, vol. 1, pp. 1074–1078, 2011.

[24] L. S. Ayeni, "Integrated application of cocoa pod ash and NPK fertilizer on soil chemical properties and yield of tomato," *American-Eurasian Journal of Sustainable Agriculture*, vol. 2, no. 3, pp. 333–337, 2008.

# 22

# DNA Barcoding for Minor Crops and Food Traceability

**Andrea Galimberti, Massimo Labra, Anna Sandionigi, Antonia Bruno, Valerio Mezzasalma, and Fabrizio De Mattia**

*ZooPlantLab, Dipartimento di Biotecnologie e Bioscienze, Università degli Studi di Milano-Bicocca, Piazza della Scienza 2, 20126 Milano, Italy*

Correspondence should be addressed to Fabrizio De Mattia; fabrizio.demattia@unimib.it

Academic Editor: Pawan L. Kulwal

This outlook paper addresses the problem of the traceability of minor crops. These kinds of cultivations consist in a large number of plants locally distributed with a modest production in terms of cultivated acreage and quantity of final product. Because of globalization, the diffusion of minor crops is increasing due to their benefit for human health or their use as food supplements. Such a phenomenon implies a major risk for species substitution or uncontrolled admixture of manufactured plant products with severe consequences for the health of consumers. The need for a reliable identification system is therefore essential to evaluate the quality and provenance of minor agricultural products. DNA-based techniques can help in achieving this mission. In particular, the DNA barcoding approach has gained a role of primary importance thanks to its universality and versatility. Here, we present the advantages in the use of DNA barcoding for the characterization and traceability of minor crops based on our previous or ongoing studies at the ZooPlantLab (Milan, Italy). We also discuss how DNA barcoding may potentially be transferred from the laboratory to the food supply chain, from field to table.

## 1. DNA Barcoding for Plant Identification

Plants as primary producers are the basis of human nutrition from time immemorial. It is estimated that about 7,000 species of plants have been cultivated for consumption in human history (FAO data) and a large number of cultivars and varieties are also recognized. The Commission on Genetic Resources for Food and Agriculture (http://www.fao.org/nr/cgrfa/cthemes/plants/en/) estimated that 30 crops are usually referred currently as major agricultural products since they provide 95% of human food energy needs (e.g., rice, wheat, maize, and potato). These resources are widely monitored and well characterized with the analysis of DNA markers specifically developed for each cultivar (see, e.g., [1–3]). On the contrary, reliable characterization tools for the minor varieties are far from being defined. Minor crops include plants for food, pharmaceutical, cosmetic, and ornamental purposes with a modest production in terms of cultivated acreage and quantity of final product [4]. There are no fixed standard values to define a minor crop; however, conventionally, all the local varieties could be placed in this category. Most of these species or varieties show peculiar traits from the alimentary, pharmaceutical, or ornamental points of view. Some examples of minor crops that are now widely cultivated and worldwide distributed are Goji (*Lycium barbarum* L. [5]), Chokeberry (*Aronia melanocarpa* (Michx.), [6]), Peach Palm (*Bactris gasipaes* Kunth [7]), Teff (*Eragrostis tef* (Zucc.) [8]), and Okra (*Abelmoschus esculentus* (L.) Moench [9]). A large number of minor crops were usually produced and consumed locally [10] but, nowadays, the continuous demand by developed countries for identifying new active metabolites for human health and nutrition has increased their diffusion at global level [11–14]. This phenomenon implies a major risk for species substitution or uncontrolled admixture of manufactured plant products. Substitution or adulteration can be deliberate (e.g., to maximize financial gains) or inadvertent (e.g., due to an insufficient knowledge by farmers) but they can have serious consequences for consumers at any rate [14–19].

Given these premises, it is clear that the definition of a reliable traceability system is an aspect of major concern when plants, parts of plants, or plant extracts are used in food industry. The need for an unequivocal identification is also essential to start quality assurance procedures for agricultural products, to authenticate their geographical provenance (in the case of protected designation of origin), and to prevent commercial frauds and adulteration cases.

Agricultural products are subjected to strong processing and manufacturing before they are released as final products to the consumer. These processes alter the plant structure, thereby impeding the use of morphological characters to identify most of the agricultural products. To overcome this limit, the analysis of proteins and/or DNA is nowadays used as the main tool for plant traceability. However, although chemical or protein-based approaches are useful in characterizing the composition of fresh products, these methods can be biased by several factors such as the strong food manufacturing processes, the limited number of detectable isozymes, or the high tissue and developmental stage specificity of the markers [20]. DNA markers are more informative than protein or chemical based methods because DNA better resists industrial processes such as shredding, boiling, pressure cooking, or transformations mediated by chemical agents (see, e.g., [18, 21, 22]). This property allows a successful identification of plant material, even when it is present in small traces [23, 24]. Moreover, the availability of advanced technologies and efficient commercial kits for DNA extraction permits obtaining an acceptable yield of genetic material from processed or degraded plant material [25].

As a consequence, DNA markers have rapidly become the most used tools in the genetic analyses of crops and cultivars, as well as in the tracking and certification of the raw materials in food industry processes [26–32]. PCR-based methods are more sensitive and faster than other technologies in characterizing agricultural products [1–3]. Among these, discontinuous molecular markers such as RAPDs, AFLPs, and their variants (e.g., ISSR, SSAP) have been successfully adopted for the characterization of crop species [24]. Moreover, sequencing-based systems such as single nucleotide polymorphisms (SNPs) and simple sequence repeats (SSRs) are also used because of their high level of polymorphism and high reproducibility [30]. However, being highly species specific, these approaches require access to the correct DNA sequence of the organisms and their application is often limited to a single species.

In the last decade, DNA barcoding was proposed as a universal DNA-based tool for species identification [33]. The name "DNA barcoding" figuratively refers to the way an infrared scanner univocally identifies a product by using the stripes of the universal product code (UPC). At the same time, this approach is based on the analysis of the variability within one or a few standard regions of the genome called "DNA barcode/s" [33]. The rationale of the method is that the DNA barcoding sequence/s univocally corresponds to each species (i.e., low intraspecific variability) but largely differs between taxa (i.e., high interspecific variability) [33, 34]. DNA barcoding has the advantage of combining three important innovations: molecularization of the identification approach (i.e., the investigation of DNA variability to differentiate taxa), standardization of the process (from sample collection to the analysis of molecular results), and computerization

(i.e., the not redundant transposition of the data using informatics) [34].

Several plastidial and nuclear regions have been proposed as barcode regions for plants [35–37] and some of them are now used for the identification of crop species, as recently reviewed by [38]. In 2009, the Plant Working Group of CBOL (consortium for the barcode of life) defined a standard core-barcode panel of markers based on the combination of portions of two coding plastidial regions: *matK* and *rbcL* [39, 40]. Despite their high universality in terms of amplification and sequencing success, the analysis of these coding regions fails in some cases due to the interspecific sharing of sequences [41]. Internal transcribed spacer regions of nuclear ribosomal DNA (ITS) were recommended as additional marker being highly variable in angiosperms [40]. ITS works well in many plant groups but, in some cases, incomplete concerted evolution and intraindividual variation make it unsuitable as universal plant barcode [40]. However, the combination of *matK* and *rbcL* with the plastidial intergenic noncoding region *trnH-psbA* increases the identification performance of DNA barcoding. As a consequence, the use of *trnH-psbA* is growing due to its easy amplification, and its high genetic variability among closely related taxa [15, 35, 42].

At the University of Milano-Bicocca (Milan, Italy), the ZooPlantLab group (http://www.zooplantlab.btbs.unimib.it/) is one of the most active centers where DNA barcoding is used as a universal traceability system. The ZooPlantLab research team investigates concrete problems dealing with agricultural production of minor crops by transferring the analytical pipeline from the laboratory to food supply chain. This approach aims to overcome technical traceability problems in order to offer solid solutions to the market.

In the following sections, we present some of the potential applications and advantages of DNA barcoding for the identification and traceability along the food supply chain of minor crops. We also examine the most innovative approaches dealing with DNA barcoding that have been recently adopted to characterize these kinds of agricultural products.

## 2. Traceability of Minor Crops in the Supply Chain: The Case of Spices

Spices represent a clear example of minor crops. Most of these belong to Lamiaceae, a large family of 264 genera and almost 7,000 described species [78] characterized by aromatic oils and secondary metabolites. Thanks to their peculiar chemical profiles, these plants are commonly used as flavor for cooking, essences for cosmetics, and active components in medicines. Given their economical importance, many members of Lamiaceae have been investigated widely with different approaches ranging from morphology to chemistry and genetics in order to characterize their variability and improve the quality of cultivated varieties [25, 26, 79, 80].

Although some species showed distinctive morphological traits, this family encompasses many critical genera such as *Thymus* [43], where differences among closely related taxa are limited to few minor morphological characters. However, morphology could be ineffective for tracing spices along the supply chain (i.e., from the crop cultivation sites to the final products) which usually encompasses strong manufacturing processes such as crushing, powdering, or aqueous/alcoholic extraction of plant material.

International agencies such as the American Spice Trade Association (ASTA, http://www.astaspice.org) and the European Spice Association (ESA, http://www.esa-spices.org/) support the characterization of the phytochemical profile to assess the quality of herbs and spices. The evaluation of chemical characteristics is essential to standardize the industrial production of spices-derived products; however, in most cases, the analysis of chemical compounds is not able to univocally identify the original plants at the species level [26]. For this reason, we proposed the DNA barcoding approach as a universal and suitable tool to characterize and trace aromatic species. DNA analyses were conducted starting from different plant portions [22] or their derived products (e.g., oils, extracts) stored at different conditions (i.e., dried, frozen). In our study [22], we investigated 6 major groups of cooking spices (i.e., mint, basil, oregano, sage, thyme, and rosemary) also including their most relevant cultivars and hybrids. We collected samples at different stages of the industrial supply chain starting from seeds and plants cultivated by private farmers or in garden centers to commercial dried spices or other manufactured products. We also tested the performances of DNA barcoding starting from plant extracts. A good yield of high quality DNA was obtained through extraction protocols from all of the considered samples and then used for the next steps of the analysis (i.e., PCR and sequencing). A sufficient amount of DNA was also extracted from several of the plant extracts (Labra M., unpublished data) by using commercial kits. This first result confirmed that the industrial processes to transform the raw plant material such as drying, crushing, and aqueous or alcoholic extractions do not excessively degrade DNA. Among the four tested DNA barcoding regions (i.e., *rbcL*, *matK*, *trnH-psbA*, and *rpoB*), the *trnH-psbA* ranked the first in genetic divergence values among species, followed by *matK* and *rbcL*. On the contrary, *rpoB* showed the lowest sequence divergence among the tested taxa (see [22] for further details).

Our results partially supported the guidelines provided by the CBOL [40]. Indeed, the two core-barcode markers (i.e., *matK* + *rbcL*) properly assigned the tested spices to the expected genus and, in most cases, they also reached the species level. However, the highest identification performances were achieved by using the additional *trnH-psbA* barcode region. A clear example is that of basil (genus *Ocimum*), a group consisting of 30–160 species with many recognized cultivars [81]. In our study, exclusive *trnH-psbA* haplotypes, were found for almost all the tested cultivars, providing a reliable system for their identification. This result deserves to be highlighted because it is one of the first pieces of evidence supporting the usefulness of DNA barcoding in discriminating organisms at a taxonomic level lower than the species one.

Other important data revealed by our analyses concerned the capability of DNA barcoding to identify parental and hybrid species in some members of Lamiaceae. An example is represented by the case of peppermint (*M. piperita* L.),

a sterile hybrid between *M. aquatica* L. × *M. spicata* L. [82, 83]. The plastidial markers used in this study confirmed that *M. spicata* L. is the maternal parental of *M. piperita* L. because both taxa showed the same DNA profile. However, to confirm definitively the hybrid origin of *M. piperita* L. and to identify the exact parental inheritance, the ITS2 codominant marker was sequenced (Labra M., unpublished data).

On the whole, the most relevant result of our work consisted in the assessment of the universality of DNA barcoding in a context of minor crops traceability. Using a single primer combination for each one of the few DNA barcoding markers and following standard laboratory protocols, it is possible to recognize the original species starting from different plant portions or derived processed materials. The same approach is also useful for validating several other herbal products commonly distributed on the market such as tea [50], saffron [44, 84], ginseng [69], black pepper [59], and many others (see also Table 1). These cases clearly emphasize the high versatility of DNA barcoding. It is an authentic functional tool for molecular traceability of agricultural products, as most of the minor crops have not yet been characterized with private markers such as SSR or SNP in order to allow a reliable DNA fingerprinting system. Moreover, DNA barcoding does not require any previous knowledge of the plant genome for the investigated species and the analytical procedures can be easily adopted by any laboratory equipped for molecular biology.

## 3. Commercial Frauds and Dangerous Substitutions

Nowadays, the global diffusion of several minor crops in the absence of suitable traceability protocols is leading to frequent cases of plant substitution and inadvertent or deliberate adulteration. There are several documented examples of commercial frauds where minor crops were substituted with related taxa showing a higher productivity or biomass but without the agronomical and nutritional characteristics of the original species/cultivars [27, 85, 86] (see also Table 1). Astounding cases of this phenomenon were observed for some of the most common spices such as the Mediterranean oregano adulterated with *Cistus incanus* L., *Rubus caesius* L. [87–89] and saffron substituted with *Crocus vernus* (L.) Hill, *Carthamus,* and *Curcuma* [19, 44, 84]. In this context, the use of DNA barcoding can be decisive because it can not only verify the presence/absence of the original species, but also identify the nature of the replaced species. One of the most striking substitution cases ever revealed by our investigations refers to fish meat (e.g., sold as slices, fillets, blocks, surimi, fish sticks, and fins). In this product category, the manufacturing processes often lead to the loss of any morphological diagnostic feature that may correctly identify the original species. In our molecular investigation [90], we documented the frequent substitutions of Palombo (i.e., the Italian vernacular name for *Mustelus mustelus* and *Mustelus asterias*) with other less valuable shark species. Our test showed that about 80% of the screened fish products did not correspond to these two species but to other species or genera,

TABLE 1: List of studies dealing with DNA barcoding identification of minor crops.

| DNA barcoding application | Minor crop/food product | Notes | References |
|---|---|---|---|
| Traceability of minor crops in the food supply chain | Aromatic plants | Identification of spices from fresh samples to manufactured or processed products | [22, 43–47] |
| | Legumes | Legume seeds traceability | [48, 49] |
| | Herbal infusions | Traceability of tea products | [50] |
| | Fruit | Identification and traceability of mango | [51] |
| | | Identification of *Citrus* species | [52] |
| | | Identification of Goji | [53] |
| | | Identification of berries | [54] |
| | Vegetables | Identification of *Capsicum* cultivars | [55] |
| | Medical plant and food supplements | Traceability of medicinal plants | [56–58] |
| Commercial frauds and dangerous substitutions | Aromatic products | Identification of spices adulterants | [59, 60] |
| | Vegetal flour | Identification of buckwheat in commercial foodstuffs | [61] |
| | Legumes | Seed admixture and adulteration | [62, 63] |
| | Fruit | Identification and adulteration of fruit-based products | [60, 64] |
| | Oil | Oil adulteration | [65] |
| | Medicinal plants/food supplements | Dangerous substitution of *Solanum lyratum* with *Aristolochia mollissima* | [66] |
| | | Adulteration of herbal products | [67] |
| | Tea | Contamination of tea products | [68] |
| Molecular identification of minor crops in complex matrices | Natural health products | Identification of pharmaceutical plants in commercial products | [69] |
| | Juice and vegetal beverages | Juice authentication | [70–72] |
| | Honey | Identification of pollen and plant residuals | [73] |
| | Jams or yogurt | Identification of fruit in commercial products | [74, 75] |
| | Food supplements | Identification of allergenic plants | [76, 77] |

some of which are fished or marketed illegally. Starting from this experience, we tested the usefulness of DNA barcoding to evaluate the contamination of plant-based products. For example, in a pilot study on spices conducted by our group, we detected contaminant DNA in commercial samples of sage (i.e., *Salvia*) produced by local farmers. This DNA corresponded to species belonging to the family Poaceae (i.e., *Festuca* sp.). We hypothesized that these contaminant plants were accidentally grown together with the sage and fragments of them were erroneously collected, shredded, and consequently admixed to the final commercial products (Labra M., unpublished data). These conditions are dangerous if the contaminant taxon is toxic or allergenic for humans. A typical example is that of nuts and almonds which cause allergies in many people [91]. Several commercial foodstuffs (e.g., bakery, pastry, and snacks) showed contamination by these plants (see, e.g., [76, 92]). Also in this case, DNA barcoding acts as a very versatile tool, allowing the detection of both species (and many other allergenic taxa) also when they were present in traces [76].

Similarly, DNA barcoding can be efficient in identifying those plant species causing intoxication or poisoning in consumers. In recent years, plant exposures are among the most

frequent poisoning cases reported by poison control centers [15, 93, 94]. Many of these are due to inadvertent misidentification as reported in [95] where the authors documented the exchange of spontaneous salad (*Lactuca alpine* (L.) Wallr.) with *Aconitum* spp. and wild garlic (*Allium ursinum* L.) with *Colchicum* sp. Both *Aconitum* and *Colchicum* contain toxic metabolites with severe consequences for human health after ingestion [96, 97]. Our analysis showed that DNA barcoding allowed us to detect the presence of poisonous plants and identify specific sequence-characterized amplified regions (SCARs) useful in a real-time PCR approach for rapid diagnosis in poison centers [60].

## 4. Plant Molecular Identification in Complex Matrices

Most food and cosmetic products are made up of a pool of plant species, major and minor crops, and spontaneous species. These are considered complex matrices [31] and, to establish traceability, the availability of universal tools able to univocally identify each plant species is needed. We underline that the assumptions for which DNA barcoding region(s) and the primers used are universal [33] imply that when the method is applied to complex matrices, PCR amplifications will produce several DNA barcoding amplicons, corresponding to different species. For this reason we tested this diagnostic method to identify the plant composition on different mixed products such as the commercial potpourris [14] and multiflower honeys (Bruni et al., submitted). For most of these herbal products, a detailed list of ingredients is not reported on the label; as a consequence, it is difficult to understand which species are used for their preparation and especially how safe these are for human health. In the case of potpourris, our results showed that the principal ingredients are simple aromatic plants (e.g., species of Lamiaceae) which are sometimes edible (e.g., *Salvia officinalis* L.; *Ocimum basilicum* L.) or ornamental (e.g., *Salvia splendens* Sellow ex J.A. Schultes, *Lavandula angustifolia* Miller) without negative effects on human health. In other cases these products revealed the presence of plants which produce natural toxic metabolites, such as alkaloids that are dangerous for human health [14, 98–100]. However, the main critical element for the identification of plant-based complex matrices is the availability of DNA barcoding reference databases [101, 102]. To date, the Barcode of Life Data System (i.e., BOLD, http://www.boldsystems.org/ [103]) contains 52,767 plant DNA sequences although several minor crops and local varieties are missing. Recent works, edited by our laboratory and other groups, highlighted the need for dedicated reference archives of DNA barcoding data for these kinds of plants [31, 67, 101, 102, 104, 105]. In another study, we demonstrated that, starting from a robust local database, it is possible to characterize the pollen composition of multiflower honey, one of the most complex food matrices. Our tests, conducted on honey samples produced in the Italian Alps, showed the conspicuous presence of endemic taxa. This result allowed us to assess not only the composition of honeys, but also their geographical origin (Bruni et al., submitted). See also Table 1 for further examples.

In comparison to agricultural products made by a single plant, the molecular characterization of complex matrices requires some technical advances, especially concerning the sequencing step. The traditional DNA-sequencing method [106] can only be adopted for direct sequencing of amplicons deriving from a single taxon. Complex matrices often contain mixtures of DNA from many individuals belonging to a certain taxonomic group (e.g., angiosperms) and DNA amplification may generate amplicons of the same size for a certain locus (e.g., a DNA barcode region for plant identification), therefore impeding direct sequencing with the Sanger approach. A possible solution could be the adoption of a preliminary cloning step to separate single DNA templates but this strategy has its own limitations (e.g., high costs) and can introduce biases (e.g., low representation of the sequenced colonies in the case of highly complex matrices [107, 108]). Recovering DNA sequences from the tens to thousands of specimens present in a complex food matrix requires the ability to read DNA from multiple templates in parallel. Since 2005, advances in the field of next-generation sequencing (NGS) technologies [109] have been helping in addressing this issue with ever-lowering costs. To date, several models of high-throughput sequencing devices have been commercially introduced based on different chemistries and detection techniques [108]. NGS technologies can generate up to tens of millions of sequencing reads in parallel and these approaches are being used in a variety of applications, including the traceability of food matrices containing agricultural products [73, 74, 110].

In conclusion, given the rapid evolution and standardization of NGS advances, we think that a universal approach such as DNA barcoding combined with them can offer a new opportunity for the traceability of minor crops from field to table.

## Conflict of Interests

The authors declare that there is no conflict of interests regarding the publication of this paper.

## References

[1] J. S. C. Smith, E. C. L. Chin, H. Shu et al., "An evaluation of the utility of SSR loci as molecular markers in maize (*Zea mays* L.): comparisons with data from RFLPS and pedigree," *Theoretical and Applied Genetics*, vol. 95, no. 1-2, pp. 163–173, 1997.

[2] F. De Mattia, G. Lovicu, J. Tardaguila et al., "Genetic relationships between Sardinian and Spanish viticulture: the case of "Cannonau" and 'Garnacha'," *Journal of Horticultural Science and Biotechnology*, vol. 84, no. 1, pp. 65–71, 2009.

[3] S. R. McCouch, K. Zhao, M. Wright et al., "Development of genome-wide SNP assays for rice," *Breeding Science*, vol. 60, no. 5, pp. 524–535, 2010.

[4] J. Womach, *Agriculture: A Glossary of Terms, Programs, and Laws*, Congressional Research Service, Library of Congress, Washington, DC, USA, 2005.

[5] H. Amagase and N. R. Farnsworth, "A review of botanical characteristics, phytochemistry, clinical relevance in efficacy and safety of *Lycium barbarum* fruit (Goji)," *Food Research International*, vol. 44, no. 7, pp. 1702–1717, 2011.

[6] S. E. Kulling and H. M. Rawel, "Chokeberry (*Aronia melanocarpa*)-a review on the characteristic components and potential health effects," *Planta Medica*, vol. 74, no. 13, pp. 1625–1634, 2008.

[7] J. M. Urpí, J. C. Weber, and C. R. Clement, *Peach Palm, Bactris Gasipaes Kunth*, vol. 20, Bioversity international, Rome, Italy, 1997.

[8] S. Ketema, *Tef-Eragrostis tef (Zucc.)*, vol. 12, Bioversity international, Rome, Italy, 1997.

[9] M. Camciuc, M. Deplagne, G. Vilarem, and A. Gaset, "Okra—*Abelmoschus esculentus* L. (Moench.) a crop with economic potential for set aside acreage in France," *Industrial Crops and Products*, vol. 7, no. 2-3, pp. 257–264, 1998.

[10] K. T. Moe, S. Kwon, and Y. Park, "Trends in genomics and molecular marker systems for the development of some underutilized crops," *Genes and Genomics*, vol. 34, no. 5, pp. 451–466, 2012.

[11] E. Ernst, "The efficacy of herbal medicine-an overview," *Fundamental and Clinical Pharmacology*, vol. 19, no. 4, pp. 405–409, 2005.

[12] H. A. Tindle, R. B. Davis, R. S. Phillips, and D. M. Eisenberg, "Trends in use of complementary and alternative medicine

by us adults: 1997–2002," *Alternative Therapies in Health and Medicine*, vol. 11, no. 1, pp. 42–49, 2005.

[13] G. Heubl, "New aspects of DNA-based authentication of Chinese medicinal plants by molecular biological techniques," *Planta Medica*, vol. 76, no. 17, pp. 1963–1974, 2010.

[14] L. Cornara, B. Borghesi, C. Canali et al., "Smart drugs: green shuttle or real drug?" *International Journal of Legal Medicine*, vol. 127, no. 6, pp. 1109–1123, 2013.

[15] I. Bruni, F. De Mattia, A. Galimberti et al., "Identification of poisonous plants by DNA barcoding approach," *International Journal of Legal Medicine*, vol. 124, no. 6, pp. 595–603, 2010.

[16] S. L. Taylor and J. L. Baumert, "Cross-contamination of foods and implications for food allergic patients," *Current Allergy and Asthma Reports*, vol. 10, no. 4, pp. 265–270, 2010.

[17] Z. P. Zeng and J. G. Jiang, "Analysis of the adverse reactions induced by natural product-derived drugs," *British Journal of Pharmacology*, vol. 159, no. 7, pp. 1374–1391, 2010.

[18] J. Costa, I. Mafra, J. S. Amaral, and M. B. P. P. Oliveira, "Detection of genetically modified soybean DNA in refined vegetable oils," *European Food Research and Technology*, vol. 230, no. 6, pp. 915–923, 2010.

[19] S. Babaei, M. Talebi, and M. Bahar, "Developing an SCAR and ITS reliable multiplex PCR-based assay forsafflower adulterant detection in saffron samples," *Food Control*, vol. 35, no. 1, pp. 323–328, 2014.

[20] Y. J. Park, J. K. Lee, and N. S. Kim, "Simple sequence repeat polymorphisms (SSRPs) for evaluation of molecular diversity and germplasm classification of minor crops," *Molecules*, vol. 14, no. 11, pp. 4546–4569, 2009.

[21] S. Soares, I. Mafra, J. S. Amaral, and M. B. P. P. Oliveira, "A PCR assay to detect trace amounts of soybean in meat sausages," *International Journal of Food Science and Technology*, vol. 45, no. 12, pp. 2581–2588, 2010.

[22] F. De Mattia, I. Bruni, A. Galimberti, F. Cattaneo, M. Casiraghi, and M. Labra, "A comparative study of different DNA barcoding markers for the identification of some members of Lamiacaea," *Food Research International*, vol. 44, no. 3, pp. 693–702, 2011.

[23] A. K. Lockley and R. G. Bardsley, "DNA-based methods for food authentication," *Trends in Food Science and Technology*, vol. 11, no. 2, pp. 67–77, 2000.

[24] I. Mafra, I. M. Ferreira, and M. B. P. Oliveira, "Food authentication by PCR-based methods," *European Food Research and Technology*, vol. 227, no. 3, pp. 649–665, 2008.

[25] J. Novak, S. Grausgruber-Gröger, and B. Lukas, "DNA-based authentication of plant extracts," *Food Research International*, vol. 40, no. 3, pp. 388–392, 2007.

[26] M. Labra, M. Miele, B. Ledda, F. Grassi, M. Mazzei, and F. Sala, "Morphological characterization, essential oil composition and DNA genotyping of Ocimum basilicum L. cultivars," *Plant Science*, vol. 167, no. 4, pp. 725–731, 2004.

[27] M. Woolfe and S. Primrose, "Food forensics: using DNA technology to combat misdescription and fraud," *Trends in Biotechnology*, vol. 22, no. 5, pp. 222–226, 2004.

[28] S. Imazio, M. Labra, F. Grassi, A. Scienza, and O. Failla, "Chloroplast microsatellites to investigate the origin of grapevine," *Genetic Resources and Crop Evolution*, vol. 53, no. 5, pp. 1003–1011, 2006.

[29] F. De Mattia, F. Grassi, S. Imazio, and M. Labra, "Chloroplast and nuclear DNA markers to characterize cultivated and spontaneous *Ribes*," *Plant Biosystems*, vol. 142, no. 2, pp. 204–212, 2008.

[30] P. Kumar, V. K. Gupta, A. K. Misra, D. R. Modi, and B. K. Pandey, "Potential of molecular markers in plant biotechnology," *Plant Omics: Journal of Plant Molecular Biology & Omics*, vol. 2, no. 4, pp. 141–162, 2009.

[31] A. Galimberti, F. De Mattia, A. Losa et al., "DNA barcoding as a new tool for food traceability," *Food Research International*, vol. 50, no. 1, pp. 55–63, 2013.

[32] H. Chuang, H. Lur, K. Hwu, and M. Chang, "Authentication of domestic Taiwan rice varieties based on fingerprinting analysis of microsatellite DNA markers," *Botanical Studies*, vol. 52, no. 4, pp. 393–405, 2011.

[33] P. D. Hebert, S. Ratnasingham, and J. R. de Waard, "Barcoding animal life: cytochrome c oxidase subunit 1 divergences among closely related species," *Proceedings of the Royal Society B: Biological Sciences*, vol. 270, supplement 1, pp. S96–S99, 2003.

[34] M. Casiraghi, M. Labra, E. Ferri, A. Galimberti, and F. de Mattia, "DNA barcoding: a six-question tour to improve users'awareness about the method," *Briefings in Bioinformatics*, vol. 11, no. 4, Article ID bbq003, pp. 440–453, 2010.

[35] J. Shaw, E. B. Lickey, E. E. Schilling, and R. L. Small, "Comparison of whole chloroplast genome sequences to choose noncoding regions for phylogenetic studies in angiosperms: the Tortoise and the hare III," *American Journal of Botany*, vol. 94, no. 3, pp. 275–288, 2007.

[36] A. J. Fazekas, K. S. Burgess, P. R. Kesanakurti et al., "Multiple multilocus DNA barcodes from the plastid genome discriminate plant species equally well," *PLoS ONE*, vol. 3, no. 7, Article ID e2802, 2008.

[37] A. J. Fazekas, P. R. Kesanakurti, K. S. Burgess et al., "Are plant species inherently harder to discriminate than animal species using DNA barcoding markers?" *Molecular Ecology Resources*, vol. 9, no. 1, pp. 130–139, 2009.

[38] D. Mathew, "Biotechnology," in *Horticulture: Methods and Applications*, K. V. Peter, Ed., chapter 2, pp. 25–50, New India Publishing Agency, New delhi, India, 1st edition, 2014.

[39] M. L. Hollingsworth, A. Andra Clark, L. L. Forrest et al., "Selecting barcoding loci for plants: evaluation of seven candidate loci with species-level sampling in three divergent groups of land plants," *Molecular Ecology Resources*, vol. 9, no. 2, pp. 439–457, 2009.

[40] P. M. Hollingsworth, S. W. Graham, and D. P. Little, "Choosing and using a plant DNA barcode," *PLoS ONE*, vol. 6, no. 5, Article ID e19254, 2011.

[41] I. Bruni, F. De Mattia, S. Martellos et al., "DNA barcoding as an effective tool in improving a digital plant identification system: a case study for the area of Mt. Valerio, Trieste (NE Italy)," *PLoS one*, vol. 7, no. 9, Article ID e43256, 2012.

[42] W. J. Kress, D. L. Erickson, N. G. Swenson, J. Thompson, M. Uriarte, and J. K. Zimmerman, "Advances in the use of DNA barcodes to build a community phylogeny for tropical trees in a puerto rican forest dynamics plot," *PLoS ONE*, vol. 5, no. 11, Article ID e15409, 2010.

[43] S. Federici, A. Galimberti, F. Bartolucci et al., "DNA barcoding to analyse taxonomically complex groups in plants: the case of Thymus (Lamiaceae)," *Botanical Journal of the Linnean Society*, vol. 171, no. 4, pp. 687–699, 2013.

[44] A. Gismondi, F. Fanali, J. M. M. Labarga, M. G. Caiola, and A. Canini, "*Crocus sativus* L. genomics and different DNA barcode applications," *Plant Systematics and Evolution*, vol. 299, no. 10, pp. 1859–1863, 2013.

[45] S. Theodoridis, A. Stefanaki, M. Tezcan, C. Aki, S. Kokkini, and K. E. Vlachonasios, "DNA barcoding in native plants of the

Labiatae (Lamiaceae) family from Chios Island (Greece) and the adjacent Çeşme-Karaburun Peninsula (Turkey)," *Molecular Ecology Resources*, vol. 12, no. 4, pp. 620–633, 2012.

[46] M. Kojoma, K. Kurihara, K. Yamada, S. Sekita, M. Satake, and O. Iida, "Genetic identification of cinnamon (*Cinnamomum* spp.) based on the trnL-trnF chloroplast DNA," *Planta Medica*, vol. 68, no. 1, pp. 94–96, 2002.

[47] M. Wang, H. Zhao, L. Wang et al., "Potential use of DNA barcoding for the identification of *Salvia* based on cpDNA and nrDNA sequences," *Gene*, vol. 528, no. 2, pp. 206–215, 2013.

[48] I. Ganopoulos, P. Madesis, N. Darzentas, A. Argiriou, and A. Tsaftaris, "Barcode High Resolution Melting (Bar-HRM) analysis for detection and quantification of PDO "fava Santorinis" (*Lathyrus clymenum*) adulterants," *Food Chemistry*, vol. 133, no. 2, pp. 505–512, 2012.

[49] P. Madesis, I. Ganopoulos, A. Anagnostis, and A. Tsaftaris, "The application of Bar-HRM (Barcode DNA-High Resolution Melting) analysis for authenticity testing and quantitative detection of bean crops (Leguminosae) without prior DNA purification," *Food Control*, vol. 25, no. 2, pp. 576–582, 2012.

[50] M. Y. Stoeckle, C. C. Gamble, R. Kirpekar, G. Young, S. Ahmed, and D. P. Little, "Commercial teas highlight plant DNA barcode identification successes and obstacles," *Scientific Reports*, vol. 1, p. 42, 2011.

[51] T. Hidayat, A. Pancoro, and D. Kusumawaty, "Utility of matK gene to assess evolutionary relationship of genus *Mangifera* (anacardiaceae) in Indonesia and Thailand," *Biotropia*, vol. 18, no. 2, pp. 74–80, 2011.

[52] J. Yu, H. X. Yan, Z. H. Lu, and Z. Q. Zhou, "Screening potential DNA barcode regions of chloroplast coding genome for *citrus* and its related genera," *Scientia Agricultura Sinica*, vol. 44, no. 2, pp. 341–348, 2011.

[53] T. Xin, H. Yao, H. Gao et al., "Super food *Lycium barbarum* (Solanaceae) traceability via an internal transcribed spacer 2 barcode," *Food Research International*, vol. 54, no. 2, pp. 1699–1704, 2013.

[54] L. Jaakola, M. Suokas, and H. Häggman, "Novel approaches based on DNA barcoding and high-resolution melting of amplicons for authenticity analyses of berry species," *Food Chemistry*, vol. 123, no. 2, pp. 494–500, 2010.

[55] R. L. Jarret, "DNA Barcoding in a crop genebank: the Capsicum annuum species complex," *Open Biology Journal*, vol. 1, pp. 35–42, 2008.

[56] S. Chen, H. Yao, J. Han et al., "Validation of the ITS2 region as a novel DNA barcode for identifying medicinal plant species," *PLoS ONE*, vol. 5, no. 1, Article ID e8613, 2010.

[57] T. Gao, H. Yao, J. Song et al., "Identification of medicinal plants in the family Fabaceae using a potential DNA barcode ITS2," *Journal of Ethnopharmacology*, vol. 130, no. 1, pp. 116–121, 2010.

[58] Y. Zuo, Z. Chen, K. Kondo, T. Funamoto, J. Wen, and S. Zhou, "DNA barcoding of panax species," *Planta Medica*, vol. 77, no. 2, pp. 182–187, 2011.

[59] V. A. Parvathy, V. P. Swetha, T. E. Sheeja, N. K. Leela, B. Chempakam, and B. Sasikumar, "DNA barcoding to detect chilli adulteration in traded black pepper powder," *Food Biotechnology*, vol. 28, no. 1, pp. 25–40, 2014.

[60] S. Federici, D. Fontana, A. Galimberti et al., "A rapid diagnostic approach to identify poisonous plants using DNA barcoding data," *Plant Biosystems*. In press.

[61] T. Hirao, S. Imai, H. Sawada, N. Shiomi, S. Hachimura, and H. Kato, "PCR method for detecting trace amounts of buckwheat (Fagopyrum spp.) in food," *Bioscience, Biotechnology and Biochemistry*, vol. 69, no. 4, pp. 724–731, 2005.

[62] I. Ganopoulos, P. Madesis, and A. Tsaftaris, "Universal ITS2 Barcoding DNA Region Coupled with High-Resolution Melting (HRM) Analysis for Seed Authentication and Adulteration Testing in Leguminous Forage and Pasture Species," *Plant Molecular Biology Reporter*, vol. 30, no. 6, pp. 1322–1328, 2012.

[63] I. Bosmali, I. Ganopoulos, P. Madesis, and A. Tsaftaris, "Microsatellite and DNA-barcode regions typing combined with High Resolution Melting (HRM) analysis for food forensic uses: a case study on lentils (*Lens culinaris*)," *Food Research International*, vol. 46, no. 1, pp. 141–147, 2012.

[64] C. C. Ng, C. Y. Lin, W. S. Tzeng, C. C. Chang, and Y. T. Shyu, "Establishment of an internal transcribed spacer (ITS) sequence-based differentiation identification procedure for mei (*Prunus mume*) and plum (*Prunus salicina*) and its use to detect adulteration in preserved fruits," *Food Research International*, vol. 38, no. 1, pp. 95–101, 2005.

[65] I. Ganopoulos, C. Bazakos, P. Madesis, P. Kalaitzis, and A. Tsaftaris, "Barcode DNA high-resolution melting (Bar-HRM) analysis as a novel close-tubed and accurate tool for olive oil forensic use," *Journal of the Science of Food and Agriculture*, vol. 93, no. 9, pp. 2281–2286, 2013.

[66] M. Li, K. Au, H. Lam et al., "Identification of Baiying (Herba Solani Lyrati) commodity and its toxic substitute Xungufeng (Herba Aristolochiae Mollissimae) using DNA barcoding and chemical profiling techniques," *Food Chemistry*, vol. 135, no. 3, pp. 1653–1658, 2012.

[67] S. G. Newmaster, M. Grguric, D. Shanmughanandhan, S. Ramalingam, and S. Ragupathy, "DNA barcoding detects contamination and substitution in North American herbal products," *BMC Medicine*, vol. 11, no. 1, p. 222, 2013.

[68] B. Dhiman and M. Singh, "Molecular detection of Cashew Husk (*Anacardium occidentale*) adulteration in market samples of dry tea (*Camellia sinensis*)," *Planta Medica*, vol. 69, no. 9, pp. 882–884, 2003.

[69] L. J. Wallace, S. M. A. L. Boilard, S. H. C. Eagle, J. L. Spall, S. Shokralla, and M. Hajibabaei, "DNA barcodes for everyday life: routine authentication of Natural Health Products," *Food Research International*, vol. 49, no. 1, pp. 446–452, 2012.

[70] M. A. Faria, A. Magalhães, M. E. Nunes, and M. B. P. P. Oliveira, "High resolution melting of *trnL* amplicons in fruit juices authentication," *Food Control*, vol. 33, no. 1, pp. 136–141, 2013.

[71] M. Li, K. Wong, W. Chan et al., "Establishment of DNA barcodes for the identification of the botanical sources of the Chinese "cooling" beverage," *Food Control*, vol. 25, no. 2, pp. 758–766, 2012.

[72] J. Han, Y. Wu, W. Huang et al., "PCR and DHPLC methods used to detect juice ingredient from 7 fruits," *Food Control*, vol. 25, no. 2, pp. 696–703, 2012.

[73] A. Valentini, C. Miquel, and P. Taberlet, "DNA barcoding for honey biodiversity," *Diversity*, vol. 2, no. 4, pp. 610–617, 2010.

[74] A. Ortola-Vidal, H. Schnerr, M. Rojmyr, F. Lysholm, and A. Knight, "Quantitative identification of plant genera in food products using PCR and Pyrosequencing technology," *Food Control*, vol. 18, no. 8, pp. 921–927, 2007.

[75] M. Arleo, F. Ruibal, J. Pereyra, E. Miquel, M. Fernández, and C. Martínez, "A DNA-based approach to discriminate between quince and apple in quince jams," *International Food Research Journal*, vol. 19, no. 4, pp. 1471–1477, 2012.

[76] T. Yano, Y. Sakai, K. Uchida et al., "Detection of walnut residues in processed foods by polymerase chain reaction," *Bioscience,*

*Biotechnology and Biochemistry*, vol. 71, no. 7, pp. 1793–1796, 2007.

[77] P. Madesis, I. Ganopoulos, I. Bosmali, and A. Tsaftaris, "Barcode High Resolution Melting analysis for forensic uses in nuts: a case study on allergenic hazelnuts (*Corylus avellana*)," *Food Research International*, vol. 50, no. 1, pp. 351–360, 2013.

[78] G. Singh, *Plant Systematics: an Integrated Approach*, Science Publishers, New York, NY, USA, 2004.

[79] H. Trindade, "Molecular biology of aromatic plants and spices. A review," *Flavour and Fragrance Journal*, vol. 25, no. 5, pp. 272–281, 2010.

[80] M. Viuda-Martos, Y. Ruiz-Navajas, J. Fernández-López, and J. A. Pérez-Álvarez, "Spices as functional foods," *Critical Reviews in Food Science and Nutrition*, vol. 51, no. 1, pp. 13–28, 2011.

[81] A. Paton, M. R. Harley, and M. M. Harley, "*Ocimum*: an overview of classification and relationships," in *Basil: The Genus Ocimum*, pp. 1–38, 1999.

[82] A. O. Tucker, "The truth about mints," *Herb Companion*, vol. 4, pp. 51–52, 1992.

[83] V. Gobert, S. Moja, M. Colson, and P. Taberlet, "Hybridization in the section *Mentha* (Lamiaceae) inferred from AFLP markers," *American Journal of Botany*, vol. 89, no. 12, pp. 2017–2023, 2002.

[84] A. Torelli, M. Marieschi, and R. Bruni, "Authentication of saffron (*Crocus sativus* L.) in different processed, retail products by means of SCAR markers," *Food Control*, vol. 36, no. 1, pp. 126–131, 2014.

[85] K. Dhanya and B. Sasikumar, "Molecular marker based adulteration detection in traded food and agricultural commodities of plant origin with special reference to spices," *Current Trends in Biotechnology and Pharmacy*, vol. 4, no. 1, pp. 454–489, 2010.

[86] P. Posadzki, L. Watson, and E. Ernst, "Contamination and adulteration of herbal medicinal products (HMPs): an overview of systematic reviews," *European Journal of Clinical Pharmacology*, vol. 69, no. 3, pp. 295–307, 2013.

[87] M. Marieschi, A. Torelli, F. Poli, A. Bianchi, and R. Bruni, "Quality control of commercial Mediterranean oregano: development of SCAR markers for the detection of the adulterants *Cistus incanus* L., *Rubus caesius* L. and *Rhus coriaria* L.," *Food Control*, vol. 21, no. 7, pp. 998–1003, 2010.

[88] M. Marieschi, A. Torelli, A. Bianchi, and R. Bruni, "Detecting *Satureja montana* L. and *Origanum majorana* L. by means of SCAR-PCR in commercial samples of Mediterranean oregano," *Food Control*, vol. 22, no. 3-4, pp. 542–548, 2011.

[89] M. Marieschi, A. Torelli, A. Bianchi, and R. Bruni, "Development of a SCAR marker for the identification of *Olea europaea* L.: a newly detected adulterant in commercial Mediterranean oregano," *Food Chemistry*, vol. 126, no. 2, pp. 705–709, 2011.

[90] M. Barbuto, A. Galimberti, E. Ferri et al., "DNA barcoding reveals fraudulent substitutions in shark seafood products: the Italian case of "palombo" (*Mustelus* spp.)," *Food Research International*, vol. 43, no. 1, pp. 376–381, 2010.

[91] Z. Hubalkova and E. Rencova, "One-step multiplex PCR method for the determination of pecan and Brazil nut allergens in food products," *Journal of the Science of Food and Agriculture*, vol. 91, no. 13, pp. 2407–2411, 2011.

[92] J. Costa, I. Mafra, I. Carrapatoso, and M. B. P. P. Oliveira, "Almond allergens: molecular characterization, detection, and clinical relevance," *Journal of Agricultural and Food Chemistry*, vol. 60, no. 6, pp. 1337–1349, 2012.

[93] F. M. Hammouda, A. M. Rizk, M. M. El-Missiry et al., "Poisonous plants contaminating edible ones and toxic substances in plant foods. IV. Phytochemistry and toxicity of *Lolium temulentum*," *International Journal of Crude Drug Research*, vol. 26, no. 4, pp. 240–245, 1988.

[94] R. Walker, "Criteria for risk assessment of botanical food supplements," *Toxicology Letters*, vol. 149, no. 1–3, pp. 187–195, 2004.

[95] M. L. Colombo, F. Assisi, T. D. Puppa et al., "Most commonly plant exposures and intoxications from outdoor toxic plants," *Journal of Pharmaceutical Sciences and Research*, vol. 2, no. 7, pp. 417–425, 2010.

[96] Y. Finkelstein, S. E. Aks, J. R. Hutson et al., "Colchicine poisoning: the dark side of an ancient drug," *Clinical Toxicology*, vol. 48, no. 5, pp. 407–414, 2010.

[97] M. A. Berdai, S. Labib, K. Chetouani, and M. Harandou, "*Atropa Belladonna* intoxication: a case report," *Pan African Medical Journal*, vol. 11, p. 72, 2012.

[98] E. Röder, "Medicinal plants in Europe containing pyrrolizidine alkaloids," *Pharmazie*, vol. 50, no. 2, pp. 83–98, 1995.

[99] C. Franz, R. Chizzola, J. Novak, and S. Sponza, "Botanical species being used for manufacturing plant food supplements (PFS) and related products in the EU member states and selected third countries," *Food and Function*, vol. 2, no. 12, pp. 720–730, 2011.

[100] H. Wiedenfeld and J. Edgar, "Toxicity of pyrrolizidine alkaloids to humans and ruminants," *Phytochemistry Reviews*, vol. 10, no. 1, pp. 137–151, 2011.

[101] K. S. Burgess, A. J. Fazekas, P. R. Kesanakurti et al., "Discriminating plant species in a local temperate flora using the rbcL+matK DNA barcode," *Methods in Ecology and Evolution*, vol. 2, no. 4, pp. 333–340, 2011.

[102] A. Sandionigi, A. Galimberti, M. Labra et al., "Analytical approaches for DNA barcoding data-how to find a way for plants?" *Plant Biosystems*, vol. 146, no. 4, pp. 805–813, 2012.

[103] S. Ratnasingham and P. D. N. Hebert, "BOLD: the barcode of life data system: barcoding," *Molecular Ecology Notes*, vol. 7, no. 3, pp. 355–364, 2007.

[104] F. De Mattia, R. Gentili, I. Bruni et al., "A multi-marker DNA barcoding approach to save time and resources in vegetation surveys," *Botanical Journal of the Linnean Society*, vol. 169, no. 3, pp. 518–529, 2012.

[105] M. L. Kuzmina, K. L. Johnson, H. R. Barron, and P. D. N. Hebert, "Identification of the vascular plants of Churchill, Manitoba, using a DNA barcode library," *BMC Ecology*, vol. 12, p. 25, 2012.

[106] F. Sanger, S. Nicklen, and A. R. Coulson, "DNA sequencing with chain-terminating inhibitors," *Proceedings of the National Academy of Sciences of the United States of America*, vol. 74, no. 12, pp. 5463–5467, 1977.

[107] M. Hajibabaei, S. Shokralla, X. Zhou, G. A. C. Singer, and D. J. Baird, "Environmental barcoding: a next-generation sequencing approach for biomonitoring applications using river benthos," *PLoS ONE*, vol. 6, no. 4, Article ID e17497, 2011.

[108] S. Shokralla, J. L. Spall, J. F. Gibson, and M. Hajibabaei, "Next-generation sequencing technologies for environmental DNA research," *Molecular Ecology*, vol. 21, no. 8, pp. 1794–1805, 2012.

[109] M. L. Metzker, "Sequencing technologies—the next generation," *Nature Reviews Genetics*, vol. 11, no. 1, pp. 31–46, 2010.

[110] M. L. Coghlan, J. Haile, J. Houston et al., "Deep sequencing of plant and animal DNA contained within traditional Chinese medicines reveals legality issues and health safety concerns," *PLoS Genetics*, vol. 8, no. 4, Article ID e1002657, 2012.

# 23

# Comparative Effects of Different Fertilizer Sources on the Growth and Nutrient Content of Moringa (*Moringa oleifera*) Seedling in a Greenhouse Trial

## S. O. Dania, P. Akpansubi, and O. O. Eghagara

*Department of Soil Science, Faculty of Agriculture, Ambrose Alli University, Nigeria*

Correspondence should be addressed to S. O. Dania; megstedania@yahoo.com

Academic Editor: Innocenzo Muzzalupo

A greenhouse experiment was conducted to investigate the effects of NPK, poultry manure, and organomineral fertilizer on the growth and nutrient concentration of *Moringa oleifera* leaves. The experimental design was completely randomized design (CRD) with four treatments replicated three times. Data collected were analysed using descriptive statistics and ANOVA at $P = 0.05$. Growth parameters measured include number of leaves per plant, plant height (cm), and stem girth (mm). The application of poultry manure increased the height, number of leaves, and stem girth of moringa compared to the application of NPK and organomineral fertilizer while the control had the least growth. Poultry manure, NPK, and organomineral fertilizer were 66%, 62%, and 39% higher in number of leaves than the control at eight weeks after planting. The application of poultry manure significantly ($P \leq 0.05$) increased the nutrient content of moringa leaves compared to other sources of fertilizer applied. The results shows that the application of poultry manure significantly ($P \leq 0.05$) improved the growth and nutrient content of moringa; however, further field trial is suggested.

## 1. Introduction

*Moringa oleifera* originated from the foothills of the Himalayas in Northwestern India and is cultivated throughout the tropics [1]. Moringa can be cultivated in a wide range of soil types but grow best in well-drained loam to clay loam soil with slightly acidic to neutral pH however, it cannot withstand prolonged water logging. Moringa is very useful in the following areas; as alley cropping, animal forage, biogas, domestic cleaning agent, green manure, gum, medicine, ornamental plants, and water purification. Moringa leaves, seeds, and roots are also use in treating diseases like lung diseases, hypertension and skin infection [2, 3].

Moringa is nutritional and rich in vitamins and minerals. Moringa leaves are the most nutritious part of the plant, being a significant source of vitamin $B_6$ vitamin C, and provitamin A as beta carotene, magnesium, and calcium [4]. However, Moringa still remains unpopular in Nigeria despite its acclaimed economic values and importance; very little research has been done on this plant, although it is widely used by the rural poor as a food resource [5].

Land degrading is one of the major impediments to agricultural productivity. This is manifested in the loss of soil fertility, desertification, and destruction of the soil structure [6]. Due to the implication of land degradation on agronomic productivity and the environment, it becomes necessary to proffer means to minimize it effects. Ultisols of tropical and subtropical regions occur in old landscapes that have a monsoon climate and are extremely weathered and leached. They have a red, brown, or yellow argillic B horizon with a base saturation of less than 50%. The soils have a low content of organic matter with ferric and hydromorphic properties. These soils are generally of low fertility and are susceptible to erosion [7]. One way to improve soil fertility is the application of fertilizer which obviously is a means required for optimum crop yield.

The use of poultry dungs has been documented to give a better result on soil amendment in degraded ultisols [8]. It has been reported that organic manure can serve as soil amendment to improve soil nutrient status and the growth of crops [9]. Organic base fertilizer such as organomineral fertilizer improves soil structure, reduces erosion, lowers the temperature at the soil surface, and increases soil water holding capacity [10]. The use of NPK fertilizer has resulted in the improvement of the growth and yield of crops. Due to increasing demand of moringa for biofuel and medicinal uses, it is therefore necessary to investigate ways to improve its growth in degraded soil. This study aimed at the responses of moringa seedling to soil amendment in degraded ultisols of Edo state under a greenhouse condition.

## 2. Materials and Methods

*2.1. Description of Experimental Site.* The experiment was conducted at the Teaching and Research Farm of Ambrose Alli University, Emaudo Annex, Ekpoma, in 2012 under greenhouse condition. The area lies between latitude North 6 degrees, 45 minutes, 34 seconds ($6° 45' 34''$) and longitude East 6 degrees, 8 minutes, 27 seconds ($6° 8' 27''$ East) with average amount of rainfall 1750 mm.

*2.2. Collection of Soil for Analysis.* Top soils (0–15 cm) were collected from the farm site; the soils were sieved with a 2 mm mesh to remove gravel and plant roots. The 5 kg polythene bags used for the experiment were filled with the sieved soil.

*2.3. Soil Physical and Chemical Analysis.* Particle size analysis was carried out using hydrometer method [11]. The pH was determined in water (ratio 1:1, soil:water). Organic carbon was determined by wet dichromate method [12] and available phosphorus by Bray extraction method [13]. Total nitrogen was determined by Kjeldahl method. Exchangeable cations (potassium, calcium, and magnesium) were extracted with ammonium acetate. Potassium was determined by flame photometer while calcium and magnesium were determined by atomic absorption spectrophotometer. Copper, zinc, manganese, and iron were also determined [14].

*2.4. Experimental Design.* The experimental design was a completely randomized design (CRD) with four treatments replicated three times. The treatments were NKP, organomineral fertilizer (OMF), poultry manure, and control.

*2.5. Planting Operation.* Moringa (*Moringa Oleifera*) seeds were first soaked in water for 24 hours to allow the seeds to absorb the moisture required for sprouting. The seeds were removed from the water, wrapped in a wet towel, and stored in a warm dark place. The towel was kept damp to allow maximum germination and prevent drought. The sprouted seeds were planted two per pot and later thinned to one stand per pot.

*2.6. Fertilizer Application.* Poultry manure (PM) was applied four weeks before planting at the rate of 100 g per pot; organomineral fertilizer (OMF) was applied two weeks before planting at the rate of 5 g per pot. NPK fertilizer was applied two weeks after planting at the rate of 2 g per pot. Ring application method was used for NPK and OMF while poultry manure was mixed with the soil.

TABLE 1: Physicochemical properties of soil before planting.

| Properties | Results |
| --- | --- |
| pH (1:1) $H_2O$ | 6.20 |
| N (g/kg) | 0.40 |
| OC (g/kg) | 3.50 |
| P (mg/kg) | 14.15 |
| Ca (cmol/kg) | 0.88 |
| Mg (cmol/kg) | 0.97 |
| K (cmol/kg) | 0.17 |
| Na (cmol/kg) | 0.73 |
| Mn (mg/g) | 63.58 |
| Fe (mg/g) | 93.34 |
| Cu (mg/g) | 0.57 |
| Zn (mg/g) | 4.77 |
| ECEC | 3.39 |
| $H^+$ | 0.64 |
| Sand | 852 |
| Silt | 34 |
| Clay | 114 |
| Textural class | |
| Sandy loam. | |

*2.7. Spacing.* The pots were arranged at a distance of 60 cm × 60 cm between and within rows. A total number of 12 pots were used for the experiment.

*2.8. Collection of Data.* Growth parameters such as plant height, number of leaves, and stem girth were measured at 4, 6, and 8 weeks after planting the sprouted seeds. Moringa leaves were collected for plant nutrient analysis at eight weeks after planting.

## 3. Results

The soil was moderately acidic with sandy loam texture. The values of the nutrient element were as follows: nitrogen: 0.40 g/kg; phosphorus: 14.15 mg/kg; calcium: 0.88 cmol/kg; magnesium: 0.97 cmol/kg; potassium: 0.17 cmol/kg; sodium: 0.73 cmol/kg. The value of ECEC was 3.39, Mn 63.58 mg/g, Fe 93.34 mg/g, Cu 0.57 mg/g, and Zn 4.77 mg/g (Table 1).

*3.1. The Chemical Properties of the Poultry Manure Used for the Experiment.* The poultry manure was alkaline with pH of 9.60, high in organic matter and other nutrient elements. The nitrogen (N) value was 3.4 g/kg, available phosphorus (P) was 59.53 mg/kg, calcium was 19.84 cmol/kg., Mg was 1.64 cmol/kg, Na was 14.32 cmol/kg, K was 7.25 cmol/kg, and $H^+$ which was 0.10 cmol/kg (Table 2).

*3.2. Organomineral Fertilizer Analysis.* Total nitrogen value was 44.0 g/kg, available phosphorus was 11.0 cmol/kg, and K and Ca were 6.8 cmol/kg and 6.8 cmol/kg, respectively. The value of Na was 0.8 cmol/kg and Mg was 10.8 cmol/kg. Fe value was 8,153.0 mg/kg, Zn was 712.7 mg/kg, Mn was 558.3 mg/kg, and Cu was 247.4 mg/kg (Table 3).

TABLE 2: Nutrient content of poultry manure.

| Parameters | Units | Values |
|---|---|---|
| pH | | 9.60 |
| Organic matter | g/kg | 31.2 |
| Total nitrogen | g/kg | 3.4 |
| Available phosphorus | mg/kg | 59.53 |
| Ca | cmol/kg | 19.84 |
| Mg | cmol/kg | 1.64 |
| Na | cmol/kg | 14.32 |
| K | cmol/kg | 7.25 |
| $H^+$ | cmo/kg | 0.10 |

TABLE 3: Nutrient content of organomineral fertilizer grade A.

| Elements | Values |
|---|---|
| Total nitrogen (g/kg) | 44.00 |
| Available phosphorus (mg/kg) | 11.00 |
| Exchangeable bases (cmol/kg) | |
| K | 6.80 |
| Na | 0.80 |
| Mg | 10.80 |
| Ca | 6.80 |
| Extractable micronutrients (mg/kg) | |
| Mn | 558.30 |
| Fe | 8153.40 |
| Cu | 247.40 |
| Zn | 712.70 |

Source: pace setter organomineral fertilizer plant, Ibadan.

TABLE 4: Responses of moringa height (cm) to different fertilizer applications.

| Treatment | Week after sprouting | | |
|---|---|---|---|
| | 4 | 6 | 8 |
| Control | $14.50^b$ | $25.00^b$ | $27.50^b$ |
| OMF | $22.40^a$ | $22.70^b$ | $27.95^b$ |
| NPK | $22.53^a$ | $48.47^a$ | $59.00^a$ |
| Poultry manure | $28.10^a$ | $52.80^a$ | $65.47^a$ |

Values followed by different letters under the same column are significantly different using Duncan's multiple range test ($P < 0.05$).

### 3.3. Growth Parameter of Moringa

*3.3.1. Plant Height (cm).* Poultry manure consistently and significantly increased the height of moringa compared to other treatments. At six weeks after planting, the height of moringa was significantly ($P < 0.05$) increased with the application of poultry manure (52.8 cm) compared to the application of NPK, organomineral fertilizer. Also at eight weeks after planting, application of poultry manure significantly ($P < 0.05$) increased the height of moringa (65.47 cm) compared to other treatments (Table 4).

*3.3.2. Stem Girth (mm).* The stem girth of moringa was not significantly different among treatments at four weeks after

TABLE 5: Responses of moringa stem girth (mm) to different fertilizer applications.

| Treatment | Week after sprouting | | |
|---|---|---|---|
| | 4 | 6 | 8 |
| Control | $2.93^a$ | $4.76^b$ | $5.04^b$ |
| OMF | $3.20^a$ | $3.88^b$ | $4.62^b$ |
| NPK | $3.16^a$ | $6.90^a$ | $8.16^a$ |
| Poultry manure | $3.46^a$ | $7.22^a$ | $8.83^a$ |

Values followed by different letters under the same column are significantly different using Duncan's multiple range test ($P < 0.05$).

TABLE 6: Responses of number of leaves of moringa to different fertilizer applications.

| Treatment | Week after sprouting | | |
|---|---|---|---|
| | 4 | 6 | 8 |
| Control | $81.00^a$ | $119.00^b$ | $127.00^c$ |
| OMF | $64.50^a$ | $135.00^b$ | $209.50^b$ |
| NPK | $99.00^a$ | $307.00^a$ | $336.00^b$ |
| Poultry manure | $102.66^a$ | $334.33^a$ | $378.33^a$ |

Values followed by different letters under the same column are significantly different using Duncan's multiple range test ($P < 0.05$).

planting. However, the application of poultry manure and NPK significantly ($P < 0.05$) increased the stem girth of moringa (7.22 mm and 6.90 mm) compared to the application of OMF and control. At eight weeks after planting, the stem girth of moringa was significantly ($P < 0.05$) higher when poultry manure and NPK were applied (8.83 mm and 8.16 mm) compared to OMF application and control (Table 5).

*3.3.3. Number of Leaves.* Application of fertilizer did not significantly ($P < 0.05$) increase the number of leaves of moringa at four weeks after planting. However, at six weeks after planting, the application of poultry manure and NPK significantly ($P < 0.05$) increased the number of leaves of moringa (334.33 and 307.007) compared to OMF and the control. It was observed that, at eight weeks after planting, poultry manure significantly ($P < 0.05$) increased the number of leaves of moringa (378.33) compared to other treatments (Table 6).

*3.3.4. Nutrient Content of Moringa Oleifera Leaves.* Nutrient content of *Moringa oleifera* leaves was significantly ($P < 0.05$) influenced by fertilizer application. Poultry manure and NPK significantly ($P < 0.05$) increased the nitrogen content of moringa leaves. Phosphorus, potassium, sodium, and manganese content in moringa leaves were significantly ($P < 0.05$) higher with the application of poultry manure compared to other treatments. OMF application had the highest calcium content with value 2.46 cmol/kg. NPK application increased the values of Cu and Fe content in the leaves of moringa while the control has higher magnesium and zinc (Zn) content compared to other treatments (Table 7).

*3.3.5. Correlation Analysis.* Positive correlation exists between phosphorus and potassium content at 0.07% and

TABLE 7: Comparative effect of PM, OMF, and NPK on the nutrient content of moringa.

| Treatments | %N (g/kg) | P (mg/kg) | Ca (cmol/kg) | Mg (cmol/kg) | K (cmol/kg) | Na (mg/g) | Mn (mg/g) | Fe (mg/g) | Cu (mg/g) | Zn (mg/g) |
|---|---|---|---|---|---|---|---|---|---|---|
| Control | 2.11[b] | 0.15[b] | 0.87[b] | 0.70[a] | 2.28[c] | 401.08[c] | 100.20[c] | 189.02[a] | 6.61[b] | 36.14[a] |
| OMF | 2.01[b] | 0.19[b] | 2.46[a] | 0.42[a] | 3.21[b] | 492.14[b] | 115.43[c] | 158.43[a] | 6.04[b] | 10.43[b] |
| NPK | 3.28[a] | 0.12[b] | 0.94[b] | 0.46[a] | 2.66[c] | 392.18[c] | 392.34[b] | 198.08[a] | 8.31[a] | 17.23[b] |
| PM | 4.13[a] | 0.26[a] | 0.94[b] | 0.53[a] | 4.78[a] | 563.05[a] | 563.00[a] | 142.17[a] | 8.22[a] | 19.78[b] |
| SE | 1.70 | 0.42 | 1.14 | 0.73 | 1.80 | 21.50 | 17.11 | 13.11 | 2.70 | 4.57 |

Values with different letters (a, b, c) are significantly ($P < 0.05$) different using Duncan's multiple range test.

TABLE 8: Correlation Analysis.

| | Nitrogen | Phos | Ca | Mg | K | Na | Mn | Fe | Cu | Zn |
|---|---|---|---|---|---|---|---|---|---|---|
| Nitrogen | 1.00000 | | | | | | | | | |
| Phos | 0.46888 | 1.00000 | | | | | | | | |
| | 0.5311 | | | | | | | | | |
| Ca | −0.54252 | 0.14043 | 1.00000 | | | | | | | |
| | 0.4575 | 0.8596 | | | | | | | | |
| Mg | −0.16828 | −0.08404 | −0.61241 | 1.00000 | | | | | | |
| | 0.8317 | 0.9160 | 0.3876 | | | | | | | |
| K | 0.73612 | 0.92446 | 0.01274 | −0.29252 | 1.00000 | | | | | |
| | 0.2639 | 0.0755 | 0.9873 | 0.7075 | | | | | | |
| Na | 0.38993 | 0.98775 | 0.28985 | −0.20127 | 0.90684 | 1.00000 | | | | |
| | 0.6101 | 0.0123 | 0.7101 | 0.7987 | 0.0932 | | | | | |
| Mn | 0.99709 | 0.45822 | −0.49191 | −0.24219 | −0.73970 | 0.38919 | 1.00000 | | | |
| | 0.0029 | 0.5418 | 0.5081 | 0.7578 | 0.2603 | 0.6108 | | | | |
| Fe | 0.91941 | 0.52196 | −0.72280 | 0.21886 | 0.68080 | 0.40206 | 0.88698 | 1.00000 | | |
| | 0.0806 | 0.4780 | 0.0772 | 0.7811 | 0.3192 | 0.5979 | 0.1130 | | | |
| Cu | 0.9560 | 0.07557 | −0.70120 | −0.11646 | 0.40534 | −0.01268 | 0.91445 | 0.81420 | 1.00000 | |
| | 0.0844 | 0.9244 | 0.2988 | 0.8835 | 0.5947 | 0.9873 | 0.0856 | 0.1858 | | |
| Zn | −0.16598 | −0.21472 | −0.67190 | 0.98880 | −0.38668 | −0.33474 | −0.23744 | 0.20113 | −0.05504 | 1.00000 |
| | 0.8340 | 0.7853 | 0.3281 | 0.0112 | 0.6133 | 0.6653 | 0.7626 | 0.7989 | 0.9450 | |

Significant at 0.001, 0.01, and 0.05.

phosphorus and sodium at 0.01%. The result showed that as phosphorus increases potassium and sodium increases as well. It was also observed that there was positive correlation between the following nutrient elements: nitrogen and iron, calcium and iron, and nitrogen and copper, and between magnesium and zinc at 0.01%. The result shows that as nitrogen and calcium increased, iron content increased correspondingly (Table 8).

## 4. Discussion

The major limiting factor of crop production in the tropics is the deficiency of soil nutrient resulting from land degradation which affects the growth, nutrient content, and uptake of the plant. Low levels of nitrogen, phosphorus, and organic carbon were observed in the soil used for the experiment and the finding corroborates with the earlier results [15]; they reported that most of Nigerian soil is deficient in nitrogen, phosphorus, and potassium even organic matter. Therefore, a sustainable method of improving the nutritional status of the soil should be employed to enhance the growth and nutrient content of the plant.

The application of NPK (15 : 15 : 15) fertilizer significantly increased the vegetative growth of moringa plant and this finding agreed with earlier work done [16]. It was reported that the application of NPK fertilizer significantly ($P < 0.05$) increased the vegetative growth of moringa which was also observed from the experiment [17].

The application of poultry manure significantly ($P < 0.05$) increased the height, stem girth, and number of leaves (vegetative growth) of moringa. This result corresponded to the earlier finding [18]. It was reported that the application of poultry manure significantly ($P < 0.05$) increased vegetative

growth of moringa. This could result from the nutritional benefits of poultry manure which include improvement of soil fertility, structure, water holding capacity, and organic matter. This will reduce the amount of inorganic fertilizers needed for the growth of moringa plant [19]. The effect of compost and other organic amendment on the growth of moringa plant may be the result of the interaction between the nutrient present and growth of moringa, as organic manure has been found to contain auxins, gibberellins, and cytokines [20].

The application of fertilizer significantly ($P < 0.05$) increased the nutrient content of moringa. Poultry manure application increased the P, K, Na, and Mn content of moringa and this result corresponded to earlier work done [21]. It has been reported that the application of organic manure increased the nutrient concentration of arable and other crops [22]. Similarly, NPK application also improved the nutrient content of moringa which has earlier been reported [23]. The improvement of calcium, potassium, and sodium content of moringa by the application of OMF agreed with the work earlier done [24]. It was reported that organic base fertilizer (OMF) improved the nutrient content of arable crops.

## 5. Conclusion

The comparative effects of NPK, poultry manure, and OMF on growth of moringa seedling and nutrient concentration was investigated. The results of this study show that the application of poultry manure significantly increased the vegetative growth of moringa. Also the nutrient concentration on the leaves of moringa was significantly improved by amending degraded soil with poultry manure. These results can be investigated further on a field trial.

## Conflict of Interests

The authors declared no conflicting interests regarding the publication of this paper.

## References

[1] M. A. Nagao, "Macadamia integrifolia: macadamia nut," in *The Encyclopedia of Fruit and Nuts*, J. Janick and R. Paull, Eds., pp. 600–610, CABI, Wallingford, UK, 2008.

[2] L. J. Fuglier, *The Miracle Tree: Moringa oleifera, Natural Nutrition for the Tropics*, Church World Service, Dakar, Senegal, 1999.

[3] Miracle Tree Foundation WHO, 2012, http://miracletrees.org/.

[4] R. Bharali, J. Tabassum, and M. R. H. Azad, "Chemomodulatory effect of moringa oleifera, lam, on hepatic carcinogen metabolising enzymes, antioxidant parameters and skin papillomagenesis in mice," *Asian Pacific Journal of Cancer Prevention*, vol. 4, no. 2, pp. 131–139, 2003.

[5] S. M. Odeyinka, D. O. Torimiro, J. O. Oyedele, and V. O. Asaolu, "Farmers' awareness and knowledge of *Moringa oleifera* in Southwestern Nigeria: a perceptional analysis," *Asian Journal of Plant Sciences*, vol. 6, no. 2, pp. 320–325, 2007.

[6] H. Eswaran, R. Lal, and P. F. Reich, "Land degradation: an overview. In responses to land degradation," in *Proceedings of the 2nd International Conference of Land Degradation and Desertification*, Oxford Press, Khon Kaen, Thailand, 2001.

[7] E. A. Fitzpatrick, *An Introduction to Soil Science*, Longman Singapore Publishers, Singapore, 1986.

[8] http://en.Wikipedia.org/wiki/ultisols.

[9] P. Y. Stoffella, M. Roe, Ozeres-Hampton, and D. A. Greatz, *Utilization of Organic Waste Compost in Vegetable Production System in Asia*, edited by R. A. Moris, Asian Vegetable Research and Development Centre, Shanhua, Taiwan, 1997.

[10] N. E. Roe, P. J. Stoffella, and D. Graetz, "Composts from various municipal solid waste feedstocks affect vegetable crops. II. Growth, yields, and fruit quality," *Journal of the American Society for Horticultural Science*, vol. 122, no. 3, pp. 433–437, 1997.

[11] G. J. Bouyoucos, "Hydrometer method improved for making particle size analyses of soil," *Agronomy Journal*, vol. 53, pp. 464–465, 1962.

[12] D. W. Nelson and L. E. Sommers, "A rapid and accurate method of estimating organic carbon in soil," *Proceeding of Indiana Academics of Science*, vol. 84, pp. 456–462, 1975.

[13] J. M. Anderson and J. S. Ingram, *Tropical Soil Biology and Fertility. A Handbook of Methods*, Information Press, Eynsham, UK, 1993.

[14] IITA, *Selected Methods for Soil and Plant Analysis*, Manual Series 1, International Institute for Tropical Agriculture Ibadan, 1979.

[15] E. A. Aduayi, V. O. Chude, L. O. Adehusuji, and S. Olayiwola, *Fertilizer Use and Management Practices for Crops in Nigeria*, Federal Ministry of Agriculture and Rural Development, Abuja, Nigeria, 2002.

[16] A. I. Makinde, "Effects of inorganic fertilizer on the growth and nutrient composition of Moringa (*Moringa oleifera*)," *Journal of Emerging Trends in Engineering and Applied Sciences*, vol. 4, no. 2, pp. 341–343, 2013.

[17] I. N. Abdullahi, K. Ochi, and A. B. Gwaram, "Plant population and fertilizer application effects on biomass productivity of *Moringa oleifera* in North-Central Nigeria," *Peak Journal of Agricultural Sciences*, vol. 1, no. 6, pp. 94–100, 2013.

[18] W. M. Imoro, I. Sackey, and A.-H. Abubakari, "Preliminary study on the effects of two different sources of organic manure on the growth performance of *Moringa oleifera* seedlings," *Journal of Biology Agriculture and Healthcare*, vol. 2, no. 10, 2012.

[19] J. M. Wapa, S. O. Ojeniyi, and J. D. OKwari, "Response of extra-early maize variety to different organic manure in the drier sub region of Northern Nigeria," *Nigeria Journal of Soil Science*, vol. 23, no. 2, pp. 52–59, 2013.

[20] K. Miezah, J. Ofosu-Anim, G. K. O. Budu, L. Enu-Kwesi, and O. Cofie, "Isolation and identification of some plant growth promoting substances in compost and co-compost," *International Journal of Virology*, vol. 4, no. 2, pp. 30–40, 2008.

[21] K. E. Law-Ogbomo, S. O. Ojeniyi, and F. E. OMazi, "Combined and sole application of compost and NPK effect on Okra yield, soil and nutrient content," *Nigeria Journal of Soil Science*, vol. 23, no. 1, pp. 130–135, 2013.

[22] C. T. Atere and A. Olayinka, "Soil chemical properties and growth of maize as affected by cocoa pod compost base N and P fertilizer," *Nigeria Journal of Soil Science*, vol. 23, no. 1, pp. 83–93, 2013.

[23] K. S. Chukwuka and O. E. Omotayo, "Soil fertility restoration potentials of tithonia green manure and water hyacinth compost on nutrient depleted soil in south western Nigeria using maize as test crop," *Research Journal of Biology*, vol. 1, pp. 20–30, 2009.

[24] F. A. Olowokere, J. K. Adesodun, C. O. Adejuyigbe et al., "Soil chemical properties, yield and nutrient uptake and animal based organic amendments," *Nigeria Journal of Soil Science*, vol. 23, no. 1, pp. 197–206, 2013.

# 24

# Nitrogen Efficiency and Nutrient Absorption by a Sorghum-Oats Forage Succession

**Rasiel Restelatto,[1] Paulo Sergio Pavinato,[2] Laércio Ricardo Sartor,[1] Suelen Maria Einsfeld,[1] and Fernanda Paula Baldicera[1]**

[1]*Department of Animal Science, Federal Technology University of Paraná, Estrada/Boa Esperança, km 04, 85660-000 Dois Vizinhos, PR, Brazil*
[2]*Department of Soil Science, Luiz de Queiroz College of Agriculture, University of São Paulo, Avenida Pádua Dias 11, 13418-900 Piracicaba, SP, Brazil*

Correspondence should be addressed to Paulo Sergio Pavinato; pavinato@usp.br

Academic Editor: Albino Maggio

Balance between nutrient uptake by the crop and the amount of applied fertilizer is desired in a production system, and this measurement can predict the nutrient use efficiency. The present study aimed to evaluate nitrogen- (N-) use and recovery efficiency and extraction of N, phosphorus (P), and potassium (K) by sorghum (*Sorghum bicolor*) and black oat (*Avena strigosa*) forages under nitrogen fertilization in three consecutive crop cultivation. The experiment was carried out at Federal Technological University of Paraná (UTFPR), Dois Vizinhos, Paraná, Brazil, in Rhodic Hapludox soil. Treatments included 0, 37.5, 75, 150, 225, 300, and 375 kg N ha$^{-1}$ applied to sorghum in the summer seasons of 2010/2011 and 2011/2012, intercalated by, respectively, 0, 40, 80, 120, 160, 200, and 240 kg N ha$^{-1}$ applied to black oat in the winter/2011. Forage production, N, P, and K extraction, and N recovery efficiency were determined through samplings during the crop cycle. The highest sorghum forage production and N and K extraction were under 230 to 300 kg N ha$^{-1}$, which was around 190 kg N ha$^{-1}$ for black oat. The values of N recovery, efficiency, and N-use physiological efficiency were higher under lower N rates for both crops. Based on this, about 150 kg N ha$^{-1}$ is recommended to obtain good sorghum forage production with higher N efficiency levels, which is about 120 kg N ha$^{-1}$ for black oat.

## 1. Introduction

Brazilian animal grazing has a significant participation in general agricultural production, composing an expressive part of the country growth income. The efficiency of animal production in a grazing system is the main concern, being necessary enough forage to feed the animals in a sustainable way [1]. The best alternative for this is improving soil and plant management, choosing forage with high yield and high quality, maintaining soil fertility in adequate levels, and determining the maximum fertilizers use efficiency [2].

Most farmers assume that grazing production systems are self-sustaining, requiring low fertilizer inputs to resupply losses by animal extraction [3]. The continued exploitation in these systems may not be able to maintain the balance for a long period of time without fertilizer amendments. Oenema et al. [4] and Martha Júnior et al. [3] suggest that removal

of nutrients through cattle grazing causes imbalance over the time, with degradation of soil reserves of nutrients.

To avoid unnecessary costs with higher amounts of fertilizers, determination of the nutrient content in plants, especially in harvested tissue, is necessary. This information would be useful to predict removal, which is essential in developing cost-effective fertilizer recommendations [5]. Following water availability, N is the second most limiting factor for biomass production in natural ecosystems [6]. However, the nutrient demand for plants is not sufficient in soil for long-term sustainability.

Selection of forage with high potential yield can respond effectively to managements [7]. Sorghum sp is characterized as $C_4$ physiology, presenting a great forage yield with a high nutritional quality, with high carbohydrates content [8, 9], a great potential for ethanol [10, 11], and biodiesel use [12], and is also an effective crop for soil protection [13]. This

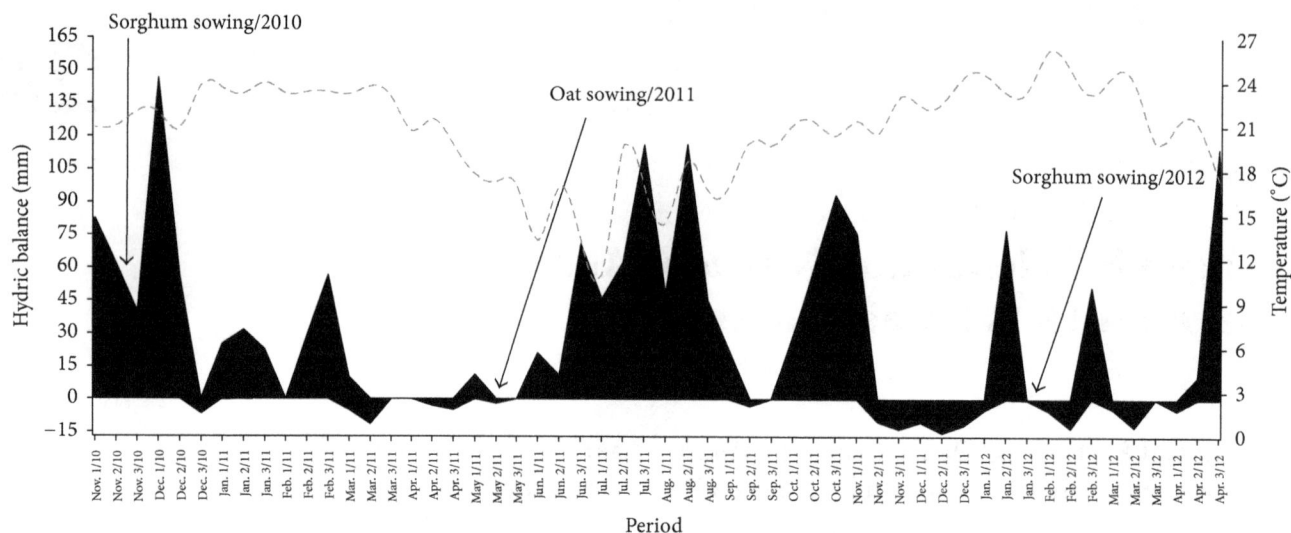

FIGURE 1: Soil water balance and mean temperatures of the experimental area during the period from November 2010 to April 2012.

TABLE 1: Chemical soil characteristics prior to sowing of sorghum in 2010.

| Depth (cm) | pH CaCl$_2$ | OM % | P-Mehlich Mg dm$^{-3}$ | K | Ca | Mg | Al | H + Al | BS* | CEC* | V* (%) |
|---|---|---|---|---|---|---|---|---|---|---|---|
| | | | | | | cmol$_c$ dm$^{-3}$ | | | | | |
| 0–10 | 4.90 | 3.48 | 8.62 | 0.30 | 5.13 | 2.34 | 0.06 | 5.35 | 7.77 | 13.12 | 59.2 |
| 10–20 | 4.50 | 3.22 | 4.20 | 0.15 | 3.59 | 1.69 | 0.28 | 5.76 | 5.43 | 11.19 | 48.5 |

*BS: basis sum; CEC: cation exchangeable capacity; V: basis saturation.

crop species is also tolerant to high temperatures and rainfall restrictions [14, 15].

The demand for cattle feeding in subtropical climates needs to be supplemented in the winter season when temperature and luminosity decrease. Including plant species adapted to the cold season in the rotation system is necessary for year-long productivity. According to Nakagawa et al. [16], black oat (*Avena strigosa*) is temperate and subtropical forage with high tillering, being adaptable to low temperature and low luminosity conditions, a good alternative to rotating with sorghum. This crop is capable of producing high quality forage, with high crude protein content (up to 18%) and low fiber levels, with a potential yield of 4 to 8 Mg of DM ha$^{-1}$ year$^{-1}$, an excellent alternative for cattle feeding.

The balance between the nutrient utilization and fertilizer amendment is desired in a production system. The objectives of this research were to evaluate the N-use and recovery efficiency by sorghum and black oat and also to evaluate the exploitation of N, P, and K by sorghum in the summer and by black oat in the winter, under nitrogen fertilization, aiming to determine the best N rates for these crops' succession.

## 2. Material and Methods

### 2.1. Experimental Site.
The study was conducted at Federal Technological University of Paraná (UTFPR), Dois Vizinhos, Paraná state, Brazil (latitude 25°44′ south, longitude 54°04′ west), with an altitude of 520 m above sea level. This region is characterized by a humid subtropical mesothermal climate

(Cfa) with no dry season, a mean annual rainfall of 2,100 mm, and temperatures averaging 22°C in summer and 17°C in winter.

Rainfall and temperature data were collected during the study (INMET-SONABRA). This information was useful to estimate the climatic water balance every 10 days, following the methodology of Thornthwaite and Mather [17] and adapted by Pereira [18] (Figure 1).

Soil was Rhodic Hapludox according to soil taxonomy [19]. Chemical soil properties prior to sorghum sowing in 2010 are presented in Table 1. The levels of available phosphorus (P) and exchangeable potassium (K), calcium (Ca), and magnesium (Mg) are considered high for this soil at 0–10 and 10–20 cm depth layers, with an adequate pH and basis saturation (V%) for the crops tested.

### 2.2. Crops Used and Establishment.
Crop sequence used in the experiment was sorghum-black oat-sorghum. For the summer crop seasons of 2010/2011 and 2011/2012 the sorghum cv. Jumbo (hybrid *S. bicolor* × *S. sudanensis*) and for the winter of 2011 black oat cv. Iapar 61 (*Avena strigosa*) were established. The first cycle of forage sorghum was conducted during November 2010 to April 2011, with sowing date on November 11. Winter oats was also established in sequence, from May to October 2011, with sowing date on May 12. The second cycle of sorghum was conducted from January to April 2012, with sowing date on January 31. This delay in the sowing was due to lack of rain, preventing earlier crop establishment (November-December 2011) (Figure 1).

2.3. *Experimental Design and Treatments.* Treatments were distributed in a completely randomized block design with three replicates. Plot size was $5 \times 5$ m, totaling 25 m$^2$. For the two sorghum cycles, the row spacing was 40 cm, with a density of 15 kg ha$^{-1}$ of viable seeds. Nitrogen rates 0, 37.5, 75, 150, 225, 300, and 375 kg N ha$^{-1}$ were distributed in three similar applications of 1/3 at sowing, 1/3 after the first cut, and 1/3 after the third cut. The application dates for the year 2010/2011 were as follows: at sowing, November 15, 2010; after the first cut, December 29, 2010; and after the third cut, March 02, 2011. For the second year (2011/2012), dates were as follows: at sowing, January 31, 2012; after the first cut, March 02, 2012; and after the third cut, March 30, 2012.

The black oat in the winter 2011 season was sown at row spacing of 20 cm, with a density of 40 kg ha$^{-1}$ of viable seeds; testing doses of 0, 40, 80, 120, 160, 200, and 240 kg N ha$^{-1}$ were also distributed in three applications: 1/3 at tillering (June 16, 2011), 1/3 after the first cut (July 09, 2011), and 1/3 after the third cut (August 13, 2011). The amount of total N applied in oats was lower than sorghum since the potential DM yield of this crop is presumed lower.

2.4. *Forage Production.* Sampling of forage sorghum plant tissue was performed manually by cutting 0.50 m above ground level of two rows per plot (0.40 m$^2$), when crops were 0.70 to 0.80 m in height. The remaining plot was also cut and removed manually, leaving a residue of about 0.20 m for regrowth and new sampling. Black oat forage sampling was performed when the crop reached about 0.25–0.30 m in height, through a manual cut to 0.10 m above ground level in 0.50 m of two rows (0.20 m$^2$) per plot. The remaining plot was handled by a mechanical cutter, in the same height of the sample collected (0.10 m). The samples were weighted to estimate green mass yield. After that, it was dried in a forced air circulation oven at a temperature of 60°C for 72 hours or until constant weight to estimate dry matter, followed by laboratory analysis to estimate nutrient uptake.

2.5. *Determination of Macronutrients.* The concentrations of the macronutrients N, P, and K in the DM were determined according to the methodology described by Tedesco et al. [20]. After sulfuric digestion, N was measured at analytical method semimicro-Kjeldahl, P was determined by analytical method of blue-molybdate in colorimeter, and K was estimated by flame photometry. All multiplied by DM yield to find the total nutrients accumulated.

2.6. *Nitrogen-Use Efficiency by Forage Production.* The recovery of N (RN), the physiology efficiency (PE), and the recovery efficiency of use of N (RAN) were calculated by the equations according to Fageria [21]. Recovery was calculated according to the following formula:

$$\text{RN} = \frac{(\text{NCT} - \text{NST})}{\text{DN}} \times 100, \tag{1}$$

where RN(%) corresponds to the recovery of nitrogen; NCT corresponds to the total nitrogen absorbed with N

application (kg ha$^{-1}$); NST is total nitrogen absorbed without N application (kg ha$^{-1}$); DN is dose of nitrogen (kg ha$^{-1}$).

The physiological efficiency (PE) was obtained by dry mass production per unit of nutrient accumulated, which is calculated by the following formula:

$$\text{PE} = \frac{Y_f - Y_0}{N_{abf} - N_{ab0}} \left( \text{kg kg}^{-1} \right), \tag{2}$$

where $Y_f$ is the total dry matter production in kg ha$^{-1}$ with fertilization; $Y_0$ is the total dry matter production in kg ha$^{-1}$ without fertilization; $N_{abf}$ is the accumulation of nutrient under fertilization in kg ha$^{-1}$; $N_{ab0}$ is the accumulation of nutrient without fertilization in kg ha$^{-1}$.

The recovery efficiency of applied nitrogen (RAN) was the accumulated amount of nutrient per unit of nutrient applied, which is calculated using the following formula:

$$\text{RAN} = \frac{QN_f - N_0}{Q_f} \left( \text{kg kg}^{-1} \right), \tag{3}$$

where $QN_f$ is the accumulation of nitrogen in kg ha$^{-1}$ with fertilization; $N_0$ is the accumulation of nitrogen in kg ha$^{-1}$ without fertilization; $Q_f$ is the amount of applied fertilizer in kg ha$^{-1}$.

All data were subjected to analysis of variance by SAS 9.2 [22] using a significance level of 0.05. When significant, treatment effects on the measured or estimated parameters were evaluated by polynomial regression analysis, using means averaged across sampling dates within each cropping season. The peak of maximum response to N application of each parameter was obtained by the derivative of the quadratic adjustment.

## 3. Results

3.1. *Forage Production.* Forage production (MV) of sorghum was influenced significantly ($P < 0.05$) by N fertilization in both seasons evaluated (Figure 2). For the first year assessed (2010/2011), the highest production was obtained under application of 288 kg N ha$^{-1}$, with peak production of 73.84 Mg MV ha$^{-1}$ and with a quadratic response (Figure 2(a)). The highest production in 2011/2012 was obtained under 267 kg N ha$^{-1}$, reaching a peak of 33.05 Mg MV ha$^{-1}$, also with a significant quadratic response (Figure 2(b)). The difference between years is explained by the rainfall deficit during November-December 2011 (Figure 1), which caused a sown delay of 77 d in second season establishment. The DM content (%) in the forage produced by sorghum was not affected by N rates (Figures 2(a) and 2(b)), reaching mean values of 17.43 and 17.09% for 2010/2011 and 2011/2012, respectively.

The rate of 190 kg N ha$^{-1}$ was responsible for the highest black oat forage production in the winter of 2011, reaching 33.91 Mg MV ha$^{-1}$ (Figure 2(c)). It demonstrates that higher N rates may not be necessary in high-yielding environments, being necessary to calibrate an adequate rate for each crop

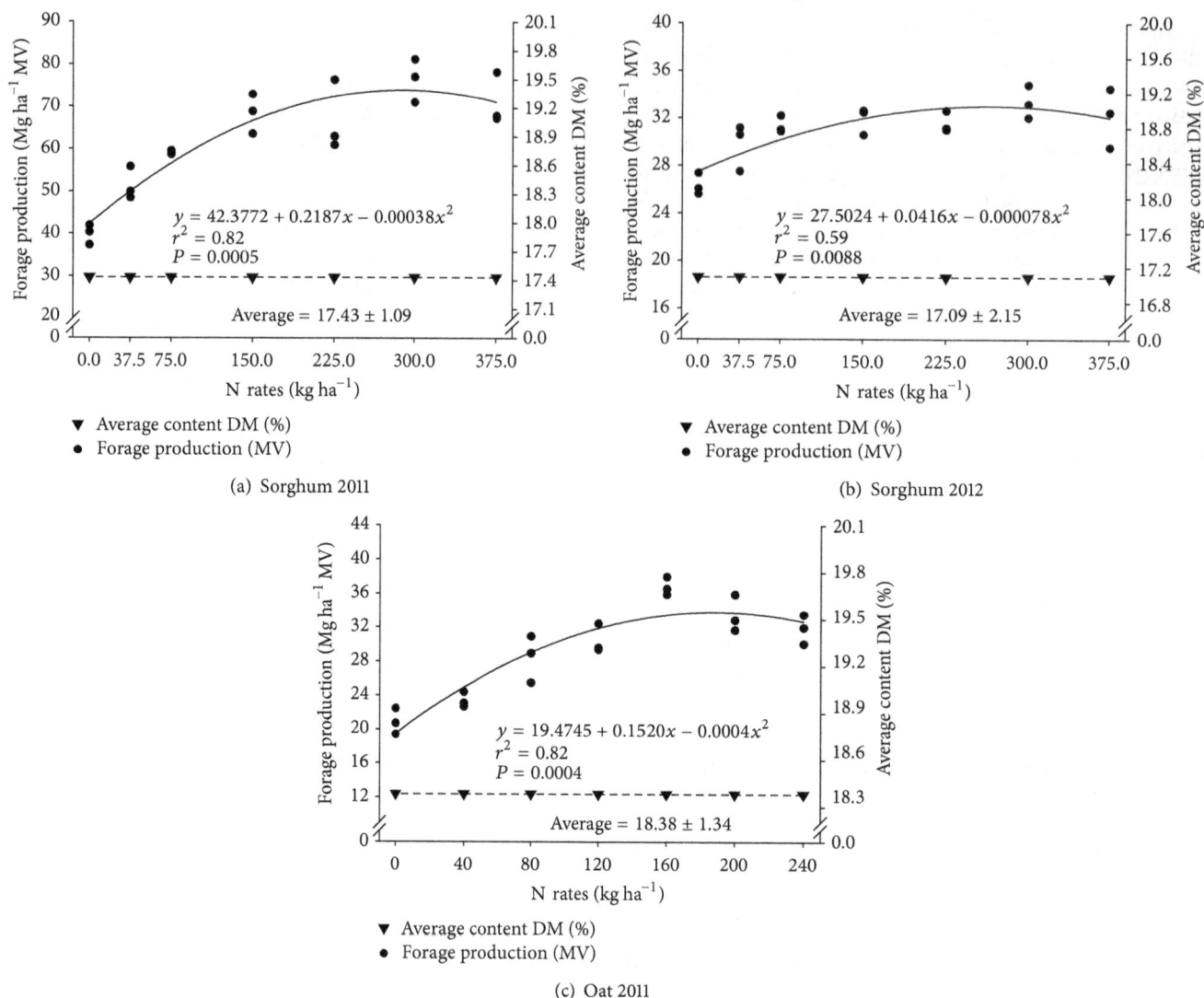

(a) Sorghum 2011

(b) Sorghum 2012

(c) Oat 2011

FIGURE 2: Cumulative production of forage ($Mg\,ha^{-1}$) and average content of dry matter (%) of sorghum in the summer seasons 2010/2011 (a) and 2011/2012 (b) and black oat in the winter of 2011 (c), under nitrogen fertilization.

species and climate condition. Also, these rates would be even lower than the maximum output obtained in this experiment when the economic viability in the recommendation is considered. Regarding the average DM of forage produced by black oat, there was no significant effect by N fertilization.

### 3.2. Nitrogen Extraction and Use Efficiency by Forage Production.

The amount of N exported through sorghum forage was affected by N rates applied ($P < 0.05$), for both years evaluated (Figures 3(a) and 3(b)), with a quadratic adjustment. In the crop season of 2010/2011, the highest N exportation was achieved under the rate of $298\,kg\,N\,ha^{-1}$, which was lower in 2011/2012, about $234\,kg\,N\,ha^{-1}$, reaching an export of 320 and $140\,kg\,N\,ha^{-1}$ each year, respectively. Comparing both seasons, exportation in 2010/2011 approached $180\,kg\,N\,ha^{-1}$ higher than in 2011/2012, explained by low rainfall and crop

establishment delay in the second cycle, as already mentioned. For black oat, the N exported by forage was also adjusted to a quadratic equation (Figure 3(c)) with the largest values under $218\,kg\,N\,ha^{-1}$.

The recovery of N by sorghum and black oat presented a significant effect of N rates ($P < 0.05$), for the three cycles evaluated (Figure 4). The N recovery by sorghum in 2010/2011 (Figure 4(a)) showed a negative linear response, reducing the amount of N recovered according to the increase in N rate. The same behavior was observed for the black oat in the winter of 2011 (Figure 4(c)). The response in N recovery of sorghum was quadratic in 2011/2012 season (Figure 4(b)). When comparing the crop years for sorghum, 2010/2011 presented a much higher N recovery than 2011/2012, explained by the low amount of rainfall in the second year.

The physiological efficiency of N means the amount of DM produced per unit of N absorbed by the crop. The

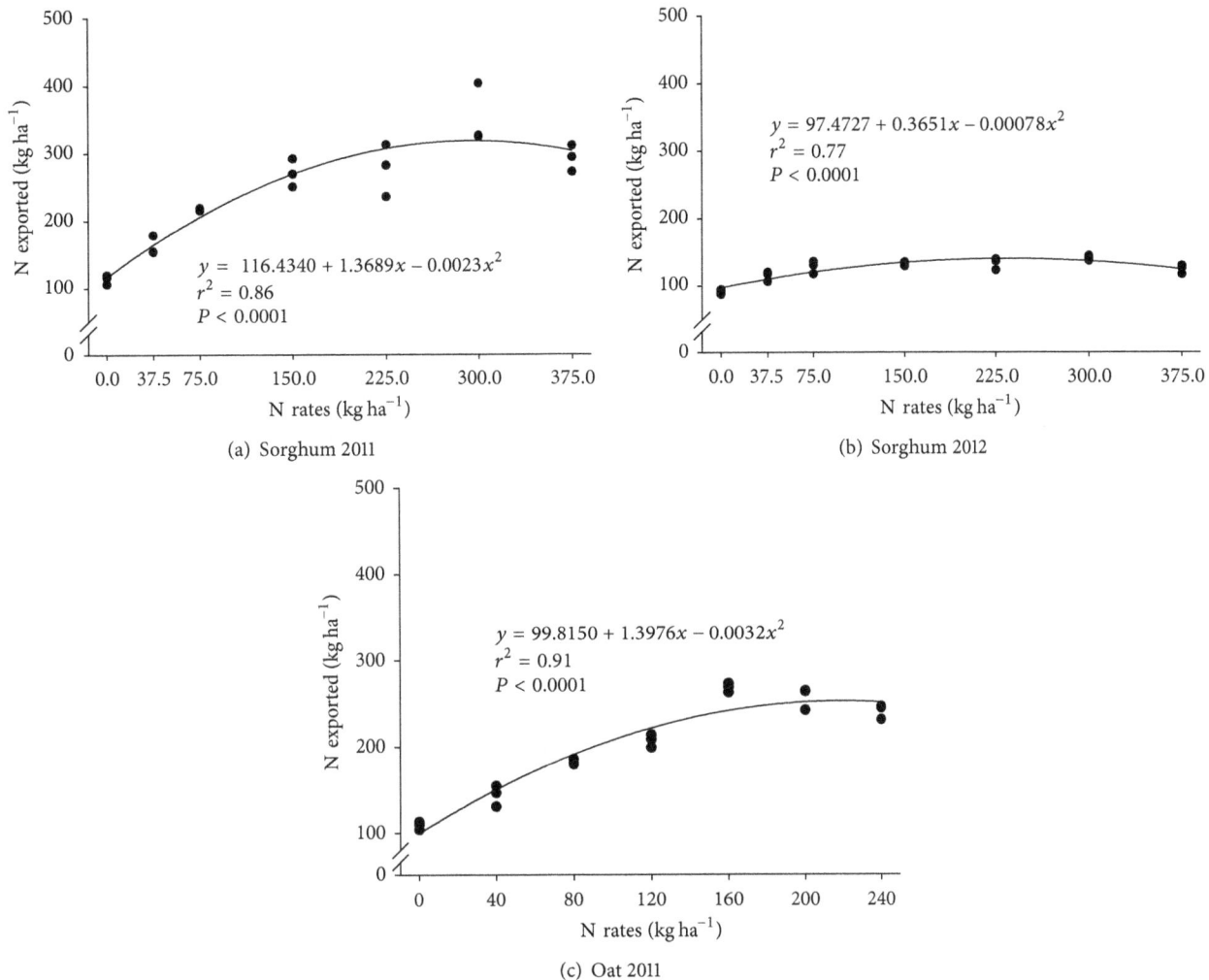

$y = 116.4340 + 1.3689x - 0.0023x^2$
$r^2 = 0.86$
$P < 0.0001$

(a) Sorghum 2011

$y = 97.4727 + 0.3651x - 0.00078x^2$
$r^2 = 0.77$
$P < 0.0001$

(b) Sorghum 2012

$y = 99.8150 + 1.3976x - 0.0032x^2$
$r^2 = 0.91$
$P < 0.0001$

(c) Oat 2011

FIGURE 3: Nitrogen exported by the forage harvest of sorghum in the summer seasons of 2010/2011 (a) and 2011/2012 (b) and black oat in the winter season of 2011 (c), under nitrogen fertilization.

physiological efficiency of N by sorghum and oats was influenced significantly ($P < 0.05$) by N fertilization (Figure 5). For sorghum, a quadratic effect was observed, reducing the efficiency according to the increases in N rates (Figures 5(a) and 5(b)). The values ranged from 40, 34, 32, 29, 28, and 27 kg DM kg$^{-1}$ N absorbed in 2010/2011 season, and 28, 20, 18, 17, 14, and 12 kg DM kg$^{-1}$ N absorbed in the season 2011/2012, under the rates of 37.5, 75, 150, 225, 300, and 375 kg N ha$^{-1}$, respectively. For black oat, there was also a quadratic effect of applied N on the physiological efficiency of N absorbed, reaching values of 22, 19, 18, 17, 17 : 16 kg DM kg$^{-1}$ N absorbed under rates of 40, 80, 120, 160, 200, and 240 kg N ha$^{-1}$, respectively (Figure 5(c)). Lower N rates (37.5 and 75 kg ha$^{-1}$) were physiologically more efficient than higher ones, for both crop species.

The nutrient use efficiency is essential for a diagnosis of the nutrient use by crops or to detect losses of this nutrient in the crop system. The efficiency of N recovery by sorghum in the years 2010/2011 (Figure 6(a)) and 2011/2012 (Figure 6(b)) and black oat in 2011 (Figure 6(c)) was highly influenced

by N fertilization ($P < 0.05$). The N levels on recovery efficiency of sorghum presented a quadratic effect, decreasing according to N increases, with values ranging from 62 to 14 kg DM kg$^{-1}$ N in the first year and 20 to 2 kg DM kg$^{-1}$ N in the second year, under 37.5 and 375 kg N ha$^{-1}$, respectively. Black oat showed a linear response, with values ranging from 24 to 9 kg DM kg$^{-1}$ N applied under 40 to 240 kg N ha$^{-1}$, respectively.

3.3. Phosphorus and Potassium Extraction. The amount of P accumulated in tissue and exported by the forage harvested is also a way to measure the effect of N fertilization in plant nutrient balance. Exported P by sorghum and black oat was directly influenced by N ($P < 0.05$) (Figure 7). For sorghum, the year 2010/2011 presented a quadratic response to N fertilization, reaching the highest P removal (46 kg ha$^{-1}$) under 375 kg N ha$^{-1}$ do not reaching the highest peak (Figure 7(a)). Otherwise, during the year 2011/2012, the response was linear, obtaining the highest P exported (24 kg ha$^{-1}$) with the higher N rate, 375 kg N ha$^{-1}$ (Figure 7(b)). Linear behavior was

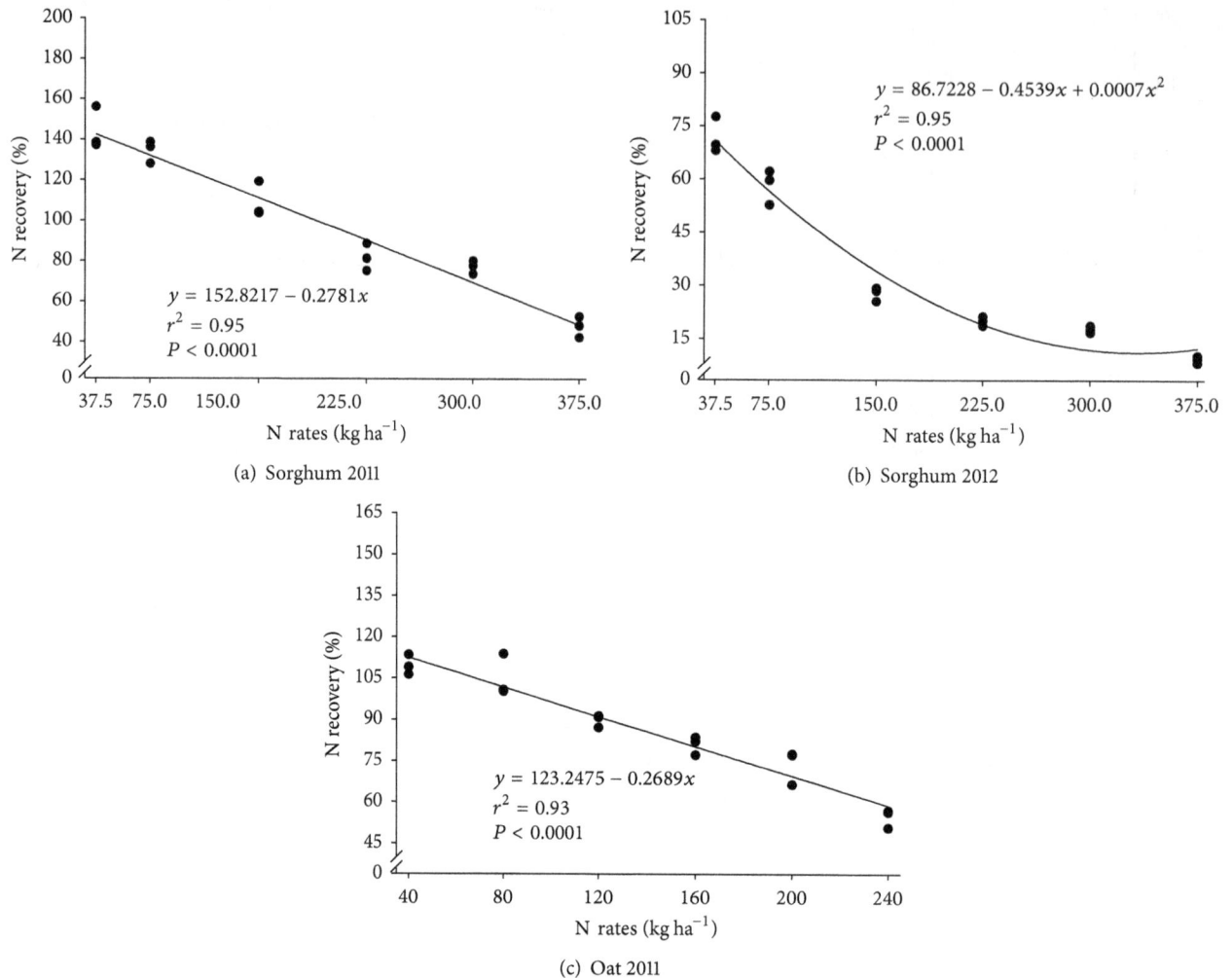

$$y = 152.8217 - 0.2781x$$
$$r^2 = 0.95$$
$$P < 0.0001$$

(a) Sorghum 2011

$$y = 86.7228 - 0.4539x + 0.0007x^2$$
$$r^2 = 0.95$$
$$P < 0.0001$$

(b) Sorghum 2012

$$y = 123.2475 - 0.2689x$$
$$r^2 = 0.93$$
$$P < 0.0001$$

(c) Oat 2011

FIGURE 4: Nitrogen efficiency (%) by the forage of sorghum in the summer seasons of 2010/2011 (a) and 2011/2012 (b) and black oat in the winter season of 2011 (c), under nitrogen fertilization.

also observed for black oat, with accumulated P exportation from 23 to 32 kg P ha$^{-1}$ under 0 to 240 kg N ha$^{-1}$, respectively (Figure 7(c)).

Potassium exported by annual pastures was influenced by N application ($P < 0.05$), since the forage production and removal are higher under fertilization. Sorghum presented a quadratic response in K exported in the year 2010/2011, reaching the peak of K removal (180 kg ha$^{-1}$) under the rate of 271 kg N ha$^{-1}$ (Figure 8(a)). Also the same behavior was observed in the year 2011/2012 but the peak of K exportation (83 kg ha$^{-1}$) was obtained under 324 kg N ha$^{-1}$ (Figure 8(b)). The difference of 97 kg K ha$^{-1}$ exported by sorghum between the years was due to lower forage production in the second season by the lack of rainfall, which was also the consequence of low crop response to N fertilization. For black oat, the K exported was also influenced by N, as shown in Figure 8(c), with a peak removal (79 kg K ha$^{-1}$) reaching up to 198 kg N ha$^{-1}$.

## 4. Discussion

Sorghum forage production (MV) in two seasons and black oat in one season were influenced by N fertilization, with quadratic responses observed (Figure 2). According to Heringer and Moojen [23], the quadratic response in MV is expected due to the limitations by other climatic and soil factors, since there is a certain limit in the available N in soil that plant can express all the productive potential. When exceeded, it causes an imbalance of other nutrients, decreasing the production and promoting unnecessary fertilizer spending.

In Brazil, there are several forage species that produce a large amount of MV. Oliveira et al. [24] observed yields of 67.18, 66.48, 82.00, and 83.90 Mg MV ha$^{-1}$ of maize (*Zea mays*), sorghum-Sudan (*Sorghum sudanensis*), sorghum, and sunflower (*Helianthus annuus*), respectively, similar to the ones observed here for sorghum in 2010/2011. Evaluating 29 sorghum genotypes, Cunha and Lima [25] found MV ranging

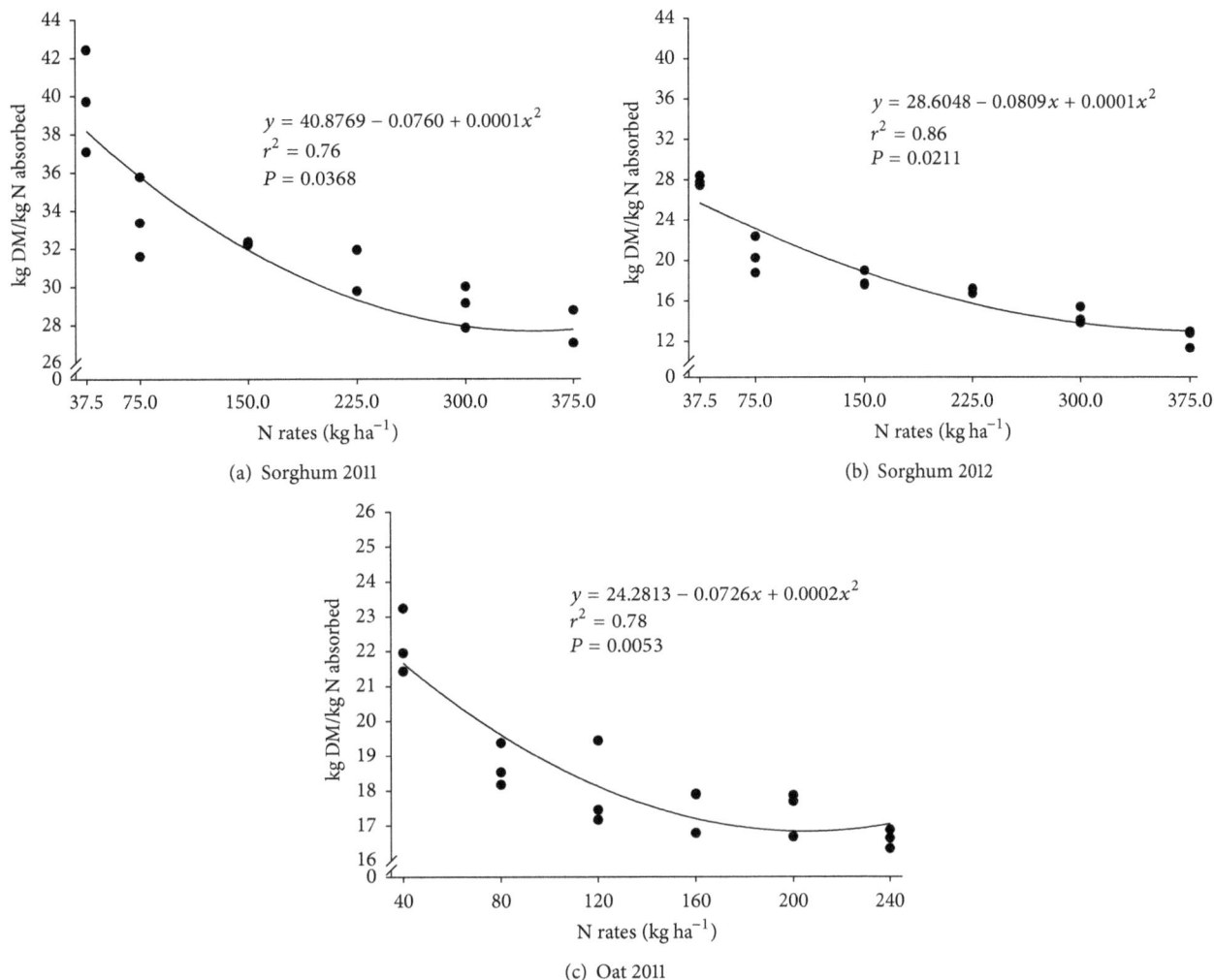

(a) Sorghum 2011

$$y = 40.8769 - 0.0760 + 0.0001x^2$$
$$r^2 = 0.76$$
$$P = 0.0368$$

(b) Sorghum 2012

$$y = 28.6048 - 0.0809x + 0.0001x^2$$
$$r^2 = 0.86$$
$$P = 0.0211$$

(c) Oat 2011

$$y = 24.2813 - 0.0726x + 0.0002x^2$$
$$r^2 = 0.78$$
$$P = 0.0053$$

FIGURE 5: Physiological N efficiency (kg DM kg$^{-1}$ N absorbed) by the forage of sorghum in the summer seasons of 2010/2011 (a) and 2011/2012 (b) and black oat in the winter season of 2011 (c), under nitrogen fertilization.

from 15.40 to 68.10 Mg ha$^{-1}$, slightly lower than in the present work. The DM content was evaluated in sorghum-Sudan cv. AG 2501C subjected to N and K rates by Simili et al. [26] who found similar results to this study, with 17.4, 16.3, and 17.3% when under 100, 200, and 300 kg N ha$^{-1}$, respectively. Cunha and Lima [25] found values ranging from 17 to 33%, higher than the present ones, probably as consequence of the different genotypes evaluated.

Black oat obviously presents lower potential of MV production than sorghum. Luz et al. [27] found yields of 18.36, 18.40, 19.49, and 18.74 Mg MV ha$^{-1}$ under irrigation and 5.72, 5.89, 5.86, and 5.92 Mg MV ha$^{-1}$ without irrigation, under 0, 50, 100, and 150 kg N ha$^{-1}$, respectively, lower than those found in the present work. Results of sorghum and black oat MV show high response to N fertilization, but it is also important to consider the genotypes selection, region climate, and especially rainfall distribution. Dry matter content is not affected by N fertilization; however, the water availability and plant phenological stage at fodder harvest are the main factors that can influence directly this character.

The amount of N exported by sorghum was affected by N rates. There was a higher N accumulation in tissue in 2010/2011 (Figure 3(a)) in relation to 2011/2012 (Figure 3(b)), which can be explained by lower forage production. However, the response of the exportation in relation to N rates applied was quadratic for both years, meaning that crop answer to N fertilization is limited by other factors. Han et al. [13], evaluating five sorghum hybrids in Beijing, China, under 120 kg N ha$^{-1}$, found exportation of 128 and 329 kg N ha$^{-1}$ with early and late hybrids, respectively, corroborating the results found in the current study.

Mateus et al. [28], evaluating the MV and nutrient accumulation in shoot of Guinea sorghum (Sorghum bicolor sp. Guinea), in the crop season of 2000/2001 in São Paulo, Brazil, found an extraction from 270 to 75 kg N ha$^{-1}$ for early and late sowing, respectively. It was slightly lower than those found here in the first season but similar to the ones observed in the second season evaluated. It is clear that, besides N rates, sowing period and cultivars (early or late) have direct effect on the amount of N extracted, which justifies the lower N extraction in the second sorghum crop season [13, 28].

(a) Sorghum 2011

(b) Sorghum 2012

(c) Oat 2011

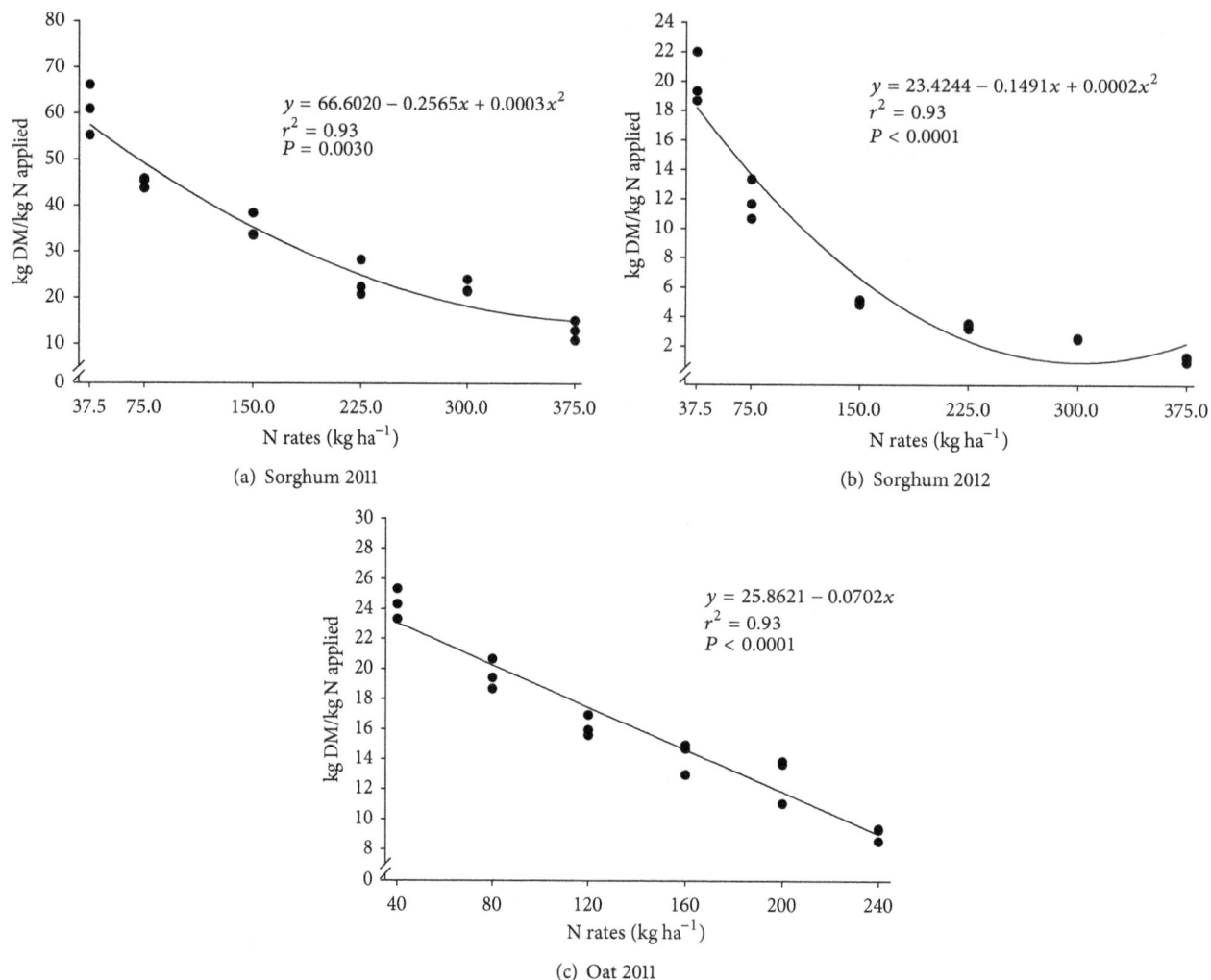

Figure 6: Nitrogen recovery efficiency (kg DM kg$^{-1}$ N applied) by the forage of sorghum in the summer seasons of 2010/2011 (a) and 2011/2012 (b) and black oat in the winter season of 2011 (c), under nitrogen fertilization.

Extraction of N by black oat was also influenced positively by N rates (Figure 3(c)). The present results are higher than those normally reported in the literature, ranging from 99 kg N ha$^{-1}$ in control to 241 kg N ha$^{-1}$ under 160 kg ha$^{-1}$ N. This difference can be explained by the cultivar planted, Iapar 61, genetically improved to assess greater vegetative cycle, produce more forage with higher feed quality, which provides a greater accumulation of N in tissue, ideal for regions of higher climatic potential [29].

Determining the recovery efficiency and the N utilization in forage production is essential to improve the system, since, in situations that the amount of N applied exceeds plant absorption capacity to convert into forage production, it can be lost by leaching, volatilization, or accumulating in plant tissues, reducing the DM conversion efficiency [30]. The recoveries of N by sorghum were higher under lower N rates for both years assessed, with levels ranging from 144 to 48% for the season 2010/2011 and 72 to 9% for the season 2011/2012, under 37.5 and 375 kg N ha$^{-1}$, respectively (Figures 4(a) and 4(b)).

The N recovery by forage may be influenced by several factors, like lack of rainfall, N rates, genetic potential of the crop, and crop management. Under adequate conditions of climate and soil, the recovery of N can exceed 100%, as observed in this study (Figure 4) and in the literature. Heringer and Moojen [23], working with millet under N fertilization, had negative linear response to N recovery from 168, 99, 94, and 79% when under rates of 150, 300, 450, and 600 kg ha$^{-1}$, respectively. Also, Sartor et al. [5], working with Alexander grass (Brachiaria plantaginea), found a recovery efficiency of 110 and 49% under 200 and 400 kg N ha$^{-1}$, respectively. For black oat, the lowest N rates were the most efficient in N recovery (Figure 4(c)). Under the rate 40 kg N ha$^{-1}$, recovery efficiency of 110% was found; however, under 240 kg N ha$^{-1}$, the efficiency decreased to 55%. Also, Kolchinski and Schuch [31] found N recovery efficiency in oats ranging from 76 to 64% under rates from 20 to 80 kg N ha$^{-1}$, respectively, lower than those found in the present work.

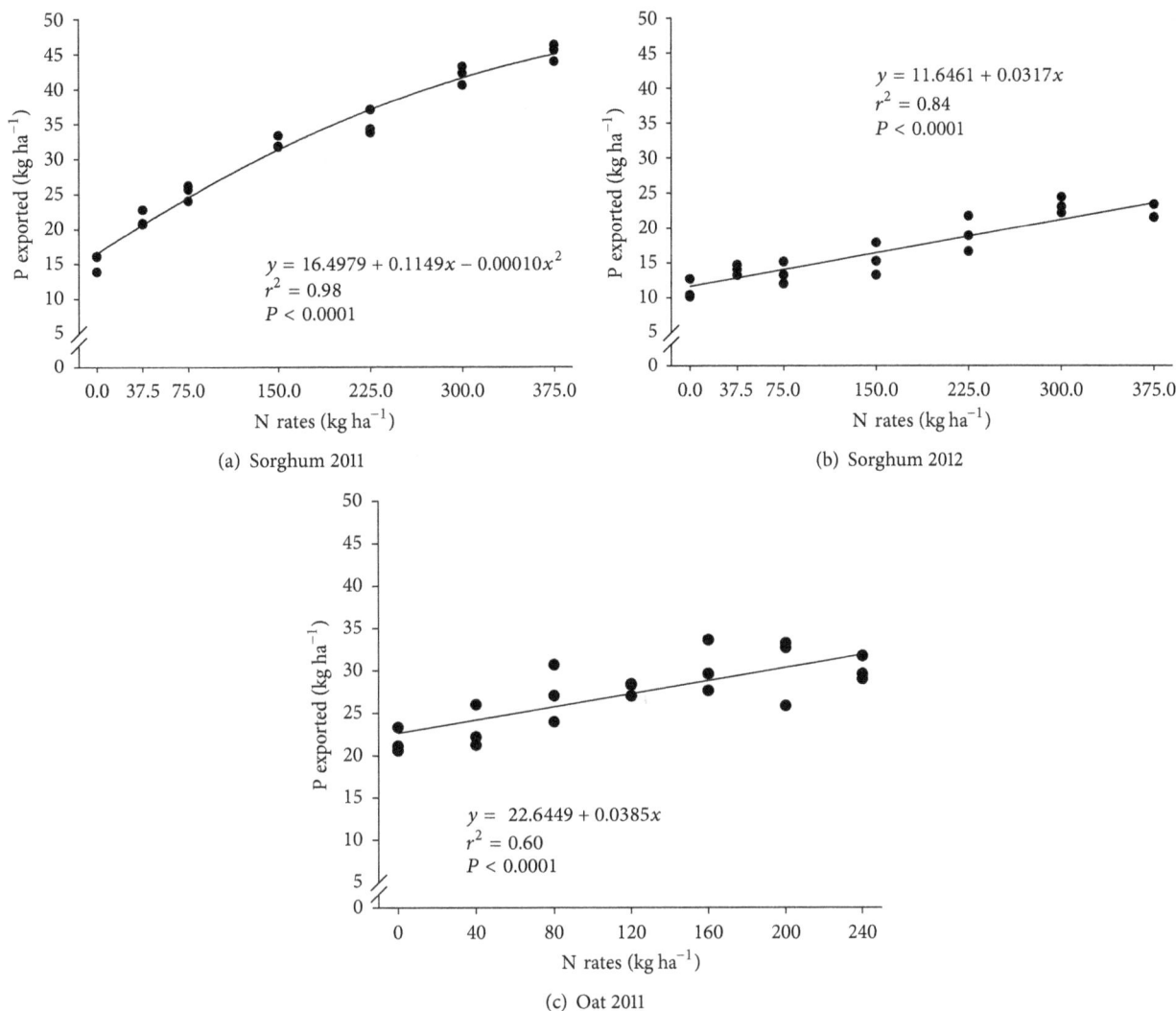

$$y = 16.4979 + 0.1149x - 0.00010x^2$$
$$r^2 = 0.98$$
$$P < 0.0001$$

(a) Sorghum 2011

$$y = 11.6461 + 0.0317x$$
$$r^2 = 0.84$$
$$P < 0.0001$$

(b) Sorghum 2012

$$y = 22.6449 + 0.0385x$$
$$r^2 = 0.60$$
$$P < 0.0001$$

(c) Oat 2011

FIGURE 7: Phosphorus exported by the forage harvested of sorghum in the summer seasons of 2010/2011 (a) and 2011/2012 (b) and black oat in the winter season of 2011 (c), under nitrogen fertilization.

The N physiological efficiency by sorghum was affected by N fertilization (Figures 5(a) and 5(b)), obtaining the highest values, 40 and 28 kg DM $kg^{-1}$ N absorbed for 2010/2011 and 2011/2012, respectively, when applying the lowest N rate, 37.5 kg $ha^{-1}$. Lupatini et al. [32], working with millet, found a physiological efficiency of 35 and 33 kg DM $kg^{-1}$ N absorbed under rates of 150 and 300 kg $ha^{-1}$ N, respectively, slightly higher than the results obtained here. Moreover, Heringer and Moojen [23] observed a progressive reduction in N physiological efficiency by millet, reaching values of 45 and 14 kg DM $kg^{-1}$ N absorbed under rates of 150 and 600 kg N $ha^{-1}$, respectively, which was also observed here, with about 13 kg DM $kg^{-1}$ N absorbed under the highest rate in the second season (2011/2012).

The N physiological efficiency of black oat was also influenced by N fertilization (Figure 5(c)), with values of 22 to 16 kg DM $kg^{-1}$ N absorbed under rates of 40 and 240 kg N $ha^{-1}$, respectively. Lupatini et al. [33] found a physiological efficiency of 31 and 25 kg DM $kg^{-1}$ N absorbed under rates of 150 and 300 kg N $ha^{-1}$, respectively, in black oat consorted to ryegrass (Lolium multiflorum Lam.), results slightly higher than the present ones. However, when N rates increased, physiological efficiency was reduced, as observed here and in most literature information. According to Pellegrini et al. [34], N fertilization may affect the production and quality of forage in ryegrass but does not alter the N physiological efficiency. Those authors found a physiological average efficiency of 29 kg DM $kg^{-1}$ N absorbed under rates of 75, 150, and 225 kg N $ha^{-1}$, contradicting the results found here, which could be explained by the way the authors evaluated the fodder, simulating grazing animals; thus it is known that animals, particularly sheep used in that research, are selective, with ability to select their diet with more than 80% of leaves.

The efficiency of N recovery by sorghum (Figures 6(a) and 6(b)) showed quadratic and negative effect in both seasons evaluated, with maximum efficiency of 61 and 20 and minimum of 13 and 1.5 kg DM $kg^{-1}$ N applied, under the rates of

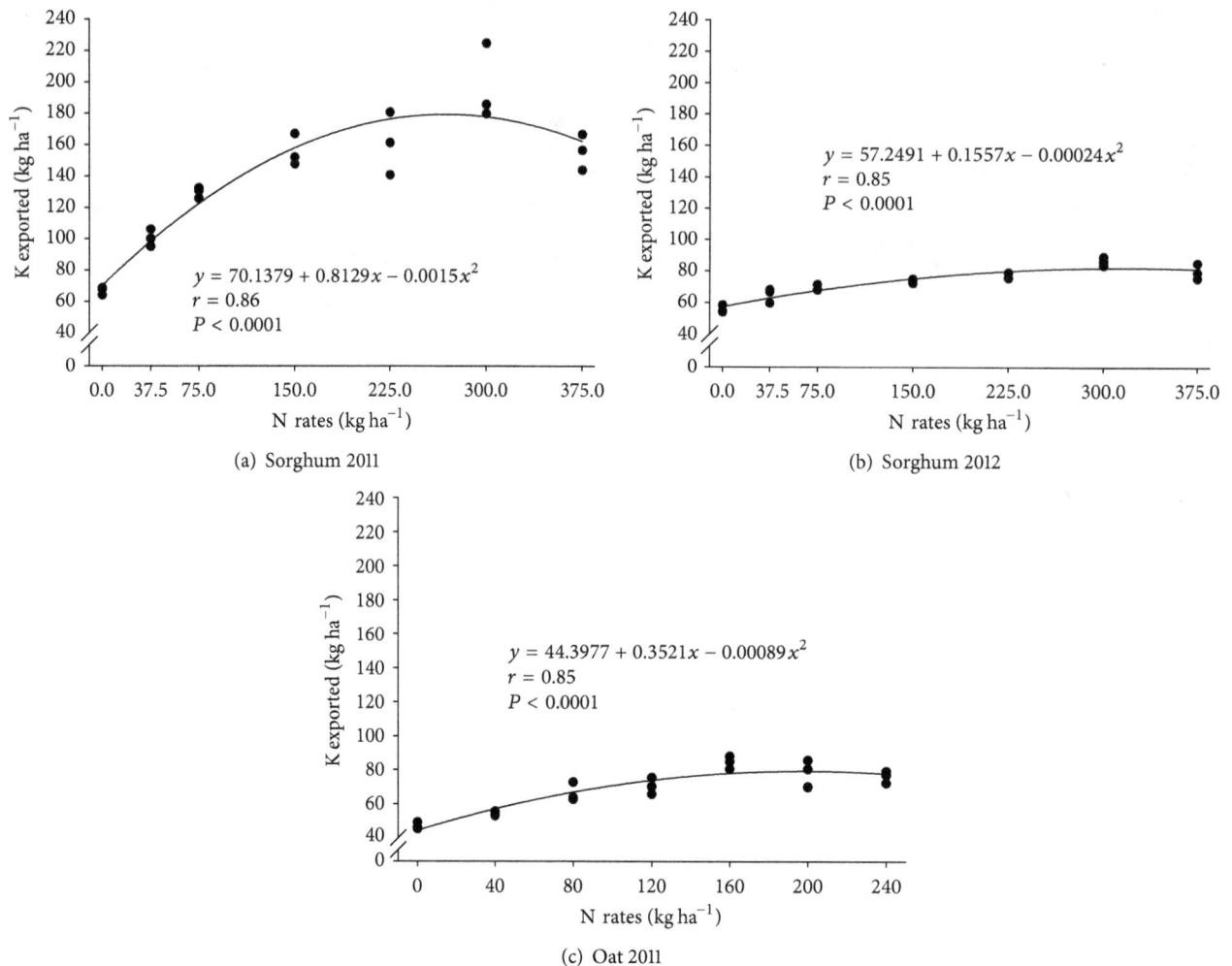

FIGURE 8: Potassium exported by the forage harvest of sorghum in the summer seasons of 2010/2011 (a) and 2011/2012 (b) and black oat in the winter season of 2011 (c), under nitrogen fertilization.

37.5 and 375 kg N ha$^{-1}$, respectively. Olanite et al. [35] found, in Nigeria, N recovery efficiency of 2, 9, and 6 kg DM kg$^{-1}$ N, respectively, under the rates of 60, 120, and 180 kg N ha$^{-1}$ in sorghum, much lower than those found here. There is huge amplitude of information in literature about efficiency of N recovery by sorghum, being credited to weather and soil variations, since it is known that N is a nutrient easily lost in soil, depending on climate changes (excess or lack of rain), fertilizer amount applied, and soil properties.

The efficiency of N recovery by black oat was also significantly influenced by N rates (Figure 6(c)), reaching a maximum of 24 and minimum of 9 kg DM kg$^{-1}$ N under the rates of 40 and 240 kg ha$^{-1}$ N, respectively. Moreira et al. [29] found 29, 17, and 11 kg DM kg$^{-1}$ N applied under the rates of 50, 100, and 200 kg N ha$^{-1}$, respectively, corroborating the results presented here.

Phosphorus is a vital component of the plant cell and has the function of stimulating plant growth and formation of the root system in early development [36]. Among the macronutrients, phosphorus is one of the elements less absorbed by sorghum and oats. The amount of P exported by sorghum

and oats was influenced by N rates (Figure 7), reaching the highest extractions in sorghum (46 and 24 kg P ha$^{-1}$) under the highest rate (375 kg N ha$^{-1}$) and lower extractions (17 and 12 kg ha$^{-1}$ P) without N fertilization, in the seasons 2010/2011 and 2011/2012, respectively. Han et al. [13] found values of P extracted ranging from 30 to 75 kg ha$^{-1}$ in sorghum hybrids, confirming the results found here, showing that N levels directly influence P extraction. Mateus et al. [28] observed that sorghum Guinea can return to the soil through biomass equivalent to 288 kg ha$^{-1}$ of superphosphate, concluding that sorghum is an excellent alternative to be used as cover crop.

The highest P extraction by black oat, 32 kg ha$^{-1}$, was observed under 240 kg N ha$^{-1}$ and reduced linearly to 23 kg P ha$^{-1}$ under no fertilization (Figure 7(c)). Evaluating four pig manure rates in single oat and intercropped with vetch, Aita et al. [37] found no significant interaction between waste rate, crop species, and N, P, and K accumulation in shoot, with mean values of P extracted of 28, 27, and 13 kg ha$^{-1}$ under single oat, oat consorted to vetch, and spontaneous vegetation, respectively. Moreover, the climatic conditions can greatly restrict nutrient uptake by plants. Working on a

Brazilian Cerrado region with rainfall restrictions, Torres et al. [38] found an extraction of only 13 kg P ha$^{-1}$ for sorghum and 4 kg ha$^{-1}$ for oats, much lower than those found here.

Potassium develops many important metabolic functions in plants, specially acting in enzyme activation reactions, moving freely within the plant. For sorghum, the K extracted by forage tissue was significantly influenced by N fertilization (Figures 8(a) and 8(b)), reaching the highest values of 180 and 83 kg ha$^{-1}$ under 271 and 324 kg N ha$^{-1}$ in the seasons 2010/2011 and 2011/2012, respectively. Oliveira et al. [24] found an amount of K extracted of 225 and 292 kg ha$^{-1}$ when evaluating sorghum-Sudan and common sorghum, respectively, higher than those found here. According to these authors, in crops under soil K deficiency, the root system will be concentrated in soil surface layers and underdeveloped, obstructing the absorption of water and nutrients. Han et al. [13] found values of 109 to 300 kg K ha$^{-1}$ extracted for early and mid/late sorghum hybrids, respectively. The average values obtained in this work are lower than those found in the literature, which can be explained by the early phenological stage of crop harvest, whose extraction would be lower compared to a reproductive stage harvest.

For black oat, K extraction was also influenced by N rates (Figure 8(c)), reaching the peak of 79 kg K ha$^{-1}$ under 198 kg N ha$^{-1}$, with a quadratic behavior. According to Borkert et al. [39], oats can be a good K recycler, since it has a deep root system, allowing the recycling of soil K previously leached to subsurface layers. Giacomini et al. [40] found an extraction by black oat of 85, 82, and 74 kg K ha$^{-1}$, respectively, for the years 1998, 1999, and 2000. The authors emphasize that oat in crop rotation could be important to reduce K losses by leaching, reducing costs in fertilizer and improving gradually the soil quality for subsequent crops.

## 5. Conclusions

In general, the highest forage production of sorghum is observed under rates from 230 to 300 kg N ha$^{-1}$, with the highest N, P, and K extraction. For black oat cv. Iapar 61, the highest forage production is up to 190 kg N ha$^{-1}$.

The values of N recovery, recovery efficiency, and N-use physiological efficiency are higher under lower N rates for both crops. Based on this, about 150 kg N ha$^{-1}$ is recommended to obtain good sorghum forage production with higher N efficiency levels, which is about 120 kg N ha$^{-1}$ for black oat.

## Conflict of Interests

The authors declare that there is no conflict of interests regarding the publication of this paper.

## Acknowledgments

Thanks are due to the Federal Technology University of Paraná (UTFPR), Dois Vizinhos, for providing the experimental area as well as the laboratory and all of the necessary material for the development and analysis of the experiment. Thanks are due to the Program of Tutorial Education (PET-Zootecnia) and the Undergraduate Program of Research Fellowship of UTFPR by grants to the students of animal science, which were dedicated to complete successfully the research work. Thanks are also due to the Education Assistance Program (PAE) for master's scholarship granted during graduate study. The paper is originated from the master's dissertation of the first author.

## References

[1] M. B. Dias-Filho, *Desafios da produção animal em pastagens na fronteira agrícola brasileira*, Embrapa Amazônia Oriental, Belém, Brazil, 2012.

[2] A. C. Primavesi, O. Primavesi, L. A. Corrêa et al., "Adubação nitrogenada em capim- coastcross : efeitos na extração de nutrientes e recuperação aparente do nitrogênio," *Revista Brasileira de Zootecnia*, vol. 33, no. 1, pp. 68–78, 2004.

[3] G. B. Martha Júnior, M. Corsi, P. C. O. Trivelin et al., "Perda de amônia por volatilização em pastagem de capim-tanzânia adubada com uréia no verão," *Revista Brasileira de Zootecnia*, vol. 33, no. 6, pp. 2240–2247, 2004.

[4] O. Oenema, H. Kros, and W. de Vries, "Approaches and uncertainties in nutrient budgets: implications for nutrient management and environmental policies," *European Journal of Agronomy*, vol. 20, no. 1-2, pp. 3–16, 2003.

[5] L. R. Sartor, T. S. Assmann, A. B. Soares, P. F. Adami, A. L. Assmann, and C. S. R. Pitta, "Nitrogen fertilizer use efficiency, recovery and leaching of an Alexandergrass pasture," *Revista Brasileira de Ciência do Solo*, vol. 35, no. 3, pp. 899–906, 2011.

[6] G. Lemaire, F. Gastal, and J. Salette, "Analysis of the effect of N nutrition on dry matter yield of a sward by reference to potential yield and optimum N content," in *Proceedings of the 16th International Grassland Congress*, pp. 179–180, 1989.

[7] D. C. Alves Filho, M. Neumann, J. Restle, A. N. M. Souza, and L. A. O. Peixoto, "Características agronômicas produtivas, qualidade e custo de produção de forragem em pastagem de azevém (*Lolium multiflorum* Lam) fertilizada com dois tipos de adubo," *Ciência Rural*, vol. 33, no. 1, pp. 143–149, 2003.

[8] S. Amaducci, A. Monti, and G. Venturi, "Non-structural carbohydrates and fibre components in sweet and fibre sorghum as affected by low and normal input techniques," *Industrial Crops and Products*, vol. 20, no. 1, pp. 111–118, 2004.

[9] Y. L. Zhao, A. Dolat, Y. Steinberger, X. Wang, A. Osman, and G. H. Xie, "Biomass yield and changes in chemical composition of sweet sorghum cultivars grown for biofuel," *Field Crops Research*, vol. 111, no. 1-2, pp. 55–64, 2009.

[10] D. R. Buxton, I. C. Anderson, and A. Hallam, "Performance of sweet and forage sorghum grown continuously, double-cropped with winter rye, or in rotation with soybean and maize," *Agronomy Journal*, vol. 91, no. 1, pp. 93–101, 1999.

[11] A. Almodares and M. R. Hadi, "Production of bioethanol from sweet sorghum: a review," *African Journal of Agricultural Research*, vol. 4, no. 9, pp. 772–780, 2009.

[12] C. Gao, Y. Zhai, Y. Ding, and Q. Wu, "Application of sweet sorghum for biodiesel production by heterotrophic microalga *Chlorella protothecoides*," *Applied Energy*, vol. 87, no. 3, pp. 756–761, 2010.

[13] L. P. Han, Y. Steinberger, Y. L. Zhao, and G. H. Xie, "Accumulation and partitioning of nitrogen, phosphorus and potassium in

different varieties of sweet sorghum," *Field Crops Research*, vol. 120, no. 2, pp. 230–240, 2011.

[14] N. Tsuchihashi and Y. Goto, "Cultivation of sweet sorghum (*Sorghum bicolor* (L.) moench) and determination of its harvest time to make use as the raw material for fermentation, practiced during rainy season in dry land of Indonesia," *Plant Production Science*, vol. 7, no. 4, pp. 442–448, 2004.

[15] C. S. Wortmann, A. J. Liska, R. B. Ferguson, D. J. Lyon, R. N. Klein, and I. Dweikat, "Dry land performance of sweet sorghum and grain crops for biofuel in Nebraska," *Agronomy Journal*, vol. 102, no. 1, pp. 319–326, 2010.

[16] J. Nakagawa, C. Cavariani, and J. R. Machado, "Adubação nitrogenada no perfilhamento da aveia-preta em duas condições de fertilidade do solo," *Pesquisa Agropecuária Brasileira*, vol. 35, no. 6, pp. 1071–1080, 2000.

[17] C. W. Thornthwaite and J. R. Mather, *The Water Balance*, vol. 8 of *Publications in Climatology*, Drexel Institute of Technology—Laboratory of Climatology, Centerton, NJ, USA, 1955.

[18] A. R. Pereira, "Simplificado o balanço hídrico de Thornthwaite-Mather," *Bragantia*, vol. 64, no. 2, pp. 311–313, 2005.

[19] Soil Survey Staff, *Soil Taxonomy: A Basic System of Soil Classification for Making and Interpreting Soil Surveys*, USDA Agricultural Handbook no. 436, Natural Resources Conservation Service, Department of Agriculture, EEUU, Washington, DC, USA, 2nd edition, 1999.

[20] M. J. Tedesco, C. Gianello, C. A. Bissani, H. Bohnen, and S. J. Volkweiss, "Análises de Solos Plantas e outros materiais," Boletim Técnico 5, Departamento de solos, Faculdade de Agronomia, Universidade Federal do Rio Grande do Sul-Porto Alegre, Porto Alegre, Brazil, 1995.

[21] N. K. Fageria, "Otimização da eficiência nutricional na produção das culturas," *Revista Brasileira de Engenharia Agrícola e Ambiental*, vol. 2, pp. 6–16, 1998.

[22] SAS Institute, *SAS/STAT. User's Guide, version 9.2*, SAS Institute Inc., Cary, NC, USA, 2008.

[23] I. Heringer and E. L. Moojen, "Potencial produtivo, alterações da estrutura e qualidade da pastagem de milheto submetida a diferentes níveis de nitrogênio," *Revista Brasileira de Zootecnia*, vol. 31, no. 2, pp. 875–882, 2002.

[24] L. B. de Oliveira, A. J. V. Pires, A. E. S. Viana, S. N. Matsumoto, G. G. P. deCarvalho, and L. S. O. Ribeiro, "Produtividade, composição química e características agronômicas de diferentes forrageiras," *Revista Brasileira de Zootecnia*, vol. 39, no. 12, pp. 2604–2610, 2010.

[25] E. E. Cunha and J. M. P. Lima, "Caracterização de genótipos e estimativa de parâmetros genéticos de características produtivas de sorgo forrageiro," *Revista Brasileira de Zootecnia*, vol. 39, no. 4, pp. 701–706, 2010.

[26] F. F. Simili, R. A. Reis, B. N. Furlan, C. C. P. Paz, M. L. P. Lima, and P. A. Bellingieri, "Resposta do híbrido de sorgo-sudão à adubação nitrogenada e potássica : composição química e digestibilidade *in vitro* da matéria orgânica," *Ciência e Agrotecnologia*, vol. 32, no. 2, pp. 474–480, 2008.

[27] P. H. C. Luz, V. R. Herling, G. J. Braga, J. C. M. Nogueira Filho, L. A. Faria, and C. G. Lima, "Resposta da aveia pret (Avena strigosa schreb ) à irrigação por aspersão e adubação nitrogenada," *Acta Scientiarum Agronomy*, vol. 30, pp. 421–426, 2008.

[28] G. P. Mateus, É. Borghi, G. S. A. Castro, R. A. Garcia, and A. C. Crusciol, "Biomass production and accumulation of nutrients in shoots of giant Guinea sorghum plants," *Revista Ciencia Agronomica*, vol. 42, no. 4, pp. 1000–1008, 2011.

[29] F. B. Moreira, U. Cecato, and I. N. do Prado, "Avaliação de aveia preta cv. Iapar 61 submetida a níveis crescentes de nitrogênio em área proveniente de cultura de soja," *Acta Scientiarum*, vol. 23, pp. 815–821, 2001.

[30] J. R. Freney, "Management practices to increase efficiency of fertilizer and animal nitrogen and minimize nitrogen loss to the atmosphere and groundwater," CSIRO Plant Industry, Australia, 2013, http://www.fftc.agnet.org/files/lib_articles/20130422100045/tb186.pdf.

[31] E. M. Kolchinski and L. O. B. Schuch, "Eficiência no uso do nitrogênio por cultivares de aveia branca de acordo com a adubação nitrogenada," *Revista Brasileira de Ciência do Solo*, vol. 27, no. 6, pp. 1033–1038, 2003.

[32] G. C. Lupatini, E. L. Moojen, J. Restle, and J. H. S. Silva, "Resposta do milheto (*Pennisetum americanum* (L) Leeke) sob pastejo à adubação nitrogenada," *Pesquisa Agropecuária Brasileira*, vol. 31, pp. 715–720, 1996.

[33] G. C. Lupatini, J. Restle, M. Ceretta, E. L. Moojen, and H. R. Bartz, "Avaliação da mistura de aveia preta e azevém sob pastejo submetida a níveis de nitrogênio," *Pesquisa Agropecuária Brasileira*, vol. 33, pp. 1939–1943, 1998.

[34] L. G. de Pellegrini, A. L. G. Monteiro, M. Neumann, A. de Moraes, A. C. R. S. de Pellegrin, and S. B. C. Lustosa, "Produção e qualidade de azevém-anual submetido a adubação nitrogenada sob pastejo por cordeiros," *Revista Brasileira de Zootecnia*, vol. 39, no. 9, pp. 1894–1904, 2010.

[35] J. A. Olanite, U. Y. Anele, O. M. Arigbede, A. O. Jolaosho, and O. S. Onifade, "Effect of plant spacing and nitrogen fertilizer levels on the growth, dry-matter yield and nutritive quality of Columbus grass (*Sorghum almum* stapf) in southwest Nigeria," *Grass and Forage Science*, vol. 65, no. 4, pp. 369–375, 2010.

[36] J. F. Karn, "Phosphorus nutrition of grazing cattle: a review," *Animal Feed Science and Technology*, vol. 89, pp. 133–153, 2001.

[37] C. Aita, O. Port, and S. J. Giacomini, "Dinâmica do nitrogênio no solo e produção de fitomassa por plantas de cobertura no outono /inverno com o uso de dejetos de suínos," *Revista Brasileira de Ciência do Solo*, vol. 30, no. 5, pp. 901–910, 2006.

[38] J. L. R. Torres, M. G. Pereira, and A. J. Fabian, "Produção de fitomassa por plantas de cobertura e mineralização de seus resíduos em plantio direto," *Pesquisa Agropecuária Brasileira*, vol. 43, no. 3, pp. 421–428, 2008.

[39] C. M. Borkert, C. A. Gaudêncio, J. E. Pereira, L. R. Pereira, and A. Oliveira Junior, "Nutrientes minerais na biomassa da parte aérea em culturas de cobertura de solo," *Pesquisa Agropecuária Brasileira*, vol. 38, no. 1, pp. 143–153, 2003.

[40] S. J. Giacomini, C. Aita, E. R. O. Vendruscolo, M. Cubilla, R. S. Nicoloso, and M. R. Fries, "Matéria seca, relação C/N e acúmulo de nitrogênio, fósforo e potássio em misturas de plantas de cobertura de solo," *Revista Brasileira de Ciência do Solo*, vol. 27, no. 2, pp. 325–334, 2003.

# 25

# Laboratory Studies on the Effects of Aqueous Extracts from *Sorghum bicolor* Stem and *Zea mays* (Roots and Tassel) on the Germination and Seedling Growth of Okra (*Abelmoschus esculentus* L.)

**Modupe Janet Ayeni and Joshua Kayode**

*Department of Plant Science, Ekiti State University, Ado Ekiti, Nigeria*

Correspondence should be addressed to Joshua Kayode; jokayode@ymail.com

Academic Editor: Albino Maggio

The allelopathic effect of the aqueous extracts from *Sorghum bicolor* stem and maize (roots and tassel) were examined on the germination and seedling growth of okra (*Abelmoschus esculentus* L.). The results showed that the extracts inhibited the germination of okra seeds which was more pronounced in seeds treated with maize (roots and tassel) extracts as no germination was recorded until 48 hours of experimental time. Also the radicle and plumule lengths were retarded. Plumule lengths were more retarded as no germination was recorded until 72 hours of experimental time. The inhibitory effects were concentration dependent as the inhibition increases with increase in concentration of the extracts. Statistical analysis ($P < 0.05$) revealed that there were significant differences in the germination of okra treated seeds most especially at higher concentration of the extracts when compared to control experiment. In the radicle lengths, statistical analysis revealed that there were significant differences in the radicle lengths of the extract treated seeds compared to the control experiment except at 24 hours of experimental time. Similarly in the plumule, significant differences abound in the extract treated seeds from 72 hrs to 144 hrs. These findings indicate that both germination and growth of okra sown in the field may be adversely affected by extracts from these residues, thus resulting in lowering yields especially by the maize root extracts.

## 1. Introduction

The chemical interference of donor plants on another receptor plants thereby affecting them negatively or positively had been established which is referred to as allelopathy [1, 2]. Allelopathic influence can have stimulatory effect on the growth of other plants thereby increasing their growth positively and inhibitory effect by suppression of neighbouring plant growth by the release of toxic compound [3].

Allelopathy plays important role in agroecosystem leading to the interaction crop to crop, crop to weed, weed to crop, and trees to crop [4], through the production of chemical compounds (allelochemicals) that escape into the environment. These allelochemicals are released from plant parts such as leaves, flowers, seeds, stems, and roots rhizomes [5, 6] from where they are released into the environment by leaching from above ground parts, root exudation, volatilization, and decomposition of plant residues in both natural and agricultural systems [7–9].

Allelopathy inhibition is complex and can involve the interaction of different classes of chemicals such as phenolic compounds, flavonoids, terpenoids, alkaloids, coumarins, glycosides, and glucosinolates. These chemicals called secondary metabolites are known to be exuded by plants to suppress emergence or growth of other plants. These substances are phytotoxic and can be suggestive of their potentials as natural herbicides [10–12]. These secondary metabolites released by plants may influence resource competition, nutrient dynamics, microbial ecology, mycorrhizae, and even soil abiotic factors [13]. When plants are exposed to allelochemicals, their growth and development are affected through inhibition of seed germination/or seedling growth decrease. The readily visible effects include inhibited or retarded germination rate [14], seeds darkening and swelling, reduced root or radicle and shoot or coleoptile extension [15, 16], swelling or necrosis of root tips, curling of the root axis, discolouration, lack of root hairs, reduced dry weight accumulation, and lowered reproductive capacity [17].

Allelopathy had been studied by many researchers [18–21]. In Nigeria, allelopathy studies so far concentrated on allelopathic potentials of weeds on crops. This includes the work of Tijani-Eniola and Fawusi [22] on Chromolaena and Kayode [23] on Aspilia africana. Recently, Kayode and Ayeni [24] and Ayeni et al. [25] reported the allelopathic potentials of crop residues on agricultural crops. The present research emphasizes the allelopathic effects of crop residues on the germination and growth of okra, an important edible fruit widely eaten in Nigeria.

Okra is an annual plant or perennial crop of the Family Malvaceae, widely grown for its edible fruits. The fruits are harvested when immature and they are widely eaten in South Western Nigeria as it provides excellent vegetable protein which is rich in tryptophan and help reduces human malnutrition. The growth of okra could be reduced due to allelochemicals released from plants in cropping.

This study aimed at examining the allelopathic potentials of aqueous extracts from residues of sorghum bicolour stem and Zea mays (tassel and roots) on the germination and growth of okra.

## 2. Materials and Methods

Laboratory experiments were conducted during June, 2010 in the Department of Plant Science, Ekiti State University, Nigeria to assess the allelopathic effects of different aqueous extracts of residues from Sorghum bicolor stem and Zea mays (root and tassel) on the germination and growth of okra.

Mature sorghum plants were harvested from the experimental farm of the Department of Plant Science, Ekiti State University, Nigeria. The sorghum stem was cut into pieces to facilitate drying. Maize (roots and tassel) was also collected from the experimental farm after the fruits had been harvested. These materials were chopped into pieces and were air-dried for three weeks after which they were pounded using pestle and mortal.

Okra seeds were obtained from Agricultural Development Project (ADP), Ado Ekiti to get improved variety.

Portions of 5 g, 10 g, 15 g, 20 g, and 25 g of each of the ground samples of the crop residues were measured out using G&G Electric Top Loading Digital balance, JJ300Y, China. Each portion was soaked in 200 mL distilled water in 500 mL conical flasks. The mixtures were shaken intermittently and for 24 hrs at $25°C ± 1°C$. The extracts for each crop residue was filtered and the filtrates were stored in a refrigerator for further usage.

In each treatment, two layers of Whatman number 1 filter papers were put in each petri dish (each with a diameter of 9 cm). Five seeds of okra were sown in the petri dish and replicated ten times for each extract concentration. The filter papers were moistened daily with different extracts concentration using syringe and needle. Control experiments

were set up for each extract and replicated ten times. All the petri dishes were arranged on germination tables at room temperature between $25–30°C$. The seeds were considered as germinated upon radicle emergence and the number that germinated was counted for six days. The radicle and plumule growth elongations were recorded at 24 hrs interval. The data obtained from the experiments were compared to those obtained from the control using Analysis of Variance (ANOVA) and using SPSS version 15 (2009) computer software. Duncan Multiple Range Test (DMRT) at $P < 0.05$ was used to separate the means.

## 3. Results and Discussion

3.1. Seed Germination. The allelopathic effects of aqueous extracts from sorghum stem and maize (roots and tassel) on the germination of okra (Abelmoschus esculentus) are shown in Table 1. The extracts brought a considerable inhibition in the germination of okra seeds. The inhibition increases with increase in the concentration of the extracts. In sorghum treated seeds at 24 hrs, no germination was observed at 20 and 25 g concentrations. At 48 hrs experimental time, the germination of okra seeds in the control was 100%, those of 5, 10, 15, 20, and 25 g/200 mL concentrations were 94%, 86%, 38%, 26%, and 24%, respectively. At 144 hrs after planting, the germination of extract treated seeds reduces as the concentration of the extracts increases. The control and 5 g concentration were 100% which reduced to 62% in 25 g concentration (Table 1A). Statistical analyses ($P < 0.05$) revealed that significant were observed in the germination of okra treated seeds when compared to the control experiment especially at higher concentrations of the extracts. At 144 hrs experimental time of sorghum treated seeds, significant differences were observed in the germination of okra seeds at 20 g and 25 g concentrations when compared to the control experiments. However, no significant differences were observed in other treatments compared to control experiment.

For maize root extracts, no germination was observed until 48 hrs (Table 1B). The germination percentage of the control experiment was 82%, germination percentages in 5, 10, 15, 20, and 25 g/200 mL concentrations were 76%, 70%, 64%, 42%, and 34%, respectively. At 144 hrs, the percentage germination of control was 96%, which decreased to 76% in 25 g concentration. Statistical analysis ($P < 0.05$) revealed that 25 g concentration showed significant difference to control experiment. Other concentrations showed no significant difference.

Similar results were observed in the maize tassel extracts. It was revealed that at after planting, 100% of the seeds germinated in the control, 5, 10, 15, 20, and 25 g/200 mL concentrations had 100%, 100%, 100%, 98%, 96%, and 94%, respectively. This tends to suggest that the effects of the extracts were concentration dependent. Statistical analysis ($P < 0.05$) also revealed that 25 g concentration showed significant difference to control experiment while other concentrations showed no significant difference (Table 1C).

Among the three different extracts, maize root showed more allelopathy and the sorghum stem showed least inhibition on okra germination. The inhibitory effect of the extracts

TABLE 1: Effects of aqueous extracts of sorghum stem and maize (roots and tassel) on the germination % of seeds of okra.

| Extracts g/200 mL | Time (hrs) | | | | | |
|---|---|---|---|---|---|---|
| | 24 | 48 | 72 | 96 | 120 | 144 |
| A | | | | | | |
| Sorghum stem 0 | 12[a] | 100[a] | 100[a] | 100[a] | 100[a] | 100[a] |
| Sorghum stem 5 | 8.00[a] | 96[a] | 98[a] | 98[a] | 98[a] | 100[a] |
| Sorghum stem 10 | 6.00[a] | 86[a] | 92[a] | 92[a] | 92[a] | 92[a] |
| Sorghum stem 15 | 6.00[a] | 38[b] | 66[b] | 72[b] | 72[a] | 82[ab] |
| Sorghum stem 20 | 0.00[a] | 26[c] | 62[b] | 66[b] | 66[b] | 66[c] |
| Sorghum stem 25 | 0.00[a] | 24[c] | 42[c] | 62[b] | 62[b] | 62[c] |
| B | | | | | | |
| Maize root 0 | 24[a] | 82[a] | 94[a] | 96[a] | 96[a] | 96[a] |
| Maize root 5 | 0.00[a] | 76[a] | 92[a] | 94[a] | 94[a] | 94[a] |
| Maize root 10 | 0.00[a] | 70[a] | 92[a] | 92[a] | 92[a] | 92[a] |
| Maize root 15 | 0.00[a] | 64[a] | 92[a] | 92[a] | 92[a] | 92[a] |
| Maize root 20 | 0.00[a] | 42[b] | 90[a] | 92[a] | 92[a] | 92[a] |
| Maize root 25 | 0.00[a] | 32[b] | 76[b] | 76[b] | 76[b] | 76[b] |
| C | | | | | | |
| Maize tassel 0 | 42[a] | 100[a] | 100[a] | 100[a] | 100[a] | 100[a] |
| Maize tassel 5 | 0.00[b] | 94[ab] | 100[a] | 100[a] | 100[a] | 100[a] |
| Maize tassel 10 | 0.00[b] | 94[ab] | 100[a] | 100[a] | 100[a] | 100[a] |
| Maize tassel 15 | 0.00[b] | 94[ab] | 98[ab] | 98[ab] | 98[ab] | 98[ab] |
| Maize tassel 20 | 0.00[b] | 88[ab] | 96[bc] | 96[bc] | 96[bc] | 96[bc] |
| Maize tassel 25 | 0.00[b] | 82[b] | 94[c] | 94[c] | 94[c] | 94[c] |

Means followed by the same letter with the column for each treatment are not significantly different at ($P < 0.05$).

was found to increase with increase in the concentration of the extracts which were in accordance with previous researches [26, 27]. Such allelopathic differences in the crop residues might be related to specific allelopathic compounds being present in each species which related to the work of Chon et al. [28].

The effects of the extracts on okra germination were similar to those obtained by Shahid et al. [29] who reported that different plant extracts significantly reduced wheat and its weeds, Kayode and Ayeni [24] on aqueous extracts of rice husk and sorghum stem on maize. Anjum et al. [30] and Monica et al. [31] reported the aqueous extracts of *Ascarum europaeum* L. inhibited the germination and growth of *Lycopersicum esculentum* Yarnia et al. [32] noted that Amaranthus seed germination was reduced by sorghum extracts. Also Oyun [33] reported that extracts from *G. sepium* caused a prolonged delay in maize seed germination.

*3.2. Radicle Length.* The effects of the aqueous extracts of sorghum stem and maize (roots and tassel) on the radicle length (cm) of okra were shown in Table 2. The results revealed that the radicle length of okra treated with both extracts seeds were retarded at 24 hrs experimental time.

In sorghum stem treated seeds, the mean radicle length in the control at 144 hrs was 3.12 cm, those of 5 g, 10 g, 15 g, 20 g, and 25 g/200 mL concentrations were 3.03 cm, 3.02 cm, 2.86 cm, 1.85 cm, and 1.65 cm, respectively. Statistical analysis ($P < 0.05$) revealed that 20 g and 25 g concentrations showed significant differences to control experiment in the

experimental times while other concentrations showed no significant difference (Table 2A).

In maize root treated seeds, the average length of radicle in the control was 3.44 cm, those of 5 g, 10 g, 15 g, 20 g, and 25 g/mL concentrations were 2.61 cm, 2.32 cm, 2.00 cm, 1.56 cm, and 1.31 cm, respectively. Similarly in the maize tassel treated seeds, the mean radicle length in the control was 3.88 cm, those of 5 g, 10 g, 15 g, 20 g, and 25 g were 3.56 cm, 2.95 cm, 2.06 cm, 1.53 cm, and 1.15 cm, respectively. The radicle lengths reduced with increase in the concentration of the extracts which tends to suggest that the effects of the extracts were concentration dependent. Statistical analysis ($P < 0.05$) also revealed that the higher concentrations (20 and 25 g) showed significant differences to control experiment in the experimental times while other concentrations showed no significant difference (Table 2B).

In maize tassel treated seeds, no radicle emerged until 72 hrs experimental time (Table 2C). The mean radicle length in the control experiment was 3.88 cm, those of 5, 10, 15, 20, and 25 g/200 mL concentrations were 3.56 cm, 2.95 cm, 2.06 cm, 1.53 cm, and 1.15 cm, respectively. The radicle lengths reduced with increase in the concentration of the extracts which tends to suggest that the effects of the extracts were concentration dependent. Statistical analysis ($P < 0.05$) at 144 hrs experimental time revealed that there were significant differences in the radicle length of okra treated seeds when compared to control experiment except 5 g concentration that showed no significant difference. The radicle length reduced with increase in the concentration of the extracts which also

TABLE 2: Effects of aqueous extracts of sorghum stem and maize (roots and tassel) on the radicle length (cm) of seeds of okra.

| Extracts g/200 mL | Time (hrs) | | | | | |
|---|---|---|---|---|---|---|
| | 24 | 48 | 72 | 96 | 120 | 144 |
| A | | | | | | |
| Sorghum stem 0 | 0.10$^a$ | 0.77$^a$ | 1.67$^a$ | **2.29$^a$** | **2.76$^a$** | **3.12$^a$** |
| Sorghum stem 5 | 0.00$^a$ | 0.74$^a$ | 1.66$^a$ | 2.26$^a$ | 2.67$^a$ | 3.03$^a$ |
| Sorghum stem 10 | 0.00$^a$ | 0.67$^a$ | 1.61$^a$ | 2.23$^a$ | 2.62$^a$ | 3.02$^a$ |
| Sorghum stem 15 | 0.00$^a$ | 0.66$^a$ | 1.18$^a$ | 1.78$^a$ | 2.29$^b$ | 2.86$^a$ |
| Sorghum stem 20 | 0.00$^a$ | 0.31$^b$ | 0.56$^b$ | **0.92$^b$** | **1.26$^b$** | **1.85$^b$** |
| Sorghum stem 25 | 0.00$^a$ | 0.10$^b$ | 0.46$^b$ | **0.82$^b$** | **1.14$^b$** | **1.65$^b$** |
| B | | | | | | |
| Maize root 0 | 0.13$^a$ | 0.58$^a$ | 1.12$^a$ | 1.99$^a$ | 2.70$^a$ | 3.44$^a$ |
| Maize root 5 | 0.00$^b$ | 0.32$^b$ | 0.81$^b$ | 1.31$^b$ | 1.78$^b$ | 2.61$^b$ |
| Maize root 10 | 0.00$^b$ | 0.310$^b$ | 0.81$^b$ | 1.23$^{bc}$ | 1.63$^{bc}$ | 2.32$^b$ |
| Maize root 15 | 0.00$^b$ | 0.27$^b$ | 0.73$^b$ | 1.07$^{bc}$ | 1.61$^{bc}$ | 2.00$^{bc}$ |
| Maize root 20 | 0.00$^b$ | 0.16$^b$ | 0.70$^b$ | 1.02$^{bc}$ | 1.27$^{cd}$ | 1.56$^c$ |
| Maize root 25 | 0.00$^b$ | 0.11$^b$ | 0.45$^c$ | 0.84$^c$ | 1.09$^d$ | 1.31$^c$ |
| C | | | | | | |
| Maize tassel 0 | 0.00$^a$ | 0.85$^a$ | 2.13$^a$ | 3.17$^a$ | 3.52$^a$ | 3.88$^a$ |
| Maize tassel 5 | 0.00$^a$ | 0.51$^b$ | 1.54$^b$ | 2.53$^b$ | 3.15$^a$ | 3.56$^a$ |
| Maize tassel 10 | 0.00$^a$ | 0.42$^{bc}$ | 1.46$^b$ | 2.19$^{bc}$ | 2.66$^{ab}$ | 2.95$^b$ |
| Maize tassel 15 | 0.00$^a$ | 0.36$^{bc}$ | 1.29$^b$ | 1.76$^c$ | 1.92$^b$ | 2.06$^c$ |
| Maize tassel 20 | 0.00$^a$ | 0.31$^{bc}$ | 0.83$^c$ | 1.03$^d$ | 1.86$^b$ | 1.53$^d$ |
| Maize tassel 25 | 0.00$^a$ | 0.15$^c$ | 0.57$^c$ | 0.71$^d$ | 0.84$^c$ | 1.15$^d$ |

Means followed by the same letter with the column for each treatment are not significantly different at ($P < 0.05$).

suggests that the effect of the extracts was concentration dependent. The roots of the crop exposed to allelochemical become brownish, leading to necrosis of the root tips, curling of root axis, lack of root hairs, and seed darkening. This might be due to the rapid inhibiting effect on respiration of root tips which might ultimately reduce its elongation. Similar results were reported by Nazim et al. [34], Monica et al. [35], and Komal [36].

3.3. *Plumule Length.* The effects of the aqueous extracts of sorghum stem and maize (roots and tassel) on the plumule length (cm) of okra were shown in Table 3. The effects of the extracts were similar to those obtained in the radicle lengths. It was revealed that no plumule emerged until 72 hrs in the three extracts. In sorghum treated seeds (Table 3A), the mean plumule length at 144 hrs was 2.35 cm, those of 5, 10, 15, 20, and 25 g/mL concentrations were 1.55 cm, 1.41 cm, 0.98 cm, 0.57 cm, and 0.42 cm, respectively. Statistical analysis revealed that 20 and 25 g concentrations showed significant differences to control experiment between 72 and 96 hrs, other treatments showed no significant difference. At 120–144 hrs, there were significant differences in okra treated seeds compared to control experiment.

In maize root aqueous treated seeds (Table 3B), the plumule length of okra reduces with increased concentration of the extracts. The mean plumule length at 144 hrs in the control was 2.10 cm which decreased to 0.46 cm in 25 g/200 mL concentration. Statistical analysis ($P < 0.05$) showed

there were significant differences in the extract treated seeds compared to the control experiment at 144 hrs.

Similar results were also observed in the maize tassel extract treated seeds (Table 3C). At 72–144 hrs, the plumule length of extract treated seed reduces with increase in the concentration of the extracts. For example, at 144 hr experiment time, the control experiment has plumule length of 2.10 cm which reduced to 0.46 cm in 25 g/mL concentration. Statistical analysis revealed that there were significant differences in the plumule length of okra treated seeds between 72–144 hrs experiment time compared to control experiment.

The effects of the different concentration of extracts in the plumule length may result from different water extracts due to manifestation of primary events caused by allelochemicals released by receiver plant which might have reduced the uptake of nutrients and may ultimately reduce shoot lengths. This findings corroborated the work of [33] who reported that the root and shoot length and seedling vigour of maize were decreased with the increasing concentration of *G. sepium*. Similarly, Salam et al. [37] reported extracts from rice hulls significantly root and shoot elongation of *E. crus galli*. Also Shahid et al. [29] reported that different plant extracts significantly reduced wheat and its weeds. The effects of the aqueous extracts on okra were also similar to those obtained by Ayeni and Kayode [38] on the effects of aqueous extracts from maize roots and sorghum stem on the germination and radicle growth of *Sphenostylis sternocarpa* Hochst ex. Rich (African Yam Bean).

TABLE 3: Effects of aqueous extracts of sorghum stem and maize (roots and tassel) on the plumule length (cm) of okra.

| Extracts g/200 mL | Time (hrs) | | | | | |
|---|---|---|---|---|---|---|
| | 24 | 48 | 72 | 96 | 120 | 144 |
| A | | | | | | |
| Sorghum stem 0 | 0.00$^a$ | 0.00$^a$ | 0.22$^a$ | 0.44$^a$ | 1.19$^a$ | 2.35$^a$ |
| Sorghum stem 5 | 0.00$^a$ | 0.00$^a$ | 0.17$^a$ | 0.41$^a$ | 0.65$^b$ | 1,55$^b$ |
| Sorghum stem 10 | 0.00$^a$ | 0.00$^a$ | 0.16$^a$ | 0.37$^a$ | 0.60$^b$ | 1.41$^{bc}$ |
| Sorghum stem 15 | 0.00$^a$ | 0.00$^a$ | 0.15$^A$ | 0.33$^a$ | 0.58$^b$ | 0.98$^c$ |
| Sorghum stem 20 | 0.00$^a$ | 0.00$^a$ | 0.03$^b$ | 0.13$^b$ | 0.23$^c$ | 0.57$^{de}$ |
| Sorghum stem 25 | 0.00$^a$ | 0.00$^a$ | 0.01$^b$ | 0.07$^b$ | 0.13$^c$ | 0.42$^e$ |
| B | | | | | | |
| Maize root 0 | 0.00$^a$ | 0.00$^a$ | 0.34$^a$ | 0.56$^a$ | 1.25$^a$ | 1.81$^a$ |
| Maize root 5 | 0.00$^a$ | 0.00$^a$ | 0.21$^{ab}$ | 0.52$^a$ | 1.02$^{ab}$ | 1.63$^{ab}$ |
| Maize root 10 | 0.00$^a$ | 0.00$^a$ | 0.15$^{ab}$ | 0.43$^a$ | 0.58$^b$ | 1.38$^{ab}$ |
| Maize root 15 | 0.00$^a$ | 0.00$^a$ | 0.11$^b$ | 0.41$^a$ | 0.72$^b$ | 1.17$^{bc}$ |
| Maize root 20 | 0.00$^a$ | 0.00$^a$ | 0.00$^c$ | 0.08$^b$ | 0.23$^c$ | 0.79$^c$ |
| Maize root 25 | 0.00$^a$ | 0.00$^a$ | 0.00$^c$ | 0.04$^b$ | 0.08$^c$ | 0.24$^d$ |
| C | | | | | | |
| Maize tassel 0 | 0.00$^a$ | 0.00$^a$ | 0.33$^a$ | 0.47$^a$ | 0.72$^a$ | 2.10$^a$ |
| Maize tassel 5 | 0.00$^a$ | 0.00$^a$ | 0.26$^{ab}$ | 0.35$^{ab}$ | 0.56$^a$ | 1.02$^b$ |
| Maize tassel 10 | 0.00$^a$ | 0.00$^a$ | 0.11$^{bc}$ | 0.32$^b$ | 0.55$^a$ | 0.99$^b$ |
| Maize tassel 15 | 0.00$^a$ | 0.00$^a$ | 0.11$^{bc}$ | 0.18$^c$ | 0.29$^b$ | 0.89$^b$ |
| Maize tassel 20 | 0.00$^a$ | 0.00$^a$ | 0.06$^c$ | 0.15$^c$ | 0.26$^b$ | 0.63$^b$ |
| Maize tassel 25 | 0.00$^a$ | 0.00$^a$ | 0.00$^c$ | 0.04$^c$ | 0.08$^c$ | 0.46$^c$ |

Means followed by the same letter with the column for each treatment are not significantly different at ($P < 0.05$).

This observation tends to suggest that the extracts from plant residues releasing allelochemical are responsible for the inhibitory effects shown on agricultural crops and weeds on farmlands.

Previous researches of Cherney et al. [39] revealed that allelochemicals in sorghum includes valinic acid, p-hydroxy-benzaldehyde, p-coumaric acid, ferulic acid, dhurin, and sorgoleone. Guenzi and McCalla (1996) had earlier asserted that the inhibitory compounds occurring in sorghum plants are mostly phenols. Einhellig and Souza (1992) reported that phenolic compounds such as dihydroquinine sorgoleone produced by *Sorghum bicolor* have been found to be extremely phytotoxic in hydroponic culture. Also Sanchez-Moreiras et al. [40] asserted that maize inflorescence allelopathy was quinine attributed to hydroxamic acid. Allelochemicals in maize root exudates are mostly saponin.

The reduction in the germination and seedling growth of okra in this study might be attributed to the allelopathic compounds present in these plants. Thus the act of leaving crop residues such as sorghum straw and maize tassels uncared for on the field might have detrimental effects on the germination and growth of subsequent crops.

## Conflict of Interests

The authors declare that there is no conflict of interests regarding the publication of this paper.

## References

[1] E. L. Rice, "Allelopathy," in *A Discipline Called Allelopathy: Basic and Applied Aspects*, S. J. H. Rizvi, H. Haque, V. K. Singh, and V. Rizvi, Eds., p. 9, Academic Press, New York, NY, USA; Chapman and Hall, London, UK, 1984.

[2] J. L. Hierro and R. M. Callaway, "Allelopathy and exotic plant invasion," *Plant and Soil*, vol. 256, no. 1, pp. 29–39, 2003.

[3] A. Fitter, "Making allelopathy respectable," *Science*, vol. 301, no. 5638, pp. 1337–1338, 2003.

[4] Z. Iqbal, H. Nasir, S. Hiradate, and Y. Fujii, "Plant growth inhibitory activity of *Lycoris radiata* Herb. and the possible involvement of lycorine as an allelochemical," *Weed Biology and Management*, vol. 6, no. 4, pp. 221–227, 2006.

[5] S. Ahmad, M. Arfan, A. L. Khan et al., "Allelopathy of Teucrium royleanum wall. Ex benth. from Pakistan," *Journal of Medicinal Plants Research*, vol. 5, no. 5, pp. 765–772, 2011.

[6] A. S. Tehmina, P. H. David, and B. Rukhasana, "Allelopathic potential of *Helianthus annus* L. (sunflower) as natural herbicides," *Allelopathy*, 2005.

[7] J. J. Ferrugson and B. Rathinasabapathi, *Allelopathy: How Plants Suppress Other Plants*, University of Florida, Institute of Food and Agriculture Sciences, UF/IFAS, Gainesville, Fla, USA, 2003.

[8] A. Inderjit and K. M. M. Dakshini, "Interference potential of *Pluchea lanceolata* (Asteraceae): growth and physiological responses of asparagus bean, *Vigna unguiculata* var. sesquipedalis," *The American Journal of Botany*, vol. 79, no. 9, pp. 977–981, 1992.

[9] M. Ben-Hammouda, H. Ghorbal, R. J. Kremer, and O. Oueslati, "Allelopathic effects of barley extracts on germination and seedlings growth of bread and durum wheats," *Agronomie*, vol. 21, no. 1, pp. 65–71, 2001.

[10] C. L. Céspedes, J. C. Marín, M. Domínguez, J. G. Avila, and B. Serrato, "Plant growth inhibitory activities by secondary metabolites isolated from Latin American flora," *Advances in Phytomedicine*, vol. 2, pp. 373–410, 2006.

[11] J. M. Herranz, P. Ferrandis, M. A. Copete, E. M. Duro, and A. Zalacaín, "Effect of allelopathic compounds produced by *Cistus ladanifer* on germination of 20 Mediterranean taxa," *Plant Ecology*, vol. 184, no. 2, pp. 259–272, 2006.

[12] T. D. Khanh, T. D. Xuan, and I. M. Chung, "Rice allelopathy and the possibility for weed management," *Annals of Applied Biology*, vol. 151, no. 3, pp. 325–339, 2007.

[13] D. A. Wardle, M.-C. Nilsson, C. Gallet, and O. Zackrisson, "An ecosystem-level perspective of allelopathy," *Biological Reviews of the Cambridge Philosophical Society*, vol. 73, no. 3, pp. 305–319, 1998.

[14] G. B. Williamson, D. R. Richardson, and N. H. Fischer, "Allelopathic mechanism in fire-prone communities," in *Allelopathy*, S. J. H. Rizvi and V. Rizvi, Eds., pp. 59–75, Chapman & Hall, London, UK, 1992.

[15] M. A. Turk and A. M. Tawaha, "Allelopathic effect of black mustard (*Brassica nigra* L.) on germination and growth of wild oat (*Avena fatua* L.)," *Crop Protection*, vol. 22, no. 4, pp. 667–673, 2003.

[16] B. P. Bhatt and N. P. Todaria, "Studies on the allelopathic effects of some agroforestry tree crops of Garhwal Himalaya," *Agroforestry Systems*, vol. 12, no. 3, pp. 251–255, 1990.

[17] A. O. Ayeni, D. T. Lordbanjou, and B. A. Majek, "*Tithonia diversifolia* (Mexican sunflower) in south-western Nigeria: occurrence and growth habit," *Weed Research*, vol. 37, no. 6, pp. 443–449, 1997.

[18] R. E. Ricklefs, *Ecology*, W.H. Freeman and Company, New York, NY, USA, 3rd edition, 1990.

[19] T. Ohno, "Oxidation of phenolic acid derivatives by soil and its relevance to allelopathic activity," *Journal of Environmental Quality*, vol. 30, no. 5, pp. 1631–1635, 2001.

[20] É. R. Alford, J. M. Vivanco, and M. W. Paschke, "The effects of flavonoid allelochemicals from knapweeds on legume—rhizobia candidates for restoration," *Restoration Ecology*, vol. 17, no. 4, pp. 506–514, 2009.

[21] A. S. Thorpe, G. C. Thelen, A. Diaconu, and R. M. Callaway, "Root exudate is allelopathic in invaded community but not in native community: field evidence for the novel weapons hypothesis," *Journal of Ecology*, vol. 97, no. 4, pp. 641–645, 2009.

[22] H. A. Tijani-Eniola and O. A. Fawusi, "Allelopathic activities of crude methanol extract of siam weed and wild poinsettia on seed germination and seedling growth in tomato," *Nigerian Journal of Weed Science*, vol. 2, no. 1-2, pp. 15–20, 1989.

[23] J. Kayode, "Allelopathic effects of aqueous extracts of *Aspillia africana* on radicle and plumule growth of *Zea mays*," *Journal of Physical and Biological Science*, vol. 2, pp. 43–46, 2004.

[24] J. Kayode and J. M. Ayeni, "Allelopathic effects of some crop residues on the germination and growth of maize (*Zea mays* L.)," *The Pacific Journal of Science and Technology*, vol. 10, no. 1, pp. 345–349, 2009.

[25] J. M. Ayeni, J. Kayode, and P. O. Tedela, "Allelopathic potentials of some crop residues on the germination and growth of *Bidens pilosa* L," *Journal of Agricultural Science and Technology*, vol. 4, no. 1, pp. 21–24, 2010.

[26] S. Sisodia and M. B. Siddiqui, "Allelopathic effect of *Lantana camara* on *Bidens pilosa*," *VEGETOS*, vol. 20, no. 1, pp. 29–32, 2008.

[27] S. Sisodia and M. Badruzzaman Siddiqui, "Allelopathic potential of rhizosphere soil of *Croton bonplandianum* on growth and establishment of some crop and weed plants," *African Journal of Agricultural Research*, vol. 4, no. 5, pp. 461–467, 2009.

[28] S.-U. Chon, Y.-M. Kim, and J.-C. Lee, "Herbicidal potential and quantification of causative allelochemicals from several Compositae weeds," *Weed Research*, vol. 43, no. 6, pp. 444–450, 2003.

[29] M. Shahid, B. Ahmad, R. A. Khattak, G. Hussan, and H. Khan, "Response of wheat and its weed to different allelopoathic plant water extracts," *Pakistan Journal of Weed Sciences Research*, vol. 12, no. 1-2, pp. 61–68, 2006.

[30] A. Anjum, U. Hussain, Z. Yousaf, F. Khan, and A. Umer, "Evaluation of allelopathic action of some selected medicinal plant on lettuce seeds by using sandwich method," *Journal of Medicinal Plants Research*, vol. 4, no. 7, pp. 536–541, 2010.

[31] M. Monica, P. Anea, M. Lucia, V. Zorica, and M. Georgeta, "Allelopathic potentials of *Ascarum Europaeum* toward *Lycopersicum esculentum*," *Analele Universitatii din Oradea, Fascicula Biologie*, vol. 18, no. 1, pp. 39–44, 2011.

[32] M. Yarnia, M. B. Khorshidi Benam, and E. Farajzadeh Memari Tabrizi, "Allelopathic effects of sorghum extracts on *Amaranthus retroflexus* seed germination and growth," *Journal of Food Agriculture and Environment*, vol. 7, no. 3-4, pp. 770–774, 2009.

[33] M. B. Oyun, "Allelopathic potentials of *Gliricidia sepium* and *Acacaia auriculiformis* on the germination and seedling vigour of maize (*Zea mays* L.)," *The American Journal of Agricultural and Biological Science*, vol. 1, no. 3, pp. 44–47, 2006.

[34] G. Nazim, A. Shabbir, R. Bajwa, and S. Bano, "Allelopathic effects of Neem (*Azadiracta indica*) on germination and mycorhizal status of Parthenium hysterophorus," in *Proceedings of the 2nd International Weed Science Conference Abstracts*, p. 29, 2005.

[35] S. M. Seyyednejad, H. Koochak, F. P. Najafabade, and M. Kolahi, "Allelopathic effect of aquatic hull extract of rice (*Oryza sativa* L.) on growth of *Silybum marianum* and *Echinochloa crus-galli*," *African Journal of Agricultural Research*, vol. 5, no. 6, pp. 2222–2226, 2010.

[36] A. S. Komal, "Allelopathic influence of aqueous extracts of *Cassia occidentalis* L.M. on *Triticum aestivum* L," *Life Science Leaflets*, vol. 18, pp. 723–725, 2011.

[37] M. A. Salam, M. Morokuma, T. Teruya, K. Suenaga, and H. Kato-Noguchi, "Isolation and identification of a potent allelopathic substance in Bangladesh rice," *Plant Growth Regulation*, vol. 58, no. 2, pp. 137–140, 2009.

[38] J. M. Ayeni and J. Kayode, "The effects of aqueous extracts from maize roots and Sorghum stem on the germination and radicle growth of *Sphenostylis sternocarpa* Hochst ex. Rich. (African Yam Bean)," *African Journal of General Agriculture*, vol. 5, pp. 117–121, 2009.

[39] D. J. R. Cherney, J. A. Petterson, J. H. Cherney, and J. D. Axtell, "Fibre and soluble phenolic monomer composition of morphological components of sorghum stover," *Journal of the Science of Food and Agriculture*, vol. 54, pp. 645–649, 1991.

[40] A. M. Sanchez-Moreiras, A. Martinez, L. Gonzalez, F. Pellisier, and M. J. Regiosa, "Mode of action of hyrdroxamic acid (BOA) and other related compounds," in *Allelopathy; Chemistry and Mode of Action of Allelochemicals*, F. A. Marcias, J. C. Galindo, J. M. Molinillo, and H. G. Cutler, Eds., pp. 239–252, CRC Press, New York, NY, USA, 2004.

# 26

# Agromorphological Traits Variability of the Ethiopian Lentil and Exotic Genotypes

**Fikru Mekonnen,**[1] **Firew Mekbib,**[1] **Shiv Kumar,**[2] **Seid Ahmed,**[3] **and Tilak R. Sharma**[4]

[1] *School of Plant Sciences, College of Agriculture and Environmental Sciences, Haramaya University, P.O. Box 138, Dire Dawa, Ethiopia*
[2] *International Center for Agricultural Research in the Dry Areas (ICARDA), Tel Hadya, P.O. Box 5466 Aleppo, Syria*
[3] *International Center for Agricultural Research in the Dry Areas (ICARDA), P.O. Box 5689, Addis Ababa, Ethiopia*
[4] *Department of Agricultural Biotechnology, Chaudhary Sarwan Kumar Himachal Pradesh Agricultural University, Palampur 176 062, India*

Correspondence should be addressed to Fikru Mekonnen; tiewoast@gmail.com

Academic Editor: Ayman Suleiman

Understanding the genetic relationships and diversity of Ethiopian lentil in relation to lentil from other countries is important in attempting to widen the genetic base of germplasm in the country. The objectives of this study were to generate information on agromorphological variability, to estimate PCV, GCV, heritability, and expected genetic advance of quantitative traits of lentil. 228 genotypes with different population types were studied for 11 agromorphological traits and rust disease severity score for two seasons (2011-2012) over three locations. The analysis of variance showed highly significant variations ($P \leq 0.01$) among genotypes for all characters studied. As per genetic parameter values, four groups of character were deduced. It is inferred that the exotic genotypes introduced from ICARDA showed rich genetic bases for 100-seed weight, number of seeds per plant, seed weight per plant, resistance source for rust, and high yielder in high yielding environment, where rainfall is not a major problem. Use the Ethiopian accessions for developing cultivars that could be used in double cropping and drought prone areas.

## 1. Introduction

Lentil (*Lens culinaris* Medik. subsp. *culinaris*) is a short, slender annual cool-season food legume. The center of origin of lentil is the Near East where it was first domesticated in the Fertile Crescent around 7000 BC [1]. According to Barulina [2], two varietal types based on seed size were recognized: the small-seeded microsperma and the macrosperma with relatively large flattened pods. The largest lentil producer is India, followed by Canada and Turkey, accounting for 68% of the global production [3]. In Africa, Ethiopia, Morocco, and Tunisia are the leading producers, but 61% of the areas and 68% of the production belong to Ethiopia [4]. Lentil is an important dietary source of macro- and micronutrients for both rural and urban dwellers. It provides sufficient amounts of the most essential amino acids to meet nutrient requirements. It is a cash crop fetching the highest price in domestic market compared to all other food legumes

and major cereal crops [5]. Global, African, and Ethiopian lentil productivity is about 887 kg/ha, 644 kg/ha [3], and 1168 kg/ha, respectively [6]. Biotic and abiotic factors limit lentil productivity and seed quality. The major yield gap contributing factors are insect pests, diseases, low yielding landraces grown by farmers, and the narrow genetic base [7, 8].

Prior knowledge of genetic variability and characterization of genetic resources within the germplasm available at a gene bank has an important implication for future utilization and collection activities, to identify areas of major priority for conservation and improvement programs [9]. Quantitative traits provide an estimate of genetic diversity, and various numerical taxonomic techniques have been successfully used to classify and measure the pattern of genetic diversity in germplasm, as in lentil [10, 11]. Morphological characterization is the first step in the classification and description of any crop germplasm [12]. The assessment of

TABLE 1: Geographical, climatic, and soil features of the experimental sites.

| Agroecological | Sirinka | Sinana | Chefe Donsa |
|---|---|---|---|
| Range of temperature (°C) | 21–32 | 9.3–20.9 | 8.9–28.3 |
| Mean annual rainfall (mm) | 876 | 808 | 851 |
| Altitude (masl) | 1850 | 2450 | 1900 |
| Latitude | 12°.11′N | 7°7′N | 8°44′N |
| Longitude | 39°62′E | 40°10′E | 39°95′E |
| Soil texture | Clay soil | Clay soil | Light soil |
| Soil type | Eutric vertisol black soil | Pellic vertisol slightly acidic | Afisols and vertisol black soil |

genetic variability present in a given crop population can be determined by using the biometrical components such as range, variance, coefficient of variation, standard error, and heritability. The relative magnitude of these components determines the genetic properties of the population, in particular, the degree of resemblance between germplasm, that is, various determinants of the phenotype [13]. Reference [14] emphasized that the uses of the genetic coefficient of variability together with heritability and genetic advance values would determine the best picture of the amount of progress to be expected from a selection and determine the selection method to improve a character.

Several researchers investigated the diversity of economically important traits of lentil germplasm and found considerable variations [10, 11, 15–20]. Some quantitative evaluations of the Ethiopian landraces revealed potential variability for morphological traits [5, 7, 8, 21–25]. Understanding the genetic relationships and diversity of the Ethiopian lentil in relation to the germplasm from other countries is important to widen the genetic base of a germplasm. However, information is not exhaustive for different population structures of local and exotic genotypes on the genetic variability of important traits, which have economic benefit. Hence, the objectives of this study were to generate information on morphological variability and to estimate the phenotypic coefficient of variation, the genetic coefficient of variation, the heritability, and genetic advance of quantitative traits in lentil.

## 2. Materials and Methods

*2.1. Description of the Study Sites.* The field experiments were conducted on a hot spot area for rust. The areas included Sirinka Agricultural Research Center (SIARC) in the northeastern part of Ethiopia for two seasons (2011 and 2012), Chefe Donsa in the central part of Ethiopia, and Sinana Agricultural Research Center (SARC) in southeastern part of Ethiopia during the 2011/12 cropping season (Table 1 and Figure 1).

*2.2. Plant Materials.* Out of the 228 genotypes considered for this study, 158 were planted for morphological evaluation at SRARC in the 2010/2011 cropping season. Of these, 104 genotypes were from the Ethiopian lentil gene pool collected from six major lentil production regions (Figure 2) and were kept by the Institute of Biodiversity Conservation (IBC).

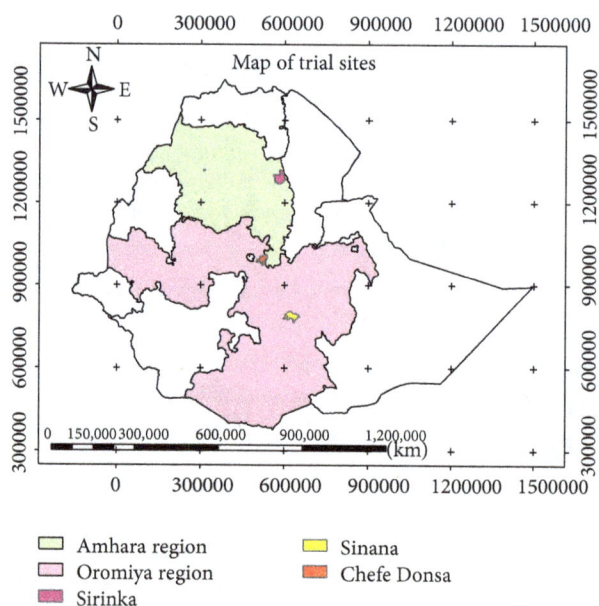

FIGURE 1: Geographical position of the experimental sites.

The accessions represent over 17% of the 618 lentil accession holdings of IBC. Rust susceptible check/EL-142/and wilt susceptible check/ILL-590/NEL 590/, as well as six commercial national released varieties (*Alemaya, Adaa, Teshale, Alem Tena, Chekol* and *Derash*), were used as a check. Ten elite breeding lines from the DebreZeit Agricultural Research Station (DZARC), Ethiopia, were included. Thirty-six parental lines and elite lines based on superiority for agronomic traits introduced from the International Center for Agricultural Research in Dry Areas (ICARDA) were included (Table 2). In the 2011/12 cropping season, 228 genotypes plus RILs were included in the study across three locations: Sirinka, Chefe Donsa, and Sinana.

*2.3. Experimental Layout and Design.* A randomized complete block design (RCBD) was used with three replications, in the 2010/2011 cropping season at SIARC. A unit plot comprised 2-meter length row with a plot size of (0.8 m²). Row-to row distance was 20 cm. The distance between two plots was 50 cm and the distance between two blocks was 100 cm. The genotypes were planted in the first week of July. In the 2011/12 cropping season, the experiment was carried out

TABLE 2: List of genotypes and their origin.

| Source of origin | Number of genotypes | Name of genotype | Type of genotypes |
|---|---|---|---|
| Tigray | 8 | Acc. no. 219957, 235383, 237503, 237504, 241785, 242604, 243447 | Landrace |
| Amhara | 54 | Acc. no. 36003, 36025, 36028, 36039, 36041, 36061, 36071, 36085, 36088, 36089, 36097, 36103, 36104, 36105, 36137, 36139, 36150, 36162, 36165, 36168, 207258, 207274, 207287, 207309, 212745, 215248, 215249, 223221, 228242, 229179, 229182, 229183, 231247, 235013, 235015, 235016, 235017, 236484, 236486, 236487, 237502, 238978, 238979, 241784, 241786, 243433, 243436, 243440, 243443, 244606, 244610, 244615, 244619, 244623 | Landrace |
| Oromya | 29 | Acc. no. 36001, 36007, 36009, 36013, 36015, 36019, 36023, 36029, 36033, 36042, 36048, 36058, 36110, 36120, 36131, 203141, 215806, 216877, 228809, 230521, 230833, 230834, 230837, 231248, 235698, 236438, 236892, 237027, 238971 | Landrace |
| SNNP | 2 | Acc. no. 36147 and Acc. No., 228243 | Landrace |
| Somali | 1 | Acc. no. 230832 | Landrace |
| DZARC | 6 | /ILL4225 x ILL4605/ /ILL 6821/ Alemaya, /ILL 1 x ILL 1169//ILL 6027/ ADAA, /ILL 7978/ Teshale, /Alemaya x FLIP88-41L/ Derash, /ILL 7981/ Aleme Tena and P160/ILL 2704/ /Chekol/ | Improved variety |
| Unknown | 10 | Acc. no. 36134, 207260, 211062, 211078, 211110, 220120, 211131, 233349, 233973, 241782 | Landrace |
| ICARDA | 22 | L-9-11, X2002S 219 /ILL 6821/, X2002S 221 /7980/, X2002S 221/FLIP 96-47 L//7979/, /ILL4965 x ILL6155/FLIP-97-16L/ILL 8078/, /ILL 883 x ILL 470/FLIP-84-95L/ILL 5722/, X2003S 233 /ILL 8009/, 2003S 235, X2003S 238 /ILL 4605/, X2005S 215/ILL 6002/, 2006S 122 /FLIP 2003-43L/ /ILL 7010 x ILL 1939/ /ILL 9932/, /ILL 2573 x ILL 7537/ /FLIP 2003-62 L/ /ILL 9951/X2006S 122, /ILL 7620 x ILL 9151/ /FLIP 2003-56L/ /ILL 9945/X2006S 127, X2006S 129/F2/, X2006S 130/FLIP 93-46L/ /ILL 547/, X2006S 130/FLIP 96-46 L//ILL 7978/, X2006S 133/FLIP87-21L/ /ILL 4349 x ILL4605//ILL 6211/, X2006S 134/ILL8174/, X2002S 219 /shehor-74/ /ILL 7554/, X2003S 222/ILL 213/, X2006S 128/ILL 5480/ | Parent |
| DARC | 11 | EL-142/ILL 5071/, 87S-93549XEL-1O3-3, 87s-93549XEL-03-4, Chekol X R-186-1, Chekol x R-186-2, EL-142 X R-186-3, EXOTIC #DZ/2008 AK, Chekol x R-186-8-0, ILL-358 X ILL-2573-2-2000, R-186X FLIP-86-38L, R-186XFLIP-86-38L-23 | Breeding line |
| ICARDA | 15 | ILL-28501, FLIP-04-26L, FLIP-2004-37L, FLIP-2004-7L, FLIP-2006-20L, FLIP-2006-60L, FLIP-87-68L, FLIP-93-63L, FLIP-97-61L, FLIP-97-68L, /ILL 6037 x ILX 87062/ FLIP2005-24L/ ILL-10045, ILL-10680, ILL-590/NEL 590/, L-830, Precoze/ILL 4605/ | Breeding line |
| ICARDA | 70 | RIL1-RIL 70 | RIL |

Genotypes represent different population groups of lentil germplasm such as landrace, elite breeding lines, putative parents, and recombinant inbred lines (RIL). Unknown = originated from Ethiopia but sites of collection are not mentioned.
RIL = recombinant inbred line.

using the augmented design with a single row of 2 m length for each genotype planted in five blocks. Eight checks were replicated within each block. Planting was done in the first week of July at SIARC, in mid-August at Chefe Donsa and in mid-September at SARC. The recommended agronomic packages were applied for raising a successful crop.

2.4. Data Collection. Data were recorded on 10 randomly selected plants on plant basis for plant height (PH), number of pods/plant (NP), number of seeds/pod (NS), number of seeds/plant (NSPP), and seed weight/plant (SWPP), whereas days to flowering (DFF), days to maturity (DM), above ground biomass (BI), 100-seed weight in gram (SW), seed

TABLE 3: Mean squares for agromorphological traits of 158 lentil genotypes at Sirinka in the 2010/11 cropping season.

| Sources of variation | df | Mean squares | | | | | | | | | | |
|---|---|---|---|---|---|---|---|---|---|---|---|---|
| | | DFF | DM | NP | NS | SWPP | NSPP | PH | SW | BI | SY | PSS |
| Replication | 2 | 561.9 | 1248.1 | 3067.3 | 1.27 | 2.1 | 8980.1 | 379.2 | 0.01 | 6475 | 46675.2 | 0.2 |
| Genotype | 157 | 70.6** | 408.3** | 359.7** | 0.3** | 0.7** | 859.2** | 48.7** | 0.99** | 35878** | 4645.4** | 1.8** |
| Error | 314 | 13.9 | 45.47 | 126.3 | 0.06 | 0.17 | 200.3 | 8.13 | 0.1 | 6140 | 994 | 0.2 |
| CV% | | 6.5 | 7 | 31.9 | 19.9 | 36.8 | 33 | 8.2 | 12.2 | 27.3 | 37.1 | 9.7 |
| Mean | | 57.2 | 95.7 | 35.3 | 1.2 | 1.1 | 42.9 | 34.6 | 2.5 | 286.8 | 84.9 | 5 |
| LSD (at 5%) | | 6 | 11.1 | 30.5 | 0.4 | 0.6 | 23.4 | 0.5 | 4.6 | 112.7 | 47.9 | 0.8 |

df = Degree of Freedom
DF = days to 50% flowering, DM = days to 90% maturity, NP = number of pods per plant, NS = number of seeds per pod, SWPP = Seed Weight per Plant, NSPP = Number of Seeds per Plant, PH = Plant height in cm, SW = 100-seed weight in gram, BI = Bio mass, PSS = Pod set Score
**, * = Significant at 5 % and 1% probability levels, respectively.

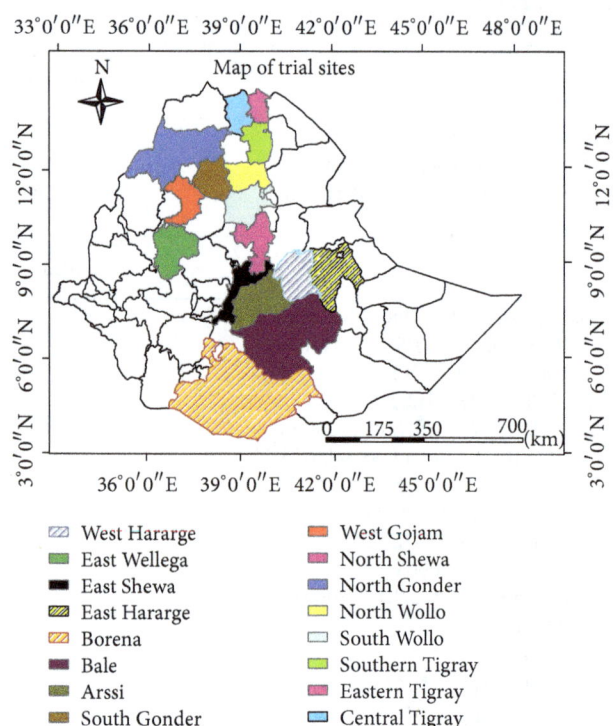

West Hararge
East Wellega
East Shewa
East Hararge
Borena
Bale
Arssi
South Gonder
West Gojam
North Shewa
North Gonder
North Wollo
South Wollo
Southern Tigray
Eastern Tigray
Central Tigray

FIGURE 2: Map of Ethiopia showing areas of collection sites (shaded region) of the landrace accessions.

yield/plant (SY), pod set score (PSS) (1–9 rating scale where 1 = very poor pod setting and 9 = excellent pod setting), and rust severity scored on a 1–9 scale (1–9 rating scale where 1 = highly resistant and 9 = highly susceptible) before flowering (RSBF) and after flowering (RSAF) were recorded on plot basis according to [26].

2.5. Data Analysis. Plot mean values were calculated for all traits and used for the analysis of variance (ANOVA). The estimation of genetic parameters was analyzed using GenStat Release 13.3. [27]. Phenotypic and genotypic variances for the augmented design were computed for all traits based on the methods of [28]. The pooled analysis for augmented design was carried out according to [29, 30] using GenStat Release

15.1 [31]. The broad sense of heritability ($h^2$) was estimated using the genotype mean according to [14]. Consider

$$h^2 = \left[ \frac{\sigma^2 g}{\sigma^2 p} \right] \times 100. \tag{1}$$

Genetic advance (GA) for selection intensity ($K$) at 5% is estimated according to [14] using the formula: GA = $(K)(\sigma p)(h^2)$.

The genetic advance as the percentage of the mean (GAM) was estimated as per [32] formula: GAM = $(GA/\overline{X}) * 100$.

## 3. Results

3.1. Analysis of Variance. Highly significant genotypic ($P \leq 0.01$) differences were observed for all parameters measured in the 2010/11 cropping season at Sirinka indicating that the variation was genetic. The mean square, CV, and mean values for traits of the genotype were presented in Table 3.

Moderate variabilities with an intermediate coefficient of variation (CV) among the genotypes were observed for days to 50% flowering, days to 90% maturity, plant height, and pod set score, whereas considerable variations were observed for the rest of the traits among the genotypes (Table 3).

The analysis of variance for the 2011/12 cropping season within the check showed significant variation at Sirinka, except for two traits: number of seeds/pods and number of seeds/plant (Table 4). At Chefe Donsa, days to 50% flowering, number of pods/plant, number of seeds/pods, number of seeds/plant, and seed weight/plant had nonsignificant variation within the checks (Table 5). At Sinana, all traits showed significant variability, except days to 50% flowering, days to 90% maturity, number of seeds/pods, and above ground biomass within the checks (Table 6). Consistent significant variations were noted across locations within the checks for plant height, 100-seed weight, and seed yield (Tables 4, 5, and 6).

The analysis of variance showed a highly significant variation ($P \leq 0.01$) for days to 50% flowering, days to 90% maturity, number of pods/plant, and number of seeds/pods within the test genotypes at Sirinka (Table 4). At Chefe Donsa, analysis depicted a highly significant variation

TABLE 4: Mean squares for agromorphological traits of 228 lentil genotypes at Sirinka in the 2011/12 cropping season.

| Sources of Variation | df | Mean squares | | | | | | | | | |
|---|---|---|---|---|---|---|---|---|---|---|---|
| | | DFF | DM | NP | NS | NSPP | SWPP | PH | SW | BI | SY |
| Block | 4 | 413.4** | 350.5** | 983.2* | 0.21** | 1693.6* | 1.16 | 35.49* | 0.8* | 113623** | 8974 |
| Within checks | 7 | 155.5** | 504.6** | 734.6* | 0.1 | 1213.8 | 1.9* | 37.3* | 2.3** | 67151* | 9542* |
| Within test genotypes | 219 | 91.9** | 182.5** | 1053.4** | 0.2** | 722.7 | 0.5 | 243.0* | 0.5* | 18463 | 2807 |
| Between check and test genotypes | 1 | 0.6 | 35.2 | 995.1 | 0.3** | 5227.6** | 13.2** | 18.2 | 4.6** | 437621** | 110311** |
| Residual | 35 | 14.1 | 34.1 | 302 | 0.04 | 546.1 | 0.7 | 13.5 | 0.3 | 24236 | 3801 |

df = degree of freedom.
DFF = days to 50% flowering, DM = days to 90% maturity, NP = number of pods per plant, NS = number of seeds per pod, SWPP = seed weight per plant, NSPP = number of seeds per plant, PH = plant height in cm, SW = 100-seed weight in gram, BI = biomass, PSS = pod set score.
**, * = significant at 5% and 1% probability levels, respectively.

TABLE 5: Mean squares for agromorphological traits of 228 lentil genotypes at Chefe Donsa in the 2011/12 cropping season.

| Sources of variation | df | Mean squares | | | | | | | | | | |
|---|---|---|---|---|---|---|---|---|---|---|---|---|
| | | DFF | DM | NP | NS | NSPP | SWPP | PH | SW | BI | SY | RSAF |
| Block | 4 | 24.4 | 327.9** | 664.7* | 0.03 | 1489.4* | 1.4* | 12.6* | 0.7** | 77552** | 3332* | 10.3** |
| Within check | 7 | 12.6 | 223.6* | 51.6 | 0.04 | 292.6 | 0.4 | 22.0** | 1.2** | 23490* | 3363* | 17.2** |
| Within test genotypes | 219 | 30.4** | 133.3 | 184.3 | 0.06 | 495.7 | 0.3 | 15.0** | 0.6** | 8941 | 1314 | 4.8** |
| Between check and test genotypes | 1 | 101.5** | 105 | 988.5* | 0.1 | 1127.8 | 2.3* | 7.7 | 0.03 | 227174** | 41227** | 27.3** |
| Residual | 35 | 12.1 | 83.4 | 188.2 | 0.06 | 560.1 | 0.4 | 4.5 | 0.1 | 10084 | 1183 | 2 |

**, * = significant at 5% and 1% probability levels, respectively.

TABLE 6: Mean squares for agromorphological traits of 228 lentil genotypes at Sinana in the 2011/12 cropping season.

| Sources of vriation | df | Mean squares | | | | | | | | | | |
|---|---|---|---|---|---|---|---|---|---|---|---|---|
| | | DFF | DM | NP | NS | NSPP | PH | SW | BI | SY | RSBF | RSAF |
| Block | 4 | 305.8** | 2224* | 50.9 | 0.04 | 319.6* | 147.8** | 1.3** | 153488** | 6897.6** | 5.1** | 16.1** |
| Within check | 7 | 224.2 | 2473.1 | 76.5** | 0.1 | 274.5* | 53.7* | 2.6** | 30505 | 6842.8** | 1.7* | 5.9 |
| Within test genotypes | 219 | 10054.6** | 36536 | 65.6** | 0.1* | 256.2** | 29.3 | 0.7** | 34051 | 1279.1* | 2.0** | 4.6 |
| Between check and test genotypes | 1 | 0.05 | 121.1 | 124.6* | 0.2 | 345.1 | 184.9** | 3.0** | 474202** | 80244.5** | 10.8** | 69.1** |
| Residual | 35 | 19.1 | 244.6 | 30 | 0.1 | 116.5 | 23 | 0.1 | 29335 | 823.2 | 0.8 | 3.8 |

RSBF; Rust disease severity score before flowering.
RSAF; Rust disease severity score after flowering.
**, * = significant at 5% and 1% probability levels, respectively.

($P \leq 0.01$) within test genotypes for days to 50% flowering, plant height, 100-seed weight, and rust disease severity score after flowering (Table 5), whereas at Sinana all traits revealed significant variation ($P \leq 0.05$) except days to 90% maturity, plant height, above ground biomass, and rust disease severity score after flowering within test genotypes (Table 6).

The analysis of variance between check and test genotypes for the 2011/12 cropping season showed highly significant differences except for days to 50% flowering, days to 90% maturity, number of pods/plant, and plant height at Sirinka (Table 4). At Chefe Donsa, significant ($P \leq 0.05$) variations were observed for days to 50% flowering, number of pods/plant, seed weight/plant, above ground biomass, seed yield, and rust disease severity score between checks and test genotypes (Table 5), whereas, at Sinana, significant variations were recorded between checks and test genotypes for number of pods/plant, plant height, 100-seed weight, seed yield, and rust severity score before and after flowering (Table 6). The test for homogeneity of error variance was significant across locations because the genotypes perform differently across the three locations for yield and yield component traits. Thus, it could not make further valid inferences for the pooled ANOVA (Table 7). Besides, rust infection was severe at Sinana, mild at Chefe Donsa, and absent at Sirinka and then no combined analysis was made for this character.

The minimum and maximum values of 11 agromorphological traits analyzed for the 2010/2011 cropping season at Sirinka were presented in (Table 8). The range for days to 50% flowering was recorded from 42 to 78 days for Acc. number 36120 and breeding line (FLIP-04-26L), respectively (Table 12). The shortest maturity period was recorded (77 days) for Acc. number 36025 and the longest (138 days) for FLIP-04-26L. Parent X2003S 223 produced the minimum (1.6) number of pods/plant while the maximum (73.3 pods/plant) was noted for breeding line (R-186XFLIP-86-38L-24). The shortest plant height of 20.6 cm was recorded for breeding line (FLIP-97-61L) and the highest was 44.7 cm for Acc. number 36001. The minimum number of seeds/pod was

TABLE 7: Mean squares of the combined analysis of variance for lentil yield and yield component traits and mean values for each location in the 2011/12 cropping season.

| Sources of variation | df | DFF | DM | NP | NS | NSPP | PH | SW | BI | SY |
|---|---|---|---|---|---|---|---|---|---|---|
| Environment mean squares | 2 | 754.0** | 1065.0** | 557.9** | 17188.6** | 162.1** | 2028.3** | 153.2** | 1031.6** | 343.1** |
| Genotype mean squares | 227 | 3.6** | 1.2$^{NS}$ | 0.4$^{NS}$ | 0.01$^{NS}$ | 0.2$^{NS}$ | 0.9** | 8.9** | 0.4** | 1.1** |
| Residual mean squares | 454 | 1.5 | 1.61 | 1.6 | 0.9 | 0.7 | 0.4 | 1.7 | 0.2 | 0.5 |
| Range | | 41–91 | 60–131 | 12.1–298 | 0.4–2.6 | 20.6–64.3 | 22.7–47 | 0.8–5.1 | 204.5–632 | 19.8–172 |
| Sirinka mean | | 56.4 | 91.4 | 37.4 | 1.2 | 43.77 | 33.9 | 2.7 | 269.8 | 83.2 |
| Chefe Donsa mean | | 56.1 | 114.5 | 29 | 1.5 | 43.8 | 27.71 | 2.8 | 269.4 | 86.8 |
| Sinana mean | | 66.3 | 104.6 | 19.2 | 1.9 | 34 | 40.5 | 2.4 | 521.3 | 44 |
| Environment mean | | 64 | 105 | 30 | 1.6 | 40 | 36 | 2.5 | 400 | 60 |

df = degree of freedom.
DFF = days to 50% flowering, DM = days to 90% maturity, NP = number of pods per plant, NS = number of seeds per pod, SWPP = seed weight per plant, NSPP = number of seeds per plant, PH = plant height in cm, SW = 100-seed weight in gram, BI = biomass, PSS = pod set score.
**, * = significant at 5% and 1% probability levels, respectively.
NS = nonsignificant.

TABLE 8: Genetic parameter estimates for agromorphological characters of lentil genotypes at Sirinka 2010/11 cropping season.

| Traits | Mean | Range | GV | PV | PCV | GCV | $h^2$ | GA | GAM |
|---|---|---|---|---|---|---|---|---|---|
| DFF | 57.2 | 42–78 | 18.7 | 32.5 | 7.6 | 10 | 57.3 | 6.8 | 11.8 |
| DM | 95.7 | 77–138 | 119.3 | 164.8 | 11.5 | 13.5 | 72.4 | 19.2 | 20.1 |
| NP | 35.3 | 1.6–73.3 | 76.4 | 202.9 | 25 | 40.5 | 37.7 | 11.1 | 31.5 |
| NS | 1.2 | 0.5–2.0 | 0.1 | 0.18 | 22.6 | 30.1 | 56 | 0.4 | 34.8 |
| PWPP | 1.5 | 0.24–8.1 | 0.7 | 0.8 | 58.4 | 60.7 | 92.6 | 1.7 | 15.8 |
| SWPP | 1.1 | 0.1–3.4 | 0.2 | 0.3 | 36.3 | 51.6 | 48.9 | 0.6 | 52 |
| NSPP | 42.9 | 5.5–103.1 | 215.4 | 416.2 | 34.6 | 47.8 | 51.8 | 21.9 | 51 |
| PH | 34.6 | 20.6–44.7 | 13.4 | 21.6 | 10.6 | 13.4 | 62.3 | 6 | 17.3 |
| SW | 2.5 | 1.38–4.9 | 0.3 | 0.4 | 21.7 | 24.9 | 75.8 | 1 | 38.9 |
| BI | 286.8 | 35–767.8 | 9806.7 | 15955 | 34.7 | 44.2 | 61.5 | 160.5 | 56 |
| SY | 84.9 | 0–347.2 | 1192.3 | 2188.7 | 41.1 | 55.4 | 54.5 | 52.8 | 62.2 |
| PSS | 5 | 2–7 | 0.5 | 0.7 | 14 | 17.1 | 66 | 1.2 | 23.2 |

Triats, DFF = days to 50% flowering, DM = days to 90% maturity, NP = number of pods per plant, NS = number of seeds per pod, NSPP = number of seeds per plant, PH = plant height in cm, SW = 100-seed weight in gram, BI = above ground biomass, SY = seed yield.
GV = genotypic variance, PV = phenotypic variance, GCV = genetic coefficient of variation, PCV = phenotypic coefficient of variation, $h^2$ = broad sense of heritability, GA = genetic advance, GAM = genetic advance as the percentage of the mean.

recorded (0.5 seed) for landrace (Acc. number 36098) and the maximum (2 seeds) for parent line (X2003S 233/8009/). Seed weight/plant ranged from 0.1 g for parent line X2002S 219/shehor-74/7554 to 3.4 g for breeding line R-186XFLIP-86-38L-24 (Table 12). The number of seeds/plant ranged from 5.5 for landrace (Acc. number 36098) to 103.1 for lines R-186XFLIP-86-38L-24. The largest seed size was observed for parent line FLIP-2004-7L (4.9 g) while the parent line X2003S 233/8009/had a minimum seed weight of 1.4 g. The lowest biomass/plant was recorded for the parent line X2003S 223 (35 g), and the highest biomass/plant was recorded for the parent line X2006S 128/5480/(767.8 g). The lowest seed yield (0 g) was recorded for landrace X2003S 223 whereas breeding line ILL-590 showed the highest seed yield (347.2 g) (Table 12).

3.2. Phenotypic and Genotypic Coefficients of Variation. Though variability in population is an indispensable prerequisite for any improvement, it is not the only criterion for deciding as to which trait is showing the highest degree of variability. Phenotype coefficient of variation (PCV) and genotype coefficient of variation (GCV) can help in this regard. Estimates of genetic parameter are shown in Tables 8, 9, 10, and 11. PCV and GCV values of approximately more than 20% are regarded as high, whereas values less than 10% are considered low and values in between are considered as medium [33]. Based on this delineation, high PCV and GCV values were recorded for number of pods/plant, number of seeds/pod, pod weight/plant, seed weight/plant, number of seeds/plant, 100 seed weight, above ground biomass, and seed yield. However, days to 50% flowering, days to 90% maturity, and plant height showed lower GCV and PCV values in the 2010/2011 cropping season. In the 2011/12 cropping season, similar GCV and PCV patterns were recorded across the three locations for number of pods/plant, seed weight/plant, number of seeds/plant, above ground biomass, seed yield, and rust disease score, whereas low PCV and GCV were noted for number of pods/plant, number of seeds/pods, seed weight/plant, and number of seeds/plant at Chefe Donsa.

TABLE 9: Genetic parameter estimates for agromorphological characters of lentil genotypes at Sirinka 2011/12 cropping season.

| Traits | Test genotype mean | Check mean | Grand mean | GV | PV | PCV | GCV | $h^2$ | GA | GAM |
|---|---|---|---|---|---|---|---|---|---|---|
| DFF | 56.3 | 56.3 | 56.3 | 77.4 | 93 | 15.6 | 17.1 | 83.3 | 16.6 | 29.4 |
| DM | 91.7 | 91.2 | 91.7 | 145.2 | 182 | 13.1 | 14.7 | 79.8 | 22.2 | 24.2 |
| NP | 35.7 | 45.2 | 35.7 | 714.7 | 1025.1 | 75 | 89.8 | 69.7 | 45.9 | 128.8 |
| NS | 1.2 | 1.3 | 1.2 | 0.1 | 0.2 | 27.6 | 34.3 | 65 | 0.6 | 49.9 |
| NSPP | 40.7 | 59.6 | 40.7 | 146.4 | 176.8 | 32.7 | 32.7 | 82.8 | 22.7 | 55.7 |
| SW | 2.7 | 3.1 | 2.7 | 0.2 | 0.5 | 18.6 | 27.6 | 45.6 | 0.7 | 24.6 |
| PH | 33.5 | 35.2 | 33.5 | 8.3 | 23.6 | 8.6 | 14.5 | 35 | 3.5 | 10.5 |
| BI | 245.9 | 392.3 | 245.9 | 5051 | 21656 | 28.9 | 59.8 | 23.3 | 70.6 | 28.7 |
| SY | 75.2 | 149.2 | 75.2 | 962 | 3429 | 41.2 | 77.8 | 28.1 | 33.9 | 45.1 |

Traits, DFF = days to 50% flowering, DM = days to 90% maturity, NP = number of pods per plant, NS = number of seeds per pod, NSPP = number of seeds per plant, PH = plant height in cm, SW = 100-seed weight in gram, BI = above ground biomass, SY = seed yield.
GV = genotypic variance, PV = phenotypic variance, GCV = genetic coefficient of variation, PCV = phenotypic coefficient of variation, $h^2$ = broad sense of heritability, GA = genetic advance, GAM = genetic advance as the percentage of the mean.

TABLE 10: Genetic parameter estimates for agromorphological characters of lentil genotypes at Chefe Donsa 2011/12 cropping season.

| Traits | Test genotype mean | Check mean | Grand mean | GV | PV | PCV | GCV | $h^2$ | GA | GAM |
|---|---|---|---|---|---|---|---|---|---|---|
| DFF | 55.9 | 57.3 | 55.9 | 14.9 | 29.3 | 6.9 | 9.7 | 50.6 | 5.6 | 10.1 |
| DM | 114.1 | 113.2 | 114.1 | 39.4 | 130.5 | 5.5 | 10.0 | 30.2 | 7.1 | 6.2 |
| NP | 26.9 | 35.3 | 26.9 | 4.6 | 185.3 | 7.9 | 50.0 | 2.5 | 0.7 | 2.6 |
| NS | 1.5 | 1.4 | 1.5 | 0.1 | 0.1 | 1.4 | 15.8 | 0.7 | 0.005 | 0.3 |
| NSPP | 41.3 | 50.2 | 41.3 | 2.2 | 502.4 | 3.6 | 53.9 | 0.4 | 0.2 | 0.4 |
| SWPP | 1.1 | 1.4 | 1.1 | 0.03 | 0.3 | 17.8 | 54 | 10.9 | 0.1 | 11.2 |
| SW | 2.8 | 2.7 | 2.8 | 0.5 | 0.6 | 23.7 | 26.2 | 81.8 | 1.3 | 46.6 |
| PH | 27.5 | 28.5 | 27.5 | 6.4 | 14.3 | 9.2 | 13.7 | 45.2 | 3.5 | 12.8 |
| BI | 262.6 | 305.6 | 262.6 | 2789.0 | 9888.0 | 20 | 37.6 | 28.2 | 57.8 | 22.0 |
| SY | 82.5 | 107.6 | 82.5 | 460.2 | 1427 | 25.7 | 45.3 | 32.3 | 25.1 | 30.5 |
| RSAF | 3.6 | 3.1 | 3.6 | 3.1 | 4.9 | 49.1 | 61.7 | 63.2 | 2.9 | 80.1 |

The lowest PCV (10%) was recorded for days to 50% flowering and the highest (60.7%) for pod weight/plant at Sirinka 2010/2011. At Sirinka 2010/2011, GCV values ranged from 7.6% for days to 50% flowering to 58.4% for pod weight/plant (Table 8). The maximum PCV and GCV recorded for number of pods/plant were 89.8 and 75, respectively, at Sirinka 2012, while the lowest PCV and GCV recorded for plant height were 14.5 and 8.6, respectively (Table 9). Days to 90% maturity and 100-seed weight showed moderate GCV. At Chefe Donsa and Sinana, values recorded for both PCV and GCV were similar with those of Sirinka (Tables 10 and 11). Traits such as number of pods/plant, seed weight/plant, 100-seed weight, above ground biomass, seed yield, and rust disease score showed relatively high PCV and GCV values across the four environments (Tables 8–11).

*3.3. Heritability and Genetic Advance.* The broad sense of heritability ($H^2$) estimates of the traits ranged from 0.4% (number of seeds/plant) to 92.6% (pod weight/plant). According to [34], heritability estimate in cultivated plants can be placed in the following categories: heritability estimate ranging from 5 to 10% low, values ranging from 10 to 30% medium, and values 30% and above as high heritability. Based on the above classification, most of the characters have shown high heritability over locations except at Chefe Donsa. At

Chefe Donsa, low heritability estimates were recorded for number of seeds/plant (0.4%), number of seeds/pods (0.7%), and number of pods/plant (2.5%) (Table 10). At Sinana, heritability estimates ranged from 18% to 88% for biomass and 100-seed weight, respectively. Days to 90% maturity, number of seeds/plant, plant height, and above ground biomass had a moderate heritability (17.7–26.1%) (Table 11), whereas seed yield, 100-seed weight, rust disease score, number of pods/plant, and number of seeds/plant scored higher heritability.

Consistently higher value of heritability was revealed for days to 50% flowering, 100-seed weight, days to 90% maturity, and plant height over locations. However, all these four characters associated with low genetic advance value (GA) because of low GCV, whereas 100-seed weight, rust severity score, and seed yield showed high heritability values coupled with high genetic advance values across locations because of high of GCV. Numbers of seeds/plant, seed weight/plant, and number of seeds/pod have intermediate GCV and heritability with a mild response to selection.

## 4. Discussion

Morphological diversity has been used to characterize germplasm from a range of plant species and allows the assessment

TABLE 11: Genetic parameter estimates for agromorphological characters of lentil genotypes at Sinana 2011/12 cropping season.

| Traits | Test genotype mean | Check mean | Grand mean | GV | PV | PCV | GCV | $h^2$ | GA | GAM |
|---|---|---|---|---|---|---|---|---|---|---|
| DFF | 66.04 | 66.8 | 66.1 | 20.7 | 66.6 | 6.9 | 12.4 | 31.1 | 7.9 | 12.1 |
| DM | 105.1 | 104.5 | 105.1 | 86.6 | 331.2 | 2.4 | 15.1 | 26.1 | 9.8 | 9.3 |
| NP | 19.2 | 20.8 | 19.3 | 33.7 | 63.6 | 30.2 | 41.4 | 52.9 | 8.6 | 44.4 |
| NSPP | 33.9 | 36.1 | 34 | 33.1 | 149.6 | 16.9 | 36 | 22.1 | 13.1 | 38.5 |
| SW | 2.38 | 2.7 | 2.4 | 0.6 | 0.7 | 32.7 | 35 | 87.1 | 1.51 | 63.2 |
| PH | 40.3 | 42.6 | 40.4 | 7.2 | 30.1 | 6.6 | 13.6 | 23.7 | 2.7 | 6.7 |
| BI | 505.1 | 619.1 | 509.1 | 6290 | 35625 | 15.6 | 37.1 | 17.7 | 69.9 | 13.8 |
| SY | 40.2 | 89.6 | 41.9 | 809.8 | 1633 | 67.9 | 96.4 | 49.6 | 45.8 | 109.2 |
| RSBF | 2 | 1.5 | 1.9 | 1.2 | 2 | 55.6 | 71.9 | 59.7 | 1.8 | 88.9 |
| RSAF | 5.8 | 4.6 | 5.7 | 1.2 | 5.1 | 19.5 | 39.4 | 24.5 | 1.2 | 21.1 |

of genetic diversity. The purpose of this study was to compare the genetic diversity within and among Ethiopian and exotic lentil genotypes of the Ethiopian possessions. In this paper, the characterizations of morphological variation of lentil germplasm using biometrical model was described.

As per ANOVA, highly significant phenotypic ($P \leq 0.01$) differences were observed for all traits measured indicating the existence of variation. The presence of highly significant variation among the genotypes for the morphological traits was indicative of the presence of high degree of genetic variation for future breeding programs through selection. Similarly, from follow-up study with the simple sequence repeat (SSR), genetic diversity analysis was a supportive evidence of the presence of a considerable variability among both local and exotic lentil genotypes of the Ethiopia germplasm (unpublished data). In agreement with this report, [21, 34–39] reported significant genetic variations among lentil landraces. On the other hand, some studies showed that no significant variation was recorded for some traits, like seed weight/plant and days to flowering [21, 38, 40].

The results recorded for coefficient of variation were comparable with [38] for days to 50% flowering (8.8), plant height (9.1), number of seeds/plant (43.2), and above ground biomass (31.5). However, [38] reported higher values vis-a-vis with this study for days to 90% maturity (19.2), pod weight/plant (47), and seed weight/plant (47.8). In agreement with this finding, [41] also noted comparable range for plant height on 3974 accessions of lentil. However, they recorded a wider range of values for 100-seed weight with an overall mean than with our records on lentil accessions. In another study, [37] reported a narrow range for yield component traits among 46 genotypes from South East Anatolia region of Turkey as opposed to our findings.

Significant variations were noted over locations among the test genotypes and within the checks with a few exceptions. In line with this study, [24] recorded the wide range of genetic variability among yield related characters in exotic lentil lines in Ethiopia. Comparable to our study, [15, 18] noted the presence of variation in characters, such as grain and straw yield, 100-seed weight, days to 50% flowering, days to 90% maturity, number of seeds/pods, plant height, pod number/peduncle, and resistance to various biotic and a biotic stresses in the lentil. Besides, the variation among

environments was significant for all morphological traits in view of diverse agro-climatic features of the test sites. Similarly, [42] also observed the greater portion of total variance was due to environment. Reference [5, 7, 23] also reported consistent regional differences among lentil landraces of Ethiopia for time to flowering and maturity, 100-seed weight, seed/pod, and plant height.

ANOVA over locations showed consistent significant variations were recorded within the checks for 100-seed weight, above ground biomass, seed yield, and rust disease score. These implied that yield increment was achieved for the last three decades, largely for 100-seed weight, above ground biomass, and seed yield of lentil genotypes. A consistent significant variation was also recorded within test genotypes over three locations for days to 50% flowering and 100-seed weight. Besides, reliable variation was noted over all locations between checks and test genotypes for above ground biomass and seed yield. Similarly, according to this (unpublished data) follow-up study, using the multivariate analysis reported that seed yield, above ground biomass, seed weight/plant, 100-seed weight, rust severity score, and plant height were the major agromorphological traits contributing to variations among the genotypes.

Estimate of PCV and GVC for days to 50% flowering, days to 90% maturity, and 100-seed weight showed narrow differences, indicating a relatively low influence of the environment on these characters. Similar observation of considerable closeness between GCV and PCV for these characters made by [43] indicated low influence of environmental factors in lentil. In general, PCV values for most of the traits at Sirinka and Sinana were higher than that record noted at Chefe Donsa. In addition, the PCV was generally higher than the GCV for all characters. Consistently higher PCV values were observed across four environments for rust disease scores, seed yield, pod weight/plant, seed weight/plant, number of pods/plant, 100-seed weight, above ground plant biomass, and number of seeds/plant. Other studies showed similar high PCV and GCV for seed yield, above ground biomass, and 100-seed weight [44, 45]. In contrast, days to 50% flowering, days to 90% maturity, and plant height showed comparatively low PCV values (less than 20%) for all locations. It was observed that GCV for number of pods/plant, seed yield, rust disease score, 100-seed

TABLE 12: Means for yield and yield components of the genotypes at Sirinka 2010/11.

| Genotype name | DF | DM | NP | NSS | PWPP | SWPP | NSPP | PHH | SW | BI | SY | PSS |
|---|---|---|---|---|---|---|---|---|---|---|---|---|
| 36001 | 57 | 89 | 39.7 | 0.9 | 1.8 | 1.2 | 39 | 44.7 | 2.6 | 345.2 | 78 | 5 |
| 36003 | 56 | 91 | 47.8 | 0.7 | 1.9 | 0.9 | 45 | 38.8 | 2.1 | 315.7 | 59.9 | 6 |
| 36007 | 58 | 90 | 38.8 | 1.8 | 2 | 1.5 | 86 | 38.4 | 2.1 | 359.3 | 79.6 | 6 |
| 36009 | 58 | 90 | 35.6 | 1.3 | 1.4 | 0.9 | 47 | 39.9 | 1.9 | 294.8 | 60.9 | 5 |
| 36013 | 58 | 90 | 39.8 | 1.2 | 3.9 | 2.3 | 52 | 39.4 | 2.1 | 440.8 | 74 | 5 |
| 36015 | 59 | 98 | 29.7 | 1.9 | 1.9 | 1.2 | 52 | 41.9 | 2.5 | 545.8 | 128.6 | 5 |
| 36019 | 55 | 94 | 41.1 | 1.2 | 1.9 | 1.5 | 69 | 37 | 2.2 | 311.9 | 71.6 | 5 |
| 36023 | 54 | 84 | 34.5 | 0.6 | 1.6 | 0.6 | 41 | 36.8 | 2.1 | 237.8 | 30.6 | 5 |
| 36025 | 46 | 77 | 34.2 | 1.1 | 2 | 1.1 | 39 | 33.6 | 2.1 | 254.7 | 62.2 | 6 |
| Alemaya | 57 | 107 | 55.2 | 1.3 | 1.5 | 2.4 | 70 | 31.6 | 3.1 | 449.6 | 160.9 | 6 |
| ADAA | 61 | 128 | 29.1 | 1.7 | 1.9 | 1.3 | 43 | 40.4 | 2.8 | 249.7 | 45.9 | 4 |
| EL-142 | 56 | 93 | 36.1 | 1.1 | 1.1 | 1 | 45 | 35.5 | 2 | 321.8 | 51.1 | 5 |
| X2003S 222/213/ | 66 | 103 | 32.7 | 1.1 | 1.7 | 1.1 | 36 | 36 | 2 | 390.8 | 95.2 | 4 |
| X2003S 238 | 55 | 92 | 32.5 | 1.1 | 2.3 | 1.4 | 40 | 32.2 | 3.8 | 377.4 | 124.8 | 6 |
| 36028 | 59 | 98 | 38.2 | 0.7 | 1.4 | 1.3 | 52 | 41.1 | 2.2 | 453 | 145.6 | 6 |
| 36029 | 57 | 89 | 39.9 | 1.3 | 2.1 | 1.1 | 49 | 31.8 | 2.3 | 331.6 | 76.7 | 5 |
| 36033 | 55 | 86 | 37.4 | 1 | 1.3 | 1 | 37 | 33.8 | 2.4 | 257.9 | 78.4 | 7 |
| 36039 | 55 | 88 | 43.5 | 1.2 | 2.1 | 1.2 | 68 | 33.6 | 2.3 | 181.4 | 60.9 | 5 |
| 36041 | 56 | 89 | 55.4 | 1.1 | 1.8 | 1.4 | 60 | 34.7 | 2.1 | 246.5 | 75.9 | 5 |
| 36048 | 57 | 88 | 52.7 | 0.9 | 1 | 1.3 | 49 | 35.1 | 1.8 | 280.4 | 89.2 | 5 |
| 36058 | 51 | 84 | 41.1 | 0.6 | 0.9 | 0.8 | 40 | 32.1 | 2.9 | 170.5 | 55.2 | 6 |
| 36061 | 54 | 82 | 36.3 | 0.6 | 0.8 | 0.6 | 22 | 30.5 | 2 | 163.3 | 30.8 | 6 |
| TESHALE | 56 | 98 | 26.9 | 1.2 | 2 | 1.3 | 34 | 33.7 | 3.6 | 372.9 | 119.7 | 5 |
| X2006S 128/5480/ | 77 | 125 | 50 | 1.2 | 4.6 | 2 | 59 | 42.9 | 2.8 | 767.8 | 153.3 | 3 |
| L-9-12 | 59 | 114 | 35.3 | 1 | 1.5 | 1.2 | 38 | 36.7 | 3.6 | 316 | 74 | 4 |
| 36071 | 57 | 91 | 33.1 | 0.9 | 1 | 0.9 | 39 | 41.5 | 2.4 | 406.5 | 104.6 | 6 |
| 36085 | 57 | 91 | 58 | 1.2 | 2 | 1.9 | 77 | 39.8 | 3 | 392.7 | 117.2 | 6 |
| 36088 | 57 | 95 | 34.4 | 1.1 | 1.2 | 0.9 | 40 | 37.5 | 3 | 399.6 | 115.3 | 5 |
| 36089 | 60 | 97 | 32.7 | 1.1 | 0.5 | 1.1 | 41 | 33.3 | 2.3 | 237.2 | 105.8 | 5 |
| 36097 | 58 | 98 | 56.4 | 1 | 1.7 | 1.5 | 75 | 34.9 | 2.4 | 288.9 | 86.5 | 4 |
| 36098 | 56 | 106 | 18.3 | 0.5 | 0.5 | 0.2 | 6 | 29.9 | 2.1 | 71.1 | 0.1 | 4 |
| 36103 | 53 | 86 | 27.5 | 1.4 | 1.3 | 1 | 36 | 35.6 | 2.5 | 274.4 | 90.9 | 5 |
| 36104 | 53 | 88 | 37.2 | 1.1 | 1.7 | 1.3 | 45 | 34.9 | 2.7 | 274.5 | 77.3 | 6 |
| 36105 | 56 | 87 | 29.7 | 1.1 | 0.3 | 0.8 | 33 | 29.8 | 3.1 | 145 | 41.6 | 5 |
| 211062 | 60 | 91 | 38.9 | 1.6 | 1.9 | 1.6 | 63 | 37 | 2 | 449.4 | 93.9 | 5 |
| X2003S 195 | 72 | 130 | 35.2 | 1.2 | 1.5 | 1.1 | 43 | 34.6 | 2.5 | 287.1 | 84.3 | 5 |
| X2002S 219/6821/ | 57 | 96 | 35.2 | 1.2 | 1.5 | 1.1 | 43 | 34.6 | 2.5 | 287.1 | 84.3 | 5 |
| X2006S 129/F2 | 51 | 94 | 28.6 | 1.2 | 1.5 | 1.1 | 43 | 34.6 | 2.5 | 287.1 | 84.3 | 5 |
| 36110 | 57 | 91 | 40.9 | 1.7 | 2.1 | 1.7 | 68 | 39.8 | 2.2 | 374.5 | 136.5 | 5 |
| 36120 | 42 | 87 | 23.3 | 0.5 | 1 | 0.4 | 12 | 25.7 | 2.2 | 174 | 30.3 | 7 |
| 36131 | 59 | 92 | 36.3 | 1.6 | 1.2 | 1.4 | 59 | 37.1 | 2.1 | 369 | 127.4 | 5 |
| 36042 | 54 | 84 | 45.5 | 0.8 | 2.2 | 0.9 | 37 | 35.4 | 1.9 | 223.4 | 48 | 6 |
| 36137 | 59 | 99 | 33.2 | 1.4 | 1.5 | 0.9 | 44 | 34 | 2.1 | 251 | 78.6 | 4 |
| 36139 | 53 | 97 | 40.7 | 0.9 | 1.8 | 0.7 | 34 | 35.8 | 2 | 243.6 | 55.8 | 6 |
| 36147 | 60 | 99 | 48 | 1.1 | 1.8 | 1.2 | 50 | 38.8 | 2.4 | 331.6 | 97.7 | 6 |
| 36150 | 56 | 89 | 34.4 | 1.3 | 1.9 | 0.8 | 46 | 35.9 | 2.4 | 261.6 | 76.3 | 5 |
| 36162 | 57 | 100 | 34.7 | 1.3 | 2 | 1 | 43 | 36.6 | 2.3 | 237.1 | 67 | 5 |
| 36165 | 57 | 95 | 30.3 | 1.4 | 0.9 | 1 | 44 | 39 | 2.4 | 313.9 | 83.4 | 6 |
| X2005S 215/6002/ | 57 | 96 | 34.1 | 0.8 | 1.6 | 1 | 25 | 40.8 | 3.9 | 329.8 | 101.7 | 5 |
| 36168 | 59 | 104 | 26.6 | 1 | 1.1 | 0.9 | 27 | 34.1 | 2.5 | 267.5 | 89.7 | 5 |
| 203141 | 59 | 90 | 41.1 | 1 | 1.5 | 1.3 | 42 | 35.4 | 2.3 | 320.5 | 118.4 | 5 |

TABLE 12: Continued.

| Genotype name | DF | DM | NP | NSS | PWPP | SWPP | NSPP | PHH | SW | BI | SY | PSS |
|---|---|---|---|---|---|---|---|---|---|---|---|---|
| 207258 | 58 | 90 | 50 | 1 | 1.3 | 1.2 | 47 | 39.1 | 2.6 | 324.7 | 104.2 | 6 |
| 207274 | 59 | 91 | 39.9 | 1.1 | 1.5 | 1.3 | 48 | 38.1 | 2.3 | 272.5 | 99.2 | 5 |
| 207260 | 58 | 94 | 40.9 | 1.2 | 0.9 | 1.1 | 52 | 38.3 | 2.3 | 340.3 | 92.7 | 5 |
| 207287 | 52 | 91 | 46.4 | 1.1 | 2.3 | 1.4 | 51 | 35.1 | 2.2 | 311.9 | 107.1 | 6 |
| 207309 | 57 | 89 | 36.9 | 1 | 1.3 | 1 | 35 | 31.8 | 2.3 | 248.1 | 57 | 5 |
| 211078 | 54 | 88 | 47.9 | 1.4 | 2.1 | 1.5 | 61 | 31 | 2.1 | 218.6 | 70.3 | 6 |
| 211131 | 53 | 87 | 40.7 | 1.7 | 1.4 | 1.8 | 65 | 32.3 | 2 | 260.4 | 102 | 6 |
| 212745 | 57 | 91 | 36.3 | 1.5 | 1.9 | 1.2 | 57 | 33.3 | 2.2 | 322.7 | 106.9 | 7 |
| X2006S 133/6211/ | 53 | 98 | 18 | 1.3 | 1.2 | 0.7 | 20 | 30.5 | 3.7 | 135.5 | 36.4 | 5 |
| 238979 | 59 | 91 | 60.1 | 1.3 | 2.3 | 1.9 | 81 | 33 | 2.1 | 259.2 | 88.6 | 5 |
| 215248 | 56 | 87 | 45.5 | 1.2 | 1.5 | 1.2 | 51 | 31.4 | 2.2 | 237.5 | 71.4 | 5 |
| 215249 | 58 | 90 | 39.7 | 1.4 | 1.1 | 1.6 | 53 | 35.7 | 2.4 | 265.7 | 87 | 5 |
| 215806 | 58 | 91 | 26 | 1.8 | 0.8 | 0.8 | 35 | 32.9 | 2.5 | 189.6 | 56.8 | 5 |
| 216877 | 58 | 92 | 29.3 | 1.3 | 0.6 | 0.8 | 35 | 33.4 | 2.1 | 226.8 | 82.4 | 4 |
| 220120 | 56 | 91 | 23.6 | 1.3 | 0.7 | 0.6 | 27 | 37.7 | 2 | 149.7 | 34.9 | 6 |
| 223221 | 58 | 89 | 23.3 | 1.5 | 0.8 | 0.7 | 35 | 34.7 | 2 | 216.9 | 72.8 | 5 |
| 228242 | 59 | 90 | 34.9 | 1.8 | 1.5 | 1.4 | 65 | 37.3 | 2.2 | 259.5 | 83.5 | 5 |
| 228243 | 59 | 87 | 24.6 | 1.9 | 0.6 | 0.5 | 43 | 33.1 | 2 | 132.5 | 35.2 | 5 |
| 228809 | 53 | 87 | 32.4 | 0.6 | 1 | 0.7 | 21 | 35.9 | 2 | 231.1 | 63.9 | 5 |
| 219957 | 53 | 97 | 31.3 | 1.2 | 0.5 | 1 | 41 | 37.3 | 2.5 | 351.3 | 107.4 | 6 |
| 211110 | 59 | 91 | 38.9 | 1.8 | 1.6 | 1.2 | 62 | 38.4 | 1.8 | 390 | 96.3 | 5 |
| 229179 | 59 | 97 | 29.7 | 1.8 | 1.1 | 0.9 | 53 | 40.3 | 2.2 | 296.5 | 72.3 | 5 |
| 229182 | 59 | 95 | 40.5 | 1 | 1.1 | 1.1 | 43 | 41 | 2.2 | 342.6 | 83.3 | 5 |
| 229183 | 59 | 92 | 29.8 | 1.3 | 1 | 0.8 | 40 | 36 | 2.2 | 265.6 | 64.8 | 5 |
| 230521 | 59 | 93 | 28.7 | 1.3 | 1.2 | 0.7 | 40 | 38.2 | 1.9 | 366.7 | 99.2 | 5 |
| 230832 | 59 | 93 | 43.3 | 1.3 | 0.9 | 1 | 54 | 38.2 | 1.7 | 270.8 | 81.8 | 5 |
| 230833 | 59 | 93 | 52.6 | 1 | 1.1 | 1.4 | 60 | 36.9 | 2 | 327.3 | 91.5 | 5 |
| 230834 | 58 | 93 | 27 | 1.5 | 1 | 0.7 | 37 | 34.7 | 1.8 | 346.2 | 94.1 | 4 |
| 230837 | 58 | 90 | 44.9 | 0.8 | 1.6 | 1 | 35 | 35.9 | 1.9 | 402.6 | 118.2 | 4 |
| 231247 | 59 | 93 | 36.4 | 1.2 | 1.1 | 1.1 | 44 | 35.6 | 2.3 | 297.5 | 103.4 | 6 |
| 231248 | 58 | 88 | 42.6 | 0.7 | 1 | 1 | 29 | 32.2 | 2.4 | 301.5 | 89.7 | 5 |
| X2006S 130/7547/ | 54 | 86 | 20.9 | 0.7 | 0.7 | 0.4 | 13 | 27.1 | 2.3 | 91.8 | 98.3 | 6 |
| FLIP-2004-7L | 51 | 99 | 17.6 | 1.6 | 1.4 | 1.3 | 28 | 32 | 4.9 | 270.8 | 92.8 | 4 |
| 233349 | 57 | 90 | 19.8 | 1.2 | 1.1 | 0.6 | 26 | 30.2 | 2.5 | 159.2 | 43 | 5 |
| 233973 | 57 | 89 | 30.6 | 1.5 | 1.2 | 1 | 40 | 31.8 | 2.4 | 252.4 | 143 | 4 |
| X2003S 223 | 66 | 124 | 1.6 | 0.8 | 8.1 | 0.4 | 6 | 25 | 2.9 | 35 | 0 | 5 |
| X2003S 195/7115/ | 57 | 96 | 35.2 | 1.2 | 1.5 | 1.1 | 43 | 34.6 | 2.5 | 287.1 | 84.3 | 5 |
| 235013 | 57 | 91 | 23.6 | 1.3 | 2.3 | 1.1 | 31 | 30.7 | 2.5 | 248.3 | 92.7 | 5 |
| 235015 | 58 | 91 | 45.4 | 1.1 | 2.2 | 1.2 | 50 | 35.5 | 2.4 | 199.6 | 66.8 | 5 |
| 235016 | 58 | 89 | 29.2 | 1.1 | 1.7 | 0.9 | 32 | 33.3 | 2.7 | 284.1 | 91.6 | 5 |
| 235017 | 58 | 91 | 38.4 | 1.1 | 1.3 | 0.8 | 42 | 33 | 2.4 | 196.6 | 66.9 | 5 |
| 235383 | 57 | 89 | 36.2 | 1.2 | 1.9 | 1.2 | 47 | 40.8 | 2.6 | 310 | 97.9 | 6 |
| 235698 | 53 | 88 | 52.1 | 0.8 | 1.4 | 1 | 40 | 33.6 | 2.2 | 196.8 | 68.3 | 5 |
| 236438 | 58 | 90 | 33.2 | 0.9 | 1.2 | 0.8 | 30 | 34.5 | 2 | 358.1 | 113.7 | 5 |
| 236484 | 57 | 88 | 52.2 | 1 | 1.7 | 1.4 | 51 | 33.6 | 2.1 | 273.1 | 87 | 6 |
| X2006S 130/7978/ | 56 | 92 | 40.8 | 1 | 1.9 | 1.6 | 39 | 35.9 | 4.1 | 334.6 | 101.3 | 5 |
| X2002S 221/7979/ | 46 | 89 | 20.5 | 0.8 | 0.9 | 0.6 | 18 | 33.3 | 2.3 | 176.6 | 40.7 | 5 |
| X2002S 221/7980/ | 58 | 99 | 28.9 | 0.9 | 1.7 | 1.5 | 30 | 36.1 | 4.2 | 310.7 | 109.1 | 5 |
| 236486 | 57 | 88 | 30.5 | 0.9 | 0.8 | 0.9 | 34 | 35 | 2.2 | 183.4 | 61.9 | 5 |
| 236487 | 53 | 88 | 36 | 1.4 | 0.8 | 1 | 39 | 33.4 | 2.3 | 207.3 | 73 | 5 |
| 236892 | 58 | 91 | 36.9 | 1.1 | 1 | 0.9 | 41 | 37.7 | 2.2 | 240.3 | 59.8 | 5 |

TABLE 12: Continued.

| Genotype name | DF | DM | NP | NSS | PWPP | SWPP | NSPP | PHH | SW | BI | SY | PSS |
|---|---|---|---|---|---|---|---|---|---|---|---|---|
| 237027 | 50 | 83 | 25.3 | 0.8 | 1.1 | 0.7 | 21 | 33.3 | 3.9 | 146.9 | 47.3 | 5 |
| 237502 | 60 | 90 | 36.8 | 1.2 | 1 | 0.9 | 44 | 33.8 | 2.1 | 195.5 | 54.1 | 5 |
| 237503 | 59 | 92 | 41.8 | 1.5 | 1.2 | 1.4 | 67 | 39.3 | 2.6 | 279.9 | 93.9 | 5 |
| 237504 | 59 | 92 | 45.4 | 0.9 | 1.8 | 1.1 | 39 | 36.4 | 2.3 | 227.3 | 76.2 | 5 |
| X2003S 224 | 57 | 96 | 35.2 | 1.2 | 1.5 | 1.1 | 43 | 34.6 | 2.5 | 287.1 | 84.3 | 5 |
| 238971 | 58 | 90 | 31 | 1.5 | 1.1 | 0.9 | 42 | 32.8 | 2.5 | 262.8 | 80.7 | 5 |
| 238978 | 60 | 97 | 31.2 | 0.9 | 0.7 | 0.8 | 33 | 35.8 | 2.4 | 247.4 | 70.7 | 4 |
| X2003S 233/8009/ | 51 | 85 | 19.8 | 2 | 1.4 | 0.4 | 29 | 27.4 | 1.4 | 75.8 | 30.3 | 2 |
| X2006S 134/8174/ | 67 | 119 | 45 | 1.2 | 3.2 | 1.8 | 54 | 40.5 | 2.9 | 413.9 | 75.6 | 5 |
| X2006S 127 | 72 | 124 | 29.1 | 0.9 | 1 | 0.7 | 28 | 31.7 | 2.2 | 188.8 | 28.7 | 3 |
| 241782 | 59 | 109 | 26.6 | 1.4 | 0.8 | 0.6 | 28 | 36.2 | 2.4 | 268.5 | 66.5 | 4 |
| 241784 | 56 | 90 | 36 | 1.2 | 1.6 | 1.2 | 41 | 36.5 | 2.6 | 282 | 98 | 6 |
| 241785 | 57 | 88 | 35.1 | 1 | 1.1 | 1.1 | 36 | 34.8 | 2.6 | 276 | 95.6 | 5 |
| 241786 | 55 | 87 | 51.3 | 1.2 | 1.4 | 1.7 | 69 | 35.4 | 2.4 | 391.6 | 145.4 | 5 |
| 242604 | 58 | 92 | 52.6 | 1.2 | 2.2 | 1.8 | 65 | 36.3 | 2.5 | 392.4 | 134.8 | 6 |
| 243433 | 59 | 94 | 32.4 | 0.9 | 1.4 | 1 | 28 | 32.8 | 2.3 | 241.3 | 73.2 | 5 |
| 243436 | 59 | 91 | 24.1 | 1.1 | 1.1 | 0.6 | 28 | 27.2 | 2.3 | 117.3 | 35 | 5 |
| 243440 | 57 | 94 | 30.7 | 0.7 | 1.1 | 1 | 20 | 35.8 | 2.3 | 244.1 | 88.5 | 5 |
| 243443 | 53 | 94 | 27.1 | 1.1 | 1.5 | 1 | 26 | 34.9 | 2.5 | 273.4 | 82.2 | 5 |
| 243447 | 53 | 94 | 26.9 | 1.1 | 2.2 | 1.1 | 26 | 34.7 | 2.7 | 257.9 | 95.7 | 6 |
| X2006S 122/9932/ | 54 | 86 | 18.1 | 1 | 0.9 | 0.5 | 20 | 30.5 | 2.1 | 163.4 | 31.4 | 5 |
| X2006S 127/9945/ | 51 | 90 | 27.6 | 0.8 | 1 | 0.5 | 21 | 28.2 | 2.4 | 206.7 | 42.1 | 4 |
| 244606 | 53 | 85 | 26.6 | 1.2 | 1 | 0.7 | 34 | 31.7 | 2.2 | 258.2 | 83.2 | 5 |
| 244610 | 53 | 85 | 33.8 | 1.3 | 1.5 | 0.9 | 50 | 34.8 | 2.2 | 222.1 | 86.8 | 5 |
| 244615 | 58 | 85 | 34.7 | 1.2 | 1.7 | 0.9 | 42 | 33.6 | 2.2 | 261.3 | 75.8 | 5 |
| 244619 | 58 | 87 | 33.8 | 1.2 | 1.2 | 0.8 | 32 | 36.7 | 2 | 227.2 | 65.1 | 5 |
| 244623 | 53 | 85 | 26.4 | 1 | 1.3 | 0.7 | 29 | 37.8 | 2.3 | 263.2 | 65.3 | 5 |
| X2006S 122/9951/ | 53 | 88 | 36.3 | 1.3 | 2.1 | 1.1 | 41 | 33 | 1.8 | 240.1 | 64.6 | 6 |
| X2002S 219/shehor-74/7554 | 46 | 82 | 10.2 | 0.8 | 0.7 | 0.1 | 8 | 28.8 | 2.5 | 59.7 | 9.1 | 6 |
| ILL-590 | 56 | 87 | 40.7 | 1.3 | 1.2 | 1.1 | 42 | 35.6 | 3.2 | 468.3 | 347.3 | 6 |
| X2003S 236 | 57 | 96 | 35.2 | 1.2 | 1.5 | 1.1 | 43 | 34.6 | 2.5 | 287.1 | 84.3 | 6 |
| EXOTIC #DZ/2008 AK | 61 | 115 | 12.2 | 1.2 | 1.5 | 1 | 16 | 31.8 | 3.3 | 564.3 | 96.4 | 5 |
| R-186XFLIP-86-38L-24 | 60 | 109 | 73.3 | 1.4 | 1.5 | 3.4 | 103 | 37.1 | 2.8 | 642.9 | 200.9 | 4 |
| FLIP-2006-60L | 54 | 126 | 23 | 1.3 | 1.5 | 1 | 31 | 24.6 | 3.4 | 200 | 54.3 | 5 |
| FLIP-97-68L | 56 | 104 | 46.1 | 1.3 | 1.5 | 1.7 | 59 | 31.1 | 3.1 | 541.5 | 165.1 | 5 |
| ILL-358 X ILL-2573-2-2000 | 64 | 112 | 32 | 1.4 | 1.5 | 1.2 | 46 | 30.4 | 2.9 | 288.1 | 99.5 | 5 |
| FLIP-04-26L | 78 | 138 | 30 | 1.1 | 1.5 | 1.2 | 26 | 38.1 | 4.2 | 578.8 | 61.2 | 5 |
| ILL-28501 | 58 | 114 | 36.2 | 1.6 | 1.5 | 1.6 | 59 | 37.7 | 3.1 | 389.1 | 98 | 5 |
| 87S-93549XEL-1O3-4 | 47 | 95 | 33.2 | 1.1 | 1.5 | 1 | 37 | 27.9 | 2.6 | 215.8 | 81.2 | 5 |
| 87s-93549XEL-103-5 | 57 | 109 | 31.6 | 1.4 | 1.5 | 0.9 | 46 | 36.4 | 2.6 | 374.7 | 109.7 | 5 |
| FLIP-2006-20L | 74 | 124 | 20.4 | 1 | 1.5 | 0.8 | 21 | 31.8 | 3.2 | 295.8 | 30.9 | 5 |
| Derash | 57 | 103 | 55.5 | 1.3 | 1.5 | 2.5 | 71 | 35.6 | 3 | 446 | 162.2 | 5 |
| ALEMETENA | 50 | 94 | 25.3 | 1.2 | 1.5 | 1.1 | 33 | 21.9 | 3.6 | 172.2 | 85.6 | 5 |
| Chekol x R-186-8-1 | 54 | 109 | 30.7 | 1.2 | 1.5 | 1.3 | 38 | 31.1 | 3.3 | 303 | 119.2 | 5 |
| R-186X FLIP-86-38L | 58 | 103 | 27.2 | 1.8 | 1.5 | 1.4 | 50 | 31.9 | 2.5 | 262.9 | 80.5 | 5 |
| FLIP-87-68L | 60 | 122 | 41 | 1.3 | 1.5 | 1.7 | 60 | 29.3 | 3.1 | 266 | 93 | 5 |
| Chekol | 57 | 98 | 38 | 1.4 | 1.5 | 1.1 | 54 | 29.4 | 2.3 | 258.1 | 104.8 | 5 |
| ILL-10045 | 60 | 115 | 8.9 | 1.1 | 1.5 | 0.3 | 10 | 35.1 | 3.1 | 305.2 | 61.6 | 5 |
| FLIP-97-16L | 61 | 102 | 17 | 1.3 | 1.5 | 0.5 | 28 | 44.7 | 2.3 | 531.7 | 60.5 | 5 |
| ILL-10681 | 63 | 116 | 41.5 | 1.3 | 1.5 | 1.9 | 53 | 32.1 | 3.3 | 202.2 | 89.7 | 5 |

TABLE 12: Continued.

| Genotype name | DF | DM | NP | NSS | PWPP | SWPP | NSPP | PHH | SW | BI | SY | PSS |
|---|---|---|---|---|---|---|---|---|---|---|---|---|
| Chekol X R-186-2 | 61 | 110 | 46.9 | 1.4 | 1.5 | 1.6 | 62 | 38.4 | 3.1 | 345.3 | 118.8 | 5 |
| FLIP-2004-37L | 63 | 109 | 32.1 | 1.9 | 1.5 | 1.3 | 50 | 29.1 | 2.5 | 181.8 | 60.1 | 5 |
| FLIP-84-95L | 59 | 129 | 12.4 | 1 | 1.5 | 0.5 | 13 | 33.2 | 4.1 | 295.2 | 48.6 | 5 |
| EL-142 X R-186-3 | 58 | 112 | 66 | 1.4 | 1.5 | 2.4 | 95 | 37.6 | 2.4 | 437.9 | 156.6 | 2 |
| FLIP-97-61L | 63 | 129 | 44 | 1.4 | 1.5 | 2.1 | 70 | 20.6 | 3.2 | 135.4 | 46.6 | 4 |
| 36134 | 55 | 90 | 33.1 | 1.2 | 1.5 | 1 | 37 | 32.7 | 2.2 | 289.2 | 108 | 5 |
| GM | 57 | 96 | 35.3 | 1.2 | 1.5 | 1.1 | 43 | 34.6 | 2.5 | 286.8 | 84.9 | 5 |
| SEM | 2.1 | 3.9 | 6.5 | 0.1 | 0.1 | 0.2 | 8.2 | 1.6 | 0.2 | 45.2 | 18.2 | 0.3 |
| SE Difference | 3 | 5.5 | 9.2 | 0.2 | 0.2 | 0.3 | 11.6 | 2.3 | 0.3 | 64 | 25.7 | 0.4 |
| LSD at 5% | 6 | 10.8 | 18.1 | 0.4 | 0.4 | 0.7 | 22.7 | 4.6 | 0.5 | 125.9 | 50.7 | 0.8 |
| CV% | 6.5 | 7 | 31.9 | 19.9 | 16.5 | 36.8 | 33 | 8.2 | 12.2 | 27.3 | 37.1 | 9.7 |

DF = days to 50% flowering, DM = days to 90% maturity, NP = number of pods per plant, NSS = number of seeds per pod, PWPP = Pod weight per plant, SWPP = seed weight per plant, NSPP = number of seeds per plant, PH = plant height in cm, SW = 100-seed weight in gram, BI = biomass, SY = seed yield, PSS = pod set score.

weight, number of seeds/plant, and biological yield/plant was high in both seasons. However, traits such as days to 50% flowering, days to 90% maturity, and plant height consistently showed lower GCV values of between 2.4% and 17%. The remaining traits lack consistency and showed low to moderate GCV for number of seeds/pods, seed weight/plant, and pod weight/plant. Similar significant and considerable variation for (phenotypic and genotypic) results was reported for biological yield/plant, seed yield/plant, and 100-seed weight by [15, 35, 43, 45–48]. However, as opposed to this finding, [49, 50] reported higher GCV and PCV for days to 50% flowering but they reported the same record for days to 90% maturity and plant height. Similarly, [35] reported the minimum PCV and GCV for days to maturity and number of seeds/pod.

According to [51], most characters have showed high heritability estimate over locations except at Chefe Donsa due to severe wilt incidence. Reference [52] reported comparable high heritability estimates for all traits except for plant height and biological yield on lentil. Reference [15] similarly recorded high heritability for 100-seed weight, days to 50% flowering, and days to 90% maturity. Reference [7] also reported high heritability for seed weight and days to 50% flowering. In line with our finding, [53] reported that those days to 50% flowering showed higher heritability estimates; however, they differed in reporting with higher genetic advance. 100-seed weight, rust severity score, and seed yield showed high heritability values coupled with high genetic advance values across locations because of high GCV. Numbers of seeds/plant, seed weight/plant, and number of seeds/pods have intermediate GCV and heritability with a mild response to selection. Similar to our findings, [45] reported higher degree of PCV, GCV, and genetic advance for seed yield/plant on lentil. However, in contrast to these reports, higher PCV, GCV, and genetic advance were recorded for biological yield/plant.

The GCV along with heritability estimates provide a reliable estimate of the amount of genetic advance expected through phenotypic selection [54]. Based on the underlying facts, four groups of traits were categorized as per the analysis: the first group included plant height, days to 50% flowering, and days to 90% maturity which have low GCV with a high heritability but they end up with low genetic advance. [42, 44] made a similar report that high heritability estimate for days to 50% flowering and days to 90% maturity with low genetic advance values on lentil. The second group of characters was above ground biomass, number of seeds/plant, and number of pods/plant with intermediate GCV, heritability, and genetic advance. However, deviated from our findings, studies by [42, 44, 45] reported high heritability and expected genetic gain estimate for above ground biomass and number of pods/plant.

The third group of characters, rust severity score, seed yield, and 100-seed weight, has higher GCV with high heritability coupled with high genetic advance as percentage of the mean. Reference [42, 44] reported high heritability estimate for seed yield and 100-seed weight coupled with higher genetic gain. High heritability for the characters indicated that these traits were less affected by environmental factors. This is indicative of the fact that these traits are mostly controlled by genetic factors and expected to respond to direct selection for traits improvement. The fourth group of characters, pod weight/plant, seed weight/plant, and number of seeds/pods, showed inconsistent GCV, heritability, and genetic advance values across locations. These findings indicated that measuring yield components in breeding program would be difficult. From our follow-up study by (unpublished data), the second and the third groups of characters responded more to selection and the most important traits that contributed to the genetic divergence of lentil. From our follow-up association study (unpublished data), these traits attributed a strong association with seed yield. Therefore, for selecting high yielding lentil genotypes, the breeder should give emphasis to higher seed weight/plant, more number of pods/plant, and plants with short to intermediate height. This observation is in accordance with [55, 56] selection based on number of pods/plant and seed weight/plant which were the most important characters that contributed to seed yield. However, days to 90% maturity, rust disease severity score, and days to 50% flowering which otherwise had a negative phenotype correlation with seed yield had substantially negative direct effect on seed yield (unpublished data).

## 5. Conclusion

This study has described a high phenotypic diversity for important agromorphological characteristics of landraces and exotic genotypes of Ethiopia gene pool. The incidence of highly significant variation between landraces and exotic line for the majority of traits considered is a sign of the presence of a high degree of genetic variation implying the great potential of the Ethiopian germplasm in future breeding programs through selection. As per the analysis, four groups of traits were categorized. Some promising landraces were identified with superior plant characteristics, such as early flowering and early maturing accessions, wide range of genetic base for plant height, number of seeds/plant, resistance to rust, and both low and high yielding lines. Seed yield, 100-seed weight in gram, rust disease score, and total number of seeds/plant consistently showed higher GCV, heritability, and genetic advance in the lentil genotypes. All these important agromorphological characteristic traits with their associated genotypes could be successfully utilized in breeding programs that are aimed at improving the yield and the yield components of lentil. The wide genetic base that is possessed by the Ethiopian germplasm would be integrated in a breeding program and used as a parent in a recombination breeding.

In conclusion, this study has demonstrated that the exotic accessions introduced from ICARDA have a rich genetic potential for 100-seed weight, number of seeds/plant, seed weight/plant, short stature, and resistant source for rust. The Ethiopian landraces were potentials for earliness, higher number of pods/plant, and long plant height. This study has showed a high phenotypic diversity of important plant characteristics of the Ethiopian lentil germplasm. However, additional search is needed to develop a specific plant idotype for major agroecological zone to provide as a platform for local adaptation and then breed for location relevant traits, such as a biotic stresses and identification source of resistance for other major important diseases.

## Conflict of Interests

The authors declare that they have no conflict of interests regarding the publication of this paper.

## Acknowledgments

The authors are most grateful for Gonder University, Haramaya University, Federal Ministry of Education, ICARDA Legume Research Program, ICARDA Ethiopia country office, Ethiopian Institute of Agricultural Research Center, National Chickpea and Lentil Research Project in Debrezeit Agricultural Research Center, Sirinka Agricultural Research Center and Sinana Agricultural Research Center, and Institute Biodiversity Conservation, for research funding supports of this study.

## References

[1] D. Zohary, "The wild progenitor and the place of origin of the cultivated lentil:*Lens culinaris*," *Economic Botany*, vol. 26, no. 4, pp. 326–332, 1972.

[2] H. Barulina, "Lentils of the USSR and other countries," in *Bulletin of Applied Botany, Genetics and Plant Breeding Supplement*, pp. 265–304, USSR Institute of Plant Industry of the Lenin Academy of Agricultural Science Leningrad, USSR, 1930.

[3] W. Erskine, "Global production, supply and demand," in *The Lentil: Botany, Production and Uses*, F. J. William Erskine, Muehlbauer, A. Sarker, and B. Sharma, Eds., pp. 4–12, CABI, London, UK, 2009.

[4] FAOSTAT, "Distribution of global lentil production among India, Canada and Turkey (the "big three" producers) and by continent," FAOSTAT, 2009, http://faostat.fao.org/.

[5] G. Bejiga, S. Tsegaye, and A. Tullu, "Stability of seed yield for some varieties of lentil grown in the Ethiopian highlands," *Crop Research*, vol. 9, pp. 337–343, 1995.

[6] CSA, *Central Statistical Agency Agricultural Sample Survey Report on: Area and Production of Crops*, Statistical Bulletin, Addis Ababa, Ethiopia, 2012.

[7] G. Bejiga, S. Tsegaye, A. Tullu, and W. Erskine, "Quantitave evaluation of Ethiopian landraces of lentil (*Lens culinaris*)," *Genetic Resources and Crop Evolution*, vol. 43, no. 4, pp. 293–301, 1996.

[8] F. Asnake and G. Bejiga, "Breeding lentil for wider adaptation," in *Forage and Food Legumes of Ethiopia: Progress and Prospects. Proceedings of the Workshop on Food and Forage Legumes*, A. Kemal, K. Gemechu, A. Seid et al., Eds., pp. 80–86, EIAR and ICARDA. International Center for Agricultural Research in the Dry Areas (ICARDA), Aleppo Syria, Addis Ababa, Ethiopia, September 2003.

[9] R. W. Allard, *Principles of Plant Breeding*, John Wiley & Sons, New York, NY, USA, 1960.

[10] W. Erskine, Y. Adham, and L. Holly, "Geographic distribution of variation in quantitative traits in a world lentil collection," *Euphytica*, vol. 43, no. 1-2, pp. 97–103, 1989.

[11] M. Ahmad, D. L. McNeil, and J. R. Sedcole, "Phylogenetic relationships in Lens species and their interspecific hybrids as measured by morphological characters," *Euphytica*, vol. 94, no. 1, pp. 101–111, 1997.

[12] J. S. C. Smith and O. S. Smith, "The description and assessment of distance between inbred lines of maize. The use of morphological traits as descriptors," *Maydica*, vol. 34, pp. 141–150, 1989.

[13] K. Mather and J. L. Jinks, *Biometrical Genetics: The Study of Continuous Variation*, Chapman and Hall, London, UK, 2nd edition, 1971.

[14] H. W. Johnson, H. F. Robinson, and R. E. Comstock, "Estimate of genetic and environmental variability in soybeans," *Agronomy Journal*, vol. 47, pp. 314–318, 1955.

[15] B. T. Bicer and D. Sakar, "Genetic variability and heritability for grain yield and other characters in lentil," *Journal Biological Science*, vol. 4, pp. 216–218, 2004.

[16] B. T. Bicer and D. Sakar, "Stability parameters in lentil," *Journal of Central European Agriculture*, vol. 7, pp. 439–443, 2006.

[17] W. Erskine, P. C. Williams, and H. Nakkoul, "Genetic and environmental variation in the seed size, protein, yield, and cooking quality of lentils," *Field Crops Research*, vol. 12, pp. 153–161, 1985.

[18] A. Tullu, I. Kusmenoglu, K. E. Mcphee, and F. J. Muehlbauer, "Characterization of core collection of lentil germplasm for phenology, morphology, seed and straw yields," *Genetic Resources and Crop Evolution*, vol. 48, no. 2, pp. 143–151, 2001.

[19] A. Sarker and W. Erskine, "Recent progress in the ancient lentil," *Journal of Agricultural Science*, vol. 144, no. 1, pp. 19–29, 2006.

[20] S. Roy, M. A. Islam, A. Sarker, M. A. Malek, M. Y. Rafii, and M. R. Ismail, "Determination of genetic diversity in lentil germplasm based on quantitative traits," *Australia Journal of Crop Science*, vol. 7, pp. 14–21, 2013.

[21] F. Edossa, T. Kassahun, and B. Endashaw, "A comparative study of morphological and molecular diversity in Ethiopian lentil (*Lens culinaris* Medikus) landraces," *African Journal of Plant Science*, vol. 4, pp. 242–254, 2010.

[22] T. Seifu, *Correlation and path coefficient analysis in lentil (Lens culinaries Medik.) and their implication for selection [M.S. thesis]*, Alemaya Agricultural University, Addis Ababa, Ethiopia, 1988.

[23] B. Geletu and A. Yadeta, "Genetic and breeding research in lentil," in *Cool-Season Food Legumes of Ethiopia. Proceedings of the 1st National Cool-Season Food Legumes Review Conference, December 1993*, A. Telaye, G. Bejiga, M. C. Saxena, and M. Solh, Eds., pp. 6416–8420, Addis AbabaInstitute of Agricultural Research. ICARDA, Aleppo, Syria, 1994.

[24] D. Tigist, *Genetic variability and associations among yield and yield related characters in exotic lentil lines (Lens culinaries Medik.) [M.S. thesis]*, Alemaya University, East Harerge, Ethiopia, 2003.

[25] D. Ketema, *Genetic Variation for Agronomic and Root Characters in Ethiopian Lentil (Lens culinaris Medikus) Landraces Grown Under Moisture Stress [M.S. thesis]*, Haramaya University, East Harerge, Ethiopia, 2007.

[26] W. Chen, *Stemphylium Blight Disease Scoring in Field Condition*, USDA-ARS, Washington State University, Pullman Wash, USA, 2007.

[27] VSN International, *GenStat for Windows*, VSN International, Hemel Hempstead, UK, 13th edition, 2010.

[28] W. T. Federer, "Augmented designs with one-way elimination of heterogeneity," *Biometrics*, vol. 17, pp. 447–473, 1961.

[29] C. S. Lin, M. R. Binns, and L. P. Lefkovitch, "Stability analysis: where do we stand?" *Crop Science*, vol. 26, pp. 894–900, 1986.

[30] S. K. Yau and J. Hamblin, "Relative yield as a measure of entry performance in variable environments," *Crop Science*, vol. 34, no. 3, pp. 813–817, 1994.

[31] VSN International, *GenStat for Windows*, VSN International, Hemel Hempstead, UK, 15th edition, 2012.

[32] H. A. Al-Jibouri, P. A. Miller, and H. F. Miller, "Genotypic and environmental variances and co variances in an upland cross of inter specific origin," *Agronomy Journal*, vol. 50, pp. 633–636, 1958.

[33] S. N. Deshmukh and P. S. Reddy, "Genetic variability, character association and path coefficients of quantitative traits in Virginia bunch varieties of ground nut," *Indian Journal Agricultural Science*, vol. 56, pp. 816–821, 1986.

[34] S. C. Agrawal, M. N. Khare, and P. S. Agrawal, "Field screening of lentil lines for resistance to rust," *Indian Journal of Phytopathology*, vol. 292, pp. 20–28, 1976.

[35] M. Chakrabotry and M. F. Haque, "Genetic variability and component analysis in lentil (*Lens culinaris* Medik.)," *Journal Lentil Research*, vol. 12, pp. 199–204, 2000.

[36] B. T. Bicer and D. Sakar, "Studies on variability of lentil genotypes in Southeastern Anatolia of Turkey," *Notulae Botanicae Horti Agrobotanici Cluj-Napoca*, vol. 36, pp. 20–24, 2008.

[37] F. Toklu, B. Tuba Biçer, and T. Karaköy, "Agro-morphological characterization of the Turkish lentil landraces," *African Journal of Biotechnology*, vol. 8, no. 17, pp. 4121–4127, 2009.

[38] T. Sultana, S. Nadeem, Z. Fatima, and A. Ghafoor, "Identification of elite pure-lines from local lentil germplasm using diversity index based on quantitative traits," *Pakistan Journal of Botany*, vol. 42, no. 4, pp. 2249–2256, 2010.

[39] A. L. A. Al-Ghzawi, E. Bsoul, F. Aukour, Z. Al-Ajlouni, M. Al-Azzam, and M. M. Ajlouni, "Genetic variation for quantitative traits in Jordanian lentil landraces," *Advances in Environmental Biology*, vol. 5, no. 11, pp. 3676–3680, 2011.

[40] K. B. Singh and S. Singh, "Evaluation of exotic germplasm in lentil," *Journal Agricultural Research*, vol. 6, pp. 304–306, 1991.

[41] M. Solh and W. Erskine, "Genetic resources of lentils," in *Genetic Resources and Their Exploitation Chickpeas, Faba Beans and Lentils*, J. R. Witcombe and W. Erskine, Eds., pp. 205–217, Martinus Nijhoff, The Hague, The Netherlands, 1984..

[42] B. N. Baidya, A. M. Eunus, and S. Sen, "Estimation of variability and correlation in yield and yield contributing characters in lentil (*Lens culinaris*)," *Environmental Ecology*, vol. 6, pp. 694–697, 1988.

[43] A. S. Rathi, J. S. Sindhu, and V. S. Singh, "Variability, heritability and genetic advance in lentil," *Legume Research*, vol. 25, pp. 113–116, 2002.

[44] O. Vir, V. P. Gupta, and O. Vir, "Variation in Macrosperma x microsperma derived gene pool of lentil under low and high fertility levels of soil at sub-tropical climate of Himalayas," *Indian Journal Agricultural Research*, vol. 32, pp. 181–184, 1998.

[45] N. Kishore and V. P. Gupta, "Early generation selection in microsperma and macrosperma derived gene pool of lentil," *Indian Journal Genetics and Plant Breeding*, vol. 62, pp. 34–37, 2002.

[46] N. I. Haddad, T. P. Bogyo, and F. J. Muehlbauer, "Genetic variance of six agronomic characters in three lentil (*Lens culinaris*Medic) crosses," *Euphytica*, vol. 31, no. 1, pp. 113–120, 1982.

[47] M. Singh, A. Sarker, and W. Erskine, "Estimation of heritability using spatial variability models: the case of lentil (*Lens culinaris* Medik.)," *Indian Journal Genetics and Plant Breeding*, vol. 65, pp. 77–83, 2005.

[48] S. D. Tyagi and M. H. Khan, "Correlation, path-coefficient and genetic diversity in lentil (*Lens culinaris* Medik) under rainfed conditions," *International Research Journal of Plant Science*, vol. 2, pp. 191–200, 2011.

[49] S. Ayaz, B. A. McKenzie, G. D. Hill, and D. L. McNeil, "Variability in yield of four grain legume species in a subhumid temperate environment. II. Yield components," *Journal of Agricultural Science*, vol. 142, no. 1, pp. 21–28, 2004.

[50] G. Singh, I. S. Singh, and R. Kumar, "Genetic variability for seed yield and its component characters in a macrosperma × microsperma cross of lentil (*Lens culinaris* Medik.)," *Crop Improvement*, vol. 31, pp. 206–209, 2004.

[51] H. F. Robinson, R. E. Comstock, and P. H. Harvey, "Genetic and phenotypic correlation in corn and their implications in selection," *Agronomy Journal*, vol. 43, pp. 283–287, 1954.

[52] S. K. Jain, S. K. Madaria, S. K. Rao, and P. K. Nigam, "Analysis of yield factors in lentil," *Indian Journal of Agricultural Research*, vol. 29, pp. 173–180, 1995.

[53] M. Singh, D. K. Maheshwari, R. K. Mittal, S. K. Sharma, and M. Singh, "Genetic variability and correlation of grain yield and other quantitative characters in lentil (*Lens culinaris* Medik)," *Annals Agricultural Biological Research*, vol. 4, pp. 121–124, 1999.

[54] G. W. Burton and E. H. DeVane, "Estimation heritability in Tall Festuca (*Festuca arundinacea*) from replicated clonal material," *Agronomy Journal*, vol. 45, pp. 478–481, 1953.

[55] S. A. M. Khattab, "Association and Path analysis in lentil under different irrigation regimes," *Egypt Journal Agronomy*, vol. 20, pp. 13–25, 1999.

[56] S. Sarvjeet, I. Singh, R. K. Gill, L. Kaur, and A. Sarker, "Evaluation of exotic lentil germplasm and studies on genetic parameters for various traits," *Journal of Lentil Research*, vol. 3, pp. 22–27, 2006.

# 27

# Multivariate Regression Analyses of Yield Associated Traits in Rapeseed (*Brassica napus* L.) Genotypes

**Valiollah Rameeh**

*Agriculture and Natural Resources Research Center of Mazandaran, Sari, Iran*

Correspondence should be addressed to Valiollah Rameeh; vrameeh@yahoo.com

Academic Editor: Nnadozie Oraguzie

The efficiency of a breeding program depends mainly on the direction of the correlation between yield and its components and the relative importance of each component involved in contributing to seed yield. The interrelationships of nine quantitative traits in 28 genotypes of spring oilseed rape (days to flowering, days to end of flowering, duration of flowering, days to maturity, pods per main raceme, pods length and pods per plant, and seed yield) were computed. Significant genotypic effects were found for phenological traits, yield components, and seed yield, indicating significant genetic differences among the genotypes. High broad sense heritability was estimated for phenological traits, seeds per pod, and seed yield, signifying high selection gain for improving these traits. Path coefficient analysis revealed that days to flowering and number of pods per plant had the highest direct effects on seed yield. Duration of flowering, number of branches, pods on main raceme, pods per plant, and seed yield had high genetic coefficient of variation. The results of factor analysis showed three factors including factor 1 (phenological traits), factor 2 (primary yield components), and factor 3 (secondary yield components). The results of stepwise regression analysis revealed that pods per plant, number of branches, and duration of flowering had considerable effects on seed yield.

## 1. Introduction

Improvement of seed yield in canola (*Brassica napus* L.) has been the main objective of canola breeders for many years [1, 2]. Seed yield is a quantitative trait, which is principally influenced by the environment and consequently has a low heritability [3, 4]. As a result, the response to direct selection for seed yield may be unpredictable, unless there is good control of environmental variation. Plant breeders are seldom interested in a single trait and therefore, there is the need to examine the relationships among different traits, especially between seed yield and other traits. As the number of independent variables influencing a particular dependent variable increases, a certain amount of interdependence is expected. In such situations, correlations may be inadequate to explain the associations in a way that will enable breeders to decide on a direct or indirect selection strategy [5]. The multivariate analyses, particularly factor and cluster analyses, are utilized for evaluation of a large number of accessions for different traits in a germplasm collection. Cluster analysis assigns genotypes into qualitative homogenous groups based on response similarities and also assists in classifying genotypes. The method produces a dendrogram showing successive fusion of individuals. Greater heterotic effect is generated when clusters are divergent. Genetic diversity among the Brassica genotypes has been assessed by Choudhary and Joshi [6] using cluster analysis. The morphological characters, namely, days to flowering, plant height, secondary branches per plant, and 1000-seed weight, had a major contribution to genetic divergence [7, 8]. Determination of correlation coefficients is an important statistical procedure to evaluate breeding programs for high yield as well as to examine direct and indirect contributions to yield variables [9–13].

The main applications of factor analytic techniques are to reduce the number of variables and to detect structure in the relationship between variables [14]. In plant breeding factor analysis is mainly applied as structure detection method, and sometimes it can be used as index selection for improving more than one trait. Factor analysis was used to determine structural factors related to growth trait and yield components and also to detect factors relating to environmental stress including drought resistance in *B. napus* [15].

The objectives of the present study were to estimate the genetic coefficient of variation, broad sense heritability, and the relationship among yield associated traits for improving seed yield in spring type rapeseed advanced lines.

## 2. Materials and Methods

The material under study consisted of 28 rapeseed (*B. napus* L.) genotypes which were selected based on different agronomic characters. The genotypes were planted in a randomized complete block design with three replications at Baykola Agriculture Research Station, located in Neka, Iran (53°13′E longitude and 36°43′N latitude, 15 m above sea level) during 2012-13. The plots consisted of four rows 5 m long and 30 cm apart. The distance between plants on each row was 5 cm resulting in approximately 400 plants per plot, which were sufficient for statistical analysis. Crop management including land preparation, crop rotation, fertilizer, and weed control followed standard practices in the local area. All plant protection measures were adopted to make the crop free from insects. Phenological traits including days to flowering, days to end of flowering, duration of flowering, and days to maturity were determined based on phenological stages of the genotypes in each plot. Pods per main raceme, pods length, and pods per plant were recorded using 10 randomly selected plants in each plot. Seed yield of two middle rows of each plot was adjusted to kg/ha.

Variance components were estimated from the mean squares (MS) in the analysis of variance [16]. The components of variance including error variance (VE), genotypic variance (VG), and phenotypic variance (VP) were estimated according to the following formula:

$$VE = MSE,$$

$$VG = \frac{(MSG - MSE)}{r}, \qquad (1)$$

$$VP = VG + VE.$$

Broad sense heritability ($h^2$) was estimated according to Khan et al. [17] as $h^2 = VG/VE$. The coefficient of variation was estimated as $CV = (\sqrt{VG})/\mu$ where $\mu$ is the mean of genotypes for each trait.

Pearson's correlation coefficients between the traits, path coefficient analysis, and factor analysis based on Varimax rotation matrix were done on the data. Principal components method analysis was used to extract factorial load of matrix and also to estimate the number of factors [14, 18]. The factors which had a root bigger than one were selected and were used to build the factorial coefficients matrix. All the calculations were performed using MS-Excel and SAS software version 9 [19].

## 3. Results and Discussions

*3.1. Analysis of Variance.* Significant mean square of genotypes were determined for the traits including days to flowering, duration of flowering, days to maturity, number of branches, pods per main raceme, pods per plant, seeds per pod, and seed yield indicating significant genetic variation for these traits (Table 1). Generally these results are similar to those reported by Aytaç and Kinaci [8] and Sabaghnia et al. [20]. Broad sense heritability estimates ranged from 0.12 to 0.98 for plant height and days to flowering, respectively. High values of broad sense heritability estimates for the phenological traits, seeds per pod, and seed yield indicate that selection gain for improving these traits will be high. Aytac et al. [3] also reported high broad sense heritability estimates for yield components. Genetic coefficient of variation, an indicator of the genetic diversity of the genotypes, varied from 18.7 to 26.8 for days to maturity and seed yield, respectively (Table 2). A high value for genetic variation was detected for duration of flowering, pods per main raceme, pods per plant, and seed yield.

*3.2. Genotypic Means.* Days to flowering ranged from 79 to 119 days (Table 2). A significant positive correlation was detected between days to flowering and days to maturity (Table 3), which suggests that selection for this trait could be done for early maturity genotypes. The genotypes including genotype 3 (G3), G16, G18, G25, and G27 with low mean values would be preferred for improving days to flowering (Table 2). These genotypes had low mean values of days for maturity. In rapeseed, since flowering and seed formation occurred about the same time, genotypes showing high mean values for duration of flowering will be preferable. High mean value for duration of flowering suggests that the plant gets enough time and opportunity for seed formation, and genotypes G3, G8, G18, and G25 with high mean values would be considered suitable for improving this trait. For reducing pest damage and also facilitate second crop cultivation after rapeseed, early maturity would be an ideal trait for breeding *B. napus* L. and other related Brassica species. Plant height ranged from 150 to 181 cm. For ideotype breeding of rapeseed, low mean value of plant height is favored, and genotypes G6, G10, G12, and G19 would be suitable parents for this trait. Number of branches varied from 3.2 to 5.6. Since pods per plant were found significantly correlated with seed yield, the genotypes including G5, G14, G15, G19, and G21 would be suitable for improving this trait. Sharma [14] reported that pods per plant, seeds per plant, and 1000-seed weight traits were positively correlated with seed yield. Khan et al. [17] also reported a positive significant correlation between seed yield and plant height, pods per plant, seeds per pod, and pod length. The genotypes G14, G19, G21, G22, G23, and G24 with high mean values for seed yield also had high mean values for some other yield components.

*3.3. Multivariate Analyses.* Path coefficient analysis revealed that days to flowering and number of pods per plant had the highest positive effects on seed yield but duration of flowering and pods on main raceme had a significant negative effect on seed yield (Table 4). Marjanović-Jeromela et al. [21] studied 30 rapeseed varieties and demonstrated that pods per plant have the highest correlation with seed yield. Sheikh et al. [22] found high heritability estimates coupled with high genetic

TABLE 1: Analyses of variance (ANOVA) and broad sense heritability estimates for the traits studied.

| S.O.V | Df | Days to flowering | Duration of flowering | Days to maturity | Plant height | Number of branches | Pods per main raceme | Pods per plant | Seeds per pod | Seed yield |
|---|---|---|---|---|---|---|---|---|---|---|
| | | | | | Mean squares | | | | | |
| Replication | 2 | 2.3 | 16.9 | 3.0 | 938.9** | 7.72** | 1074** | 4005** | 216.3** | 973157* |
| Treatments | 27 | 161.1** | 219.9** | 51.6** | 156.1 | 1.03** | 98** | 561** | 16.0** | 810934** |
| Error | 54 | 1.0 | 13.9 | 3.7 | 111.2 | 0.31 | 32.2 | 175.1 | 3.6 | 202709 |
| Broad sense heritability | | 0.98 | 0.83 | 0.81 | 0.12 | 0.44 | 0.41 | 0.42 | 0.53 | 0.50 |

SOV: source of variation.

*,**Significant at $P < 0.05$ and 0.01, respectively.

TABLE 2: Mean comparison of the rapeseed genotypes for phenological traits, plant height, yield components, and seed yield.

| Genotypes | Days to flowering | Duration of flowering | Days to maturity | Plant height (cm) | Number of branches | Pods per main raceme | Pods per plant | Seeds per pod | Seed yield (kg/ha) |
|---|---|---|---|---|---|---|---|---|---|
| | | | | | Traits | | | | |
| 1-Zar08-1 | 106 | 68 | 213 | 168 | 4.9 | 52 | 90 | 24 | 1952 |
| 2-Zar01-3 | 106 | 64 | 213 | 176 | 4.9 | 41 | 72 | 22 | 2171 |
| 3-Zar01-7 | 97 | 77 | 206 | 169 | 5.3 | 37 | 77 | 24 | 1893 |
| 4-Zar01-8 | 104 | 61 | 206 | 160 | 4.6 | 47 | 72 | 23 | 2493 |
| 5-22B01-14 | 108 | 61 | 212 | 163 | 5.1 | 38 | 110 | 24 | 3094 |
| 6-22B01-15 | 111 | 54 | 211 | 154 | 4.3 | 31 | 87 | 18 | 3053 |
| 7-22B01-16 | 109 | 65 | 211 | 168 | 3.9 | 39 | 102 | 19 | 3070 |
| 8-Forn08-1 | 103 | 69 | 206 | 157 | 4.1 | 35 | 101 | 18 | 2194 |
| 9-Forn01-7 | 106 | 68 | 208 | 165 | 4.3 | 36 | 70 | 19 | 1846 |
| 10-Forn08 | 111 | 51 | 215 | 150 | 4.6 | 42 | 103 | 18 | 2704 |
| 11-Mod04 | 108 | 68 | 213 | 160 | 3.8 | 34 | 71 | 19 | 2445 |
| 12-Mod05 | 106 | 58 | 209 | 154 | 3.9 | 35 | 82 | 19 | 2484 |
| 13-Mod06 | 117 | 51 | 214 | 181 | 5.6 | 47 | 81 | 22 | 2322 |
| 14-Mod08 | 102 | 68 | 208 | 172 | 4.6 | 41 | 109 | 22 | 3113 |
| 15-Mod09 | 104 | 68 | 212 | 158 | 5.2 | 34 | 107 | 22 | 2859 |
| 16-Mod010 | 99 | 67 | 205 | 167 | 4.7 | 40 | 93 | 22 | 2274 |
| 17-Mod011 | 103 | 67 | 209 | 163 | 5.1 | 34 | 83 | 19 | 2126 |
| 18-Mod014 | 97 | 68 | 204 | 165 | 5.1 | 42 | 90 | 20 | 2195 |
| 19-Mod01 | 109 | 50 | 210 | 152 | 4.6 | 32 | 107 | 21 | 3338 |
| 20-Mod02 | 110 | 54 | 211 | 166 | 3.2 | 32 | 82 | 20 | 2830 |
| 21-Zar01 | 110 | 54 | 212 | 163 | 4.8 | 40 | 114 | 20 | 3415 |
| 22-Zar03 | 105 | 61 | 209 | 161 | 3.3 | 32 | 69 | 18 | 3104 |
| 23-Zar025 | 103 | 61 | 213 | 167 | 4.5 | 43 | 102 | 21 | 3664 |
| 24-Zar029 | 109 | 58 | 215 | 170 | 4.8 | 36 | 101 | 19 | 3582 |
| 25-OKP08 | 79 | 82 | 201 | 172 | 4.9 | 39 | 84 | 17 | 2152 |
| 26-Sarigol | 119 | 49 | 221 | 171 | 4.4 | 45 | 100 | 19 | 2427 |
| 27-Hyola401 | 97 | 60 | 210 | 167 | 5.2 | 47 | 91 | 26 | 2764 |
| 28-RGS003 | 104 | 47 | 213 | 163 | 4.6 | 49 | 85 | 24 | 2487 |
| LSD$_{(\alpha=0.05)}$ | 0.94 | 3.52 | 1.81 | 9.94 | 0.52 | 5.35 | 12.48 | 1.79 | 424.48 |
| LSD$_{(\alpha=0.01)}$ | 1.25 | 4.68 | 2.41 | 13.22 | 0.70 | 7.12 | 16.59 | 2.38 | 564.56 |
| Genetic coefficient of variation | 19.8 | 23.0 | 18.7 | 19.1 | 22.4 | 23.4 | 23.8 | 21.6 | 26.8 |

TABLE 3: Pearson's correlations between pairs of traits in rapeseed.

| Traits | (1) Days to flowering | (2) Duration of flowering | (3) Days to maturity | (4) Plant height | (5) Number of branches | (6) Pods per main raceme | (7) Pods per plant | (8) Seeds per pod | (9) Seed yield |
|---|---|---|---|---|---|---|---|---|---|
| (1) Days to flowering | 1 | | | | | | | | |
| (2) Duration of flowering | $-0.75^{**}$ | 1 | | | | | | | |
| (3) Days to maturity | $0.80^{**}$ | $-0.68^{**}$ | 1 | | | | | | |
| (4) Plant height | $-0.11$ | 0.23 | 0.06 | 1 | | | | | |
| (5) Number of branches | $-0.21$ | 0.16 | $-0.04$ | 0.35 | 1 | | | | |
| (6) Pods per main raceme | $-0.02$ | $-0.14$ | 0.21 | $0.44^{*}$ | $0.46^{*}$ | 1 | | | |
| (7) Pods per plant | 0.16 | $-0.22$ | 0.25 | $-0.16$ | 0.23 | 0.04 | 1 | | |
| (8) Seeds per pod | $-0.07$ | $-0.04$ | 0.05 | 0.28 | $0.52^{**}$ | $0.56^{**}$ | 0.04 | 1 | |
| (9) Seed yield | 0.29 | $-0.45^{*}$ | 0.33 | $-0.21$ | $-0.21$ | $-0.23$ | $0.58^{**}$ | $-0.08$ | 1 |

$^{*,**}$ Significant at $P < 0.05$ and 0.01, respectively.

TABLE 4: Direct (along the diagonal) and indirect (above and below the diagonal) effects of yield components on seed yield.

| Traits | 1 | 2 | 3 | 4 | 5 | 6 | 7 | 8 |
|---|---|---|---|---|---|---|---|---|
| (1) Days to flowering | $0.40^{*}$ | 0.41 | 0.16 | $-0.02$ | 0.06 | 0.01 | 0.09 | $-0.01$ |
| (2) Duration of flowering | 0.30 | $-0.54^{**}$ | $-0.14$ | 0.04 | $-0.05$ | 0.06 | $-0.13$ | $-0.01$ |
| (3) Days to maturity | $-0.32$ | 0.36 | 0.20 | 0.01 | 0.01 | $-0.09$ | 0.15 | 0.01 |
| (4) Plant height | 0.04 | $-0.12$ | 0.01 | 0.18 | $-0.11$ | $-0.19$ | $-0.09$ | 0.05 |
| (5) Number of branches | 0.08 | $-0.09$ | $-0.01$ | 0.06 | $-0.29$ | $-0.19$ | 0.13 | 0.09 |
| (6) Pods per main raceme | 0.01 | 0.07 | 0.04 | 0.08 | $-0.14$ | $-0.42^{*}$ | 0.02 | 0.11 |
| (7) Pods per plant | $-0.06$ | 0.12 | 0.05 | $-0.03$ | $-0.07$ | 0.02 | $0.58^{**}$ | 0.01 |
| (8) Seeds per pod | 0.03 | 0.02 | 0.01 | 0.05 | 0.05 | 0.08 | 0.11 | 0.17 |

$^{*,**}$ Significant at $P < 0.05$ and 0.01, respectively.

TABLE 5: The results of stepwise regression analysis in which 3 out of 9 studied traits were selected.

| Step | Variable entered | Partial $R$-square | Model $R$-square | $F$-test |
|---|---|---|---|---|
| 1 | $X7$: pods per plant | 0.34 | 0.34 | $13.4^{**}$ |
| 2 | $X5$: number of branches | 0.12 | 0.46 | $5.7^{*}$ |
| 3 | $X2$: duration of flowering | 0.07 | 0.56 | 3.3 |
| Final regression model: $Y$ (seed yield) $= 2835.9 - 16.4X7 - 269.6X5 + 22.5X7$. | | | | |

$^{*,**}$ Significant at $P < 0.05$ and 0.01, respectively.

advance for seed yield per plant, pods per plant, and seed weight in rapeseed (*B. campestris*) genotypes.

The results of stepwise regression analysis indicated that pods per plant, number of branches, and duration of flowering had considerable effects on seed yield (Table 5).

The results of factor analysis revealed three factors for the nine traits studied (Table 6). The eigenvalues for factors 1, 2, and 3 were 2.96, 2.35, and 1.33, respectively. The cumulative variation for these factors was 0.74 while the proportion of variation contributed by each factor was 0.32, 0.26, and 0.15, respectively. Factor 1 "could be regarded as the main factor (phenological traits)" and include days to flowering, duration of flowering, and days to maturity which had high coefficients on the factor loading. The second factor includes yield components such as number of branches, pods on

main raceme, and seeds per pod. The third factor includes second yield components such as pods per plant and seed yield. In earlier studies [15, 18] factor analyses were used to determine structural factors related to growth trait and yield components and also for detecting factors relating to environmental stress including drought resistance in *B. napus* L.

In general, phenological traits, seeds per pod, and seed yield were more heritable than the other traits. Due to the significant positive correlation of yield components with seed yield these traits can be used as indirect selection criteria for improving seed yield. The low value of genetic coefficient of variation for days to maturity suggests that for improving this trait the correlated trait including days to flowering can be used as indirection selection criterion.

TABLE 6: Factor analysis of all traits studied in rapeseed genotypes.

| Traits | Factor loadings | | |
|---|---|---|---|
| | 1 | 2 | 3 |
| (1) Days to flowering | **0.92** | −0.11 | 0.05 |
| (2) Duration of flowering | **−0.86** | 0.04 | −0.23 |
| (3) Days to maturity | **0.90** | 0.13 | 0.13 |
| (4) Plant height | −0.03 | 0.62 | −0.32 |
| (5) Number of branches | −0.23 | **0.79** | 0.19 |
| (6) Pods per main raceme | 0.18 | **0.83** | −0.13 |
| (7) Pods per plant | 0.09 | 0.13 | **0.91** |
| (8) Seeds per pod | 0.01 | **0.78** | 0.07 |
| (9) Seed yield | 0.31 | −0.23 | **0.77** |
| Eigen value | 2.96 | 2.35 | 1.33 |
| Portion | 0.32 | 0.26 | 0.15 |
| Cumulative | 0.32 | 0.59 | 0.74 |

## Conflict of Interests

The author declares that there is no conflict of interests regarding the publication of this paper.

## Acknowledgments

The author wishes to thank Agricultural and Natural Resources Research Center of Mazandaran and Seed and Plant Improvement Institute (SPII) for providing genetic materials and facility for conducting this experiment.

## References

[1] G. G. Khachatourians, A. K. Summer, and P. W. B. Phillips, *An Introduction to the History of Canola and the Scientific Basis for Innovation*, CABI, London, UK, 2001.

[2] M. J. Mahasi and J. W. Kamundia, "Cluster analysis in rapeseed (*Brassica napus* L.)," *African Journal of Agricultural Research*, vol. 2, no. 9, pp. 409–411, 2007.

[3] Z. Aytac, G. Kinaci, and E. Kinaci, "Genetic variation, heritability and path analysis of summer rapeseed cultivars," *Journal of Applied and Biological Science*, vol. 2, no. 3, pp. 35–39, 2008.

[4] M.-J. Ana, K.-S. Ankica, S. Dejana, M. Radovan, and H. Nikola, "Phenotypic and molecular evaluation of genetic diversity of rapeseed (*Brassica napus* L.) genotypes," *African Journal of Biotechnology*, vol. 8, no. 19, pp. 4835–4844, 2009.

[5] I. Ofori, "Correlation and path-coefficient analysis of components of seed yield in bambara groundnut (*Vigna subterranea*)," *Euphytica*, vol. 91, no. 1, pp. 103–107, 1996.

[6] B. R. Choudhary and P. Joshi, "Genetic diversity in advanced derivatives of Brassica interspecific hybrids," *Euphytica*, vol. 121, no. 1, pp. 1–7, 2001.

[7] A. A. Leilah and S. A. Al-Khateeb, "Yield analysis of canola (*Brassica napus* L.) using some statistical procedures," *Saudi Journal of Biological Sciences*, no. 12, pp. 103–113, 2005.

[8] Z. Aytaç and G. Kinaci, "Genetic variability and association studies of some quantitative characters in winter rapeseed (*Brassica napus* L.)," *African Journal of Biotechnology*, vol. 8, no. 15, pp. 3547–3554, 2009.

[9] F. A. Khan, S. Ali, A. Shakeel, A. Saeed, and G. Abbas, "Correlation analysis of some quantitative characters in *Brassica napus* L.," *Journal of Agricultural Research*, no. 44, pp. 7–14, 2006.

[10] S. Ivanovska, C. Stojkovski, Z. Dimov, A. Marjanovic-Jeromela, M. Jankulovska, and L. J. Jankuloski, "Interrelationship between yield and yield related traits of spring canola (*Brassica napus* L.) genotypes," *Genetika*, vol. 39, pp. 325–332, 2007.

[11] D. Basalma, "The correlation and path analysis of yield and yield components of different winter rapeseed (*Brassica napus* ssp. *oleifera* L.) cultivars," *Research Journal of Agriculture and Biological Sciences*, vol. 4, pp. 120–125, 2008.

[12] H. A. Sadat, G. Ali Nematzadeh, N. B. Jelodar, and O. G. Chapi, "Genetic evaluation of yield and yield components at advanced generations in rapeseed (*Brassica napus* L.)," *African Journal of Agricultural Research*, vol. 5, no. 15, pp. 1958–1964, 2010.

[13] Y. Semahegn Belete, "Genetic variability, correlation and path analysis studies in Ethiopian mustard (*Brassica carinata* A. Brun) genotypes," *International Journal of Plant Breeding and Genetics*, vol. 5, no. 4, pp. 328–338, 2011.

[14] S. Sharma, *Applied Multivariate Techniques*, John Wiley & Sons, New York, NY, USA, 1st edition, 1996.

[15] R. Naderi and Y. Emam, "Interrelationships among grain yield and related characters of four oilseed rape (*Brassica napus* L.) cultivars under drought stress conditions," *Desert*, vol. 15, no. 2, pp. 133–138, 2010.

[16] M. Singh, S. Ceccarelli, and J. Hamblin, "Estimation of heritability from varietal trials data," *Theoretical and Applied Genetics*, vol. 86, no. 4, pp. 437–441, 1993.

[17] S. Khan, I. Farhatullah, and H. Khallil, "Phenotypic correlation analysis of elite F3:4 Brassica populations for quantitative and qualitative traits," *ARPN, Journal of Agriculture and Biological Sciences*, no. 3, pp. 38–42, 2008.

[18] V. Rameeh, "Combining ability and factor analysis in $F_2$ diallel crosses of rapeseed varieties," *Plant Breeding and Seed Science*, vol. 62, pp. 73–83, 2010.

[19] SAS Institute Inc, *SAS/STAT User's Guide, Version 9*, Statistical Analysis Institute Inc., Cary, NC, USA, 4th edition, 2004.

[20] N. Sabaghnia, H. Dehghani, B. Alizadeh, and M. Mohghaddam, "Interrelationships between seed yield and 20 related traits of 49 canola (*Brassica napus* L.) genotypes in non-stressed and water-stressed environments," *Spanish Journal of Agricultural Research*, vol. 8, no. 2, pp. 356–370, 2010.

[21] A. Marjanović-Jeromela, R. Marinković, A. Mijić, M. Jankulovska, and Z. Zdunić, "Interrelationship between oil yield and other quantitative traits in rapeseed (*Brassica napus* L.)," *Journal of Central European Agriculture*, vol. 8, no. 2, pp. 165–170, 2007.

[22] F. A. Sheikh, A. G. Rather, and S. A. Wani, "Genetic variability and inter-relationship in toria *Brassica campestris* L. var. toria," *Advances in Plant Sciences*, vol. 12, no. 1, pp. 139–143, 1999.

# 28

# Tolerance of Glyphosate-Resistant Maize to Glyphosate Plus MCPA Amine Is Influenced by Dose and Timing

**Nader Soltani,**[1] **Kris McNaughton,**[1] **Chris L. Gillard,**[1]
**Robert E. Nurse,**[2] **and Peter H. Sikkema**[1]

[1]*University of Guelph Ridgetown Campus, Ridgetown, ON, Canada N0P 2C0*
[2]*Agriculture and Agri-Food Canada, 2585 Country Road 20, Harrow, ON, Canada N0R 1G0*

Correspondence should be addressed to Nader Soltani; soltanin@uoguelph.ca

Academic Editor: Kassim Al-Khatib

There is little information on tolerance of glyphosate-resistant maize to glyphosate plus MCPA amine as influenced by dose and timing under Ontario environmental conditions. A total of seven field trials were conducted at various locations in Ontario, Canada, in 2011–2013 to evaluate tolerance of field maize to tank mixes of glyphosate (900 g a.e./ha) plus MCPA amine (79, 158, 315, 630, 1260, 2520, or 5040 g a.e./ha) at either the 4- or 8-leaf stage. The predicted dose of MCPA amine that caused 5, 10, and 20% injury was 339, 751, and 1914 g a.e./ha when applied to 4-leaf maize but only 64, 140, and 344 g a.e./ha when applied to 8-leaf maize, respectively. The predicted dose of MCPA amine that caused 5, 10, and 20% reduction in shoot dry weight of maize was 488, 844, and 1971 g a.e./ha when applied to 4-leaf maize and only 14, 136, and 616 g a.e./ha when applied to 8-leaf maize, respectively. The predicted dose of MCPA amine that caused 5, 10, and 20% yield reduction was 2557, 4247, and >5040 g a.e./ha when applied to 4-leaf maize and 184, 441, and 1245 g a.e./ha when applied to 8-leaf maize, respectively. Based on these results, glyphosate plus MCPA amine applied at the manufacturer's recommended dose of 630 g a.e./ha applied to 4-leaf maize has potential to cause injury but the injury is transient with no significant reduction in yield. However, when glyphosate plus MCPA amine is applied to 8-leaf maize it has the potential to cause significant injury and yield loss in maize.

## 1. Introduction

Canada produces nearly 11,000,000 metric tonnes of maize annually making it the 11th largest maize producer in the world [1]. Maize (*Zea mays* L.) is the third most valuable crop produced in Canada [1]. Greater than 95% of maize produced in Canada is grown in Eastern Canada, with Ontario producing 62% of the national production [1, 2]. Effective weed management is critical for maize producers to be competitive in the global market. Growers need weed management options that provide grass and broadleaved weed control with a wide margin of crop safety thereby resulting in minimal yield loss due to weed interference and higher net returns for producers.

In Ontario, most weed management studies in maize have focused on common weeds such as common lambsquarters (*Chenopodium album* L.), common ragweed (*Ambrosia artemisiifolia* L.), redroot pigweed (*Amaranthus retroflexus* L.), velvetleaf (*Abutilon theophrasti* Medicus), and foxtails (*Setaria* spp.). In contrast, very little research has been conducted on weeds that affect only small percentage of the maize hectarage such as field horsetail (*Equisetum arvense* L.). Field horsetail is a competitive weed that has adapted to current agronomic practices and has moved into fertile grain fields [3]. It is increasing in prevalence in areas where growers rely on glyphosate for weed management since glyphosate is not very effective for the control of field horsetail. Field horsetail normally grows to a height of 30 to 40 cm and can be very competitive in crops as its rhizomes can form a thick mat underground during the summer growing season [4]. Significant yield losses have been reported in maize with heavy field horsetail stands that can reach densities of 400 shoots m$^{-1}$ [3].

Currently registered postemergence (POST) herbicides in field maize in Ontario including nicosulfuron/rimsulfuron, flumetsulam, and nicosulfuron/rimsulfuron + flumetsulam do not provide adequate control of field horsetail in maize [5].

However, studies have shown that MCPA amine (dimethylamine salt of 2-methyl-4-chlorophenoxyacetic acid) applied alone or in tank mix with nicosulfuron/rimsulfuron or flumetsulam applied after emergence has potential to control field horsetail [4–8].

MCPA amine is a phenoxy herbicide that in addition to field horsetail controls many broadleaved weeds including pigweed spp., *Polygonum* spp., *Ambrosia* spp., *Capsella bursa-pastoris* (shepherd's purse), *Taraxacum officinale* (dandelion), *Delphinium bicolor* (buttercup), and *Galeopsis tetrahit* (hempnettle) [9, 10]. MCPA amine is a growth regulating herbicide that affects respiration, translocation, and cell division which results in chlorosis of the growing points, growth inhibition, wilting, necrosis, and eventual death of susceptible plants within 3–5 weeks [10, 11].

MCPA amine is a desirable compliment to the current weed management programs in glyphosate-resistant maize. It provides an alternative mode of action thereby reducing the selection intensity for glyphosate-resistant weeds and can effectively control troublesome weeds such as field horsetail in maize. MCPA amine is currently registered in Ontario at 380–630 g a.e./ha applied to maize up to 15 cm in height. MCPA amine is more injurious to maize than either 2,4-D or dicamba and therefore farmers are reluctant to use it. Earlier studies have shown as much as 5% injury in maize with MCPA amine applied after emergence at 630 g a.e./ha [5]. However, MCPA is the best herbicide available in Ontario for the control of field horsetail in maize. There is little information on the sensitivity of glyphosate-resistant maize to glyphosate plus MCPA amine applied after emergence at various doses and application timings under Ontario environmental conditions. Determining the appropriate MCPA amine dose and application timing will help maize growers avoid crop injury and associated yield loss and provide an additional option for control of troublesome, glyphosate tolerant weeds such as field horsetail.

The objective of this research was to evaluate tolerance of field maize to tank mixes of glyphosate (900 g a.e./ha) plus MCPA amine applied after emergence at 79, 158, 315, 630, 1260, 2520, or 5040 g a.e./ha at either the 4- or 8-leaf stage of maize.

## 2. Materials and Methods

A total of seven field trials were conducted at various locations in Ontario, Canada, during 2011–2013. There were 3 experiments at Ridgetown, ON, Canada, in 2011 and 2012 and one experiment at Harrow, ON, Canada, in 2013. Seedbed preparation at all sites consisted of autumn moldboard plowing followed by two passes with a field cultivator with rolling basket harrows in the spring.

Field trials were established as a randomized complete block design with four replications. Treatments included glyphosate 900 g a.e./ha (weed-free control) and glyphosate at 900 g a.e./ha plus MCPA amine at 79, 158, 315, 630, 1260, 2520, or 5040 g a.e./ha. Plots consisted of four rows of maize that were spaced 0.75 m apart and were 8.0 m long. Maize was planted at the population of approximately 80,000 seeds ha$^{-1}$ in May of each year. Herbicide treatments were applied

with a $CO_2$-pressurized backpack sprayer equipped with Hypro ULD120-02 nozzle tips (Hypro, New Brighton, MN) calibrated to deliver 200 L ha$^{-1}$ of water at 200 kPa. Herbicide applications were made with a 1.5 m boom with four nozzles spaced 50 cm apart. Herbicide treatments were applied at 4-leaf or 8-leaf maize (each timing to a separate set of plots). All plots including the nontreated control were kept weed-free by hand weeding as required.

Maize injury was visually estimated on a scale of 0 (no injury) to 100% (complete plant death) at 1, 2, 4, and 8 weeks after the corresponding treatment application (WAT). Maize shoot dry weight was determined 4 WAT by cutting 10 plants per plot at the soil surface, drying at 60°C to constant moisture. At 8 WAT, maize height was measured for 10 plants per plot and brace root malformation was rated on a scale of 0 (no malformation) to 10 (severe malformation). Percent lodging (goose-necked or falling stalks) was rated prior to harvest, approximately 16 WAT. The maize crop was harvested with a small plot combine, weight and seed moisture content were recorded, and yields were adjusted to 15.5% seed moisture content.

Data were analyzed using nonlinear regression (PROC NLIN) in SAS 9.2 [12]. Maize dry weight, height, and yield were converted to a percent of the glyphosate-only check prior to analysis.

All parameters were regressed against MCPA amine dose, designated as RATE in the equations. The equation used for percent crop injury, brace root injury, and lodging (rectangular hyperbola based on Cousen's yield loss model) was

$$Y = \frac{(i * \text{RATE})}{(1 + (i * \text{RATE}/a))},\qquad(1)$$

where $a$ is the upper asymptote and $i$ is the initial slope. For percent shoot dry weight, a descending rectangular hyperbola based on Cousen's model was used:

$$Y = d\left(1 - \frac{(b * \text{RATE})}{100(1 + b * \text{RATE}/c)}\right),\qquad(2)$$

where $b$ is the initial slope, $c$ is the lower asymptote, and $d$ is the $Y$-intercept. The equation for percent maize height and yield was an inverse exponential:

$$Y = f + g * (\exp[-h * \text{RATE}]),\qquad(3)$$

where $f$ is the lower asymptote, $g$ is the magnitude of the response, and $h$ is the slope of the response.

## 3. Results and Discussions

Regression equations were used to calculate predicted MCPA rates (g a.e./ha) that would result in a 5, 10, or 20% crop/brace root injury or lodging or a 5, 10, or 20% reduction in percent crop dry weight, height, or yield, relative to the glyphosate-only check. If any MCPA rate was predicted to be higher than 5040 g a.e./ha, it was simply expressed as ">5040" since it would be improper to extrapolate outside the range of rates evaluated in these experiments.

TABLE 1: Parameter estimates and predicted MCPA rates from regression models of visible percent field maize injury 1, 2, 4, and 8 WAT, brace root injury 8 WAT, and percent lodging prior to harvest[a].

| Crop injury | Application timing | Parameter estimates[b] (±SE) | | Predicted MCPA rate[c] | | |
| | | a | i | $R_5$ | $R_{10}$ | $R_{20}$ |
| | | % | | | g a.e./ha | |
|---|---|---|---|---|---|---|
| 1 WAT | 4-leaf corn | 45 (6) | 0.015 (0.002) | 380 | 867 | 2415 |
| | 8-leaf corn | 90 (5) | 0.052 (0.005) | 101 | 215 | 492 |
| 2 WAT | 4-leaf corn | 56 (9) | 0.016 (0.002) | 339 | 751 | 1914 |
| | 8-leaf corn | 78 (4) | 0.056 (0.005) | 95 | 204 | 479 |
| 4 WAT | 4-leaf corn | 63 (22) | 0.009 (0.002) | 630 | 1380 | 3403 |
| | 8-leaf corn | 65 (4) | 0.081 (0.012) | 67 | 146 | 357 |
| 8 WAT | 4-leaf corn | 21 (6) | 0.008 (0.003) | 820 | 2414 | >5040 |
| | 8-leaf corn | 63 (5) | 0.085 (0.014) | 64 | 140 | 344 |
| Brace root | 4-leaf corn | 3 (1) | 0.002 (0.001) | 267 | 671 | 2740 |
| | 8-leaf corn | 9 (0) | 0.068 (0.009) | 8 | 17 | 38 |
| Lodging | 4-leaf corn | 0 (0) | 0 (0) | — | — | — |
| | 8-leaf corn | 37 (7) | 0.034 (0.013) | 168 | 400 | 1283 |

[a]WAT, weeks after treatment.
[b]Regression parameters ((1), rectangular hyperbola): a, upper asymptote; i, initial slope.
[c]$R_5$, $R_{10}$, and $R_{20}$ are the MCPA rates predicted to result in 5, 10, or 20% crop injury.

### 3.1. Crop/Brace Root Injury or Lodging.

For all injury ratings MCPA amine caused more injury when applied to 8-leaf maize compared to when it was applied to 4-leaf maize. At 1 WAT, the predicted dose of MCPA amine that caused 5, 10, and 20% injury was 380, 867, and 2415 g a.e./ha when applied to 4-leaf maize but only 101, 215, and 492 g a.e./ha when applied to 8-leaf maize, respectively (Table 1), indicating that MCPA amine is far more injurious to maize when the application timing is delayed. At 2 WAT, the predicted dose of MCPA amine that caused 5, 10, and 20% injury was 339, 751, and 1914 g a.e./ha when applied to 4-leaf maize but only 95, 204, and 479 g a.e./ha when applied to 8 leaf maize, respectively.

Generally beyond 2 WAT, MCPA amine applied at 630 g a.e./ha or lower doses to 4-leaf maize caused no visible injury in maize. At 4 WAT, the predicted dose of MCPA amine that caused 5, 10, and 20% injury was 630, 1380, and 3403 g a.e./ha when applied to 4-leaf maize and 67, 146, and 357 g a.e./ha when applied to 8-leaf maize, respectively (Table 1). At 8 WAT, the predicted dose of MCPA amine that caused 5, 10, and 20% injury was 820, 2414, and >5040 g a.e./ha when applied to 4-leaf maize and 64, 140, and 344 g a.e./ha when applied to 8-leaf maize, respectively.

The predicted dose of MCPA amine that caused 5, 10, and 20% brace root malformation was 267, 671, and 2740 g a.e./ha when applied to 4-leaf maize and only 8, 17, and 38 g a.e./ha when applied to 8-leaf maize, respectively (Table 1). There was no lodging injury when MCPA amine was applied at doses evaluated to 4-leaf maize but the predicted dose of MCPA amine that caused 5, 10, and 20% lodging injury was 168, 400, and 1283 g a.e./ha when applied to 8 leaf maize, respectively (Table 1).

In other studies, Buhler in 1991 [13] found no visible injury or reduction in plant stand in maize with other grass herbicides such as metolachlor and atrazine. However, other studies have shown as much as 28% injury in maize with herbicides such as alachlor and metolachlor depending on environmental conditions [14–17]. Other herbicides such as rimsulfuron, nicosulfuron, nicosulfuron plus rimsulfuron, and primisulfuron have been shown to cause no injury in maize [18]. However, Swanton et al. [19] found as much as 19% injury in maize with sulfonylurea herbicides such as nicosulfuron and rimsulfuron.

### 3.2. Shoot Dry Weight and Height.

The predicted dose of MCPA amine that caused 5, 10, and 20% reduction in shoot dry weight of maize compared to the control was 488, 844, and 1971 g a.e./ha when applied to 4-leaf maize and 14, 136, and 616 g a.e./ha when applied to 8-leaf maize, respectively (Table 2). MCPA amine reduced shoot dry weight much more when applied to 8-leaf maize compared to when it was applied to 4-leaf maize (Table 2).

There was no effect on plant height when MCPA amine was applied to 4-leaf maize but the predicted dose of MCPA amine that reduced height by 5% was only 396 g a.e./ha when applied to 8-leaf maize (Table 2). In other studies Buhler [13] found no reduction in maize height with other herbicides such as metolachlor when applied with atrazine. However, Swanton et al. [19] found significant height reduction with sulfonylurea herbicides such as nicosulfuron and rimsulfuron in maize.

### 3.3. Yield.

Maize was more tolerant to MCPA amine when applied at the 4-leaf stage than the 8-leaf stage. The predicted dose of MCPA amine that caused 5, 10, and 20% yield reduction compared to the control was 2557, 4247, and >5040 g a.e./ha when applied to 4-leaf maize and only 184,

TABLE 2: Parameter estimates and predicted MCPA rates from regression models of percent maize dry weight (4 WAT), height (8 WAT), and yield[a].

| Corn measurement % of check | Application timing | Parameter estimates[b] (±SE) | | | Predicted MCPA rate[c] | | |
|---|---|---|---|---|---|---|---|
| | | $b$ | $c$ | $d$ | $R_5$ | $R_{10}$ | $R_{20}$ |
| | | | % | | | g a.e./ha | |
| Dry weight | 4-leaf corn | 0.024 (0.015) | 48 (17) | 105 (4) | 488 | 844 | 1971 |
| | 8-leaf corn | 0.054 (0.030) | 33 (5) | 96 (3) | 14 | 136 | 616 |
| | | $f$ | $g$ | $h$ | | | |
| | | | % | | | | |
| Height | 4-leaf corn | 0 (0) | 100 (0) | 0.00001 (0.000002) | >5040 | >5040 | >5040 |
| | 8-leaf corn | 84 (1) | 15 (1) | 0.0008 (0.0002) | 396 | 1116 | — |
| Yield | 4-leaf corn | 0 (0) | 103 (1) | 0.00003 (0.000006) | 2557 | 4247 | >5040 |
| | 8-leaf corn | 70 (3) | 29 (4) | 0.0009 (0.0003) | 184 | 441 | 1245 |

[a] WAT, weeks after treatment.
[b] Regression parameters ((2), descending rectangular hyperbola): $b$, initial slope; $c$, lower asymptote; $d$, Y-intercept. ((3), inverse exponential): $f$, lower asymptote; $g$, magnitude of response; $h$, slope of response.
[c] $R_5$, $R_{10}$, and $R_{20}$ are the MCPA rates predicted to result in a 5, 10, or 20% reduction in percent corn dry weight, height, and yield.

441, and 1245 g a.e./ha when applied to 8-leaf maize, respectively (Table 2).

In other studies, Buhler [13] found significant yield reduction with herbicides such as metolachlor with atrazine in maize. Other herbicides such as nicosulfuron plus rimsulfuron have been shown to cause as much 16% yield reduction in maize [19].

## 4. Conclusions

Based on these results, glyphosate plus MCPA amine applied at the manufacturer's recommended dose of 630 g a.e./ha to 4-leaf maize has the potential to cause injury in maize but the injury is transient with no significant effect on maize yield. In contrast, glyphosate plus MCPA amine applied at the manufacturer's recommended dose of 630 g a.e./ha to 8-leaf maize has the potential to cause significant injury resulting in substantial yield losses at harvest time. Glyphosate plus MCPA amine applied early after emergence is a desirable compliment to the current weed management programs in maize, especially for weeds that are naturally tolerant to glyphosate such as field horsetail. In addition, this tank mix will increase the spectrum of annual, biennial, and perennial broadleaf weeds control and will reduce the selection intensity for glyphosate-resistant biotypes.

## Conflict of Interests

The authors have no conflict of interests to report.

## Acknowledgments

Funding for this project was provided by the Grain Farmers of Ontario and the GF2 Program of Agricultural Adaptation Council.

## References

[1] Statistics Canada, *Corn: Canada's Third Most Valuable Crop*, 2015, http://www.statcan.gc.ca/pub/96-325-x/2014001/article/11913-eng.htm.

[2] K. Kulasekera, *Grain Corn: Area and Production, by County*, Ministry of Agriculture, Food and Rural Affairs, Ontario, Canada, 2014, http://www.omafra.gov.on.ca/english/stats/crops/estimate_new.htm#metric.

[3] W. J. Cody and V. Wagner, "The biology of Canadian weeds. 49. *Equisetum arvense* L.," *Canadian Journal of Plant Science*, vol. 61, pp. 123–133, 1980.

[4] T. K. James and A. Rahman, "Chemical control of field horsetail (*Equisetum arvense*)," *New Zealand Plant Protection*, vol. 63, pp. 102–107, 2010.

[5] P. H. Sikkema, *Weed Control Trials Research Report*, University of Guelph Ridgetown Campus, Ridgetown, Canada, 2013.

[6] C. C. Bell and A. I. Popay, "Chemical control of field horsetail," in *Proceedings of the 41st New Zealand Weed and Pest Control Conference*, vol. 41, pp. 66–69, 1988.

[7] J. Doll, "Biology and control of field horsetail (*Equisetum arvensis* L., Horsetail Family)," 2001, http://fyi.uwex.edu/weedsci/2001/05/02/biology-and-control-of-field-horsetail/.

[8] P. B. Hoyt and A. C. Carder, "Chemical control of field horsetail," *Weeds*, vol. 10, no. 2, pp. 111–115, 1962.

[9] Ontario Ministry of Agriculture; Food and Rural Affairs (OMAFRA), "Guide to weed control," Publication 75, Ontario Ministry of Agriculture, Food and Rural Affairs (OMAFRA), Toronto, Canada, 2014.

[10] D. L. Shaner, *Herbicide Handbook*, Weed Science Society of America, Champaign, Ill, USA, 10th edition, 2014.

[11] D. Lemerle, R. B. Hinkley, C. R. Kidd, and A. R. Leys, "Symptoms of injury caused by herbicides in wheat and barley," *Advisory Bulletin, Department of Agriculture, New South Wales*, vol. 4, pp. 1–14, 1986.

[12] Statistical Analysis Systems (SAS), *The SAS System for Windows, Release 9.2.*, Statistical Analysis Systems Institute, Cary, NC, USA, 2008.

[13] D. D. Buhler, "Early preplant atrazine and metolachlor in conservation tillage corn (*Zea mays*)," *Weed Technology*, vol. 5, no. 1, pp. 66–71, 1991.

[14] G. A. Dixon, E. W. Stoller, and M. D. McGlamery, "Acetanilide herbicides for yellow nutsedge (*Cyperus esculentus*) control in corn (*Zea mays*)," *Weed Science*, vol. 28, no. 5, pp. 593–598, 1980.

[15] G. A. Dixon and E. W. Stoller, "Differential toxicity, absorption, translocation, and metabolism of metolachlor in corn (*Zea mays*) and yellow nutsedge (*Cyperus esculentus*)," *Weed Science*, vol. 30, pp. 225–230, 1982.

[16] J. R. Leavitt and D. Penner, "Protection of corn (*Zea mays*) from acetanilide herbicidal injury with the antidote R-25788," *Weed Science*, vol. 26, no. 6, pp. 653–659, 1978.

[17] C. J. Peter and B. Weber, "Adsorption, mobility, and efficacy of alachlor and metolachlor as influenced by soil properties," *Weed Science*, vol. 33, pp. 874–881, 1985.

[18] P. H. Sikkema, C. Kramer, J. D. Vyn, J. J. Kells, D. E. Hillger, and N. Soltani, "Control of *Muhlenbergia frondosa* (wirestem muhly) with post-emergence sulfonylurea herbicides in maize (*Zea mays*)," *Crop Protection*, vol. 26, no. 10, pp. 1585–1588, 2007.

[19] C. J. Swanton, K. Chandler, M. J. Elmes, S. D. Murphy, and G. W. Anderson, "Postemergence control of annual grasses and corn (*Zea mays*) tolerance using DPX-79406," *Weed Technology*, vol. 10, no. 2, pp. 288–294, 1996.

# Permissions

All chapters in this book were first published in AAG, by Hindawi Publishing Corporation; hereby published with permission under the Creative Commons Attribution License or equivalent. Every chapter published in this book has been scrutinized by our experts. Their significance has been extensively debated. The topics covered herein carry significant findings which will fuel the growth of the discipline. They may even be implemented as practical applications or may be referred to as a beginning point for another development.

The contributors of this book come from diverse backgrounds, making this book a truly international effort. This book will bring forth new frontiers with its revolutionizing research information and detailed analysis of the nascent developments around the world.

We would like to thank all the contributing authors for lending their expertise to make the book truly unique. They have played a crucial role in the development of this book. Without their invaluable contributions this book wouldn't have been possible. They have made vital efforts to compile up to date information on the varied aspects of this subject to make this book a valuable addition to the collection of many professionals and students.

This book was conceptualized with the vision of imparting up-to-date information and advanced data in this field. To ensure the same, a matchless editorial board was set up. Every individual on the board went through rigorous rounds of assessment to prove their worth. After which they invested a large part of their time researching and compiling the most relevant data for our readers.

The editorial board has been involved in producing this book since its inception. They have spent rigorous hours researching and exploring the diverse topics which have resulted in the successful publishing of this book. They have passed on their knowledge of decades through this book. To expedite this challenging task, the publisher supported the team at every step. A small team of assistant editors was also appointed to further simplify the editing procedure and attain best results for the readers.

Apart from the editorial board, the designing team has also invested a significant amount of their time in understanding the subject and creating the most relevant covers. They scrutinized every image to scout for the most suitable representation of the subject and create an appropriate cover for the book.

The publishing team has been an ardent support to the editorial, designing and production team. Their endless efforts to recruit the best for this project, has resulted in the accomplishment of this book. They are a veteran in the field of academics and their pool of knowledge is as vast as their experience in printing. Their expertise and guidance has proved useful at every step. Their uncompromising quality standards have made this book an exceptional effort. Their encouragement from time to time has been an inspiration for everyone.

The publisher and the editorial board hope that this book will prove to be a valuable piece of knowledge for researchers, students, practitioners and scholars across the globe.

# List of Contributors

**Majid Talebi and Nima Jaafari**
Graduate School of Horticulture, Islamic Azad University, Karaj Branch, Karaj 31876 44511, Iran

**Ebrahim Hadavi**
Department of Horticulture, Islamic Azad University, Karaj Branch, Karaj 31876 44511, Iran

**Abena O. Adjapong, Kwame D. Ansah, Faustina Angfaarabung, and Henry O. Sintim**
Department of General Agriculture, School of Applied Science & Technology, Sunyani Polytechnic, Sunyani, Ghana

**Asghar Farajollahi**
Combating Desertification, Gorgan University of Agricultural Science and Natural Resources, Gorgan 49138 15739, Iran

**Bahram Gholinejad and Hamed Jonaidi Jafari**
Faculty of Natural Resources, University of Kurdistan, Sanandaj 31585 4314, Iran

**Erik Delaquis and Philippe Seguin**
Department of Plant Science, McGill University, Macdonald Campus, 21111 Lakeshore Road, Sainte-Anne-de-Bellevue, QC, Canada H9X 3V9

**Roger Samson**
REAP Canada, 21111 Lakeshore Road, Sainte-Anne-de-Bellevue, QC, Canada H9X 3V9

**Arif Mustafa**
Department of Animal Science, McGill University, Macdonald Campus, 21111 Lakeshore Road, Sainte-Anne-de-Bellevue, QC, Canada H9X 3V9

**Huguette Martel**
MAPAQ, Direction Régionale de l'Estrie, Sherbrooke, QC, Canada J1N 2A5

**Arfan Ali**
Centre of Excellence in Molecular Biology, University of the Punjab, Lahore, Pakistan

**Shelley Jansky**
USDA Agricultural Research Service and University ofWisconsin-Madison, 1575 Linden Drive, Madison, WI, USA

**Mitchell K. Williams, RonnieW. Heiniger, Wesley J. Everman and David L. Jordan**
Department of Crop Science, North Carolina State University, P.O. Box 7620, Raleigh, NC 27695-7620, USA

**S. M. Shirazi, Zulkifli Yusop and N. H. Zardari**
Institute of Environmental andWater Resource Management (IPASA), Universiti Teknologi Malaysia (UTM), 81310 Skudai, Johor, Malaysia

**Z. Ismail**
Department of Civil Engineering, Faculty of Engineering, University of Malaya, 50603 Kuala Lumpur, Malaysia

**Balayogan Sivasankari and Marimuthu Anandharaj**
Department of Biology, Gandhigram Rural Institute-Deemed University, Gandhigram, Dindigul, Tamil Nadu 624 302, India

**Joann K. Whalen**
Department of Natural Resource Sciences, McGill University, Macdonald Campus, 21111 Lakeshore Road, Ste-Anne-de-Bellevue, QC, Canada H9X 3V9

**A. Bah, M. H. A. Husni and C. B. S. Teh**
Department of Land Management, Faculty of Agriculture, Universiti Putra Malaysia, 43400 Serdang, Selangor, Malaysia

**M. Y. Rafii**
Institute of Tropical Agriculture, Universiti Putra Malaysia, 43400 Serdang, Selangor, Malaysia

**S. R. Syed Omar**
Diversatech (M) Fertilizer Sdn. Bhd., Bandar Baru Bangi, 43650 Selangor, Malaysia

**O. H. Ahmed**
Department of Crop Science, Faculty of Agriculture and Food Sciences, Universiti Putra Malaysia, Bintulu Campus Sarawak, 97008 Bintulu, Sarawak, Malaysia

**Ashok Kumar Yadav**
CSIR-Institute of Himalayan Bioresource Technology, Palampur, Himachal Pradesh 176061, India

**Rajesh Kumar Arya and M. S. Narwal**
Department of Genetics and Plant Breeding, CCS HAU, Hisar 125004, India

**Aniela Pilar Campos deMelo**
Federal University of Goiás, Sector of Horticulture, Brazil

**PauloMarçal Fernandes**
Federal University of Goiás, Sector of Phytosanitary, Brazil

**Fábio Venturoli and Carlos de Melo Silva-Neto**
Federal University of Goiás, Sector of Forestry, Brazil

**Aurélio Rubio Neto**
Federal Institute Goiano, Brazil

**G. J. Silva, Tatiane Medeiros Souza, Rosa Lía Barbieri and Antonio Costa de Oliveira**
Plant Genomics and Breeding Center, Federal University of Pelotas, 96001-970 Pelotas, RS, Brazil

**Nader Soltani, Christy Shropshire and Peter H. Sikkema**
University of Guelph Ridgetown Campus, 120 Main Street East, Ridgetown, ON, Canada N0P 2C0

**Robert E. Nurse**
Agriculture and Agri-Food Canada, Harrow, ON, Canada N0R 1G0

**Guy R. Knudsen**
Soil & Land ResourcesDivision, University of Idaho, Moscow, ID83844,USA

**Louise-Marie C. Dandurand**
Department of Plant, Soil, & Entomological Sciences, University of Idaho, Moscow, ID 83844, USA

**Ayeni Leye Samuel and Adeleye Omotayo Ebenezer**
Department of Agricultural Science, Adeyemi College of Education, PMB 520, Ondo State, Ondo City, Nigeria

**Savita Jandaik**
Shoolini University of Biotechnology and Management Sciences, Solan, Himachal Pradesh 173212, India

**Preeti Thakur and Vikas Kumar**
Shoolini Institute of Life Sciences and Business Management, Solan, Himachal Pradesh 173212, India

**Shilpi Khare Saikia and Rakesh Pandey**
Microbial Technology and Nematology Department, CSIR-Central Institute of Medicinal and Aromatic Plants, P.O. CIMAP, Lucknow 226015, India

**Hossain Sohrawardy**
Department of Crop Production Systems in the Tropics, Faculty of Agricultural Sciences, Georg-August-University, 37077 Göttingen, Germany

**Md. Lokman Hossain**
Global Change Ecology (International Elite Graduate Programme), Faculty of Biology, Chemistry and Geosciences, University of Bayreuth, 95440 Bayreuth, Germany

**Rahul Tripathi, A. K. Nayak, R. Raja,Mohammad Shahid, Anjani Kumar, Sangita Mohanty, B.B. Panda, B. Lal and Priyanka Gautam**
Crop Production Division, Central Rice Research Institute, Cuttack, Odisha 753006, India

**Olowoake Adebayo Abayomi and Ojo James Adebayo**
Department of Crop Production, Kwara State University, PMB 1530, Ilorin, Nigeria

**Andrea Galimberti,Massimo Labra, Anna Sandionigi, Antonia Bruno, Valerio Mezzasalma and Fabrizio DeMattia**
ZooPlantLab, Dipartimento di Biotecnologie e Bioscienze, Università degli Studi di Milano-Bicocca, Piazza della Scienza 2, 20126 Milano, Italy

**S. O. Dania, P. Akpansubi and O. O. Eghagara**
Department of Soil Science, Faculty of Agriculture, Ambrose Alli University, Nigeria

**Rasiel Restelatto, Laércio Ricardo Sartor, Suelen Maria Einsfeld and Fernanda Paula Baldicera**
Department of Animal Science, Federal Technology University of Paraná, Estrada/Boa Esperanc¸a, km 04, 85660-000 Dois Vizinhos, PR, Brazil

**Paulo Sergio Pavinato**
Department of Soil Science, Luiz de Queiroz College of Agriculture, University of São Paulo, Avenida P´adua Dias 11, 13418-900 Piracicaba, SP, Brazil

**Modupe Janet Ayeni and Joshua Kayode**
Department of Plant Science, Ekiti State University, Ado Ekiti, Nigeria

**Fikru Mekonnen and Firew Mekbib**
School of Plant Sciences, College of Agriculture and Environmental Sciences, Haramaya University, P.O. Box 138, Dire Dawa, Ethiopia

**Shiv Kumar**
International Center for Agricultural Research in the Dry Areas (ICARDA), Tel Hadya, P.O. Box 5466 Aleppo, Syria

**Seid Ahmed**
International Center for Agricultural Research in the Dry Areas (ICARDA), P.O. Box 5689, Addis Ababa, Ethiopia

**Tilak R. Sharma**
Department of Agricultural Biotechnology, Chaudhary Sarwan Kumar Himachal Pradesh Agricultural University, Palampur 176 062, India

**Valiollah Rameeh**
Agriculture and Natural Resources Research Center of Mazandaran, Sari, Iran

**Nader Soltani, Kris McNaughton, Chris L. Gillard and Peter H. Sikkema**
University of Guelph Ridgetown Campus, Ridgetown, ON, Canada N0P 2C0

**Robert E. Nurse**
Agriculture and Agri-Food Canada, 2585 Country Road 20, Harrow,ON, Canada N0R 1G0

www.ingramcontent.com/pod-product-compliance
Lightning Source LLC
Chambersburg PA
CBHW080513200326
41458CB00012B/4194